# Principles of Astronomy

# Principles of Astronomy

Editor
**Donald R. Franceschetti, PhD**
*The University of Memphis*

SALEM PRESS

A Division of EBSCO Information Services, Inc.
Ipswich, Massachusetts

GREY HOUSE PUBLISHING

Copyright © 2016, by Salem Press, A Division of EBSCO Information Services, Inc., and Grey House Publishing, Inc.

All rights in this book are reserved. No part of this work may be used or reproduced in any manner whatsoever or transmitted in any form or by any means, electronic or mechanical, including photocopy, recording, or any information storage and retrieval system, without written permission from the copyright owner. For permissions requests, contact proprietarypublishing@ebsco.com.

∞ The paper used in these volumes conforms to the American National Standard for Permanence of Paper for Printed Library Materials, Z39.48–1992 (R2009).

**Publisher's Cataloging-In-Publication Data
(Prepared by The Donohue Group, Inc.)**

Names: Franceschetti, Donald R., 1947- editor.
Title: Principles of astronomy / editor, Donald R. Franceschetti, PhD, the University of Memphis.
Description: [First edition]. | Ipswich, Massachusetts : Salem Press, a division of EBSCO Information Services, Inc. ; [Amenia, New York] : Grey House Publishing, [2016] | Series: Principles of | Includes bibliographical references and index.
Identifiers: ISBN 978-1-61925-948-5 (hardcover)
Subjects: LCSH: Astronomy.
Classification: LCC QB15 .P75 2016 | DDC 520--dc23

PRINTED IN THE UNITED STATES OF AMERICA

# CONTENTS

| | |
|---|---|
| Publisher's Note . . . . . . . . . . . . . . . . . . . . . . . . vii | Corona Australis . . . . . . . . . . . . . . . . . . . . . . . . . 90 |
| Editor's Introduction . . . . . . . . . . . . . . . . . . . . . ix | Corona Borealis . . . . . . . . . . . . . . . . . . . . . . . . . . 93 |
| Contributors . . . . . . . . . . . . . . . . . . . . . . . . . . . . xiv | Cosmic inflation . . . . . . . . . . . . . . . . . . . . . . . . . . 96 |
| | Crux . . . . . . . . . . . . . . . . . . . . . . . . . . . . . . . . . . . 98 |
| Ablation . . . . . . . . . . . . . . . . . . . . . . . . . . . . . . . . 1 | Crystal spectroscopy . . . . . . . . . . . . . . . . . . . . . 100 |
| Accretion disks . . . . . . . . . . . . . . . . . . . . . . . . . . . 2 | *Curiosity* . . . . . . . . . . . . . . . . . . . . . . . . . . . . . . 103 |
| Aerospace design . . . . . . . . . . . . . . . . . . . . . . . . . 4 | |
| Aerospace propulsion . . . . . . . . . . . . . . . . . . . . . 8 | Dactyl: (243) Ida I Dactyl . . . . . . . . . . . . . . . . . 106 |
| Agamemnon: 911 Agamemnon . . . . . . . . . . . . . 11 | Dark energy . . . . . . . . . . . . . . . . . . . . . . . . . . . 107 |
| Amor asteroid group: 1221 Amor . . . . . . . . . . . 12 | Dark matter . . . . . . . . . . . . . . . . . . . . . . . . . . . 110 |
| Amor asteroids . . . . . . . . . . . . . . . . . . . . . . . . . . 14 | *Dawn* mission . . . . . . . . . . . . . . . . . . . . . . . . . 112 |
| Annefrank: 5535 Annefrank . . . . . . . . . . . . . . . 16 | Deep-space navigation . . . . . . . . . . . . . . . . . . 115 |
| Apollo asteroid group: 1862 Apollo . . . . . . . . . 18 | Dysnomia: (136199) Eris I Dysnomia . . . . . . . 117 |
| Apollo asteroids . . . . . . . . . . . . . . . . . . . . . . . . . 20 | |
| Aquarius . . . . . . . . . . . . . . . . . . . . . . . . . . . . . . . 22 | Earth . . . . . . . . . . . . . . . . . . . . . . . . . . . . . . . . . 119 |
| Aries . . . . . . . . . . . . . . . . . . . . . . . . . . . . . . . . . . 25 | Earth-imaging satellites . . . . . . . . . . . . . . . . . 122 |
| Asteroids . . . . . . . . . . . . . . . . . . . . . . . . . . . . . . 28 | 136199 Eris . . . . . . . . . . . . . . . . . . . . . . . . . . . 124 |
| Astrochemical reaction models . . . . . . . . . . . . 31 | 433 Eros . . . . . . . . . . . . . . . . . . . . . . . . . . . . . . 127 |
| Astronomical image processing . . . . . . . . . . . . 33 | |
| Astrophysics . . . . . . . . . . . . . . . . . . . . . . . . . . . 35 | Fireball . . . . . . . . . . . . . . . . . . . . . . . . . . . . . . . 129 |
| Aten asteroids . . . . . . . . . . . . . . . . . . . . . . . . . . 38 | Fuel systems . . . . . . . . . . . . . . . . . . . . . . . . . . 131 |
| Aten asteroids: 2062 Aten . . . . . . . . . . . . . . . . . 40 | |
| Atmospheric opacity . . . . . . . . . . . . . . . . . . . . . 42 | Galaxy types . . . . . . . . . . . . . . . . . . . . . . . . . . 133 |
| | Galileo (ESA) . . . . . . . . . . . . . . . . . . . . . . . . . 136 |
| Ball of light particle model . . . . . . . . . . . . . . . . 46 | Gas-grain models . . . . . . . . . . . . . . . . . . . . . . 139 |
| Black holes . . . . . . . . . . . . . . . . . . . . . . . . . . . . 48 | Gas-phase models . . . . . . . . . . . . . . . . . . . . . . 141 |
| Blazars . . . . . . . . . . . . . . . . . . . . . . . . . . . . . . . . 50 | Gemini . . . . . . . . . . . . . . . . . . . . . . . . . . . . . . . 144 |
| Bolide . . . . . . . . . . . . . . . . . . . . . . . . . . . . . . . . . 52 | GLONASS . . . . . . . . . . . . . . . . . . . . . . . . . . . . 146 |
| Bremsstrahlung radiation . . . . . . . . . . . . . . . . . 54 | GPS . . . . . . . . . . . . . . . . . . . . . . . . . . . . . . . . . 149 |
| | |
| Cancer . . . . . . . . . . . . . . . . . . . . . . . . . . . . . . . . 56 | Hertzsprung gap . . . . . . . . . . . . . . . . . . . . . . . 152 |
| Canis Major . . . . . . . . . . . . . . . . . . . . . . . . . . . . 58 | High-mass stars . . . . . . . . . . . . . . . . . . . . . . . 154 |
| Canis Minor . . . . . . . . . . . . . . . . . . . . . . . . . . . 61 | Hubble's galaxy classification . . . . . . . . . . . . 156 |
| Capricornus . . . . . . . . . . . . . . . . . . . . . . . . . . . . 63 | Hydrogen cloud . . . . . . . . . . . . . . . . . . . . . . . 159 |
| Cassiopeia . . . . . . . . . . . . . . . . . . . . . . . . . . . . . 66 | |
| Celestial mechanics . . . . . . . . . . . . . . . . . . . . . 69 | International Space Station (ISS) . . . . . . . . . . 161 |
| Centaur . . . . . . . . . . . . . . . . . . . . . . . . . . . . . . . 71 | Interstellar chemistry . . . . . . . . . . . . . . . . . . . 164 |
| Ceres: 1 Ceres . . . . . . . . . . . . . . . . . . . . . . . . . . 73 | Inverse Compton scattering . . . . . . . . . . . . . . 167 |
| Ceto: 65489 Ceto . . . . . . . . . . . . . . . . . . . . . . . 75 | |
| CNO Cycle . . . . . . . . . . . . . . . . . . . . . . . . . . . . 77 | (216) Kleopatra . . . . . . . . . . . . . . . . . . . . . . . . 169 |
| Coma (Comet anatomy) . . . . . . . . . . . . . . . . . . 79 | Kleopatra's moons: (216) Kleopatra I Alexhelios |
| Combustion in space . . . . . . . . . . . . . . . . . . . . 80 | and II Cleoselene . . . . . . . . . . . . . . . . . . . . 171 |
| Comet anatomy . . . . . . . . . . . . . . . . . . . . . . . . 82 | Koronis asteroids: 243 Ida . . . . . . . . . . . . . . . 172 |
| Compton scattering . . . . . . . . . . . . . . . . . . . . . 84 | Kuiper Belt objects (KBOs) . . . . . . . . . . . . . . 174 |
| Constellations . . . . . . . . . . . . . . . . . . . . . . . . . . 86 | |
| Core navigation system . . . . . . . . . . . . . . . . . . 88 | |

## Contents

Langmuir-Hinshelwood model . . . . . . . . . . . . . . 176
Laser bees . . . . . . . . . . . . . . . . . . . . . . . . . . . . . 177
Launch vehicle aerodynamics . . . . . . . . . . . . . . 180
Leo . . . . . . . . . . . . . . . . . . . . . . . . . . . . . . . . . . 183
Libra . . . . . . . . . . . . . . . . . . . . . . . . . . . . . . . . 185

Mars . . . . . . . . . . . . . . . . . . . . . . . . . . . . . . . . 188
Mass attenuation coefficients . . . . . . . . . . . . . . 191
Mercury . . . . . . . . . . . . . . . . . . . . . . . . . . . . . 193
Messier catalog . . . . . . . . . . . . . . . . . . . . . . . . 195
Milky Way's structure . . . . . . . . . . . . . . . . . . . 197
*Mir* . . . . . . . . . . . . . . . . . . . . . . . . . . . . . . . . . 200
Moon impacts (General) . . . . . . . . . . . . . . . . . 203
Morgan-Keenan classification
  system (MK or MKK) . . . . . . . . . . . . . . . . . 205

*Near Shoemaker* . . . . . . . . . . . . . . . . . . . . . . . 209
NEOWISE . . . . . . . . . . . . . . . . . . . . . . . . . . . 211
Neptune . . . . . . . . . . . . . . . . . . . . . . . . . . . . . 214
Neutrino astrophysics . . . . . . . . . . . . . . . . . . . 216
Neutron star . . . . . . . . . . . . . . . . . . . . . . . . . . 218
Nucleosynthesis . . . . . . . . . . . . . . . . . . . . . . . 220
Nucleus (Comet anatomy) . . . . . . . . . . . . . . . 224

OB stars . . . . . . . . . . . . . . . . . . . . . . . . . . . . . 226
Orbit plotting . . . . . . . . . . . . . . . . . . . . . . . . . 228
Orion . . . . . . . . . . . . . . . . . . . . . . . . . . . . . . . 230

4450 Pan . . . . . . . . . . . . . . . . . . . . . . . . . . . . . 233
Particle acceleration . . . . . . . . . . . . . . . . . . . . 234
Phorcys: (65489) Ceto I Phorcys . . . . . . . . . . . 237
Photometry . . . . . . . . . . . . . . . . . . . . . . . . . . . 238
Pisces . . . . . . . . . . . . . . . . . . . . . . . . . . . . . . . 242
Planet spin-orbit coupling . . . . . . . . . . . . . . . . 244
Plasma physics . . . . . . . . . . . . . . . . . . . . . . . . 246
Pluto . . . . . . . . . . . . . . . . . . . . . . . . . . . . . . . . 249

Quark star . . . . . . . . . . . . . . . . . . . . . . . . . . . . 252
Quasar (quasi-stellar radio source) . . . . . . . . . . 254

Red giant . . . . . . . . . . . . . . . . . . . . . . . . . . . . 257
Rocket engines and rocketry . . . . . . . . . . . . . . 259

Sagittarius . . . . . . . . . . . . . . . . . . . . . . . . . . . . 263
Scattered disk objects (SDOs) . . . . . . . . . . . . . 266
Scorpius . . . . . . . . . . . . . . . . . . . . . . . . . . . . . 267
90377 Sedna . . . . . . . . . . . . . . . . . . . . . . . . . . 270
Space navigation systems . . . . . . . . . . . . . . . . 272
Space robotics . . . . . . . . . . . . . . . . . . . . . . . . . 275

Space transportation systems . . . . . . . . . . . . . . 278
Space-time properties . . . . . . . . . . . . . . . . . . . 281
*Sputnik* space program . . . . . . . . . . . . . . . . . . 284
Star structure . . . . . . . . . . . . . . . . . . . . . . . . . 287
Star-forming regions . . . . . . . . . . . . . . . . . . . . 290
Stellar classification . . . . . . . . . . . . . . . . . . . . . 292
Stellar population mapping . . . . . . . . . . . . . . . 295
Stellar remnants . . . . . . . . . . . . . . . . . . . . . . . 297
Sub-brown dwarfs . . . . . . . . . . . . . . . . . . . . . . 299
Supernova remnants . . . . . . . . . . . . . . . . . . . . 301

Tail (Comet anatomy) . . . . . . . . . . . . . . . . . . . 303
Taurus . . . . . . . . . . . . . . . . . . . . . . . . . . . . . . . 304
Theoretical astrophysics . . . . . . . . . . . . . . . . . 307
Three-body problem . . . . . . . . . . . . . . . . . . . . 310
Trans-Neptunian objects . . . . . . . . . . . . . . . . . 312
Trojan asteroids . . . . . . . . . . . . . . . . . . . . . . . . 314

Ursa Major . . . . . . . . . . . . . . . . . . . . . . . . . . . 317
Ursa Minor . . . . . . . . . . . . . . . . . . . . . . . . . . . 319

Venus . . . . . . . . . . . . . . . . . . . . . . . . . . . . . . . 323
4 Vesta . . . . . . . . . . . . . . . . . . . . . . . . . . . . . . 325
Virgo . . . . . . . . . . . . . . . . . . . . . . . . . . . . . . . 327

Wormholes . . . . . . . . . . . . . . . . . . . . . . . . . . . 330

*XMM-Newton* . . . . . . . . . . . . . . . . . . . . . . . . 333

### Appendices
Nobel Notes . . . . . . . . . . . . . . . . . . . . . . . . . . 339
Glossary . . . . . . . . . . . . . . . . . . . . . . . . . . . . . 340
Bibliography . . . . . . . . . . . . . . . . . . . . . . . . . . 355
Web Resources . . . . . . . . . . . . . . . . . . . . . . . . 371
Timeline of Space Exploration . . . . . . . . . . . . 373
Famous Figures and Events in Astronomy . . . . . . 378
Subject Index . . . . . . . . . . . . . . . . . . . . . . . . . 382

# Publisher's Note

Grey House Publishing is pleased to add *Principles of Astronomy* to its Salem Press collection, the third of four titles in the *Principles of* series: Chemistry, Physics, Astronomy, and Computer Science. This new resource introduces students and researchers to the fundamentals of astronomy using easy-to-understand language, giving readers a solid start and deeper understanding and appreciation of this complex subject.

The 140 entries range from Ablation to Quarks to XMM-Newton, and are arranged A to Z, making it easy to find the topic of interest. All entries include the following:

- Related fields of study that illustrate connections between various branches of astronomy, such as sub-planetary astronomy, theoretical astrophysics, astrochemistry, and planetary astronomy;
- Brief summary of the topic and how the entry is organized;
- Principal terms fundamental to the topic and to understanding the concepts presented;
- Illustrations, diagrams, and charts that clarify difficult topics such as solar systems, white dwarf and red giant stars, orbital plotting, rocket propulsion, and near-Earth objects;
- Star charts and photographs taken from space observatories;
- Photographs of significant contributors to the study of physics;
- Further reading lists that relate to the entry.

Ancient Greeks and Romans lent their names to many of the planets, stars, and constellations we study. Entries related to these celestial bodies explore how they were named, as well as the significance these figures held in the ancient world. Astronomy in the modern era has more to do with math and physics than with deities and myth. What we know about Black holes, Blazars, Quarks, and Supernovas is due to the work of the world's most renowned mathematicians and physicists, including Charles Messier, whose catalog is used to this day to describe and categorize objects visible in the sky, Edwin Hubble and his ground-breaking work in the discovery of universes beyond our own, and Albert Einstein and how his theory of general relativity elegantly describes the way that gravity can affect space and time.

This reference work begins with a comprehensive introduction by editor Donald R. Franceschetti, PhD, that discusses the significance of naked eye observations by the earliest astronomers; the use of optical, radio, and neutron telescopes; and the contributions derived from space probes, manned missions, and space stations. The goal of this material is to advance our understanding of our own planet, the celestial bodies around us, and the ways in which astronomy and physics have become intertwined.

The book's backmatter is another valuable resource:

- Nobel Notes explain how prizes awarded in areas such as chemistry and physics have advanced work in the area of astronomy;
- List of important figures in astronomy, including Ptolemy, Copernicus, Newton, and Hubble;
- Timeline of important advances in the field of astronomy;
- Timeline of space exploration;
- Glossary;
- General bibliography;
- Web Resources, and
- Subject index.

Salem Press extends its appreciation to all involved in the development and production of this work. The entries have been written by experts in the field, whose names follow the Editor's Introduction.

*Principles of Astronomy,* as well as all Salem Press titles, is available in print and as an e-book. Please visit www.salempress.com for more information.

# Editor's Introduction

Astronomy is the oldest of the natural sciences, and human beings have looked to the skies since before recorded time for answers to some of their most important questions. It is believed that the first people to observe the skies and make calculations based on the movement of the stars were priests, so it is natural that they would associate heavenly bodies with gods and the divine, as well as with the events in nature which they associated with the stars—night changing to day, the procession of seasons, the tides, and annual weather events from rains to winds to floods.

One of the earliest practical tasks of astronomy was to set the length of the year and determine when the sun would pass from the northern to the Southern hemisphere. Over the last ten to twenty years, archaeologists have determined that those priests were quite sophisticated when it came to applying their skills as observers and as mathematicians. They devised both lunar (moon-based) and solar (sun-based) calendars that allowed them to determine such important events as planting seasons. Some of these early calendars date back as far as 4000 BCE. It was the need to improve the calendar and better determine the dates of religious feasts that led Copernicus to first suggest that the sun be treated as the center of the solar system.

Early on, though, it was also taken as truth that one's horoscope, based on the precise position of the stars and planets at the moment of one's birth, held all the details of one's fate. Since the sky might not be clear enough to "read" when important personages like kings or emperors were born, it was important to have some device for predicting the positions of the heavenly bodies. Ptolemy's book of tables, the *Almagest*, was the first of these; it was used for 2000 years, though it became inaccurate over time. While today, we draw a clear distinction between astronomy and astrology, at first there was no distinction between the activity of the stars and the activities of the gods or spirits thought to rule over men, which meant that astrology–an attempt to decipher how the positions, movements, and properties of the stars can affect human beings–was not truly distinct from astronomy. In fact the names of the stars and constellations shows quite clearly the powerful influence this believe in the connection between deities and spirits on human destiny.

## NAKED EYE OBSERVATIONS

Despite the fact that the ancients may have had a more "mystical" approach to some of their observations, the fact remains that they were intent upon observing the behaviors of the bodies they saw in the sky. They built structures known as observatories specifically to observe the heavens, including one of the earliest known observatories, Stonehenge, located on Salisbury Plain in England. Researchers have demonstrated that Stonehenge is aligned quite precisely with sunrise on the summer solstice and sunset on the winter solstice. The pyramids in Egypt also represent important astronomical observatories; they are thought to be aligned with the pole star. Chinese astronomers created star charts and used their recorded observations to predict the movements of the stars as well as lunar and solar eclipses, and also to set the dates and times for important festivals. All of these observations were made with the naked eye, well before the invention of the telescopes used by modern astronomers.

## TELESCOPES AND A MODEL OF THE SOLAR SYSTEM

While Galileo did not invent the telescope, he was one of the first scientists to turn the telescope skyward rather than using it solely as a way to keep track of sailing vessels. He discovered marvelous things: The moon was covered with craters; the planet Venus went through phases just like the moon; the planet Saturn had rings which disappeared when viewed edge on; and the planet Jupiter had its own little solar system of at least four moons. All of these discoveries contradicted the prevailing Aristotelian orthodoxy. The story of Galileo's battles with the Pope are legend—his treatment at the hands of the Church and the fact that he died under house arrest discouraged scientists in the Catholic countries of Europe from following in his footsteps. Espousing the understanding that the Earth revolved around the Sun was considered heresy and led to the persecution of such important figures as Galileo and Copernicus, and even Frederick Kepler.

Tycho Brahe (1546-1601) was the last of the naked eye observers of astonomy. He was renowned for the precision and accuracy of his observations, made from an observatory built for him by by King Frederick of Denmark, on the island of Hven, where he recorded planetary positions night after night. It was Brahe's observations that led to geocentric (Earth centered) understanding of our solar system, with Earth at the center and the stars and sun revolving around Earth. It took many years for a different system to come into wide acceptance. Although there were errors in some of the transcriptions of these records by scribes and students who flocked to work with Brahe, the data concerning the movements of the planet Mars were accurate enough to allow Johannes Kepler to devise his laws of planetary motion. Kepler's mathematical models of the movement of the planets and stars were able to join physics with astronomy and so became the basis of Newton's more general laws of motion.

Newton, a Protest and Non-Conformist in religion, completed the work of dismantling of Aristotle's mechanics. His three laws of motion did away with the distinction between celestial and terrestrial motion, as well as the need for a divinity to set things in motion. His law of universal gravitation explained why things seemed to be attracted to the center of the Earth, stating that every mass in the universe attracted every other mass with a force proportional to the product of the two masses and inversely proportional to the square of the distance between them. Thus, there was nothing physically special about the Earth.

By the nineteenth century, the Copernican model of the solar system had been general accepted. The question then became how accurately did Newton's laws predict the time evolution of planetary positions? Astronomy turned to computational mathematics for the answers. In classical physics there are generally no exact solutions for systems of three or more interacting bodies. To handle the solar system with its many interacting bodies, one approach was to begin with the solution for two interacting bodies, the sun and Jupiter for instance, and to treat the effects of the additional bodies as perturbations on a readily solvable problem.

**DISCOVERING NEW PLANETS**
William Herschel built Observatory House in Slough, England, to house his 48-inch reflector. In 1791, he discovered the planet Uranus, the first new member of the solar system to be discovered since classical times and the first to be located on the basis of Newton's laws and mathematical computations.

Then, after 50 years of observation, it became apparent that Uranus was deviating from the orbit predicted by Newton's laws, so it was hypothesized there might be another planet whose gravitational effect had not been taken into account. Two pairs of mathematicians, one in Berlin and the other in Paris, computed the coordinates of the new planet—Neptune—which was eventually found just where it was expected to be. The situation was repeated in 1930 when observations of Neptune led to the discovery of Pluto. In recent years, observations of the motions of nearby stars have made possible the identification of some 2000 objects outside the solar system.

Newton's theory of gravitation provided a framework for understanding the structure of the universe. Before Newton, celestial objects were thought to move in circles because circular movement was considered perfect. Newton's theory meant that objects were understood to respond to the gravitational force and, therefore, one could extract from the motion of any object the vector sum of all the other objects influencing it. Gravitation is one of the fundamental forces in the universe, and certainly the weakest of the long-range forces. Nonetheless planets have been identified through their gravitational effects on other planet's orbits. Neptune was first identified this way and then in 1930, Pluto.

The optical telescope, which gathers and focuses light from visible light from electromagnetic spectrum, (with wavelengths between 400 and 750 nm) has been used since before Galileo and is still used up to the present day by astronomers and star-gazers alike. The optical telescope reached its high point with the construction of the 200-inch Hale telescope in 1939, set atop Mt. Palomar in CalTech's Palomar Observatory. This telescope, augmented by radio telescopes and then adaptive optics programs that can rapidly change the mirror's shape, make it possible to correct for atmospheric distortions so that it can produce images comparable to those received by space telescopes taking direct images of extra-solar objects.

Radio astronomy is the study of the universe through the analysis of radio emissions from celestial objects. The first radio telescope was built in 1937. Radio telescopes are capable of capturing information from the microwave region of the electromagnetic spectrum that ranges between a meter and a centimeter in wavelength. Radio telescopes typically involve a large reflector and a detector at the focus of the reflector. Many radio telescopes operate at a wavelength of 21 cm, which is associated with a proton spin-flipping transition in atomic hydrogen.

**GRAVITY AND GENERAL RELATIVITY**
Astronomers have studied the orbit of Mercury for centuries. This planet, nearest to the sun, describes an elliptical orbit, as all planets do, but its orbit appears to move ever so slightly in space. Even allowing for the orbit being perturbed by the gravitational pulls of other planet, Mercury's perihelion (point of closest approach to the sun) precesses by about 43 seconds of arc per century. This residual precession inspired the search for a planet within the orbit of Mercury, given the tentative name of Vulcan, but no such planet was found.

The explanation was found in the general theory of relativity put forth by Albert Einstein in 1915 just over a hundred years ago. The sun is massive enough to distort space by a slight amount in the vicinity of Mercury and this small deviation from Euclidean geometry was just enough to account for the precession. In a similar way the sun's gravity should result in a slight deviation of starlight when it passes near the sun's surface. This deviation should be observed during a solar eclipse and was captured on photographic plates by a British expedition in 1919 as predicted. It was at this point that Albert Einstein became famous world-wide. Since that time small perturbations in the orbits of the sun's nearer neighbors have led to the discovery of about 2,000 extra-solar planets with more discoveries expected each year. The search not only demonstrated the validity of the laws of physics far from the Earth, but also led to new discoveries

Once it was understood that interstellar distances could extend for hundreds or thousands of light years, astronomers began to wonder whether the Milky Way galaxy was in fact the entire universe or whether it was just one of a great many objects at incredible distances from the Earth. Interestingly, a number of cloud like objects had been cataloged by Charles Messier as a aid to comet hunters at the end of the eighteenth century. Harlow Shapley, noted the presence of large variable stars in such clouds. Their size put them beyond our galaxy. That fact meant that the universe was larger than anyone had thought.

Heinrich Olbers is often given credit for pointing out that the darkness of the sky at night is itself a clue to the distribution of matter in the universe. The basic argument is that if the sky were full of stars, randomly distributed, then looking in any direction, the line of sight would sooner or later end on a stellar surface. If the stars had been shining forever then the night sky should be ablaze with light. That it was not implied that the stars had finite lifetimes or were not randomly distributed or both.

While it is now common knowledge that the stars shine because they are releasing their mass energy, this was not known at the beginning of the twentieth century. Edwin Hubble used the Doppler shift of the hydrogen lines in the spectrum of other galaxies to determine how rapidly and in which direction they were moving. Hubble's conclusion was that galaxies were moving away from each other at a speed that roughly increased with distance. If you were to "run the movie backwards," all the galaxies would converge at a single point about 13 billion years ago, indicating that the universe had not existed forever but rather had begun in a cosmic explosion: the big bang.

Examining Einstein's work on general relativity, astronomers realized that Einstein's equations allowed for an expanding or contracting universe. The expansion was ongoing but slowing down. The open question was then: Is there enough matter that, at some time in the future, the universe would begin contracting? It turned out that the answer to this question was much more involved than anyone might have guessed.

Confirmation of the big bang cosmology would come not from astronomy but from communications engineers. Arno Penzias and Robert Wilson were working for the Bell Telephone system, developing a microwave horn antenna to communicate through the earth's atmosphere. They found a mysterious signal which they could not attribute to any source. Eventually they realized that it was a radiation residue from the big bang, now cooled to 2.7 kelvins.

**THE SPACE AGE**
The "space-age" began in October 1958 when, to the great surprise of the American public, the first artificial earth satellite, *Sputnik I,* was launched by the Soviet Union. The satellite epoch was ushered in by rocket-launched, space-based telescopes, which

offered two advantages over the terrestrial variety: Far more accurate images and the ability to function in frequency regions in which earth's atmosphere was opaque.

Since 1958 some thousands of space vehicles have been launched. Some of those probes include telescopes that can make observations in the microwave, infrared, ultraviolet, and x-ray regions. Prominent among them is the Hubble astronomical telescope, an optical telescope launched at a cost of over one billion dollars. This telescope afforded earth-based observers a factor of 10 improvement in resolution. Images from the Hubble provided evidence that not only was the universe's expansion continuing, it seemed to be accelerating. Further measurements of galactic rotational speeds indicate that galaxies have much more mass than would be estimated on the basis of their luminosity. Estimates of the amount of dark matter in the universe vary but may exceed 50 percent.

As we review the various epochs of astronomy fron naked eye observations to optical telescopes, radio telescopes, and neutrino telescopes, we can see our understanding of the universe grow, starting with Ptolemy's world, through the computational models of Copernicus and on to the period where the Milky Way was the "island" universe, to Hubble and Shapley's world of many hundreds of galaxies, and into Einstein's world, where we cannot uniquely separate space and time, and ultimately, into the highly speculative worlds of the far future.

**DETERMINING LOCATIONS OF ASTRONOMICAL BODIES**

One perennial challenge to astronomers was to locate the various astronomical objects, that is, determine their distance from us. The absence of parallax of the stars was one of the strongest arguments against the heliocentric model of Copernicus. Stellar parallax is quite small, only 0.76 seconds of arc for the nearest star Centauri, and the first accurate determination of stellar parallax was not made until 1828 when parallaxes could be determined for only the closest stars.

Herzsprung and Russell provided the next clue for determining interstellar distances. They found that stars could be classified by their optical spectra, which itself was a clue to the chemical composition of the stellar atmosphere. Determining the spectrum of a star let the astronomers classify it. They found that for the vast majority of stars the brightness correlated with their spectral type. Thus from the spectrum we would know in absolute terms how bright a star is and thus how far away it is.

Next came the realization that for certain variable stars, the period of their variation was directly related to their luminosity. Finding some of these stars in nearby galaxies led Shapley to the conclusion that the some of the nebulae were in fact galaxies like our own, but millions of light years older.

It is a truism that in spite of our ability to map out the universe, the human race has been to one planet and one moon—both ours. Everything else has been learned by observation.

Since 1958 space probes have surveyed the solar system, visited all of the planets and returned pictures, Rocket observations have also been made of the sun, planets, and a comet and space probes have only just departed into deep space.

Some of the strangest of astronomical instruments include the neutrino telescope and the gravitational wave observatory. Our principal source of information about all but our immediate neighborhood in space is the electromagnetic radiation we receive although we are able to draw some inferences from other forms of radiation and materials on Earth of extraterrestrial life.

The neutrino was proposed by Wolfgang Pauli in 1930, and given its current name in 1934 by Enrico Fermi. The particle only interacts via the nuclear weak force and presumably gravitation. As a result neutrinos are very hard to detect. The basic event of neutrino emission from the care of stars remained conjectural until the 1950's. Every time four protons in the sun core combine to form a helium nucleus two neutrinos are released into space. Most neutrino detectors are based on the weak force driven reaction

$$\text{Proton} + \text{electron} \rightarrow \textbf{neut}_{\text{ron}} \rightarrow \textbf{neut}_{\text{ron}} + \text{neutrino}$$

Thus if a chlorine nucleus absorbs a neutrino, it becomes a nucleus of the inert gas argon. Technology now exists that can cycle a gigantic tank of CCl4 and count the number of argon atoms that have been created in this way. Neutrinos from the Sun have been detected in this way in a neutrino "telescope" in an abandoned gold mine in Lead, N.D. In 1987 two supernovas were recorded in the Magellenic clouds, satellites of our own Milky Way galaxy, Analysis of the detector fluid was done and the results confirmed a detectable increase in neutrino flux.

The only problem with neutrino detectors is that the number of neutrinos detected is about only one-third of the number that was expected from the known energy output of the sun. Current theory says that the neutrino released in solar physics actually oscillated between three different neutrino species. The seven-minute journey was enough to allow the beam of neutrinos to become an equal mixture of electron neutrinos, muon neutrinos, and taon neutrinos. This implied that, rather than being massless particles traveling at the speed of light, neutrinos actually had a small rest mass.

**CONCLUSION**
As this essay is being written, on February 11, 2016, the National Science Foundation has just announced that on September 14, 2015, the Laser Interferometer Gravitational Wave Observatory (LIGO) with detectors in locations in the United States—one in Louisiana and one in Washington—detected gravitational waves coming from the collision of two black holes. In what can only be described as an observational tour de force, this confirms Albert Einstein's 1915 prediction that detectable waves in the gravitational field could come about from a sufficiently violent cosmic collision. The detector is basically a pair of massive tuning forks, some 2000 miles underground. By looking for vibrations of the forks in phase with each other, one can identify their response to the collision. It is worth noting that there is still some uncertainty as to the source of the universe's cosmic expansion. In fact, there are quite a few mysteries associated with gravitation so, after all, there is still much work to be done.

*Donald R. Franceschetti, PhD*

# Contributors

Tyler Biscontini

Cait Caffrey

Josephine Campbell

Nancy Comstock

Mark Dziak

Donald R. Franceschetti, PhD

Angela Harmon

Marie Keenan, MS

Adrienne A. Kennedy

Marianne M. Madsen, MSc

Michael Mazzei

Elizabeth Mohn

Richard Renneboog, MSc

Randa Tantawi, PhD

Janine Ungvarsky

# Principles of Astronomy

# A

# ABLATION

### FIELDS OF STUDY

Astrophysics; Astronautics

### SUMMARY

Ablation is a phenomenon during which material vaporizes, sublimates, melts, or otherwise erodes from a surface. In space science, it refers to a cooling effect by mass transfer during aerodynamic heating. A famous example of ablation is a meteor or spacecraft entering Earth's atmosphere. Ablation has important implications for humans, including the way it protects Earth from large meteors.

### PRINCIPAL TERMS

- **ablation shield:** the material on spacecraft that is meant to ablate to protect the craft during reentry.
- **sublimation:** the process of a frozen solid becoming a gas without passing through the liquid phase.

### FUNDAMENTALS OF ABLATION

Ablation is a phenomenon during which materials are eroded from the exterior of an object. This may occur through sublimation, vaporization, melting, chipping, or other processes. In the context of space, ablation is a form of cooling through mass transfer.

### EXAMPLES OF ABLATION

Many different examples of ablation exist. One of the most common examples is that of meteors that enter Earth's atmosphere. Friction ablates the surface of the falling meteor, and the eroded particles form a tail behind the object.

Another example of ablation is the ablation shield on some spacecraft. Scientists have designed shields made from special material that is meant to ablate upon reentry into Earth's atmosphere. This helps keep the spacecraft cooler while it enters the atmosphere. Such shields are not reusable.

### IMPLICATIONS OF ABLATION

Ablation has important implications for humans on Earth. The natural ablation of meteors in the atmosphere helps protect Earth's surface from large impacts. As meteors ablate, they become smaller. Most disintegrate completely before reaching the surface. Without ablation, meteor impacts could devastate life on Earth.

Another way in which ablation is important is its use in aerospace technology. Ablation shields, commonly called heat shields, allow astronauts to land

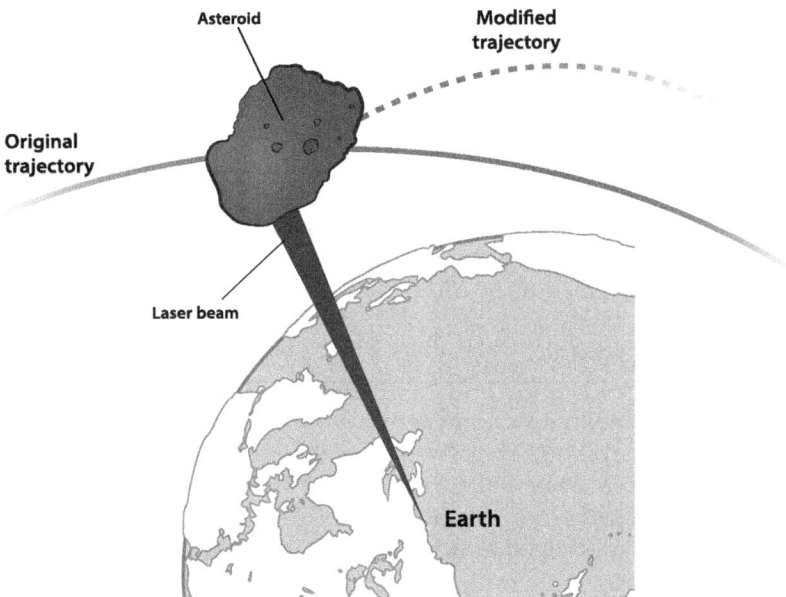

A high-intensity laser impacting with an object, such as space debris or a near-Earth asteroid, causes vaporization of surface material. This vaporization is called ablation, which produces thrust and can change the trajectory of the target object.

1

safely back on Earth, making space exploration possible. Ablation may also provide protection from asteroids large enough to pose a threat to Earth. Near-Earth objects have impacted Earth's surface in the past, and similar objects will likely impact it in the future. However, some scientists believe they can deflect such threats using laser ablation. This process would use solar-powered lasers to sublimate an asteroid's surface. The gas created by the sublimation could propel the asteroid enough to change its course so it does not impact Earth.

—Elizabeth Mohn

**FURTHER READING**

"Ablation." *Cosmos*. Swinburne University of Technology, n.d. Web. 17 Mar. 2015.

"Ablation Shield." *Oxford Index*. Oxford University Press, 2014. Web. 17 Mar. 2015.

Angelo, Joseph A. "Ablation." *Encyclopedia of Space and Astronomy*. New York: Facts On File, 2006. Print.

Cammorata, Nicole. *Words You Should Know 2013: The 201 Words from Science, Politics, Technology, and Pop Culture that Will Change Your Life This Year*. Avon: Adams, 2013. Print.

"Investigating Heat Shield Physics May Improve Design of Space Capsules." *Aerospace Engineering and Engineering Mechanics*. Cockrell School of Engineering, University of Texas at Austin, 13 Apr. 2013. Web. 20 Mar. 2015.

Jones, Daniel C., and Iwan P. Williams. "High Inclination Meteorite Streams Can Exist." *Advances in Meteoroid and Meteor Science*. Ed. J. M. Trigo-Rodríguez. New York: Springer, 2008. Print.

"Sublimation." *Cosmos*. Swinburne University of Technology, n.d. Web. 17 Mar. 2015.

# ACCRETION DISKS

**FIELDS OF STUDY**

Astrophysics; Observational Astronomy; Sub-planet Astronomy

**SUMMARY**

Accretion disks are relatively flat, rotating collections of debris and gas that form around black holes, young stars, or other large celestial objects with great gravitational force. Accretion disks can produce large amounts of energy and light, and they are one way scientists can identify the location of a black hole.

**PRINCIPAL TERMS**

- **angular momentum:** the momentum, or force of motion, of a rotating object or system, determined by the object's mass, speed, and distance from the point about which it is rotating.
- **binary star:** two stars that orbit around a common center of mass.
- **black hole:** a region of space with a gravitational pull so strong that even light cannot escape it.
- **galactic nucleus:** the center of a galaxy, relatively small but with a high concentration of stars.
- **quasar:** short for "quasi-stellar"; an extremely bright celestial object that produces very large amounts of energy.
- **white dwarf:** a small, very dense star that has exhausted nearly all of its fuel and is nearing the end of its life cycle.
- **x-ray:** high-energy electromagnetic radiation with a very short wavelength.

**BLACK HOLE MARKERS**

Accretion disks are one way that scientists detect the presence of black holes. Some black holes form when a massive star dies in a supernova explosion and then collapses in on itself. For a star to collapse into a black hole, its mass must be about twenty times that of the sun or greater. Other black holes are formed by the collision of dense objects, such as neutron stars. The tremendous gravitational power of a black hole is the result of a massive amount of matter being compressed into a relatively small area. The very center of a black hole is called a "singularity." A singularity is a point in space where mass density, and thus gravitational force, is essentially infinite.

The enormous size and power of a black hole pulls in everything within range, including light. Because of this, scientists cannot locate or study black holes

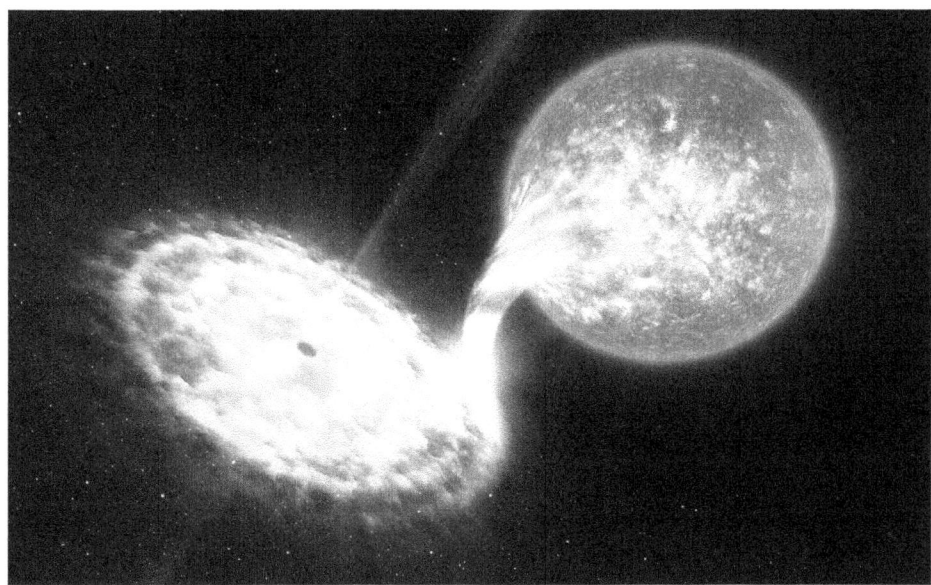

An artist's rendition of an accretion disk with a black hole at its center and a star, similar to the sun, orbiting it; the GX 339-4 binary system. By ESO/L. Calçada (eso.org), via Wikimedia Commons.

directly using any of the usual instruments that detect electromagnetic radiation. Instead, they look for objects and phenomena that accompany black holes. One such object is an accretion disk. Accretion disks are made up of the dust, gas, and other debris that massive objects such as black holes attract. Because of the law of conservation of angular momentum, this material continues to orbit the massive object as it approaches, following a spiral path and forming a flat, rotating mass. The friction produced by this rotation causes the material in the accretion disk to heat up and generate x-rays. The x-rays then propagate through space. Scientists look for these x-rays in order to identify the probable location of a black hole.

### Other Accretion Disks

The largest black holes can be more than a billion times as massive as the sun. These supermassive black holes are believed to be at the center of the galactic nuclei of most massive galaxies, including the Milky Way. If the accretion disk of a supermassive black hole produces enough friction, it generates a tremendous amount of heat and light, more than is produced by the rest of the galaxy combined. This phenomenon is called a quasar. The term "quasar" comes from "quasi-stellar radio source," as quasars were first thought to be individual starlike objects that emitted strong radio waves. In fact, a quasar is a type of active galactic nucleus (AGN), which is a galactic nucleus that emits more radiation than can be produced by the stars it contains. Quasars are believed to be the brightest objects in the universe. They are often more than a hundred times brighter than their surrounding galaxies. Quasars last as long as there is matter in the accretion disk to fuel them.

Accretion disks also form around young stars as they exert their gravitational forces on the space around them. Some of the material in such a disk may eventually form planets that orbit around the new star. On the other end of the stellar life span, accretion disks can also form when a white dwarf, a type of dying star, is one half of a binary star. The white dwarf compresses and grows both smaller in size and heavier in mass, increasing its gravitational power. This can cause it to draw material from its partner in the binary system, forming an accretion disk.

### Bringing Accretion Disks into View

In 2011, scientists at the National Aeronautics and Space Administration and the European Space Agency were able to use the Hubble Space Telescope to study a quasar accretion disk. Because quasars are at the center of distant galaxies, too far to study in detail with even the most powerful telescopes, the scientists developed a new technique: they used a galaxy between Earth and the quasar as a gravitational lens. When the stars in this galaxy passed in front of the quasar, their gravitational fields amplified the quasar's light and made it easier to see. This was the first time scientists were able to gather direct information about a quasar and its accretion disk instead of relying on theoretical data.

—*Janine Ungvarsky*

### Further Reading

"Angular Momentum." *Cosmos: The SAO Encyclopedia of Astronomy.* Swinburne University of Technology, n.d. Web. 30 Mar. 2015.

"Binary Star." *Cosmos: The SAO Encyclopedia of Astronomy.* Swinburne University of Technology, n.d. Web. 30 Mar. 2015.

"Black Holes." *NASA Science.* NASA, 30 Mar. 2015. Web. 31 Mar. 2015.

"Galactic Nuclei." *Cosmos: The SAO Encyclopedia of Astronomy.* Swinburne University of Technology, n.d. Web. 30 Mar. 2015.

"Hubble Directly Observes the Disc around a Black Hole." *ESA Science & Technology.* European Space Agency, 4 Nov. 2011. Web. 31 Mar. 2015.

Lemonick, Michael D. "Winds Blasting from Black Holes Shut Down Star Growth." *National Geographic.* Natl. Geographic Soc., 19 Feb. 2015. Web. 31 Mar. 2015.

Wanjek, Christopher. "Ring around the Black Hole." *Solar System Exploration.* NASA, 21 Feb. 2011. Web. 6 Mar. 2015.

# AEROSPACE DESIGN

### FIELDS OF STUDY

Aerospace Engineering, Astronautics; Space Technology

### SUMMARY

Aerospace design is crucial to the development and manufacturing of aircraft and spacecraft. Since the introduction of aircraft in the nineteenth century, aerospace design has led to the production of countless flight vehicles, including spacecraft used for space exploration. Aerospace design involves a design process, which may take several years to complete. The phases of this process typically are system/mission requirements, conceptual design, preliminary design, and critical design. With spacecraft, the design generally includes major components such as the mission payload and the platform. The design of the Space Shuttle is one good example.

### PRINCIPAL TERMS

- **attitude control:** the process of obtaining and sustaining the proper orientation in space.
- **mission payload:** the extra equipment carried by a craft for a specific mission. For a launch vehicle, payload usually refers to scientific instruments, satellites, probes, and spacecraft attached to the launcher.
- **platform:** all parts of a spacecraft that are not part of the payload; also known as the bus.
- **telemetry:** the process of transmitting measurement data via radio to operators on the ground. Telemetry is used to improve spaceflight performance and accuracy. It provides important information about standard operational health and status of a craft as well as mission-specific payload data.
- **thermal control:** the system aboard a spacecraft that controls the temperature of various components to ensure safety and accuracy during a mission.
- **tracking and commanding:** tracking takes account of a craft's position in relation to the ground base with transponders, radar, or other systems. Commanding refers to the ground station sending signals to a craft to change settings such as ascent and orbit paths.

### LEARNING TO FLY

The term "aerospace" refers to Earth's atmosphere and the space beyond. Aerospace design or aerospace engineering is the branch of science and technology that focuses on creating and manufacturing effective aircraft and spacecraft. Aerospace design is divided into two subfields: aeronautic design and astronautical design. Aeronautic design refers to the creation of machines that fly in Earth's atmosphere. Astronautical design deals with designing and developing spacecraft and their launch vehicles, generally powered by highly powerful rockets. The aerospace industry caters to military, industrial, and commercial consumers.

The history of aerospace design begins with the history of aviation. The first powered lighter-than-air craft existed as early as the 1850s. Jules Henri Giffard (1825–82) invented a steam-powered airship in 1852. Ferdinand von Zeppelin (1838–1917) introduced rigid airships in 1900; they came to be known as "zeppelins" and later as "blimps." Brothers Orville (1871–1948) and Wilbur Wright (1867–1912) designed a heavier-than-air, piloted airplane that is generally credited as the first of its kind to execute a successful powered flight, in 1903. Air flight designs continued to progress throughout the twentieth century, paving the way for the development of rotary winged aircraft such as helicopters, which use revolving wings or blades to lift into the air. Powerful air flight engines that fuel modern aircrafts did not emerge until the mid-twentieth century.

At the same time that engine design was advancing, aerospace engineers also began developing machinery that would take humans to the upper atmosphere and into space. Konstantin Tsiolkovsky (1857–1935), known as the Russian father of rocketry, was a pioneer of astronautics with his insightful studies in space travel and rocket science. American engineer Robert Goddard launched the first liquid-fueled rocket in 1926. Goddard continually improved his rocket design over the years. His calculations contributed to the development of other rocket-powered devices such as ballistic missiles. Tsiolkovsky and German scientist Hermann Oberth (1894–1989) independently made similar breakthroughs in the same time period.

During World War I and World War II, aerospace design saw rampant progress as military engineers pursued higher performance aircraft design. In particular, advances in rocket research during World War II laid the foundation for astronautics. Continued advances in aerospace design were spurred by the Cold War and the space race between the United States and the Soviet Union. The competition to reach outer space led to the launch of the first spacecraft, *Sputnik I*, by the Soviet Union in 1957. The United States established the National Aeronautics and Space Association (NASA) in 1958. Aerospace engineers designed a wide range of spacecraft to explore space, incorporating new technologies and capabilities as they became available. For example, in 1969, an American-made spacecraft successfully sent astronauts to the moon, which marked the first time humans landed on the moon. Many other spacecraft have been used to carry out missions elsewhere in the solar system. Artificial satellites have proven critical to twenty-first-century communications systems and become widespread with a variety of designs and functions.

**DESIGN PROCESS**
The design process of a spacecraft generally occurs in the following phases: system/mission requirements, conceptual design, preliminary design, and critical design. Some of these phases can take years to complete. In the system/mission requirements phase, the requirements of the spacecraft are addressed. The type of mission the spacecraft will be used for helps determine these requirements, as specific tasks may dictate certain design elements. For example, a deep space probe would require a significantly different design than a weather monitoring satellite.

The conceptual design phase deals with several possible system concepts that could fulfill the requirements of the mission. These concepts are first conceived and then analyzed. After the most suitable concept is chosen, costs and risks are examined and schedules are made.

The preliminary design phase involves several tasks. Variations of the selected concept are identified, examined, and improved. Specifications for each subsystem and component level are identified. The projected performance of the systems and subsystems is analyzed. Documents are composed, and an initial parts list is put together. This phase may run for several years depending on the novelty and complexity of the mission.

Lastly, the critical design phase, or detailed design phase, takes place. During this phase, the detailed characteristics of the structural design of the spacecraft are established. Equipment, payload, the crew, and provisions are all taken into account. Plumbing, wiring, and other secondary structures are reviewed. Various tests involving design verification are performed, including tests of electronic circuit models and software models. Design and performance margin estimates are improved. Test and evaluation plans are settled. Like the preliminary design phase, the critical design phase may take several years to complete.

Once the design process has been completed, the spacecraft can finally be built. It is then tested before being delivered for use.

**TYPICAL COMPONENTS**

Most spacecraft share two key components: the mission payload and the platform, or the bus. The mission payload includes all of the equipment that is specific to the mission, such as scientific instruments and probes, rather than general operation. The platform comprises all other parts of the spacecraft, used to deliver the payload. It consists of several subsystems, including the structures subsystem, thermal control subsystem, electrical power subsystem, attitude control subsystem, and telemetry, tracking, and commanding (TT&C) subsystem.

The structures subsystem serves various functions such as enclosing, supporting, and protecting the other subsystems, as well as sustaining stresses and loads. It also provides a connection to the launch vehicle. Two main types of structure subsystems exist: open truss and body mounted. The open-truss type typically has the shape of a box or cylinder, while the body-mounted type does not have a definite shape. The choice of structural materials is an important consideration in aerospace design. Light, durable, and heat-resistant materials, such as aluminum, titanium, and some plastics, are typically used.

The thermal control subsystem regulates the temperature of the spacecraft's components. This helps guarantee that the components function properly throughout the mission. Different components require different temperatures. Thermal control systems may be active or passive. Active thermal control involves the use of electrical heaters, cooling systems, and louvers. Passive thermal control includes the use of heat sinks, thermal coatings and insulations, and phase-change materials (PCM). With passive thermal control, electrical power is not needed and there are no moving parts or fluids.

The electrical power subsystem provides the power the spacecraft needs for the duration of the mission, which can last for years. In most cases, the subsystem includes the following components: a source of energy; a device that converts the energy into electricity; a device that stores the electrical energy; and a system that conditions, charges, discharges, distributes, and regulates the electrical energy. The source of energy is generally solar radiation, nuclear power, or chemical reactions.

The attitude control subsystem deals with the process and hardware necessary for obtaining and sustaining the proper orientation in space, or attitude. It has several functions, including maintaining an orbit (station keeping), adjusting an orbit, and stabilization. That subsystem typically includes navigation sensors, a guidance section, and a control section. As with thermal control, the attitude control may be either active or passive. Active attitude control uses continuous decision making and hardware that are closed loop. This includes the use of thrusters, electromagnets, and reaction wheels. Passive attitude control uses open-loop environmental torques to sustain attitude, such as gravity gradient and solar sails.

The telemetry, tracking, and commanding (TT&C) subsystem involves communication with operators on the ground. Telemetry uses a radio link to

System/Mission requirements → Conceptual design → Preliminary design → Critical design ↔ Secondary design

The process of designing aerospace technology begins with reviewing system and mission requirements. From this a conceptual design is developed. This phase is followed by the preliminary design phase in which concept designs are refined and preliminary parts are identified. The critical design phase follows and identifies finer details of the design. During this stage preliminary structure details are finalized and secondary structures are evaluated for inclusion in the critical design.

transmit measurement data to those operators. It is typically used for improving spacecraft performance and for monitoring the health of the spacecraft, including the payload. Tracking and commanding deals with the spacecraft's position. Tracking is used to report the spacecraft's position to the ground station, while commanding is used to change the spacecraft's position. Some common tracking methods are the use of a beacon or a transponder, Doppler tracking, optical tracking, interferometer tracking, and radar tracking and ranging. Commanding is achieved through coded instructions that the ground station sends to the aircraft.

## Practical Example

A good example of aerospace design is that of the Space Shuttle, officially called the Space Transportation System (STS). In 1969, US president Richard Nixon established the Space Task Group to study the United States' future in space exploration. Among other things, the group envisioned a reusable spaceflight vehicle. It was not long before NASA, along with industry contractors, began the design process of such a vehicle. The process involved numerous studies, including design, engineering, cost, and risk studies. Some of the studies focused on the concepts of an orbiter, dual solid-propellant rocket motors, a reusable piloted booster, and a disposable liquid-propellant tank.

In 1972, the design of the Space Shuttle was moved forward. It would feature an orbiter, three main engines, two solid rocket boosters (SRBs), and an external tank (ET). The orbiter would house the crew, the SRBs would provide the shuttle's lift at the beginning of its flight, and the ET would hold fuel for the main engines. All of the components would be reusable, except for the ET, which would be jettisoned after launch. Refinements continued to be made as the project continued and systems were tested.

The first orbiter spacecraft, named *Enterprise*, was completed in 1976 and underwent several flight tests. However, *Enterprise* was merely a test vehicle and was not used for any actual space missions. In 1981, the first Space Shuttle mission took place. *Columbia* lifted off from the Kennedy Space Center and became the first orbiter in space. Over the next several decades, several shuttles successfully performed many space missions.

—*Michael Mazzei*

## Further Reading

*Aerospace Design Lab.* Stanford U, Dept. of Aeronautics and Astronautics, 28 July 2013. Web. 9 June 2015.

Garino, Brian W., and Jeffrey D. Lanphear. "Spacecraft Design, Structure, and Operations." *AU-18 Space Primer.* Air U, 2009. Web. 9 June 2015.

Lucas, Jim. "What Is Aerospace Engineering?" *Live Science.* Purch, 4 Sept. 2014. Web. 9 June 2015.

"Part I. Historical Context." *Space Transportation System.* Denver: Historic American Engineering Record, Natl. Park Service, US Dept. of the Interior, Nov. 2012. PDF file.

"Space Shuttle History." *Human Space Flight.* NASA, 27 Feb. 2008. Web. 9 June 2015.

"Spacecraft Design, Structure, and Operations." *AU Space Primer.* Air U, 2003. *Air University.* Web. 9 June 2015.

# AEROSPACE PROPULSION

## FIELDS OF STUDY

Spacecraft Propulsion; Aerospace Engineering; Space Technology

## SUMMARY

Propulsion is the system creating force for movement. Aerospace propulsion encompasses everything involving the movement of aerospace vehicles. This includes aircraft engine design as well as launching an object into outer space and directing travel through space. Aerospace propulsion is based on the laws of motion developed by English physicist Isaac Newton. While different types of propulsion systems exist, each has the same purpose: to create a thrust to move the object. Scientists are developing new technology to create more efficient, ecofriendly propulsion systems.

## PRINCIPAL TERMS

- **chemical propulsion:** the use of a chemical reaction to propel an object.
- **duct propulsion:** a form of propulsion that uses a duct to direct air or fluid into an engine, increase its momentum, and expel it to generate thrust.
- **jet propulsion:** the means of propelling an object by discharging a jet of ejected matter in the opposite direction of the intended motion.
- **nuclear propulsion:** the use of a nuclear reactor to propel an object.
- **propulsion:** the way of creating force to move an object.
- **rocket propulsion:** a class of jet propulsion in which the thrust is produced by ejecting a propellant such as fuel; used as the main form of spacecraft propulsion because it does not need air.
- **solar propulsion:** propulsion using solar energy, whether collected by solar panels to power an electric engine (ion propulsion) or harnessed directly to push a solar sail.
- **turbomachinery:** devices that use an engine with rotating blades to generate or transfer energy from a fluid.

## THE BEGINNINGS OF AEROSPACE PROPULSION

Propulsion is the way in which a force is created to cause movement. The word "propulsion" is derived from the Latin words *pro*, which means before or forward, and *pellere*, which means to drive. There are many forms of propulsion, from biological processes to complex mechanical inventions. Aerospace propulsion involves those systems used to move aircraft and spacecraft.

The birth of modern aerospace propulsion technology dates back to the early twentieth century with the advent of powered airplanes. These early flying devices used engines and one or more propellers to generate thrust and gain lift off. Most were powered by combustion, or burning fuel, although electric motors and other systems were also developed. Propeller technologies were refined over the years, including the use of turbomachinery such as turboprops, and remained in use with certain types of aircraft.

The creation of more powerful propulsion technology accelerated through the 1930s, pushed by the formation of what in 1943 would become the Jet Propulsion Laboratory (JPL) in the United States. It built on the early development of rocket technology by engineer Robert Goddard (1882–1945). JPL applied and adapted rockets to help aircraft achieve liftoff and refined rocket designs. World War II further spurred advances in rocketry in Germany, the Soviet Union, and the United States. The first fully rocket-powered aircraft were tested and expanded the capabilities of flight.

Rocket technology opened up the possibility of space travel. After World War II, the space race broke out between the United States and the Soviet Union as the superpowers competed technologically. The Soviet Union was the first to reach space with the launch of *Sputnik* in 1957. A year later, the United States launched the unmanned satellite *Explorer 1*. The same year, the United States formed the National Aeronautics and Space Administration (NASA) to continue its focus on aerospace research. JPL joined NASA in 1958, and both groups continued to work on new propulsion methods to allow further exploration. As of 2015, rockets remained the chief form of spacecraft propulsion technology, but alternative

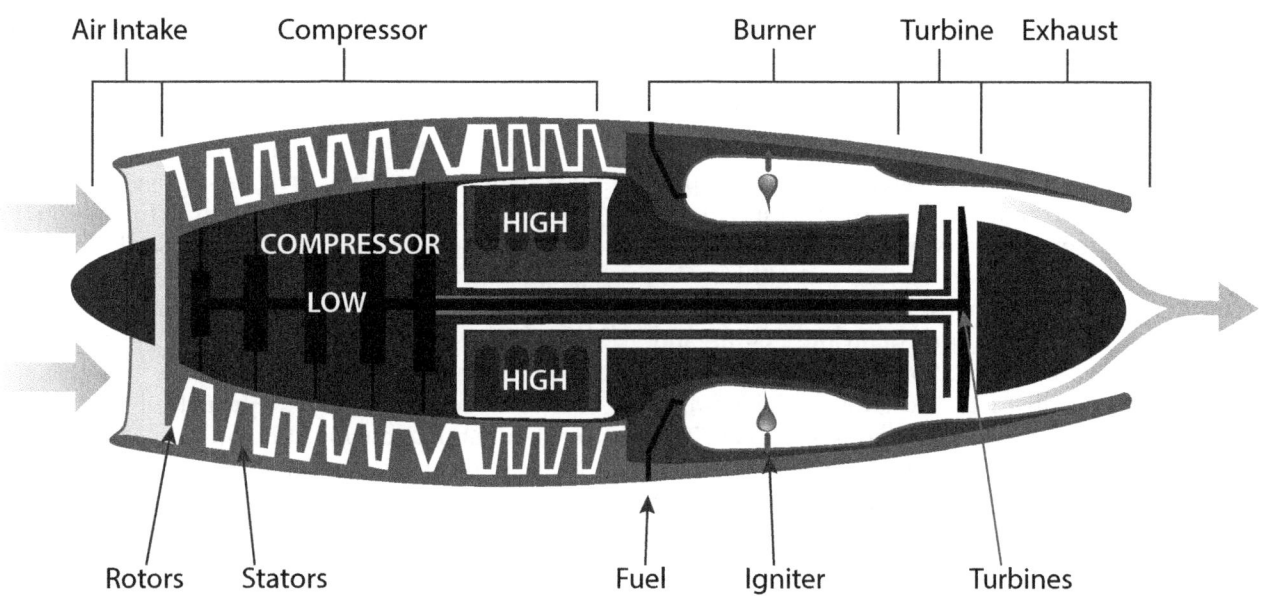

Simplified model of a cross section of a turbojet engine. Air flows in through the air inlet (or intake) section, then passes through the compressor to the burner section where fuel is ignited. After pushing through the turbine section, air exits the engine through the exhaust section, providing thrust.

forms have the potential to allow cheaper, more efficient space travel.

**How Does Aerospace Propulsion Work?**
All propulsion is based on the work of English physicist Sir Isaac Newton (1642–1727), specifically his second and third laws of motion. The third law states that every action has an equal and opposite reaction. This means that a pair of forces (push and pull) exist that act when any two objects interact. The sizes of the two forces are always the same. The direction of the force on the first object is always directly opposite the direction of the force on the second object. If a person or thing pushes on something, it pushes back on the person or thing.

For example, if a person sits in a chair, the person exerts a downward force on the chair, while the chair exerts an upward force on the person. The result from the interaction is two forces: one pushing on the chair and one pushing on the person. The chair and person stay in one place only because of the restrictive force of friction and comparatively powerful force of Earth's gravity acting on both.

In propulsion, acceleration and mass also play a role. Newton's second law states that the acceleration of an object is directly proportional to the magnitude of the net force exerted on the object. It is inversely proportional to the object's mass. A bird flapping its wings exerts a force on the air, and the air exerts a force against the bird. Because the mass of the bird is relatively small, the acceleration produced by the interaction overcomes friction and allows the bird to take off.

These examples can be applied to how aircraft and spacecraft move. Engines power the propulsion system, whether a propeller or jet. This creates a thrust force, typically by accelerating a gas's or other propellant's mass. The engine forces the propellant out one end, forcing the craft itself in the opposite direction. The expelled matter and the craft exert equal and opposite forces on each other. If the thrust force is large enough compared to the mass of the craft and surrounding friction, or drag, the craft will move.

Airplanes and launch rockets operate in the atmosphere, and many aerospace propulsion systems, such as propellers, rely on air. They also must overcome the air's friction to remain in flight. For propulsion in space, where there is no air to push against, rockets are suitable because they push against their own exhaust gases.

## Propulsion Systems

Several types of aerospace propulsion systems exist, but each basically has the same purpose. They use different sources and move to propel an object through air or space. The two main categories are propellers and jet propulsion, with subcategories of each providing different advantages. Some spacecraft employ other systems, generally relying on solar power. The major difference between various propulsion systems is the amount of thrust generated. The thrust depends on the mass flow through the engine and the exit velocity of the gas. Some aircraft and spacecraft need high thrust to accelerate quickly, while others need less to fly along at a constant speed. The amount of thrust also dictates fuel efficiency.

A spinning propeller provides propulsion through the air and was the first means of propelling aircraft. Propellers may be powered by combustion or electric engines and may take several forms. A propeller may be powered by turbomachinery to create a turboprop or a propfan. One form of duct propulsion involves a fan propeller surrounded by a duct, which enhances certain operating qualities.

Jet propulsion involves an engine that expels a jet of matter to achieve thrust. Some jets use turbines and duct propulsion to rely on air. The oxygen gained from the air flow is used to burn the fuel and create exhaust propellant. Some examples of duct propulsion jet engines include turbojets and ramjets. Duct propulsion engines give a high thrust-to-weight ratio and a high thrust to the frontal area of an aircraft or spacecraft. This allows longer flight ranges in low altitudes. The turbojet is the most common of type of duct propulsion engine, while the ramjet is used for high-speed flight within the atmosphere.

Rocket propulsion systems are a class of jet propulsion in which the thrust is produced by ejecting a stored propellant without taking in air. Rockets are the most powerful but least efficient type of jet engine. They are used to launch spacecraft into orbit as well as to direct most types of spacecraft. They can be classified in different ways, such as by size, propellant type (liquid, solid, or gas), basic function, vehicle type (aircraft or spacecraft), or energy source (chemical, nuclear, or electric).

Chemical propulsion systems use a chemical reaction to propel an object. They are classified by the type of propellant used: liquids, solids, gases, or a combination of liquids and solids. Bipropellant systems contain two liquids, an oxidizer such as liquid oxygen and a fuel such as kerosene. Monopropellants use a single liquid that contains a mixture of both an oxidizer and a fuel. Solid propellant engines use solids called grain. Gaseous propellant engines use high-pressure gas such as nitrogen or helium. Because of the heaviness of the tanks for gaseous propellants, however, few spacecraft use these. Hybrid propellant engines use both liquid and solid propellants, but they are less common than liquid or solid propellant engines.

Nuclear propulsion engines use a nuclear reactor to generate heat to propel an object. They work much like chemical liquid propellant engines. Nuclear propulsion engines are a type of electric propulsion engine. In electric propulsion engines, energy comes from a separate source than the propellants. (In chemical propulsion systems, the power comes from the propellants themselves.) Several theoretical types of nuclear spacecraft propulsion have been proposed.

Solar propulsion refers to the use of solar energy to propel a craft. The simplest form is an airplane with an electric engine powered by solar panels. However, some spacecraft use solar-generated electricity to power electromagnetic propulsion. These include electrostatic ion propulsion (EIP) and Hall effect drives, which accelerate ions to generate thrust. Other electromagnetic systems, such as the variable specific impulse magnetoplasma rocket (VASIMR), are under development. Solar sails may also be considered a type of solar propulsion.

## Solar Sails

The advancement of aerospace propulsion technology not only applies to aeronautics but also extends to other fields. This technology could help the environment in the long term. Scientists have the potential to devise propellants that have less of a negative impact on Earth. They even have the ability to work on new technologies that could eliminate the need for propellants. One of these is the solar sail, which has been in development for many years. It uses the pressure of solar wind and solar radiation to propel a vehicle through space without ejecting any mass.

A solar sail was first successfully used by the Japanese *Interplanetary Kite-Craft Accelerated by Radiation of the Sun (IKAROS)* satellite that launched in May 2010. The United States launched the NanoSail-D in November 2010, which initially had some issues. However, by January 2011, it had released from its larger satellite and deployed its solar sail. In May 2015, the Planetary Society successfully launched the solar sail satellite *LightSail* aboard the *Atlas V* rocket and deployed into space.

In future, advancements may increase the speed of space vehicles while reducing fuel consumption. This would allow humans to probe deeper into outer space and travel greater distances to celestial bodies.

—*Angela Harmon*

**FURTHER READING**

Desain, John D. "Green Propulsion: Trends and Perspectives." *Aerospace*. Aerospace Corp., 2011. Web. 5 June 2015.

Goebel, Greg. "Spaceflight Propulsion." *Vectors*. Greg Goebel, 1 Feb. 2014. Web. 5 May 2015.

"How Do DS1 and Other Spacecraft Propel Themselves through Space?" *Qualitative Reasoning Group*. Northwestern U, n.d. Web. 5 June 2015.

Howell, Elizabeth. "Ikaros: First Successful Solar Sail." *Space*. Purch, 7 May 2014. Web. 5 June 2015.

"JPL History." *Jet Propulsion Laboratory California Institute of Technology*. NASA, n.d. Web. 5 June 2015.

"LightSail Launch Success!" *EarthSky*. Earthsky Communications, 20 May 2015. Web. 5 June 2015.

"NASA Solar Sail Prototype Gets Longer Life in Orbit." *Space*. Purch, 26 Apr. 2011. Web. 5 June 2015.

"Newton's Third Law." *Physics Classroom*. Physics Classroom, 2015. Web. 5 June 2015.

Sutton, George P., and Oscar Biblarz. *Rocket Propulsion Elements*. Hoboken: Wiley, 2010. Print.

"Welcome to the Beginner's Guide to Propulsion." *Glenn Research Center*. NASA, 5 May 2015. Web. 5 June 2015.

# AGAMEMNON: 911 AGAMEMNON

**FIELDS OF STUDY**

Astronomy; Observational Astronomy; Cosmology

**SUMMARY**

911 Agamemnon is a Trojan asteroid near Jupiter. Trojan asteroids are asteroids that share the orbital periods of other planets. They travel just ahead of or behind the planet's orbit. 911 Agamemnon is one of the largest Trojan asteroids. German astronomer Karl Wilhelm Reinmuth discovered it in 1919.

**PRINCIPAL TERMS**

- **asteroid:** a small, rocky space object that orbits the sun or another celestial body.
- **Trojan asteroid:** an asteroid that shares an orbital period with a planet and maintains a consistent distance from that planet.

**TROJAN ASTEROIDS AND 911 AGAMEMNON**

911 Agamemnon is a Jupiter Trojan asteroid. A Trojan asteroid orbits the sun at about the same rate of time as its planet does and just ahead of or behind the planet's orbit. Trojan asteroids are sometimes called minor planets, and they are left over from when the solar system formed billions of years ago. Most of the known Trojan asteroids share Jupiter's orbit, but Mars and Neptune Trojans also exist. Scientists even discovered an Earth Trojan asteroid in 2011.

Researchers believe that 911 Agamemnon is the second-largest Jupiter Trojan asteroid, with a diameter of about 167 kilometers (104 miles). In 2012, researchers recorded 911 Agamemnon as it passed over North America. From their observations, researchers determined that it most likely has a small satellite, or moon, that orbits it.

**HISTORY OF 911 AGAMEMNON**

Karl Wilhelm Reinmuth (1892–1979) was a scientist who searched for undiscovered asteroids. Reinmuth started working at the Heidelberg Observatory in Germany in 1912 and remained there until 1957. During that time, Reinmuth discovered 395 asteroids. On March 19, 1919, Reinmuth discovered Trojan asteroid 911 Agamemnon.

From the 1960s to the early 1990s, scientists studied the movements of 911 Agamemnon and determined that it had a rotational period of about seven hours.

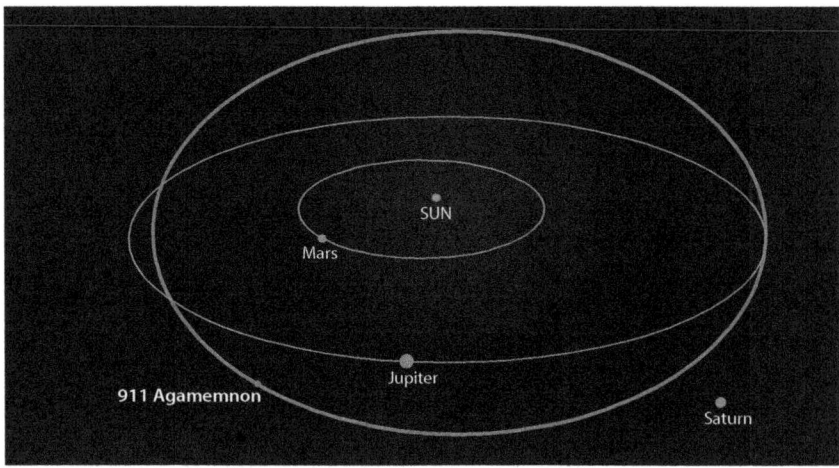

Orbit of 911 Agamemnon, asteroid.

However, a 2009 study gave a more precise measurement of 6.59 hours per rotation.

Studying Trojan asteroids, such as 911 Agamemnon, is important for several reasons. The first is that those asteroids were most likely formed when the solar system was formed. Therefore, they could give scientists information about the very early solar system. Furthermore, Trojan asteroids and other asteroids could be mined in the future to bring valuable resources back to Earth.

—Elizabeth Mohn

**FURTHER READING**

"Asteroids: Overview." *Solar System Exploration*. NASA, 13 Jan. 2015. Web. 13 Mar. 2015.

Hughes, Stefan. *Catchers of the Light*. Cyprus: ArtDeCiel, 2013. Print.

Khan, Amina. "Trojan Asteroid Tags along on Earth's Orbit." *Los Angeles Times*. Los Angeles Times, 28 July 2011. Web. 11 Mar. 2015.

Mottola, Stefano, et al. "Rotational Properties of Jupiter Trojans." *Astronomical Journal* 141.5 (2011): 170–201. Print.

"911 Agamemnon." *NASA*. NASA, n.d. Web. 11 Mar. 2015.

Timerson, Bradley, et al. "Occultation Evidence for a Satellite of the Trojan Asteroid (911) Agamemnon." *Planetary and Space Science* 87 (2013): 78–84. Print.

# AMOR ASTEROID GROUP: 1221 AMOR

## FIELDS OF STUDY

Astronomy; Observational Astronomy; Asteroid Impact Avoidance

## SUMMARY

1221 Amor is the namesake asteroid of the Amor asteroid group. It was discovered in 1932 at an orbit about 16 million kilometers (about 9.9 million miles) from Earth. As the first asteroid seen orbiting close to Earth, it showed astronomers that asteroids exist in the solar system in places other than the asteroid belt. It also helped them realize that asteroids have the potential to collide with Earth and other planets.

## PRINCIPAL TERMS

- **Amor asteroid group:** a group of near-Earth asteroids that come within 0.3 astronomical units (AU) of Earth but do not cross its orbit, although most do cross Mars's orbit.
- **aphelion:** the point in a space object's orbital path that is farthest from the sun.
- **asteroid:** a small, irregularly shaped celestial body that orbits the sun and is composed of rock, metal, and silicate.
- **near-Earth asteroid:** an asteroid that travels within 0.3 astronomical units (AU) of Earth's orbit and within 1.3 AU of the sun.
- **perihelion:** the point in a space object's orbital path that is closest to the sun.

## NEAR-EARTH ASTEROIDS

1221 Amor is the namesake asteroid of the Amor asteroid group. Amor asteroids are small, irregularly shaped space objects that travel an elliptical path around the sun. They often cross the orbit of Mars. Amor asteroids are a type of near-Earth asteroid (NEA). This means that they come within 0.3

astronomical units (AU)—about 44.9 million kilometers, or 27.9 million miles—of Earth. An NEA's perihelion, or closest point to the sun, is between 1 and 1.3 AU. When 1221 Amor was first discovered, it was orbiting at about 0.1 AU (16 million kilometers, or 9.9 million miles) from Earth. It was later determined that this is Amor's closest point to Earth.

There are millions of asteroids in Earth's solar system. Most of these can be found within the asteroid belt, a region of space between the orbits of Mars and Jupiter. However, some asteroids are ejected from the asteroid belt when they collide with other space objects or encounter the immense gravity around Jupiter. When these asteroids end up orbiting close to Earth and the sun, they are called NEAs. Amor asteroids are unique among NEAs because while they travel close to Earth, they do not cross its orbit. Most NEAs are among the category of space objects known as near-Earth objects (NEOs). Other NEOs include meteoroids and comets that have been pulled by gravity into orbit near the four inner planets (Mercury, Venus, Earth, and Mars). Amor asteroids cross the orbit of Mars. Some have orbital paths so eccentric (elongated) that their aphelions, or farther points from the sun, are beyond Jupiter.

### Defining Characteristics

Of all the asteroid groups, Amor asteroids vary the most in composition. Some are mainly rock, while others are composed of some combination of rock, mineral, and metal. This variation in composition suggests that Amor asteroids originated in a number of different places in the asteroid belt. Planetary scientists believe that Amor asteroids were forced into their current orbits by encounters with the gravity of either Jupiter or Mars. Many also believe that a number of Amor asteroids may be comets that have burned off their volatile substances and can no longer form a comet's characteristic coma or tail.

In terms of appearance, Amor asteroids resemble other asteroids, with surfaces full of visible pits and craters. These are most likely the result of collisions with other objects in space. Amor asteroids also vary in size. The largest Amor asteroid is 1036 Ganymed, with an estimated diameter between 31 and 35 kilometers (19 and 22 miles).

In contrast, 1221 Amor is on the smaller side, at about 1 kilometer (0.6 mile) in diameter. Its semimajor axis, or the most distant point of its orbital radius, lies between Mars and the asteroid belt. It crosses the orbit of Mars from the outside. Amor has been observed orbiting close to Earth numerous times since its discovery. It reaches its closest point to Earth (about 0.1 AU) every eight years.

Amor is a faint object with an absolute magnitude of 17.7. Absolute magnitude measures how bright an object appears from a distance of 10 parsecs (about 309 trillion kilometers, or 192 trillion miles). The smaller the number, the brighter the object is. The brightest objects in the universe have a negative absolute magnitude.

### Discovery of 1221 Amor

Belgian astronomer Eugène Delporte (1882–1955) discovered sixty-six asteroids during his lifetime. He achieved this by taking countless photographs of the night sky and studying them closely. Delporte found 1221 Amor in March 1932 while working at the Royal Observatory of Belgium. Amor was the first asteroid to be found so close to Earth. It was originally designated 1932 EA1 but was later renamed. Since then, astronomers have identified more than 4,700 Amor asteroids in the solar system. In fact, they make up about 38 percent of all known NEAs.

The gravitational fields of nearby planets may capture some Amor asteroids and change their orbital paths. Some planetary scientists hypothesize that Mars's two moons, Phobos and Deimos, are actually Amor asteroids that were pulled in by Mars's gravity.

### Significance of Discovery

The discovery of 1221 Amor was significant for planetary scientists because it made them realize that some asteroids travel close to Earth. This helped scientists to uncover truths about life on Earth, but it also raised new concerns about a possible collision between Earth and an asteroid that might get too close. Most planetary scientists agree that planet-asteroid collisions were probably fundamental to the formation of Earth as it exists today. Such collisions caused asteroids to deposit substances on Earth that are necessary for life to occur, including water and organic material.

More recent collisions have had more catastrophic effects. An immense crater near Chicxulub, Mexico, is believed to be evidence of a collision between Earth and a large asteroid about sixty-five million years ago. This event would have released energy equivalent to a massive nuclear explosion and may have been a major factor in the Cretaceous-Tertiary mass extinction, which resulted dinosaurs and other large lifeforms dying out.

Amor asteroids such as 1221 Amor are generally not considered an immediate threat because they do not cross Earth's orbit in space. However, some Amor asteroids have been known to become caught in the gravity around Mars and evolve into Apollo asteroids. Apollo asteroids are asteroids that cross the orbits of both Earth and Mars.

### NASA AND THE THREAT OF NEAS

The National Aeronautic and Space Administration (NASA) created the Near-Earth Object Program to search for unknown NEOs and keep careful track of known NEOs. The goals of the program are to learn more about these objects and to develop sophisticated technologies that will someday allow scientists to divert NEOs away from Earth. Teams of international experts use state-of-the art technologies, such as massive radars that bounce radio waves off asteroids, to conduct their work.

The Asteroid Redirect Mission (ARM) is NASA's first attempt to physically capture an NEA so scientists can study it directly. The ARM team is developing technologies that will allow scientists to redirect NEAs into the orbit around Earth's moon. Once the NEA is moved into the moon's orbit, astronauts plan to land on it and explore it firsthand.

—*Marie Keenan, MS*

### FURTHER READING

"Amor Asteroids." *Cosmos: The SAO Encyclopedia of Astronomy.* Swinburne University of Technology, n.d. Web. 27 Mar. 2015.

Chapman, Clark R. "The Hazard of Near-Earth Asteroid Impacts on Earth." *Earth and Planetary Science Letters* 222.1 (2004): 1–15. Print.

Harris, Alan W., et al. "Near-Earth Objects." *Encyclopedia of the Solar System.* Ed. Tilman Spohn, Doris Breuer, and Torrence V. Johnson. 3rd ed. Waltham: Elsevier, 2014. 603–23. Print.

"Near-Earth Objects—NEO Segment." *European Space Agency.* ESA, n.d. Web. 27 Mar. 2015.

*Near-Earth Object Program.* Caltech, 6 Feb. 2015. Web. 27 Mar. 2015.

"1221 Amor (1932 EA1)." *JPL Small-Body Database Browser.* Caltech, 9 Dec. 2014. Web. 27 Mar. 2015.

"Responding to Potential Asteroid Redirect Mission Targets." *Jet Propulsion Laboratory.* Caltech, 14 Feb. 2014. Web. 27 Mar. 2015. <http://www.jpl.nasa.gov/news/news.php?release=2014-052>

Trachet, Tim. "Delporte, Eugène-Joseph." *Biographical Encyclopedia of Astronomers.* Ed. Thomas Hockey et al. New York: Springer, 2007. 288–89. Print.

# AMOR ASTEROIDS

### FIELDS OF STUDY

Astronomy; Observational Astronomy; Cosmology

### SUMMARY

Amor asteroids are a type of near-Earth asteroid (NEA) that often cross the orbit of Mars as they travel an elliptical path around the sun. Their orbits are sometimes large enough to extend beyond Jupiter. Although Amor asteroids travel close to Earth, they do not cross its orbit.

### PRINCIPAL TERMS

- **near-Earth asteroid:** a small, irregularly shaped celestial body with an orbital path that brings it within 0.3 astronomical units (AU) of Earth's orbit and within 1.3 AU of the sun.

### NEAR-EARTH ASTEROIDS

Amor asteroids are small, irregularly shaped space objects that follow an elliptical orbit around the sun. Their orbital paths often cross that of Mars. Amor asteroids are a type of near-Earth asteroid (NEA), which are asteroids that come within 0.3 AU (about 44.8 million kilometers, or 28 million miles) of Earth. Amor asteroids are unique among NEAs because although they travel close to Earth, they do not cross its orbit.

The solar system contains millions of asteroids. Most of these are found within the asteroid belt, a doughnut-shaped ring located between the orbits of Mars and Jupiter, about 2 to 4 AU (300 to 600 million kilometers) from the sun. However, not all asteroids stay within the asteroid belt. For example, if two asteroids collide, one or both of them can be pushed out of the belt and into the larger solar system. Similarly, an encounter with the immense gravity around Jupiter, the largest planet, can also fling an asteroid out into space. When such asteroids end up orbiting close to Earth and the sun, they are referred to as NEAs.

NEAs belong to the category of space objects known as near-Earth objects (NEOs). NEOs include both comets and asteroids that have been pulled near the solar system's four inner planets—Mercury, Venus, Earth, and Mars—by the force of gravity.

Asteroids are composed mainly of rock but may also include clay, minerals, and metals. Their outer surfaces are full of visible pits and craters, which scientists believe are the result of collisions with other objects in space. Asteroids vary greatly in size. The largest known asteroid, Ceres, measures about 940 kilometers (nearly 600 miles) in diameter. Most asteroids are much smaller; in fact, many are only the size of small rocks.

NEAs can be distinguished from one another by characteristics such as their average distance from the sun and the size and shape of their orbital paths. Amor asteroids can be identified through their distinct orbital paths, which often cross the orbital path of Mars but do not cross that of Earth. Some Amor asteroids have such wide orbits that they travel beyond Jupiter.

Astronomers have identified about fifteen hundred Amor asteroids in the solar system. These objects make up about 32 percent of all known NEAs.

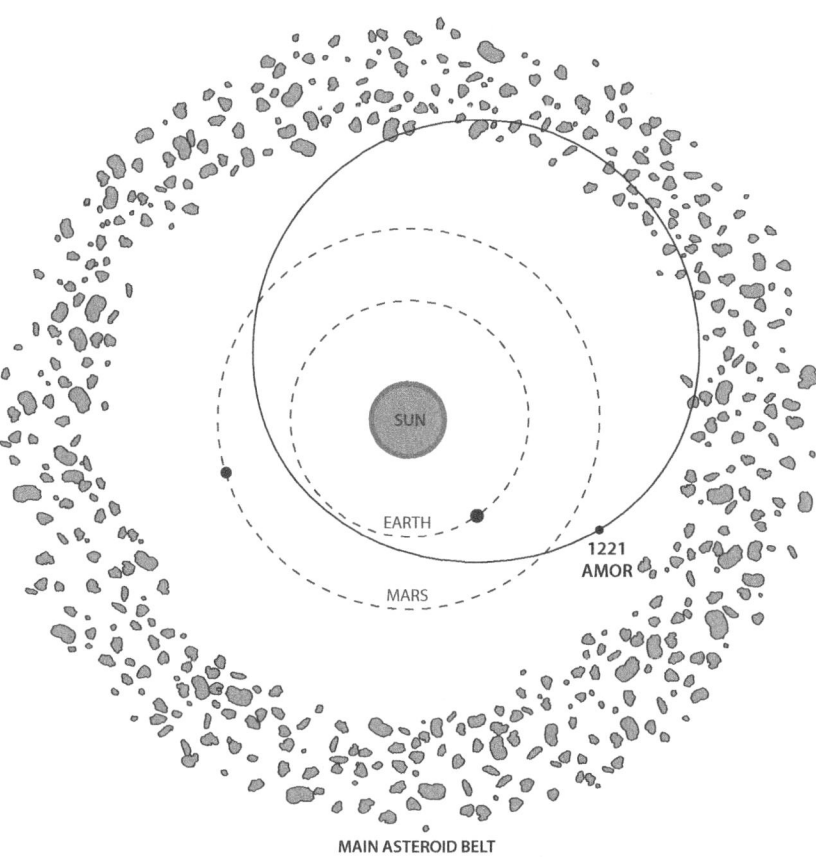

The orbit of an Amor asteroid circles the sun and extends into the asteroid belt, crosses Mars's orbit, but does not cross Earth's orbit.

Additionally, the gravitational forces that surround Earth and Mars may capture some Amor asteroids. When this occurs, scientists believe these asteroids change paths and evolve into Apollo asteroids, which are asteroids that cross the orbits of both Earth and Mars.

## NEA IMPACTS ON EARTH

Scientists are interested in Amor asteroids and other NEAs because they are thought to be leftover debris from the formation of the solar system about 4.6 billion years ago. They can provide valuable information about what happened at this time. Additionally, about 20 percent of NEAs come close enough to Earth to raise concerns about potential collisions. Although smaller space objects such as meteors collide with Earth frequently, these objects are usually small enough to burn up in Earth's atmosphere before reaching the planet's surface.

However, major collisions between Earth and NEAs have occurred. For example, a huge crater near Chicxulub, Mexico, is believed to be evidence of a past collision between Earth and a large NEA. Many experts believe this collision, which probably had the force of a nuclear explosion, played a major role in the Cretaceous-Tertiary mass extinction sixty-five million years ago, during which dinosaurs and other large life-forms on the planet died out.

Other NEAs have collided with Earth more recently. For example, in February 2013, a relatively small NEA exploded over Russia, releasing the energy of a large atomic bomb.

### ASTEROID MISSIONS

In the late twentieth and early twenty-first centuries, the National Aeronautics and Space Administration (NASA) and its international counterparts have focused on robotic missions. These missions involve sending unmanned spacecraft into the solar system to learn more about space objects. The NEAR Shoemaker mission, for example, landed on Amor asteroid 433 Eros in February 2001. This spacecraft sent valuable data back to Earth, furthering scientific knowledge of asteroids and other space objects.

NASA also created the Near Earth Object Search Program (NEOP) to detect and track NEOs and to develop new technologies for both robotic and manned space research. The Asteroid Redirect Mission (ARM) is part of this program. It would be NASA's first attempt to redirect a NEA in space so that it can be studied directly by human astronauts.

—*Marie Keenan, MS*

### FURTHER READING

"Amor Asteroids." *Cosmos: The SAO Encyclopedia of Astronomy*. Swinburne University of Technology, n.d. Web. 9 Mar. 2015.

Chapman, Clark R. "The Hazard of Near-Earth Asteroid Impacts on Earth." *Earth and Planetary Science Letters* 222.1 (2004): 1–15. Print.

*Discovery Is NEAR: Clues to Our Origins*. Laurel: Johns Hopkins U Applied Physics Laboratory, 2001. *Near Earth Asteroid Rendezvous Mission*. Web. 27 Mar. 2015.

"Near Earth Asteroids (NEAs): A Chronology of Milestones." *International Astronomical Union*. IAU, 7 Oct. 2013. Web. 27 Mar. 2015.

"Near-Earth Objects—NEO Segment." *European Space Agency*. ESA, n.d. Web. 9 Mar. 2015.

*Near-Earth Object Program*. National Aeronautics and Space Administration, 6 Feb. 2015. Web. 27 Mar. 2015.

"Responding to Potential Asteroid Redirect Mission Targets." *Jet Propulsion Laboratory*. California Inst. of Technology, 14 Feb. 2014. Web. 9 Mar. 2015.

"The Helin Commemorative Exhibit: Searching the Sky for Dangerous Neighbors; Eleanor Helin and the 18-Inch Telescope." *Palomar Observatory*. California Inst. of Technology, 8 May 2014. Web. 27 Mar. 2015.

# ANNEFRANK: 5535 ANNEFRANK

### FIELDS OF STUDY

Sub-planetary Astronomy; Observational Astronomy; Cosmology

### SUMMARY

5535 Annefrank is an asteroid that is part of the asteroid belt in the solar system. Discovered in 1942, this asteroid was named for the famous World War II diarist Anne Frank. The NASA *Discovery Stardust* spacecraft captured images of 5535 Annefrank during a flyby in 2002.

### PRINCIPAL TERMS

- **asteroid:** a small, irregularly shaped space object that is made of rock, silicates, and metals, and orbits the sun in the solar system.

### ASTEROID BELT

5535 Annefrank is an asteroid. Like most asteroids in the solar system, it is found in the asteroid belt. The asteroid belt is an oval-shaped ring between the orbits of Mars and Jupiter. It is about 2 to 4 astronomical units (186 to 370 million miles) from the sun. Scientists estimate that the belt contains more than

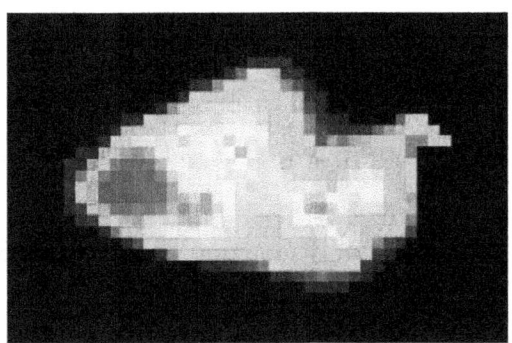

5535 Annefrank, asteroid. STARDUST Team, JPL, NASA.

two hundred asteroids larger than 97 kilometers (60 miles) in diameter and thousands more of smaller sizes. 5535 Annefrank is about 6 kilometers (3.5 miles) long and 5 kilometers (3 miles) wide.

Asteroids are believed to be rocky debris left over from the formation of the solar system about 4.6 billion years ago. Scientists theorize that after the gas giant Jupiter formed, its massive gravity caused many small bodies nearby to collide. This created the objects known as asteroids.

As an S-class asteroid, 5535 Annefrank is made mainly of silicates (stone) and metals like nickel and iron. It is one of the brighter S-class asteroids. This suggests that it may be a newer space object.

### CHARACTERISTICS OF 5535 ANNEFRANK

German astronomer Karl Reinmuth (1892–1979) discovered 5535 Annefrank in 1942. It was the 5535th asteroid found. The asteroid was later given the name of the famous World War II diarist Anne Frank, who died in a German concentration camp.

Images of 5535 Annefrank were taken by the NASA *Discovery Stardust* spacecraft in 2002. They show a highly irregular object that appears to be shaped like a triangular prism with three sides. Its surface looks quite angular with many flat planes.

### STARDUST MISSION

The NASA Stardust mission photographed the asteroid in November 2002 on its way to the comet 81P/Wild 2. Stardust took more than seventy images of 5535 Annefrank during a fifteen-minute period. These images captured about 40 percent of the asteroid's surface. They highlight features such as the asteroid's irregular triangular shape, few visible craters, and relative brightness.

—*Marie Keenan, MS*

### FURTHER READING

"Asteroids: Overview." *Solar System Exploration.* NASA, 13 Jan. 2015. Web. 23 Mar. 2015.

Duxbury, T. C., et al. "Asteroid 5535 Annefrank Size, Shape, and Orientation: Stardust First Results." *Journal of Geophysical Research* 109.E2 (2004): n. pag. Web. 13 Mar. 2015.

"5535 Annefrank Overview." *NASA.* NASA, 13 Jan. 2015. Web. 11 Mar. 2015.

Lang, Kenneth R. "Asteroids and Meteorites." *The Cambridge Guide to the Solar System.* Cambridge: Cambridge University Press, 2011. 365–90. Print.

"Missions to Comets: Stardust, Deep Impact." *NASA Discovery Program.* NASA, 13 Jan. 2015. Web. 11 Mar. 2015.

"Small Worlds: The Neighborhood." *Marshall Space Flight Center Discovery Program.* NASA, n.d. Web. 11 Mar. 2015.

# APOLLO ASTEROID GROUP: 1862 APOLLO

## FIELDS OF STUDY

Asteroid Impact Avoidance; Astronomy; Observational Astronomy

## SUMMARY

1862 Apollo is the namesake object of the Apollo asteroid group. These are asteroids whose perihelion (minimum distance from the sun) lies within Earth's orbit and whose aphelion (maximum distance from the sun) lies outside it, causing them to cross Earth's orbit as they travel around the sun. 1862 Apollo is a near-Earth asteroid (NEA) that is also classified as a potentially hazardous object (PHO) because of its size and close orbit.

## PRINCIPAL TERMS

- **Apollo asteroid group:** the group of asteroids whose perihelion lies within Earth's orbit and whose aphelion lies outside it, causing them to cross Earth's orbit as they travel around the sun.
- **asteroid:** a small, irregularly shaped space rock that orbits the sun.
- **potentially hazardous object:** a space object greater than 140 meters (about 460 feet) in diameter with an orbit that passes within 0.05 astronomical units (about 7.5 million kilometers, or 4.6 million miles) of Earth's orbit.
- **Q-type asteroid:** an asteroid that is made up of olivine (a rare iron silicate), pyroxene, and metals.
- **YORP effect:** a phenomenon in which objects in space change their rate of spin when exposed to rays from the sun.

## NEAR-EARTH ASTEROIDS (NEAs)

1862 Apollo is the original member of the Apollo asteroid group. Apollo asteroids are a particular type of asteroid whose perihelion, or minimum distance from the sun, lies within Earth's orbit, while their aphelion, or maximum distance from the sun, lies outside it. This causes the asteroids to cross Earth's orbit twice as it completes one full revolution around the sun. Apollo asteroids may cross the orbits of other planets as well. 1862 Apollo crosses the orbits of three planets in its travels: Earth, Mars, and Venus. Completing this orbit takes 1862 Apollo about two years.

Millions of asteroids orbit within the solar system. Astronomers believe that asteroids are debris left over from the formation of the solar system about 4.6 billions of years ago. Most of these are located in the asteroid belt, a doughnut-shaped ring of space between the orbits of Mars and Jupiter. However, some asteroids orbit in other parts of the solar system. Although scientists have different theories about where these asteroids came from, they know that many of them were ejected from the asteroid belt after colliding with one another or getting caught by Jupiter's or Mars's gravity. Additionally, scientists hypothesize that some asteroids may have originated as comets in the outer reaches of the solar system. After these comets burned off their volatile substances, they left behind rocky space bodies that were no longer able to form the distinctive comet's tail.

Asteroids sometimes travel close to Earth and the sun. Apollo asteroids are a type of near-Earth asteroid (NEA), which is an asteroid that comes within 0.3 astronomical units (AU)—about 44.9 million kilometers, or 27.9 million miles—of Earth's orbit, and thus 1.3 AU of the sun. (One AU is equal to about 149.6 million kilometers, or 92.9 million miles—approximately the distance from Earth to the sun.) NEAs are part of a larger group of space objects called near-Earth objects (NEOs). This group includes comets and asteroids that have been pulled close to the four inner planets—Mercury, Venus, Earth, and Mars—by gravitational forces.

## DISCOVERY OF 1862 APOLLO AND ITS MOON

German astronomer Karl Reinmuth (1892–1979) discovered 1862 Apollo in 1932. He noticed the asteroid tumbling past Earth at very high speed. However, since the asteroid appeared small in size, it did not seem important. Reinmuth noted the observation but did not carefully track the asteroid. It was lost shortly thereafter.

Astronomers at the Harvard Observatory rediscovered Reinmuth's lost asteroid in 1973. They noted that its diameter was about 1.5 kilometers (about 0.9 mile) and it approached the sun at a distance of about 0.65 AU (97.2 kilometers, or 60.4 million miles). It was designated HA 1932. Later, it was officially named 1862 Apollo, after the mythological Greek god who is sometimes said to be the sun's chariot driver. Astronomers determined that 1862 Apollo followed an orbit that brought it within 0.3 AU of Earth every two years.

In 2005, astronomers at the Arecibo Observatory in Puerto Rico announced the discovery of a satellite, or moon, orbiting 1862 Apollo. The satellite was detected via radar. It is believed to be about 80 meters (262.5 feet) in length. It travels a three-kilometer (about two miles) orbit around 1862 Apollo.

## Apollo's Composition and Rotation

1862 Apollo is classified as a Q-type asteroid. The composition of Q-type asteroids is similar to that of common meteorites. They are made up of rock, olivine, pyroxene, and several other metals. While Q-types are rarely observed in the asteroid belt, they are quite common among small NEAs. Apollo asteroids vary in size. The largest, 1866 Sisyphus, measures about 9 kilometers (5.6 miles) in diameter.

In 2007, astronomers confirmed that 1862 Apollo is subject to a phenomenon known as the YORP effect. The YORP (which stands for Yarkovsky-O'Keefe-Radzievskii-Paddack) effect states that exposure to sunlight affects asteroids' rates of spin and orientation. When photons from sunlight reach an asteroid, it absorbs this energy and eventually reradiates it away from itself as heat. This process both increases the asteroid's rate of spin and changes its orientation. This can greatly impact how the asteroid evolves over millions of years. The YORP effect is especially impactful on small asteroids, which can begin to spin so quickly that they can break into many pieces. Scientists believe that the YORP effect can explain why some asteroids have satellites: a large chunk spun off the asteroid's body and then was caught in its orbit.

## Potentially Hazardous Objects (PHOs)

The National Aeronautics and Space Administration (NASA) defines potentially hazardous objects (PHOs) as asteroids and comets that have the potential to collide with Earth. These are objects that pass within 0.05 AU (about 7.5 million kilometers, or 4.6 miles) of Earth's orbit. PHOs are also large enough to survive the passage through Earth's atmosphere. Objects of this size—about 50 meters (164 feet) or larger—could cause significant damage to Earth on impact.

Astronomers keep a close eye on known PHOs, including 1862 Apollo, because of the potential for collisions with Earth. Earth has experienced major collisions with space objects in the past. A huge crater near Chicxulub, Mexico, is likely evidence of a collision between Earth and a large asteroid about sixty-five million years ago. This collision may have played a major role in the Cretaceous-Tertiary mass extinction of dinosaurs and other large life-forms. More recently, in 2013, a relatively small NEA entered Earth's atmosphere and exploded over Russia, releasing about the same amount of energy as a large atomic bomb.

NASA created the Near-Earth Object Program (NEOP) to detect and track PHOs and other NEOs. This program studies these objects and is exploring technologies that will someday allow scientists to divert NEOs from Earth.

Additionally, the Asteroid Redirect Mission (ARM) will be NASA's first attempt to capture an NEA so it can be studied directly. ARM is developing technologies that will allow scientists to redirect an NEA into the orbit around Luna, Earth's moon. Astronauts will be able to land on this object to explore it. This is expected to occur around 2025.

—*Marie Keenan, MS*

Model of 1862 Apollo, asteroid. Astronomical Institute of the Charles University: Josef Ďurech, Vojtěch Sidorin, via Wikimedia Commons.

### Further Reading

Chapman, Clark R. "The Hazard of Near-Earth Asteroid Impacts on Earth." *Earth and Planetary Science Letters* 222.1 (2004): 1–15. Print.

de Pater, Imke, and Jack J. Lissauer. *Planetary Sciences.* Updated 2nd ed. Cambridge: Cambridge University Press, 2015. Print.

Ďurech, J., et al. "New Photometric Observations of Asteroids (1862) Apollo and (25143) Itokawa: An Analysis of YORP Effect." *Astronomy & Astrophysics* 488.1 (2008): 345–50. Print.

Gold, Lauren. "Asteroids Spin at YORP Speed, Thanks to the Effects of Sunlight, Cornell and Belfast Astronomers Discover." *Cornell Chronicle.* Cornell U, 7 Mar. 2007. Web. 7 Apr. 2015.

Johnston, William Robert, comp. "(1862) Apollo." *Johnston's Archive.* Johnston, 16 Nov. 2014. Web. 7 Apr. 2015.

Leverington, David. *Babylon to Voyager and Beyond: A History of Planetary Astronomy.* Cambridge: Cambridge University Press, 2003. Print.

Nelson, Marcia L., Daniel T. Britt, and Larry A. Lebofsky. "Review of Asteroid Compositions." *Resources of Near-Earth Space.* Ed. John S. Lewis, Mildred Shapley Matthews, and Mary L. Guerrieri. Tucson: U of Arizona P, 1993. 493–522. Print.

"Responding to Potential Asteroid Redirect Mission Targets." *Jet Propulsion Laboratory.* California Inst. of Technology, 14 Feb. 2014. Web. 7 Apr. 2015.

Rozitis, B., et al. "A Thermophysical Analysis of the (1862) Apollo Yarkovsky and YORP Effects." *Astronomy & Astrophysics* 555 (2013): N. pag. Web. 7 Apr. 2015.

United States. Natl. Aeronautics and Space Administration. *Near-Earth Object Survey and Deflection Analysis of Alternatives: Report to Congress.* N.p.: Author, 2007. *NASA.* Web. 7 Apr. 2015.

# APOLLO ASTEROIDS

### FIELDS OF STUDY

Astronomy; Observational Astronomy

### SUMMARY

Apollo asteroids are a type of near-Earth asteroid (NEA). These space objects orbit around the sun close enough to cross the orbits of both Earth and Mars. Planetary scientists study NEAs because they have the potential to collide with Earth. Additionally, NEAs can provide information about the early solar system.

### PRINCIPAL TERM:

- **near-Earth asteroid (NEA):** a small, irregularly shaped celestial body with an orbital path that brings it close to Earth. NEAs travel within 0.3 astronomical units (AUs) of Earth's orbit and within 1.3 AUs of the sun. An AU is equal to about 149.6 million kilometers (about 93 million miles).

### Near-Earth Asteroids

Apollo asteroids are small, irregularly shaped space objects that travel around the sun in the solar system. Like other objects in the solar system, they orbit the sun in a roughly elliptical (oval) shape. However, Apollo asteroids are of particular interest because their paths of travel cross the orbits of both Earth and Mars.

The solar system contains millions of asteroids, most of which can be found within the asteroid belt, an oval-shaped ring located between the orbits of Mars and Jupiter, about 2 to 4 AU (about 186 to 370 million miles) from the sun. However, some asteroids are ejected from the asteroid belt into the larger solar system. This can occur when asteroids collide or become trapped in the massive gravitational force surrounding Jupiter.

Asteroids that are pushed into the solar system sometimes end up orbiting close to Earth and the sun. near-Earth asteroids (NEAs) are asteroids that come within 0.3 AU (about 44.8 million kilometers, or about 31 million miles) of Earth. NEAs are categorized by distinguishing characteristics, such as their average distance from the sun (both near and far) and the size and shape of their orbital paths.

NEAs are part of a larger grouping of space objects called near-Earth objects (NEOs). This group includes comets and asteroids that have been pulled near the four inner planets (Mars, Mercury, Earth,

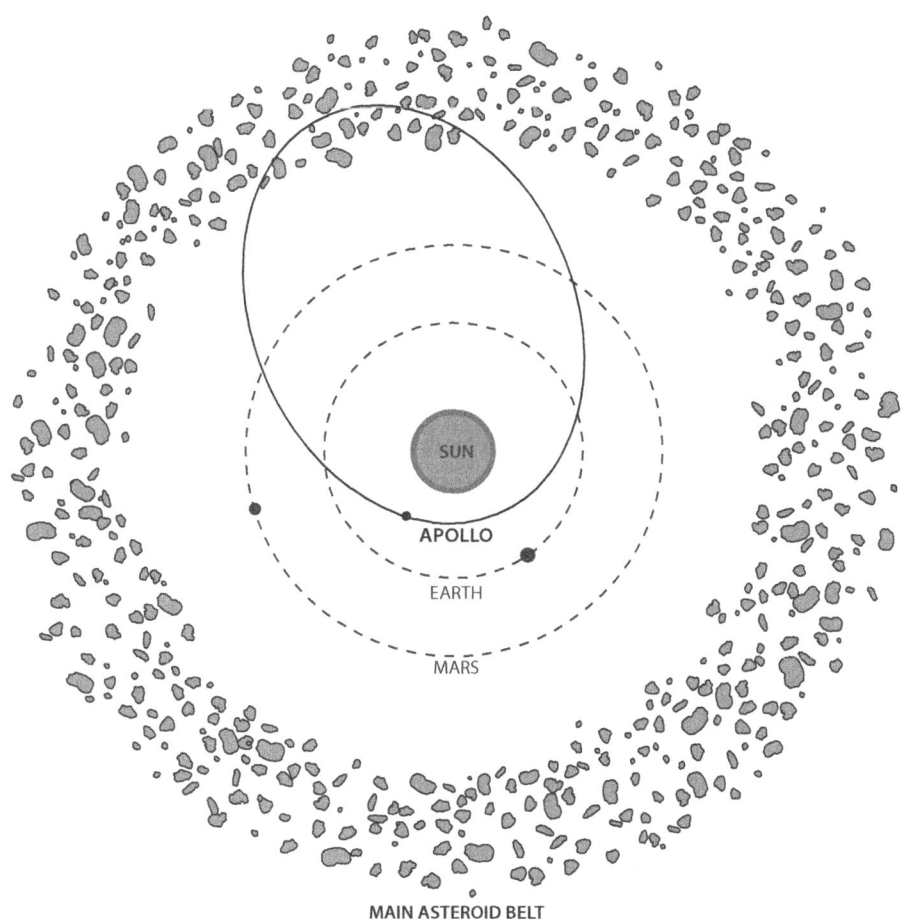

The orbit of an Apollo asteroid forms an ellipse around the sun, crosses both Earth's and Mars's orbits, and extends into the asteroid belt.

Astronomers keep a close eye on known Apollo asteroids, because they come close enough to Earth to collide with the planet.

## What NEAs Tell Us

Scientists are interested in asteroids and other NEOs because they are thought to be remaining debris from the formation of the solar system about 4.6 billion years ago. They can provide valuable information about the early solar system. Additionally, about 20 percent of NEAs come close enough to Earth to raise concerns about potential collisions with Earth. Although smaller space objects such as meteors collide with Earth frequently, these objects are usually so small that they burn up in the atmosphere before they can reach the planet's surface.

Major collisions between Earth and NEAs have occurred in the past. For example, a huge crater near Chicxulub, Mexico, is believed to be evidence of a past collision between Earth and a large NEA. Many experts believe this collision, which probably had the force of a nuclear explosion, played a major role in the Cretaceous-Tertiary mass extinction 65 million years ago. During this extinction, dinosaurs and other large life forms on the planet died out. Researchers believe that this NEA was about the same size as Apollo asteroid 1866 Sisyphus.

Other NEAs have collided with Earth more recently. For example, in February 2013, a relatively small NEA exploded over Russia, releasing the energy of a large atomic bomb.

## NASA's Near-Earth Object Program

The National Aeronautic and Space Administration (NASA) created the Near-Earth Object Program to detect and track NEOs. This program uses teams

and Venus) by gravitational forces. Asteroids are made up mainly of rock. Their surfaces are covered with visible craters. Asteroids vary greatly in size. The largest known asteroid, Ceres, measures about 966 kilometers (600 miles) in diameter. However, most asteroids are considerably smaller. Apollo asteroids are on the smaller side, measuring no more than about 6 miles (10 kilometers) in diameter. At 8.5 kilometers wide, 1866 Sisyphus is the largest known Apollo asteroid. Astronomers have identified more than 1,600 Apollo asteroids in the solar system. They are the most common type of NEA, comprising more than 60 percent of known NEAs. They are distinguished by their unique orbital paths, which cross the orbits of both Earth and Mars.

around the world to learn more about these objects and to develop technologies that will someday allow scientists to divert NEOs away from Earth. These teams use state-of-the art technologies such as massive radars that bounce radio waves off asteroids in order to identify, track, and learn about NEOs.

Additionally, NASA has successfully launched many robotic spacecraft missions to study NEOs. For example, the NEAR-Shoemaker mission was launched in 1996 to study the asteroid Eros, while the Dawn mission was launched in 2007 to study asteroids Vesta and Ceres.

Finally, the Asteroid Redirect Mission (ARM) is NASA's first attempt to capture an NEA so it can be studied directly. ARM is developing technologies that will allow scientists to redirect an NEA into the orbit around Earth's moon. Astronauts will then be able to land on this object to explore it.

—Marie Keenan, MS

**FURTHER READING**

"Apollo Asteroids." *Cosmos: The SAO Encyclopedia of Astronomy.* Swinburne University of Technology, 2014. Web. 9 Mar. 2015.

Chapman, Clark. "The Hazard of Near-Earth Asteroid Impacts on Earth." *Earth and Planetary Science Letters* 222 (2004): 1–15. PDF file.

Hannes, Alfvén, and Gustaf Arrhenius. "Evolution of the Solar System: 4. The Small Bodies." *NASA History Office.* NASA, 1976. Web. 9 Mar. 2015.

"Near Earth Object Program." *NASA.* National Aeronautics and Space Administration, n.d. Web. 9 Mar. 2015.

"Near Earth Objects: NEO Segment." *ESA.* European Space Agency, n.d. Web. 9 Mar. 2015.

"Responding to Potential Asteroid Redirect Mission Targets." *Jet Propulsion Laboratory.* California Institute of Technology, 14 Feb. 2014. Web. 9 Mar. 2015.

"Searching the Sky for Dangerous Neighbors: Eleanor Helin and the 18-Inch Telescope." *Palomar Observatory.* California Institute of Technology, 8 May 2014. Web. 9 Mar. 2015.

Van der Hucht, Karel. "Near Earth Asteroids: A Chronology of Milestones." *IAU.* International Astronomical Union, 7 Oct. 2013. Web. 9 Mar. 2015.

# AQUARIUS

**FIELDS OF STUDY**

Stellar Astronomy; Observational Astronomy

**SUMMARY**

Aquarius is a constellation that is visible in the night sky from Earth. As part of the zodiac, Aquarius is one of the oldest and most well known of the constellations. Although Aquarius is a faint constellation with relatively dim stars, it can be seen in both the Northern and Southern Hemispheres at different times of the year. Constellations such as Aquarius help astronomers locate, track, and study the stars and other significant celestial bodies to better understand the universe.

**PRINCIPAL TERMS**

- **celestial equator:** the imaginary line above Earth's equator that halves the celestial sphere; it is equally distant from the celestial poles.
- **constellation:** a region of space defined by a pattern of stars that can be seen in the night sky from Earth.
- **declination:** the north-south position of a celestial body relative to the celestial equator expressed in degrees of arc.
- **International Astronomical Union:** an association of professional astronomers from all over the world who define astronomical constants while promoting research, education, and discussion on important astronomical topics.
- **right ascension:** the east-west position of a celestial body defined in relation to the celestial equator and expressed in hours and minutes, not degrees of arc.

**THE WATER BEARER**

Aquarius is a constellation. A constellation is an area of space defined by a grouping of stars that appears to form a pattern when viewed in the night sky from Earth. Aquarius is one of the oldest and best-known constellations. Its name comes from the Latin word *aquarius*, which means "water carrier."

Ancient texts and artifacts show that people have been trying to make sense of the stars for thousands of years. Early cultures mapped the constellations in patterns that resembled people, animals, and objects, and they used these patterns for a variety of practical purposes. For example, star patterns helped them to navigate their travels by land or sea, explain nature, and differentiate the seasons for planting and harvesting. The Greeks described more than half of the total constellations that have been identified. In modern times, Eugène Delporte assigned official boundaries to the eighty-eight recorded constellations on behalf of the International Astronomical Union (IAU) to formalize their positions and designations.

Ancient people also believed that constellations had mystical powers. This was especially true for constellations like Aquarius, which is part of the zodiac. The zodiac is a group of thirteen constellations located in a specific section of the sky. Because Earth's orbit around the sun creates an illusion of the sun passing through these constellations, ancient people placed great significance upon them.

Most modern scientists dismiss the more mystical explanations of the zodiac. However, they still see the constellations as convenient maps to the sky, and they study them to learn more about the universe.

**ATTRIBUTES OF AQUARIUS**

Aquarius is the tenth largest of the constellations and covers about 980 square degrees of sky. It has a declination of about -15 degrees from the celestial equator and a right ascension of twenty-three hours. However, despite this impressive size, Aquarius is generally difficult to find due to the low magnitude (brightness) of the stars, galaxies, and planetary nebulas within it.

In the Northern Hemisphere, Aquarius can be seen in the fall, with greatest visibility in October. In the Southern Hemisphere, Aquarius can be seen in the spring (February to March).

Aquarius is filled with a variety of fixed stars, but they are low-magnification objects located hundreds of thousands of miles from Earth, making them appear quite dim. This constellation is also home to several globular clusters—densely packed collections of stars at least ten billion years in age. Globular cluster Messier 2 (M2), one of the oldest objects in the Milky Way galaxy, is located in Aquarius. M2 is about thirty-eight thousand light-years from Earth but is so densely packed with expired stars that it can be seen with binoculars under the right circumstances.

Two well-known planetary nebulas are also found within Aquarius. Named for their resemblance to gas giants, these objects are actually the remains of stars once had very similar makeup to the sun's. The Saturn Nebula, one of the brightest planetary nebulas familiarized by its Saturn-like ring, and the Helix Nebula, which is one of the closest planetary nebulas to Earth, both reside in Aquarius.

One important star system in Aquarius is Gliese 876, a small red star known as an M dwarf. M dwarfs are the most common type of star in the galaxy. Gliese 876 is significantly only fifteen light-years away from Earth but still too faint to be seen without a telescope. Research indicates that despite Gliese 876's relatively small size (about one-third the size of the sun), it is orbited by at least four planets.

Two of these planets are gas giants like Jupiter. The first is about twice the size of Jupiter, while the second is about half of Jupiter's size. Their orbits around Gliese 876 take about thirty and sixty days to complete. The third planet orbiting Gliese 876 is of special interest to astronomers because it appears to be a rocky, terrestrial planet like Earth. However, it is around seven times the size of Earth and located so close to the surface of Gliese 876 that its surface temperature probably hovers between 200 and 400 degrees Celsius (400 and 750 degrees Fahrenheit). The planet's short distance from Gliese 876 also makes its orbit incredibly short. Astronomers estimate that it takes the planet a mere two days to orbit the star.

**GOOD LUCK AND WATER**

The constellation Aquarius is associated with luck and good fortune. Several of the stars in Aquarius demonstrate this connection. For example, the original Arabic names for many of its stars have to do with luck or fortune. Three of the brightest stars in Aquarius—alpha Aquarii, beta Aquarii, and gamma Aquarii—are also known as Sadalmelik, Sadalsuud, and Sadachbia. These names roughly translate to "lucky one of the king," "luckiest of the lucky," and "lucky star of hidden things." Astronomers are not entirely sure why Aquarius is associated with luck and fortune, but it may be due to the time of year that Aquarius rises. As the sun appears to pass through the constellation, winter ends and the early spring season begins, ushering in a time of fertility and new life.

Aquarius is also identified with water. This may have been due to Aquarius's ascension in the sky at the start of the rainy season. The shape of the constellation is also associated with water. Its pattern is said to resemble a young man bending and pouring water from a jar. Additionally, Aquarius is located in the section of the sky referred to as "water" or "sea." It is near many other water-related constellations, including Pisces (the fish), Delphinus (the dolphin), and Eridanus (the river), among others.

### Mythological Connections

Aquarius was one of the first constellations to be identified in recorded history. The ancient astronomer Ptolemy (ca. 100–170 CE) listed it among the forty-eight constellations he and his contemporaries recognized. It also became entwined in the study of the zodiac. Over the centuries, many cultures applied their own interpretations to the meaning of the constellation.

Aquarius is most often associated with the Greek myth of Ganymede, the son of King Tros of Troy. Ganymede was a handsome young man whose appearance captured the attention of Zeus, the chief Greek god. Zeus had Ganymede brought to the gods' home on Mount Olympus, where he served them by bringing them water and acting as cupbearer to Zeus. According to the myth, Zeus honored Ganymede's service by placing Aquarius (the water carrier constellation) among the stars.

The Greeks were not the only ancient culture to associate Aquarius with water. The ancient Babylonians, who lived in the Middle East near modern-day Iraq, related Aquarius to Ea, the god of the waters. The Egyptians associated Aquarius with the waters of the Nile, which overflowed its banks each year.

### Studying Aquarius and Other Constellations

Modern astronomers use constellations such as Aquarius to make sense of the stars in the sky. Although the constellations are only imaginary pictures, they do provide a standard way for astronomers to locate and keep track of the thousands of named stars scattered in the sky. Aquarius provides a useful map for far distant regions of the sky.

Aquarius's globular clusters and planetary nebulas are also of interest to astronomers seeking to better understand the formation of celestial bodies. For instance, by studying the Gliese 876 star system,

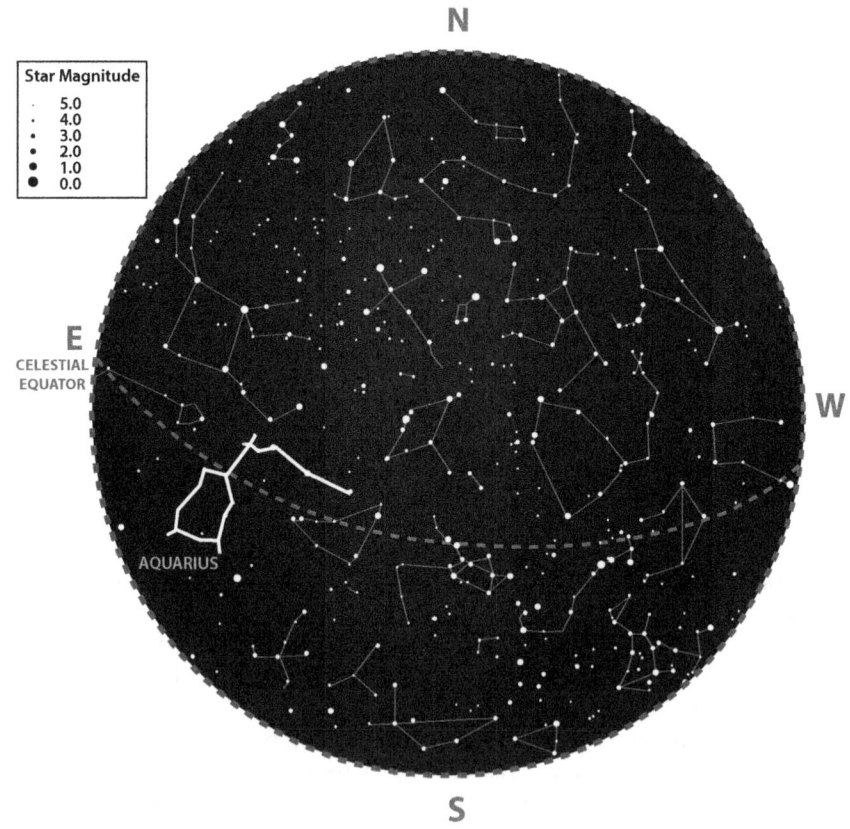

Star chart of the Northern and Southern Hemispheres in June. Lines connect the stars of the constellation Aquarius.

scientists can learn more about small exoplanets that could be similar to Earth.

—*Marie Keenan, MS*

**FURTHER READING**

"Aquarius." *Peoria Astronomical Society*. Peoria Astronomical Soc., n.d. Web. 13 Apr. 2015.

"Aquarius." *StarDate*. University of Texas McDonald Observatory, n.d. Web. 13 Apr. 2015.

"Aquarius (the Water Bearer)." *Chandra X-Ray Observatory*. Harvard-Smithsonian Center for Astrophysics, 2 Dec. 2013. Web. 13 Apr. 2015.

"The Constellations." *International Astronomical Union*. IAU, n.d. Web. 13 Apr. 2015.

"Constellations in the Zodiac." *National Aeronautics and Space Administration*. NASA, n.d. Web. 13 Apr. 2015.

Sanders, Robert, and Tim Stephens. "Astronomers Announce the Most Earth-Like Planet Yet Found outside the Solar System." *Currents Online*. UC Santa Cruz, 13 June 2005. Web. 13 Apr. 2015.

Zimmerman, Kim Ann. "Aquarius Constellation: Facts about the Water Bearer." *Space. com*. Purch, 10 June 2012. Web. 13 Apr. 2015.

# ARIES

**FIELDS OF STUDY**

Stellar Astronomy; Observational Astronomy

**SUMMARY**

The constellation Aries is a group of stars best seen in the Northern Hemisphere. Its main stars appear to form a line with a bent end, a pattern some ancient people thought resembled a ram. Aries contains a few bright stars, double stars, and variable stars, as well as distant galaxies and nebulae. It is the source of numerous meteors seen entering Earth's atmosphere. Long ago, constellations such as Aries guided agriculture and religion; today, astronomers study them to learn about the universe.

**PRINCIPAL TERMS**

- **celestial equator:** the imaginary line in the night sky that follows Earth's equator, dividing the celestial sphere in half.
- **constellation:** a region of space defined by a pattern of stars that can be seen in the night sky from Earth.
- **declination:** the north-south position of a celestial body relative to the celestial equator, expressed in degrees of arc.
- **International Astronomical Union:** an association of professional astronomers from all over the world who define astronomical constants while promoting research, education, and discussion on important astronomical topics.
- **right ascension:** the east-west position of a celestial body when viewed from the Earth's equator, defined in relation to the vernal equinox (one of two points at which the ecliptic intersects the celestial equator) and expressed in hours and minutes.

**THE RAM CONSTELLATION**

Since ancient times, people have scanned the sky and searched for images among the stars. These images helped shape myths and spiritual narratives as well as mark the changing seasons of the year. Since 1930, the International Astronomical Union (IAU) has recognized eighty-eight constellations, the boundaries of which were set by astronomer Eugène Delporte (1882–1955). Among the best known of these is Aries, which is shaped like a line with a bent end. Ancient cultures interpreted this shape in many different ways, most famously as a ram.

Aries is also famous for its inclusion in the zodiac, a group of twelve or thirteen constellations intersected by the path the sun appears to travel with respect to the stars over the course of a year. This path, called the ecliptic, is an illusion caused by Earth's orbit around the sun. Since ancient times, many cultures have ascribed mystical or spiritual significance to the zodiac.

Modern astronomers locate constellations by measuring their position relative to the celestial equator, the projection of Earth's equator onto the night sky. Aries has a declination of approximately +20 degrees and a right ascension of 2.66 hours. Declination and right ascension correspond to latitude and longitude on Earth.

**ATTRIBUTES OF ARIES**

Aries is a Northern Hemisphere constellation that is best seen during the late winter months of November and December. It is located between fellow zodiacal constellations Pisces to the east and Taurus to the west. Below Aries is Cetus, the whale; above it is Musca Borealis, the northern fly, a constellation no longer officially recognized by astronomers. Other constellations nearby include Perseus, named for the ancient Greek mythological hero, and Triangulum (the triangle). Aries is roughly average in size among the constellations, ranking thirty-ninth largest of the eighty-eight major stellar groups recognized by modern astronomers. Despite its considerable span, Aries can be challenging to see because it is a relatively faint constellation with only a few bright stars.

Three of the brightest stars in the constellation make up the head and horn of the ram. These stars are alpha Arietis (known as Hamal), beta Arietis (Sheratan), and gamma Arietis (Mesarthim). Hamal is an orange giant, twice as massive and fifteen times as wide as the sun. It is the brightest star in the constellation. Observers in many areas can see Hamal with the naked eye, even though it is about sixty-six light-years from Earth. The name Hamal comes from an Arabic word meaning "the lamb." Sheratan is slightly dimmer, and viewing it may require the use of binoculars in some areas.

Sheratan and Mesarthim are both notable for being binary stars. A binary star is a system in which two stars in close proximity revolve around a common center of mass. There are several other binary stars in Aries, including epsilon Arietis, lambda Arietis, iota Arietis, and tau1 Arietis. Other stars in the constellation are variable stars, which is a star whose apparent brightness varies, sometimes greatly and unpredictably.

Like many other constellations, Aries also features an array of deep-sky objects such as nebulae (clouds of gas and dust) and galaxies. Some of the galaxies in or near Aries exhibit widely varying shapes and behaviors. These galaxies include elliptical (oval), spiral, and interacting forms. (The gravitational forces of interacting galaxies act on one another.) Two of the most notable spiral galaxies in Aries are New General Catalog 697 (NGC 697) and New General Catalog 772 (NGC 772). The galaxy New General Catalog 1156 (NGC 1156) is considered an irregular galaxy because its form cannot be classified as any of the regular galactic shapes. The galaxy New General Catalog 877 (NGC 877) is the brightest in a cluster of eight interacting galaxies.

Aries is also the point of origin for several meteor showers, mostly in the months of May and June. Some of these meteor showers are unusual because they can be seen during the daytime.

**THE FLYING RAM AND GOLDEN FLEECE**

The star pattern now identified as Aries has a long history reaching back through many cultures. Although the constellation is now mostly known as a ram, various civilizations long ago created their own interpretations. For some people in the Marshall Islands, the stars represented a porpoise. To the Maya people, they formed the shape of an ocelot, a wildcat native to Central and South America.

Despite the varied interpretations, most of the stories that contribute to the popular mythology of Aries originated in ancient Greece. These stories differ in their details, but most deal with an extraordinary ram that rescued two children from danger. In these myths, the Theban king Athamas took a new queen, who schemed to have the king's son Phrixus and daughter Helle killed. The air nymph Nephele took pity on the endangered children and, with the help of Hera, goddess of marriage, arranged for their rescue. The gods dispatched a flying ram to Athamas's kingdom. The ram, known as Chrysomallus, picked up Phrixus and Helle to deliver them to safety. Along the way, Helle fell from the flying ram and drowned in a strait near modern-day Istanbul, Turkey, which was subsequently named Hellespont in her honor. Phrixus survived, arriving with the ram in the land of Colchis. There, Phrixus showed his gratitude by sacrificing the ram to the chief god, Zeus, in a sacred grove.

In a later myth, the fleece of the sacrificed ram turned to gold and became a prized possession of the king Aeetes. A hero named Jason and a group of adventurers known as the Argonauts embarked on a quest to capture the coveted fleece. Meanwhile, the spirit of Chrysomallus the ram had flown into the sky to assume a place among the stars, becoming the constellation Aries. Ancient astronomers claimed that

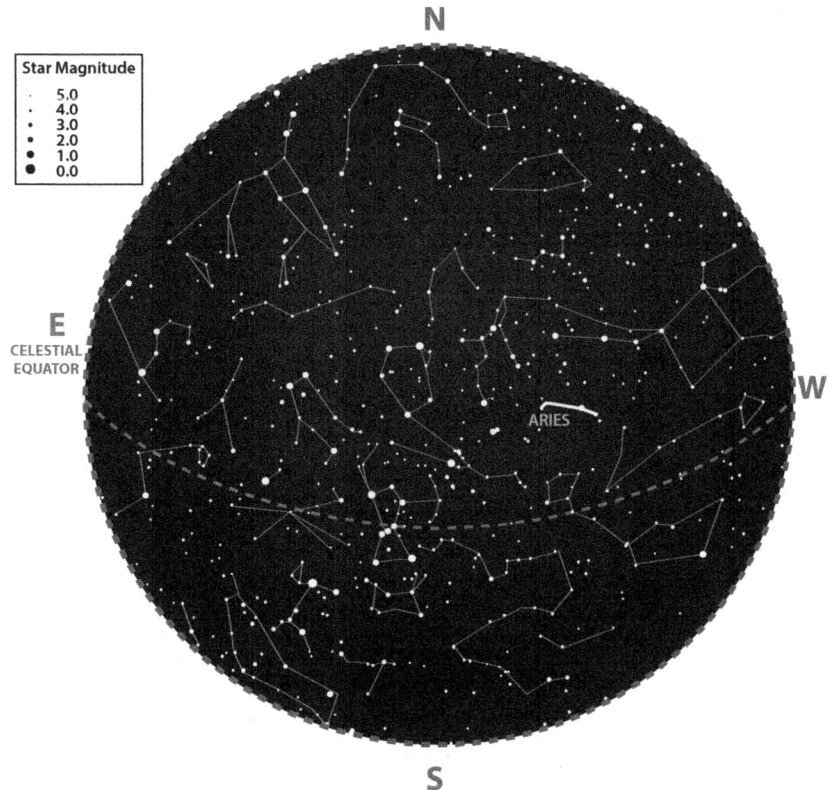

Star chart of the Northern and Southern Hemispheres in November. Lines connect the stars of the constellation Aries.

the sun touched Aries on the first day of spring, the vernal equinox. People rejoicing in the new year would celebrate the appearance of Aries in the night sky. Even today, the location of the sun at vernal equinox is called the "First Point of Aries." That name is outdated now, however, since the motion of Earth's axis has shifted the position of the sun at vernal equinox into the Pisces constellation.

One of the many stars in Aries, HD 19445, provided an important lesson for modern scientists. Astronomers studying this star in 1951 determined that it had relatively little iron, calcium, and other materials once believed common in all stars. Since HD 19445 is an extremely old star, astronomers were able to deduce that in the earliest days of the galaxy, iron and calcium were rare. Only over time did the stars create these vital materials that helped form not only Earth but also the human body.

the constellation is fairly dim because the ram had lost its sparkling golden coat.

Another, lesser-known Greek myth provided an alternate origin story for the constellation Aries. In this tale, the god of revelry, Dionysus, was lost in the desert. Just when he seemed doomed to die of thirst and hunger, Dionysus spotted a ram that led him to a well full of life-giving water. The grateful god then honored the ram by placing it among the stars.

## Aries's Importance Today

Modern astronomers and stargazers value the constellations for their unique beauty as well as the scientific information they can provide about the universe beyond Earth. Constellations are important for mapping the stars and planets of the universe as well as studying the different space objects within the stellar patterns. Like the other constellations, Aries contains several objects of astronomical interest.

Since early times, Aries has been important for tracking the progress of the seasons. In ancient times,

—*Mark Dziak*

**Further Reading**

"Aries." *Constellations.* Peoria Astronomical Soc., n.d. Web. 13 Apr. 2015.

"Aries, the Ram." *StarDate.* University of Texas McDonald Observatory, n.d. Web. 13 Apr. 2015.

Bratton, Mark. *The Complete Guide to the Herschel Objects: Sir William Herschel's Star Clusters, Nebulae and Galaxies.* New York: Cambridge University Press, 2011. Print.

"Greek Mythology and the Constellations." *Skywise Unlimited.* Western Washington U, n.d. Web. 13 Apr. 2015.

Kronberg, Christine. "Aries." *The Munich Astro Archive.* U Observatory Munich / Leibniz Supercomputing Centre, 3 July 1997. Web. 13 Apr. 2015.

McClure, Bruce. "Aries? Here's Your Constellation." *EarthSky.* Earthsky Communications, 2 Dec. 2014. Web. 13 Apr. 2015.

Plotner, Tammy. "Aries." *Universe Today.* Universe Today, 13 Oct. 2008. Web. 13 Apr. 2015.

Simpson, Phil. *Guidebook to the Constellations: Telescopic Sights, Tales and Myths.* New York: Springer, 2012. Print.

# ASTEROIDS

### FIELDS OF STUDY

Astronomy; Sub-planet Astronomy; Asteroid Impact Avoidance

### SUMMARY

Asteroids are small, airless, cratered astronomical objects that are made of mostly rock. Most are irregularly shaped, but some have regular spherical shapes. Most asteroids can be found orbiting the sun between the planets Mars and Jupiter. Ceres is the largest known asteroid, at approximately 950 kilometers (590 miles, or 3,115,000 feet) wide, and 1991 BA is among the smallest known, at 6 meters (20 feet) wide. Asteroids routinely collide with Earth. They can be potentially dangerous, and scientists are working on developing technology to explode such asteroids or change their course to protect Earth.

### PRINCIPAL TERMS

- **asteroid belt:** an area between the planets Mars and Jupiter where the majority of asteroids can be found.
- **dwarf planet:** a relatively large astronomical object that orbits the sun and has enough mass to be roughly spherical in shape but has not cleared its orbit of other astronomical objects.
- **small solar system body:** an astronomical object other than a planet, dwarf planet, or satellite that orbits the sun.

### WHAT ARE ASTEROIDS?

Asteroids are celestial bodies located in the solar system. While they share some characteristics with planets, they are much smaller in size. To put their size into perspective, the total mass of all known asteroids combined is less than the mass of Earth's moon.

Asteroids are sometimes referred to as planetoids or minor planets, which are defined as astronomical objects other than planets or comets that orbit the sun. However, new discoveries in the twenty-first century began to blur the differences between planets, comets, asteroids, tiny meteoroids, and distant trans-Neptunian objects (TNOs). In 2006, the International Astronomical Union (IAU) created the umbrella term small solar system bodies (SSSBs). The designation includes all celestial objects except for planets, satellites, and dwarf planets. The term "asteroid" is still widely used, however, especially for objects in the inner solar system. All inner solar system asteroids are classified as SSSBs except for Ceres, which is a dwarf planet.

Asteroids were formed about 4.6 billon years ago during the formation of the solar system. Basically, asteroids are the leftover pieces that did not form into planets. When the planet Jupiter formed, it prevented other planets from forming between itself and the planet Mars. Small objects between these planets collided with one another until they broke apart into asteroids.

The area between Mars and Jupiter, where the majority of asteroids can be found, is called the asteroid belt. Billions of asteroids can be found in the asteroid belt. Of these, more than 200 are over 100 kilometers (62 miles) wide, and more than 750,000 are over 1 kilometer (0.62 mile) wide. Other celestial objects, such as comets and dwarf planets, also can be found in the asteroid belt.

Some asteroids lie beyond the asteroid belt. Trojan asteroids are a group of asteroids that share the orbital path of Jupiter. Three other types of asteroids—Aten, Amor, and Apollo asteroids—orbit close enough to Earth to pose a potential threat.

Asteroids are irregularly shaped objects composed mainly of rock. Some are made of solid rock, while others are merely balls of dust and rubble held together by gravity. They are often categorized based on their composition. C-type (carbon-rich) asteroids are the most common and have remained largely unchanged since their formation. S-type (stony or siliceous) asteroids have melted and reformed. M-type (metal-rich) asteroids are a reddish color due to their nickel-iron content. Most asteroids fall into these three categories, although other types exist. All types are typically airless, dust covered, and pitted or cratered on the surface. Scientists have also found asteroids with characteristics similar to those of other celestial bodies, such as comets and planets. Rings, tails, moons, and water vapor have all been found on asteroids.

Asteroids rotate and orbit around the sun. Most follow an elliptical path, although some asteroids spin erratically due to their irregular shapes. Some asteroids are binary systems, in which two asteroids of similar size orbit a common center of mass. A few triple asteroid systems exist.

Sometimes asteroids get caught up in a planet's gravitational pull and become moons. Astronomers believe that Mars's moons, Phobos and Deimos, were once asteroids before they were pulled into Mars's orbit. Many of the distant moons of Jupiter, Saturn, Uranus, and Neptune are thought to be captured asteroids as well.

Asteroids are very cold, which makes them inhospitable to life. Their average surface temperature is –73 degrees Celsius (–99 degrees Fahrenheit). Some have conditions that have not changed since they formed billions of years ago, making them valuable for learning about the solar system's history.

## DISCOVERY OF ASTEROIDS

Astronomers in the eighteenth century used the Titius-Bode law, or simply Bode's law, to predict the spacing of planets in the solar system. German astronomer Johann Titius (1729–96) first proposed the theory in 1766, and Johann Elert Bode (1747–1826) formulated it as a mathematical expression in 1778. The law relates the distances of the planets from the sun to a progression of numbers. In 1781 Sir William Herschel (1738–1822) discovered the seventh planet, Uranus, which seemed to confirm Bode's law. However, the law also predicted that a planet should exist between Mars and Jupiter. Near the end of the eighteenth century, a group of astronomers in Germany set out to find the missing planet. But the Italian priest and astronomer Giuseppe Piazzi (1746–1826) beat the group and found what was thought to be the missing planet on January 1, 1801.

Scientists named the new object Ceres, for the Roman goddess of agriculture. However, they quickly determined that Ceres (later officially dubbed 1 Ceres), which was only about 950 kilometers (590 miles) wide, should not be considered a planet. As astronomers continued to search for the predicted planet, they found more objects even smaller than Ceres. The asteroids 2 Pallas, 3 Juno, 4 Vesta, and 5 Astraea were observed and named. It was established that no large body lay between Jupiter and Mars. Scientists determined that Jupiter's gravitational pull prevented the loose material in the area from forming a planet. This loose material was called the asteroid belt, and scientists began to study the new class of objects.

Ceres was the first asteroid discovered and is the largest known. It accounts for about one-quarter of the total mass of all asteroids in the asteroid belt, yet it is fourteen times less massive than Pluto. In 2006 Ceres was classified as a dwarf planet. It is the only asteroid in the inner solar system to be so designated. Pluto and 136199 Eris are other examples of dwarf planets.

## EXPLORING ASTEROIDS

Since asteroids were discovered in the nineteenth century, scientists have continued to study them and search for new ones. New technology in the twentieth century enabled closer study and images of asteroids. In 1971, the National Aeronautics and Space Administration (NASA) spacecraft *Mariner 9* took the first detailed images of Phobos and Deimos, which are believed to be captured asteroids.

The first close-up images of non-satellite asteroids were taken by the NASA spacecraft *Galileo*. In 1991, *Galileo* photographed 951 Gaspra, which is located on the inner edge of the asteroid belt. The spacecraft took images of a second asteroid, 243 Ida, in 1993. These images revealed the presence of a moon orbiting Ida. The moon, which was named (243) Ida I Dactyl, was the first confirmed asteroid moon.

In the beginning of the twenty-first century, NASA's *NEAR* spacecraft studied 433 Eros, a near-Earth asteroid, for nearly a year. *NEAR* landed on Eros in 2001, becoming the first spacecraft to successfully land on an asteroid. In 2005 the Japanese spacecraft *Hayabusa* landed on the asteroid 25143 Itokawa, took samples, and then relaunched. The spacecraft returned to Earth in 2010 with the first samples taken from an asteroid.

In 2007 NASA sent the space probe *Dawn* to explore Vesta. It reached the asteroid in 2011 and spent

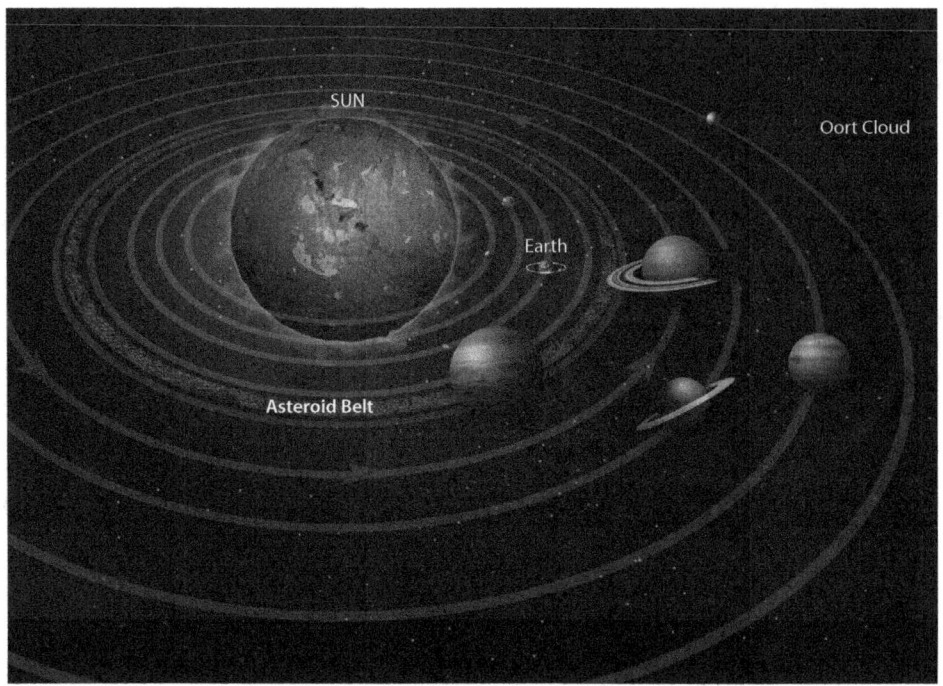

A diagram of the solar system, with the asteroid belt highlighted.

a year in orbit, collecting information. It then left for Ceres, entering the dwarf planet's orbit in March 2015. *Dawn* was the first spacecraft to successfully visit either of these asteroids.

In addition to space probe missions, astronomers on Earth continue to study asteroids. In 2013 they spotted an asteroid with six tails located in the asteroid belt. Dubbed P/2013 P5, the asteroid is thought to have gained the six tails as result of intense spinning. The following year scientists discovered 10199 Chariklo, an asteroid with a set of rings. Considered small at 250 kilometers (155 miles) wide, Chariklo is located between Neptune and Uranus and is also thought to have an undiscovered moon. Researchers believe more asteroids such as these exist, and they continue to search the skies for them.

### Earth Impacts

Even though asteroids are not very large in comparison to Earth, they have the potential to be dangerous. Many asteroids have hit Earth and other planets since their formation, and they will continue to do so in the future. Scientists study the orbits and physical characteristics of asteroids to determine if they pose a danger to Earth.

According to NASA, asteroids more than a quarter-mile wide pose a risk of disrupting life on Earth. These large asteroids have the potential to collide with Earth once every one thousand centuries. Smaller asteroids can still cause tsunamis or destroy large areas. Such an asteroid is estimated to strike Earth every one thousand to ten thousand years.

Asteroid impacts are believed to be responsible for a number of major extinctions throughout Earth's history, including that of the dinosaurs. Less powerful impact events have been recorded by humans. On June 30, 1908, an asteroid exploded in the atmosphere near the Podkamennaya Tunguska River in Siberia, Russia. It caused a shock wave that destroyed a large area of the Siberian forest and registered effects throughout Europe and Asia. On February 15, 2013, another asteroid crashed into the atmosphere over Chelyabinsk, Russia. The explosion caused a shock wave that injured more than one thousand people. The asteroid was estimated to be twenty meters (nearly sixty-six feet) wide when it entered the atmosphere.

Scientists spend much time tracking asteroids and have cataloged numerous asteroids as "potentially hazardous" to the Earth. These near-Earth objects (NEOs) could crash into the planet if their orbits shift. In 2001 a United Nations team was formed to explore joint actions to respond to an asteroid threat. As of 2015, NASA was developing technology to redirect asteroids. In 2013, NASA's Asteroid Initiative began developing plans for the Asteroid Redirect Mission, a robotic mission to visit an NEO, collect materials from its surface, and redirect it.

—*Angela Harmon*

## FURTHER READING

"Asteroids: The Discovery of Asteroids." *Space for Europe*. European Space Agency, n.d. Web. 11 May 2015.

"Bode's Law." *Astro 2201: Our Home in the Universe*. Cornell U, n.d. Web. 11 May 2015.

"Ceres: Overview." *Solar System Exploration*. NASA, 6 Jan. 2015. Web. 6 May 2015.

Cessna, Abby. "Planetoid." *Universe Today*. Fraser Cain, 10 Aug. 2009. Web. 6 May 2015.

Choi, Charles Q. "Asteroids: Fun Facts and Information about Asteroids." *Space.com*. Purch, 21 Nov. 2014. Web. 6 May 2015.

Gannon, Megan. "Bizarre Asteroid with Six Tails Spotted by Hubble Telescope." *Space. com*. Purch, 7 Nov. 2013. Web. 6 May 2015.

Phillips, Tony, ed. "The Tunguska Impact: 100 Years Later." *NASA Science*. NASA, 30 June 2008. Web. 6 May 2015.

Redd, Nola Taylor. "Asteroid Found with Rings! First-of-Its-Kind Discovery Stuns Astronomers." *Space. com*. Purch, 26 Mar. 2014. Web. 6 May 2015.

"Small Worlds: The Neighborhood." *Discovery Program*. NASA, n.d. Web. 6 May 2015.

"What Is NASA's Asteroid Redirect Mission?" *Asteroid Redirect Mission*. NASA, 30 Apr. 2015. Web. 11 May 2015.

# ASTROCHEMICAL REACTION MODELS

## FIELDS OF STUDY

Astrophysics; Theoretical Astronomy

## SUMMARY

Astrochemical reaction models include laboratory experiments and computer models that simulate the reactions and interactions of the chemical components of interstellar dust. Essentially invisible, interstellar dust exists in the dark areas of space between star systems. Astrochemistry research offers insight into questions about the fundamental role these chemical components have played in the evolution of life.

## PRINCIPAL TERMS

- **astrochemistry:** the interdisciplinary field that studies the chemical composition and evolution of the universe, including the chemical reactions that occur between the elements of interstellar dust.
- **hot core region:** an area of space where stars form.
- **interstellar dust:** solid particles of matter that exist in the vast regions of space between star systems.
- **wavelength:** the distance between two peaks of a light wave.

## The Interstellar Medium

For thousands of years, people thought that the dark spaces between stars in the sky were empty of matter. However, astronomical research in the 1950s began to tell a different story. Astronomers discovered that these dark areas between star systems are actually filled with matter known as the interstellar medium. The vast majority of the interstellar medium—about 99 percent—consists of various forms of gas, mainly hydrogen and helium. The remaining 1 percent or so is made up of tiny, frozen particles called interstellar dust, also known as cosmic dust.

Hydrogen and helium are the two simplest chemical elements. Scientists believe that these gases were left behind in the interstellar medium after the universe was formed. Trace amounts of other gases are present in the interstellar medium, including carbon, oxygen, nitrogen, and dimethyl ether, an organic compound. These gases may have either formed inside the cloud or originated in nearby stars. They are especially abundant in stars found in hot core regions. When these stars collapse in supernova explosions, the matter within them, including these gases, is scattered throughout the interstellar medium.

Scientists are not sure where the frozen particles of interstellar dust originated or what specific role they play in the universe. However, they do know that the reactions and interactions of the elements that make up the interstellar medium were crucial to the formation of the universe, particularly the individual star systems and planets. Modern astronomers agree that

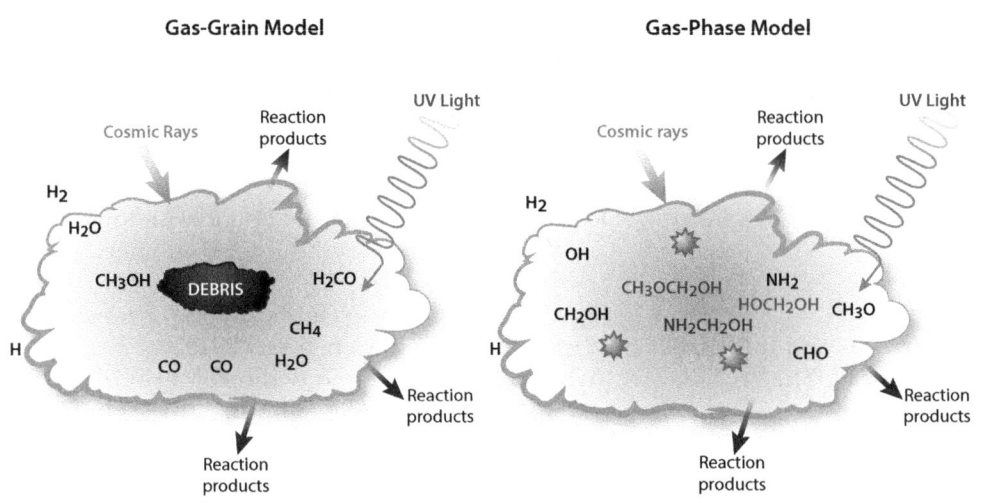

Gas-grain astrochemistry model describes gas-dust interactions—chemical reactions among gases that occur around and on the surface of some grains of debris. Gas-phase astrochemistry model describes gas-phase chemical reactions among gases released from ices into the clouds surrounding protostars.

these elements were delivered to Earth from space via meteorites, creating the conditions that would later allow life to evolve.

## Studying the Interstellar Medium

Clouds of interstellar medium are invisible to the human eye. This makes them difficult to study. However, astronomers have discovered that interstellar medium can be studied using particular types of light.

When most people think of light, they think of visible light, such as that which is emitted from a lamp or the sun. However, visible light is only a small portion of the light that exists in the universe. Many different types of light exist along the electromagnetic spectrum. Not all of them are visible to the human eye.

What people call "light" is actually a form of energy that can behave as either a wave or a particle. Different types of light are classified according to their wave properties, specifically their wavelength. Visible light is somewhere in the middle of the electromagnetic spectrum. It has a longer wavelength than do ultraviolet light, x-rays, and gamma rays and a shorter wavelength than infrared, microwaves, and radio waves.

Light waves are important to astronomers because they can be used to study invisible elements of the universe. For example, when interstellar dust clouds absorb visible and ultraviolet light, the dust particles heat up and then reradiate a portion of that energy in the form of infrared. Astronomers can then use an infrared telescope to locate and study these clouds.

Scientists study interstellar medium to learn more about its composition and evolution. They believe that the interactions between the gases and dust in these molecular clouds played a key role in the evolution of life on Earth and in the formation of new stars and planetary systems.

## Research Models

Studying interstellar dust in space is difficult. Therefore, scientists re-create space conditions in laboratories to learn how these elements interact. They design laboratory experiments and create innovative computer models to simulate the reactions and interactions that occur among the elements of interstellar dust. This helps them understand how these elements affect each other and the universe around them. Mass spectrometry, low-energy electron irradiation, and quartz-crystal microbalancing are some of the experimental techniques scientists use.

Research into interstellar dust is part of an interdisciplinary branch of science called astrochemistry. Astrochemistry is the study of the chemical components of the universe and how they interact with one another. It can provide insight into important questions that have long interested scientists, such as how the universe began, how stars and planets formed, how life formed and evolved on Earth, and whether or not life might exist elsewhere in the universe.

—*Marie Keenan, MS*

## Further Reading

"The Electromagnetic Spectrum." *Imagine the Universe!* NASA, Mar. 2013. Web. 14 Apr. 2015.

Hermans-Killam, Linda. "Infrared Astronomy." *Ask an Infrared Astronomer.* California Inst. of Technology, n.d. Web. 15 Apr. 2015.

Marchione, Demian. *The Heriot-Watt Astrochemistry Research Group.* Heriot-Watt U, 26 Mar. 2014. Web. 14 Apr. 2015.

Peeters, Z., et al. "Astrochemistry of Dimethyl Ether." *Astronomy & Astrophysics* 445.1 (2006): 197–204. Print.

Viti, Serena, et al. "The Making of *Stars 'R' Us!*" *Astronomy & Geophysics* 45.6 (2004): 6.22–24. Print.

Williams, David A., and Thomas W. Hartquist. "The Chemistry of Star-Forming Regions." *Accounts of Chemical Research* 32.4 (1999): 334–41. Print.

# ASTRONOMICAL IMAGE PROCESSING

## FIELDS OF STUDY

Astronomy; Space Technology

## SUMMARY

The instruments used to gather space data are designed to gather scientific information and not necessarily to produce high-quality photographic images. Astronomical image processing allows scientists and amateur enthusiasts alike to produce high-quality images of distant stellar objects. Astronomical image processing uses specialized equipment and software to filter and combine multiple images from space telescopes. These images often provide a dramatic look at stellar objects in distant parts of the universe.

## PRINCIPAL TERMS

- **grayscale:** a photographic image produced in shades of gray rather than in color, commonly referred to as a black-and-white image.
- **matrix:** the arrangement or grid of pixels that make up a digital image.
- **pixel:** a small square of a single color that is the smallest discrete component of a digital image.

## REFINING RAW DATA

It can be difficult to produce images of stellar objects. First, space telescopes are not designed to produce the kind of images taken by a traditional camera. Second, stellar objects are far from Earth and are often too large to fit in a single image. In addition, some space objects, such as new stars, are hidden by clouds of dust and gas. Astronomical image processing allows researchers to overcome these problems by creating composite

A mosaic of over a hundred thousand infrared images are combined to create a single picture of the Large Magellanic Cloud near our Milky Way galaxy. NASA/JPL-Caltech/STScI.

images of stellar objects to use in scientific study. This process is often used to create the images of galaxies, planets, and other space objects that appear in popular media.

To detect and study celestial objects, researchers use instruments that detect electromagnetic radiation outside the visible light spectrum, such as infrared, ultraviolet, or x-rays. The images produced by these instruments are often colored differently from the actual objects themselves. Researchers use filters, software, and other photographic techniques to refine such images. The process can involve combining multiple images of the same object to create a single, more complete picture.

### How Astronomical Images Are Made

All digital images are formed from a matrix, or grid, of pixels. The placement and color of each pixel in relation to those alongside it create the image. Even grayscale images are made of pixels. Each pixel has an assigned value that determines its color or shade. Depending on the color model being used, the pixels might represent shades in the red, green, and blue (RGB) spectrum, or they might represent shades in the cyan, magenta, yellow, and black (CMYK) spectrum. The stacking of these colors creates all the visible colors in an image.

While some telescopes use color filters to collect images, most produce grayscale images. These images are taken with a charge-coupled device (CCD). CCDs are more sensitive to photons than standard photography equipment and therefore yield much more detailed images. They capture the data digitally, making it easier to access, transmit, and share.

An image processor can highlight areas of interest to researchers or minimize items, such as dust and gas clouds, that obstruct the image. This is done using specialized software designed for the purpose. A processor working with raw data from a CCD would first load the data into a program to produce the desired image type. The type of image varies depending on the image's intended use. A processor who wants to release a photograph of a newly discovered galaxy to the media will produce a dramatic, colorful image. On the other hand, a researcher who wants to determine the number and types of stars in an area will create an image that makes counting and categorizing the stars easier, without much care for its attractiveness.

The processor might choose to stack the image using the Drizzle algorithm. An algorithm is a step-by-step problem-solving process. Drizzling stacks multiple images of the same object, allowing a fuller view than could be captured in any one image. This algorithm could be used to fill in areas that are blocked by cosmic rays, for example. It could also be used to render a complete image of a large object, such as a galaxy, that would impossible to capture completely in a single image.

If the final color of the image is important, the processor will apply filters that assign color values to the grayscale pixels. In some cases, colors other than the object's actual colors are used. For instance, if a dust cloud surrounding a new star is nearly the same color as the star, it might be shown in a different color. Researchers call these representative color images.

### Uses for Astronomical Images

The photos made from the data collected by giant telescopes have given researchers great insight into the universe. For example, by studying a particular type of star called a Cepheid variable using data and images from the Hubble Space Telescope, researchers from the National Aeronautics and Space Agency (NASA) were able to determine that the universe is about 13.7 billion years old, give or take 200 million years.

Other discoveries made with Hubble images include proof that dark matter exists, proof that black holes are probably the forces that create galaxies, and proof that the universe has been expanding since its creation. In early 2015, images from the Hubble captured the rarely seen phenomenon of an exploding star and provided a glimpse of three of Jupiter's moons against the planet's face.

NASA's Wide-Field Infrared Survey Explorer (WISE) telescope focuses on infrared data. Among other tasks, it has provided information on the location of a number of failed stars, known as brown dwarves.

### Not for Professionals Only

Access to first-hand scientific information is often limited to researchers working for governments, agencies, or universities. However, the raw data gathered from the giant space telescopes is available to anyone. The image archive for the Hubble telescope has more than twenty years' worth of data. The government agencies responsible for this data have made it available to the public.

Numerous volunteer image processors, including academics, amateur photographers, and astronomy enthusiasts, have produced images of star-forming regions in the Large Magellanic Cloud, the center of the Messier 77 (M77) galaxy, and a newly formed star in a spray of gas. In 2009, astronauts renewed the array on the Hubble, it to keep gathering data for new deep-space discoveries by scientists and enthusiasts alike.

—*Janine Ungvarsky*

**FURTHER READING**

"Breakthroughs in Cosmology: Taking the Universe's Baby Pictures." *HubbleSite.* NASA, n.d. Web. 7 Apr. 2015.

Ferguson, Harry, and Andy Fruchter. "Drizzle." *Space Telescope Science Institute.* STScI, 1997. Web. 7 Apr. 2015.

Hall, Geoffrey. "Astronomical Image Processing A1." *Workspace: Physics Undergraduate Laboratories.* Imperial Coll. London, 6 Mar. 2015. Web. 7 Apr. 2015.

"Hubble Image Processors: Raiders of the Hubble Archive." *HubbleSite.* NASA, n.d. Web. 7 Apr. 2015.

Netting, Ruth. "Infrared Waves." *Mission:Science.* NASA, 13 Aug. 2014. Web. 7 Apr. 2015

Peterson, Kit A. "Introduction to Basic Measures of a Digital Image for Pictorial Collections." *Library of Congress Prints & Photographs Division.* LOC, June 2005. Web. 7 Apr. 2015.

"A Short Introduction to Astronomical Image Processing." *Hubble Space Telescope.* European Space Agency, n.d. Web. 7 Apr. 2015.

Zuckerman, Catherine. "Hubble Pictures: Top Five Hidden Treasures." *National Geographic.* Natl. Geographic Soc., 4 Sept. 2012. Web. 7 Apr. 2015.

# ASTROPHYSICS

**FIELDS OF STUDY**

Astrophysics; Astronomy; Cosmology

**SUMMARY**

Astrophysics is the study of the physics of planets, stars, and other space objects. It involves presenting theoretical ideas and technical data and collecting supporting measurements and materials from space. Authorities such as Isaac Newton and Albert Einstein provided firm theoretical and mathematical foundations for planetary movement, light, and relativity, while Edwin Hubble and George Gamow contributed further understanding of the origin of the universe and its age. Ideas such as the big bang theory and the life cycle of stars also contributed knowledge and insight to the study of astrophysics.

**PRINCIPAL TERMS**

- **big bang:** the theoretical massive explosion that created the universe.
- **cosmology:** the study of the origin, evolution, and structure of the universe.
- **luminosity:** in astronomy, the amount of radiant energy emitted by a celestial body, such as a star.
- **nucleosynthesis:** the process of creating new atomic nuclei by fusing together protons and neutrons.
- **redshift:** an apparent increase in the wavelength of electromagnetic radiation, caused by the radiation source moving away from the observer.
- **stellar evolution:** the process by which a star moves through its life cycle, from its formation to its eventual end as a white dwarf, neutron star, or black hole.

**THE SCIENCE OF SPACE**

Like other branches of space science, astrophysics involves the study of the universe, its origins, and its evolution. In particular, it studies the nature of space objects and how they interact with forces and other matter in space. Because most objects in the universe cannot be seen from Earth, research relies on data such as radiation emitted from distant stars and the light frequencies of substances in

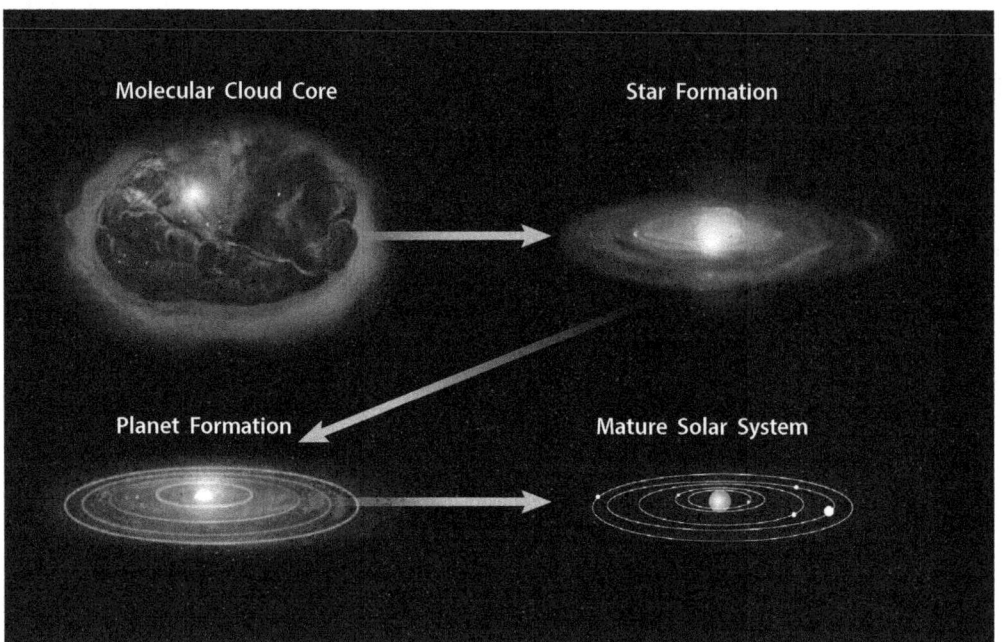

Phases of solar-system formation with a single star and surrounding planets. A molecular cloud core collapses, forming a protostar, which then accretes material to form a rotating disk. Other protostars are pushed out, leaving only a single star and protoplanet debris in the accretion disk. Over time, the debris condenses into planets to form a mature solar system.

space. Astrophysicists use both theory and applied technology to understand such data and to make new discoveries.

Sir Isaac Newton (1645–1727) developed a single theory on the movements of the planets. Before that, astronomers could only observe and describe what they saw. Newton's three laws of motion were unique because they applied both on Earth and in outer space. He not only accurately described the orbits of planets and moons but also identified and defined gravity as the force that controls them. Newton's ideas allowed researchers to predict how planets would move and where they would be in the future.

Other researchers built on what Newton had discovered to expand humankind's understanding of astrophysics. In 1905, Albert Einstein (1879–1955) proposed his theory of special relativity, which was based on two fundamental postulates: that the speed of light is always constant, and that the laws of physics remain the same everywhere on Earth and in space. Einstein's theory of general relativity, published ten years later, introduced the idea that gravity can bend light. He also described how heavy objects bend space, creating a gravity well. Edwin Hubble (1889–1953) identified spiral "nebulae" as galaxies similar to the Milky Way, helped discover the expansion of the universe, and calculated the universe's approximate age. George Gamow (1904–68) helped develop the big bang theory in 1948. He believed that if this theory were correct, astronomers would find leftover radiation from the big bang throughout the universe, a prediction that was later proved true.

Astrophysics is related to cosmology, another branch of astronomy that concentrates on large celestial objects and the study of the universe. Both subjects focus on the origin and development of the cosmos.

## THE BIG BANG THEORY

Astrophysicists have long wondered how the universe began. It was not until the 1920s that evidence was found to support a viable theory. Previously, several scientists had observed that some spiral nebulae displayed significant redshift, indicating that they were moving away from Earth at high speeds. Early in the 1920s, Hubble proved for the first time that these "nebulae" were in fact other galaxies outside the Milky Way. In 1929, by comparing the redshifts of different galaxies, he discovered that those farthest away from Earth were moving the fastest. This suggested that the universe was not only expanding but still accelerating.

Hubble's observations supported a theory proposed two years earlier by Belgian physicist and Catholic priest Georges Lemaître (1894–1966). In 1931, armed with Hubble's evidence, Lemaître proposed that if the universe is constantly expanding, it must have started as a single point, what he called a "single quantum" or a "unique atom." This was the first articulation of what would later be called the big bang theory.

Lemaître's expanding-universe theory was not widely accepted at first. In the 1940s, however, Gamow and his student Ralph Alpher (1921–2007) built on this theory to introduce the idea of big bang nucleosynthesis (BBN). According to this theory, several seconds after the big bang, once the universe had cooled somewhat, protons and neutrons began to combine to form deuterium and helium nuclei. (Deuterium is an isotope of hydrogen that has one neutron in its nucleus.) Gamow and Alpher suggested that BBN created all of the different elements. While this was later disproved, the BBN model does account for the creation of hydrogen, helium, and a small amount of lithium. Heavy elements, such as nickel and iron, were synthesized later, once stars had formed. The stars acted as thermonuclear reactors, fusing atomic nuclei within their cores. This process is called stellar nucleosynthesis.

Based on their research on BBN, Gamow, Alpher, and Robert Herman (1914–97) predicted the existence of cosmic microwave background radiation (CMB) in 1948. Essentially, CMB is the photons left over from the big bang. The early universe was extremely hot, producing thermal radiation in the form of photons. When the first atoms began to form, these photons started to spread throughout the universe. As the universe expanded, it gradually cooled. Based on this, Alpher and Herman attempted to calculate the current temperature of the universe. Although their calculations were incorrect, their reasoning was sound.

CMB was first observed in 1965 by American physicists Arno Penzias (b. 1933) and Robert Wilson (b. 1936). This discovery aligned with several predictions of the big bang theory. The existence of CMB, its uniformity, and its current temperature—about 2.7 kelvins (K)—support the theory that the universe began in an explosion approximately 13.8 billion years ago.

## Stellar Evolution

Stars have a natural life cycle from their formation to their end. This progression is called stellar evolution. Stars begin as huge clouds of gas and dust molecules. Turbulence causes knots in these clouds, which can then collapse as a result of the gravitational attraction within them. This causes the mass of materials to heat up. Heat from the core of the developing star creates pressure that counters the gravitational force, maintaining the equilibrium of the mass and preventing it from collapsing completely. At this stage, the mass is called a protostar. If its mass is great enough, thermonuclear reactions will begin in the core, fusing hydrogen into helium and building up heat. Then the fully developed star enters the main sequence of its life cycle. In some cases, the remaining debris will eventually condense into a solar system with the new star at its center.

Many protostars can form from a single molecular cloud, but not all protostars develop into stars. If a protostar's core cannot generate enough heat, gravity causes it to collapse and die, resulting in what is called a brown dwarf. Those that succeed burn for billions of years, until the star's core begins to run out of fuel. As the hydrogen is depleted, the core loses pressure and is unable to withstand the force of gravity, causing it to contract. As a result, both temperature and pressure increase in the layer above the core.

At this point, the star can follow one of several evolutionary tracks, depending on its mass. Some low-mass stars expand outward to become red giants. A star can burn as a red giant for up to a billion years. When it finally burns out, only the core remains, which is called a white dwarf. Other low-mass stars become white dwarfs directly, without becoming red giants first. A white dwarf is extremely dense, with a surface gravity more than one hundred thousand times as strong as Earth's. One cubic centimeter of white dwarf matter would weigh a ton.

Stars of intermediate mass burn much longer than large stars. The luminosity of a star reflects the rate at which it consumes fuel. High-mass stars consume their fuel much faster than smaller ones, so their luminosity is higher and they burn out faster. Some massive stars burn through several red giant stages before they begin to cool off. When they collapse again, they become even more dense and heavy. Finally, the energy from the core is released as light and heat as it explodes in a supernova, blasting heavy metals into space.

The largest stars end much as smaller ones do, except that nothing supports their enormous mass as the cores collapse in on themselves. Massive amounts of matter are compressed into a smaller and smaller area. Eventually, the star collapses into a single point of infinite density and thus infinite gravity. This point, called a singularity, forms the center of a black hole.

**ASTROPHYSICS IN CONTEXT**

Throughout the twentieth century, new discoveries and technological advances enabled scientists to study the cosmos in greater depth, giving birth to the field of astrophysics. Using phenomena such as redshift, cosmic microwave background radiation, and the life cycles of stars, astrophysicists hope to learn more about the universe—not just how it works now, but also how it began and how, eventually, it may one day end.

—*Nancy Comstock*

**FURTHER READING**

Anderson, Scott R. "Lecture 6: The Laws of Motion and Gravity." *Open Course Info*. Author, 10 Apr. 2003. Web. 6 May 2015.

"High Redshift Quasars." *Astronomy Picture of the Day*. NASA, 11 Dec. 1998. Web. 6 May 2015.

Howell, Elizabeth. "What Is the Big Bang Theory?" *Space.com*. Purch, 19 Mar. 2014. Web. 6 May 2015.

Krebs, Robert. "Fowler's Theory of Stellar Nucleosynthesis." *Encyclopedia of Scientific Principles, Laws, and Theories*. Vol. 1. Westport: Greenwood, 2008. 195–96. Print.

Lemaître, Georges. "The Beginning of the World from the Point of View of Quantum Theory." *Nature* 127 (1931): 706. Rpt. in *The Book of the Cosmos: Imagining the Universe from Heraclitus to Hawking*. Ed. Dennis Richard Danielson. Cambridge: Perseus, 2000. 407–8. Print.

"Stars." *Imagine the Universe!* NASA, 18 Nov. 2014. Web. 6 May 2015.

"Stellar Evolution: The Birth, Life, and Death of a Star." *NASA Education*. NASA, 10 Apr. 2009. Web. 6 May 2015.

Tate, Karl. "Einstein's Theory of Relativity Explained (Infographic)." *Space.com*. Purch, 5 Mar. 2015. Web. 6 May 2015.

"Tests of Big Bang: The CMB." *Wilkinson Microwave Anisotropy Probe*. NASA, 12 May 2014. Web. 6 May 2015.

White, Martin. "Big Bang Nucleosynthesis." *Martin White, Professor of Physics, Professor of Astronomy, UC Berkeley*. Regents of the University of California, 27 Mar. 2006. Web. 6 May 2015.

Wilcox, Roger M. "Stellar Evolution." *The Internet Stellar Database*. Author, n.d. Web. 6 May 2015.

# ATEN ASTEROIDS

**FIELDS OF STUDY**

Astronomy; Observational Astronomy

**SUMMARY**

Aten asteroids are a type of near-Earth asteroid (NEA). These objects orbit around the sun close enough to Earth to cross its orbit. NEAs are important because of the information they can offer about the early solar system as well as their potential to collide with Earth.

**PRINCIPAL TERMS**

- **near-Earth asteroid (NEA):** a small, irregularly shaped celestial body with an orbital path that brings it close to Earth. NEAs travel within 0.3 astronomical units (AU) of Earth's orbit and within 1.3 AU of the sun. An AU is equal to about 149.6 million kilometers (about 93 million miles).

**NEAR-EARTH ASTEROIDS**

Aten asteroids are a type of asteroid that travels in the solar system close to Earth, named after 2062 Aten. The first known asteroid of this type, it was discovered in 1976 by scientist Eleanor Helin (1932–2009) at Palomar Observatory in San Diego, California. Asteroids are small, irregularly shaped space bodies. Like other space objects, they travel around the sun in elliptical (oval-shaped) paths. Asteroids are composed mainly of rock but may also include minerals and metals. Their rocky surfaces are full of visible

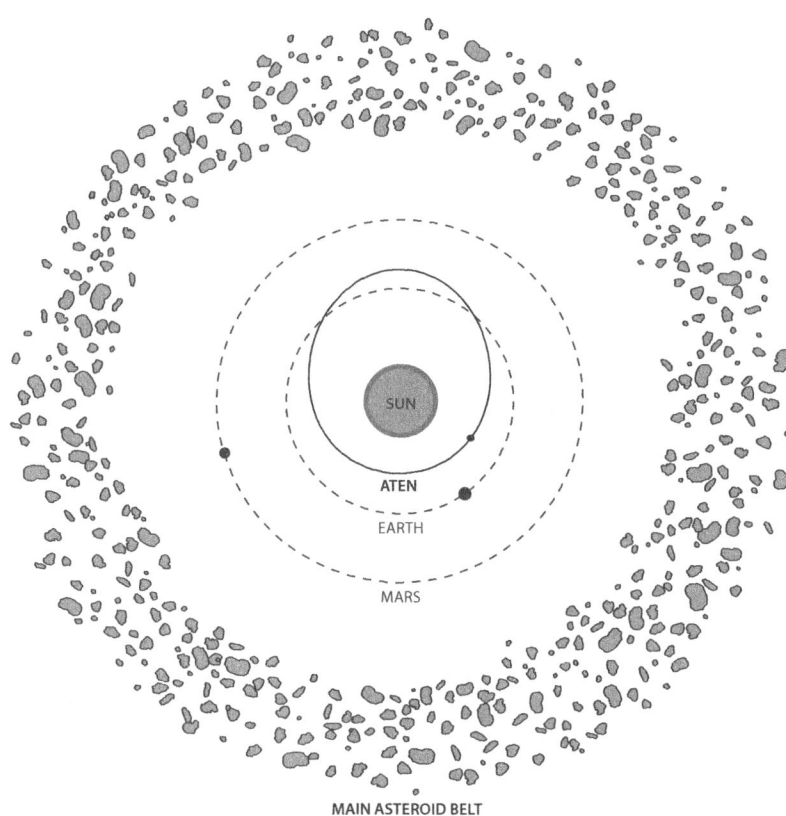

The orbit of Aten asteroids ellipses the sun and crosses Earth's orbit, but does not cross Mars's orbit.

craters and pits, which show where the asteroids have been struck by other objects in space.

Asteroids vary greatly in size. The largest known asteroid, Ceres, measures about 966 kilometers (600 miles) in diameter and is so large that it is considered a dwarf planet. However, most asteroids are much smaller; in fact, many asteroids are only the size of small rocks.

The solar system contains millions of asteroids. Most are found within the asteroid belt, a doughnut-shaped ring located between the orbits of Mars and Jupiter, about 2 to 4 AU from the sun, a distance of about 300 million to 600 million kilometers (186 to 370 million miles). Sometimes asteroids within the asteroid belt are pushed into the solar system. For example, if two asteroids collide, one or both of them can be flung into space. Similarly, an encounter with Jupiter's massive gravity can have the same result. These asteroids sometimes end up orbiting close to Earth and the sun. Those that come within 0.3 AU, or around 45 million kilometers (28 million miles), of Earth are called near- Earth asteroids (NEAs).

Several categories of NEAs exist. These categories are distinguished by characteristics such as the asteroids' average distance from the sun (near and far) as well as the orbital paths they take through space. As of 2015, about 6 percent of known NEAs are considered Aten asteroids. Because their orbit is smaller than Earth's, Aten asteroids take an orbital path mainly inside of Earth's orbit but also cross it. Astronomers keep a close eye on known Aten asteroids because they come close enough to Earth to collide with it. NEAs are also known as near-Earth objects (NEOs). NEOs include both comets and asteroids that have been pulled closer to Earth's orbit by gravitational forces, mainly near the four inner planets (Mars, Mercury, Earth, and Venus).

## NEA Impacts on Earth

Scientists are interested in asteroids and other NEOs because they are thought to be remaining debris from the formation of the solar system about 4.6 billion years ago. They can provide valuable information about the early solar system.

Additionally, thousands of NEAs come close enough to Earth to raise concerns about potential collisions. Although smaller space objects such as meteors collide with Earth frequently, these objects are usually so small that they burn up in the atmosphere before they reach the planet's surface and do much damage.

However, major collisions between Earth and NEAs have occurred. For example, a huge crater near Chicxulub, Mexico, is believed to be evidence of a past collision between Earth and a large NEA. Many experts believe this collision, which probably had the force of a nuclear explosion, played a major role in the mass extinction that occurred over 65 million years ago. During this extinction, dinosaurs and other large life forms on the planet died out.

Other NEAs have collided with Earth in more modern times. For example, in February 2013, a relatively small NEA exploded over Russia, releasing more energy into the atmosphere than a large atomic bomb.

### NEAR-EARTH OBJECT PROGRAM
The National Aeronautics and Space Administration (NASA) created the Near-Earth Object Program (NEOP) to detect and track NEOs. This program uses teams around the world to learn more about NEOs and to eventually develop technologies to divert NEOs away from Earth. These teams, such as Deep Space Network (Goldstone, California) and Arecibo Observatory (Puerto Rico), use state-of-the-art technologies to identify and track NEOs. They use radar to bounce radio waves off asteroids in space to provide valuable information. Radar is even capable of generating surface images.

Additionally, NASA has successfully launched many robotic spacecraft missions to study NEOs. For example, the NEAR Shoemaker mission was launched in 1996 to study the asteroid Eros, while the Dawn mission was launched in 2007 to study asteroids Vesta and Ceres.

The Asteroid Redirect Mission (ARM) is NASA's first attempt to capture and redirect an NEA in order to allow it to be studied directly by humans. ARM is developing technologies that can redirect an NEA to orbit around Earth's moon. Astronauts would then be able to land on and explore the asteroid. This mission is scheduled for the 2020s.

—*Marie Keenan, MS*

### FURTHER READING
Chapman, Clark. "The Hazard of Near-Earth Asteroid Impacts on Earth." *Earth and Planetary Science Letters* 222.1 (2004): 1–15. Print.

Dymock, Roger. *Asteroids and Dwarf Planets and How to Observe Them.* New York: Springer, 2010. Print.

"Near Earth Asteroids: A Chronology of Milestones." *International Astronomical Union.* IAU, 7 Oct. 2013. Web. 4 Mar. 2015.

"Near-Earth Object Program." *National Aeronautics and Space Administration.* NASA, 6 Feb. 2015. Web. 5 Mar. 2015.

"Responding to Potential Asteroid Redirect Mission Targets." *Jet Propulsion Laboratory.* California Inst. of Technology, 14 Feb. 2014. Web. 5 Mar. 2015.

Yeomans, Donald K. *Near-Earth Objects: Finding Them before They Find Us.* Princeton: Princeton University Press, 2013. Print.

# ATEN ASTEROIDS: 2062 ATEN

### FIELDS OF STUDY
Astronomy; Cosmology; Asteroid Impact Avoidance

### SUMMARY
2062 Aten is in a class of space objects known as Aten asteroids, which for the most part orbit between the sun and Earth. Atens are near-Earth asteroids (NEAs), which are important because they both provide insight into the early solar system and are close enough to strike Earth. 2062 Aten was the first Aten asteroid to be identified.

### PRINCIPAL TERMS
- **asteroid:** small, rocky space objects that orbit other larger celestial bodies such as the sun and planets.
- **Aten objects:** asteroids that orbit between Earth and the sun and are classified as near-Earth asteroids. Aten asteroids have a semimajor axis, or half the greatest diameter of its orbit, of less than 1 astronomical unit (AU).

### NEAR-EARTH ASTEROIDS
2062 Aten was the first of the Aten objects to be identified. Asteroids are irregular objects made of rock, metal, and minerals that are larger than meteors and travel in a circular or oval orbit around another celestial body, such as the sun or Earth. Studying the material in asteroids is valuable because asteroids are thought to be made up of remnants of the early solar system.

Asteroids are defined by their orbits and fall into three main groups. The largest of these is the asteroid belt. This irregular ring of asteroids is located between the orbits of Jupiter and Mars in an area ranging from about 2.12 to 3.3 AU from the sun (about 317 million to 493 million kilometers, or 197 to 306 million miles). Another group of asteroids in the same vicinity orbit Jupiter and are known as the Trojans. The group of asteroids that orbit in the area between the sun and Earth are known as near-Earth asteroids (NEA). Scientists believe that NEAs started out in the asteroid belt but were pushed out by collisions with other asteroids and pulled into a closer orbit with the sun by gravitational forces. NEAs belong to a broader category of space bodies called near-Earth objects (NEOs), which also includes comets. NEAs can orbit as close as about 0.3 AU (about 45 million kilometers or 29 million miles) from Earth.

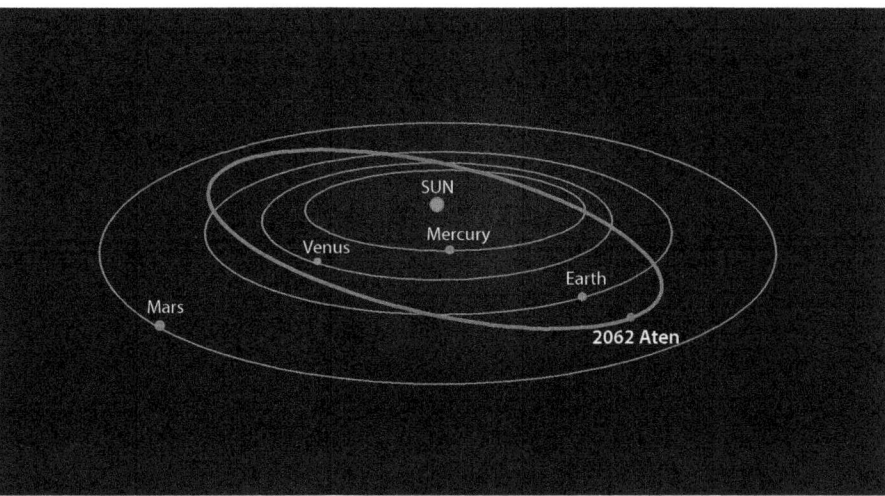

Orbit of 2062 Aten, asteroid.

Several types of NEAs are differentiated by their orbit paths and distances from the sun. NEA families include Aten, Apollo, Amor, and Atira. An individual asteroid can change categories if its orbit is altered by forces such as collisions and gravity. Aten asteroids orbit in a relatively close and nearly circular path that is smaller than Earth's orbit. At its closest, an Aten asteroid comes within 1 AU (about 149.6 million kilometers or 93 million miles) of the sun; at the aphelion, or greatest distance from the sun, of its orbit, it is more than 1 AU away. This means its orbit both takes less than a year and crosses over Earth's orbit. One of the reasons that scientists are so interested in Aten asteroids is because they come so close to Earth.

### Discovery of 2062 Aten

Planetary scientist Eleanor Francis Helin discovered 2062 Aten, the first Aten asteroid to be identified, at the Palomar Observatory in California, on January 7, 1976. The name Aten comes from the name of the Egyptian sun god. Helin discovered 2062 Aten using a Schmidt photographic telescope. Helin was a leader in the discovery and study of NEAs and was instrumental in establishing the Palomar Planet-Crossing Asteroid Survey (PCAS). Between 1973 and its end in 1995, PCAS identified sixty-five NEAs, enabling scientists to track these potentially dangerous objects.

### Potential NEA Impacts

NEOs that come close enough to pose a threat to life on Earth are called potentially hazardous asteroids (PHAs). Scientists believe that one of these struck Earth 65 million years ago in Chicxulub, Mexico, which may have led to the Cretaceous-Tertiary extinction that wiped out the dinosaurs.

More recently, in Russia, the explosion of small NEAs in Earth's atmosphere resulted in significant damage to an uninhabited area of Tunguska in 1908. In 2013, a similar event shattered windows and rocked buildings in the city of Chelyabinsk. Because of the potential for Earth impact, scientists from the National Aeronautics and Space Administration (NASA) established the Near-Earth Object Program (NEOP).

Working with teams worldwide and using state-of-the-art equipment such as radar and robotic spacecraft, NEOP searches for and monitors asteroids such as 2062 Aten. NASA has also established the Asteroid Redirect Mission (ARM) to more closely study PHAs. Scientists hope that by the 2020s, ARM will be able to redirect an NEA to orbit the moon, which would allow hands-on exploration by astronauts.

—*Janine Ungvarsky*

### Further Reading

"Aten Asteroids." *Cosmos: The SAO Encyclopedia of Astronomy*. Swinburne U of Technology. Site, n.d. Web. 19 Feb. 2015.

"Near Earth Asteroids." *Cosmos: The SAO Encyclopedia of Astronomy*. Swinburne University of Technology, n.d. Web. 19 Feb. 2015.

"Near Earth Asteroids: A Chronology of Milestones." *International Astronomical Union*. IAU, 7 Oct. 2013. Web. 19 Feb. 2015.

"Near Earth Objects: NEO Segment." *Space Situational Awareness*. European Space Agency, n.d. Web. 19 Feb. 2015.

"Near Earth Object Program." *NASA*. NASA, 10 Mar. 2015. Web. 10 Mar. 2015.

"Potentially Hazardous Asteroids." *Cosmos: The SAO Encyclopedia of Astronomy*. Swinburne University of Technology, n.d. Web. 19 Feb. 2015.

Schmadel, Lutz D. *Dictionary of Minor Planet Names*. 6th ed. 2 vols. Heidelberg: Springer, 2012. Print.

"Searching the Sky for Dangerous Neighbors: Eleanor Helin and the 18-Inch Telescope." *Palomar Observatory*. California Inst. of Technology, 8 May 2014. Web. 19 Feb. 2015.

# ATMOSPHERIC OPACITY

### FIELDS OF STUDY

Observational Astronomy; Theoretical Astrophysics

### SUMMARY

Atmospheric opacity is the degree to which electromagnetic radiation penetrates the layers of atmosphere surrounding a celestial body. Electromagnetic radiation in the form of light or radio waves reaches the surface in differing amounts, depending on the type of wave, its frequency and energy, and the level of opacity of the surrounding atmosphere. Atmospheric opacity affects surface conditions on Earth and provides information about the potential for life on other celestial bodies.

### PRINCIPAL TERMS

- **electromagnetic waves:** the classical form of electromagnetic radiation, produced when electric and magnetic fields come together and interact; can be in the form of radio waves, microwaves, infrared, optical, ultraviolet, x-rays, or gamma rays, depending on their frequency, energy, and wavelength.
- **opacity:** the degree to which a substance or object lets various forms of electromagnetic radiation pass through it.

### PROPERTIES OF THE ATMOSPHERE

Earth's atmosphere is a complex, layered field of gases, water vapor, and dust surrounding the planet. It provides and protects the conditions necessary for life. The atmosphere holds in the oxygen and other gases that make up breathable air and allows the light and heat of the sun to reach Earth's surface. It also deflects or absorbs most harmful forms of electromagnetic waves, such as ultraviolet, gamma rays, and x-rays.

Earth is not the only celestial object with an atmosphere. While all atmospheres serve as filters for light, heat, and radiation, each differs in the amount of waves it allows to reach the surface. The degree to which an atmosphere does this is called atmospheric opacity. Opacity is affected by both the conditions of the atmosphere and the properties of the electromagnetic waves. Relevant atmospheric conditions include temperature, cloud cover, and amount of water vapor. Scientists have been learning about atmospheric opacity for as long as they have been studying Earth's atmosphere. The earliest known experiments in the seventeenth and eighteenth centuries were limited to studying temperature and air pressure on Earth's highest mountains. The invention of hot air balloons in the late eighteenth century allowed scientists to study the higher levels of the atmosphere, but the effects of extreme cold and air pressure on humans hampered these experiments. At very high altitudes, the scientists got frostbite and lost consciousness. French meteorologist Léon Teisserenc de Bort (1855–1913) became one of the first to launch an unmanned weather balloon. Instruments attached to the balloon allowed him to record information at altitudes higher than could be safely reached by humans.

In 1901, Italian physicist and inventor Guglielmo Marconi (1874–1937) proved that one layer of the atmosphere—the ionosphere—reflects radio waves. He used this property to bounce the first wireless radio signal across the Atlantic Ocean, from England to Canada. The next year, de Bort presented his research on the upper atmosphere, identifying the troposphere and stratosphere. In 1932, American physicist Karl Guthe Jansky (1905-50) found that the static heard in radio signals is caused by radio waves from deep in the Milky Way galaxy. This was the first proof that radio waves could also pass through the atmosphere.

Improved technology in the twentieth century enabled closer study of atmospheric layers and their properties. The development of spaceflight in the late 1950s and 1960s represented a huge leap forward. Scientists began using satellites and special telescopes to study Earth's atmosphere, both from space and from within its layers. Several key research instruments, including the International Space Station (ISS), orbit in the part of Earth's atmosphere known as the thermosphere. The outer boundary of this layer is 690 kilometers (429 miles) above Earth's surface. Other satellites and telescopes have allowed scientists to study the atmospheres of other planets.

**EARLY ATMOSPHERIC STUDY**

Scientists have been learning about atmospheric opacity for as long as they have been studying Earth's atmosphere. The earliest known experiments in the seventeenth and eighteenth centuries were limited to studying temperature and air pressure on Earth's highest mountains. The invention of hot air balloons in the late eighteenth century allowed scientists to study the higher levels of the atmosphere, but the effects of extreme cold and air pressure on humans hampered these experiments. At very high altitudes, the scientists got frostbite and lost consciousness. French meteorologist Léon Teisserenc de Bort (1855-1913) became one of the first to launch an unmanned weather balloon. Instruments attached to the balloon allowed him to record information at altitudes higher than could be safely reached by humans.

In 1901, Italian physicist and inventor Guglielmo Marconi (1874-1937) proved that one layer of the atmosphere—the ionosphere—reflects radio waves. He used this property to bounce the first wireless radio signal across the Atlantic Ocean, from England to Canada. The next year, de Bort presented his research on the upper atmosphere, identifying the troposphere and stratosphere. In 1932, American physicist Karl Guthe Jansky (1905-50) found that the static heard in radio signals is caused by radio waves from deep in the Milky Way galaxy. This was the first proof that radio waves could also pass through the atmosphere.

Improved technology in the twentieth century enabled closer study of atmospheric layers and their properties. The development of spaceflight in the late 1950s and 1960s represented a huge leap forward. Scientists began using satellites and special telescopes to study Earth's atmosphere, both from space and from within its layers.

Several key research instruments, including the International Space Station (ISS), orbit in the part of Earth's atmosphere known as the thermosphere. The outer boundary of this layer is 690 kilometers (429 miles) above Earth's surface. Other satellites and telescopes have allowed scientists to study the atmospheres of other planets.

**PROPERTIES OF ELECTROMAGNETIC RADIATION**

The nature of electromagnetic waves is determined by their frequency and wavelength. These properties are related through the following wave equation:

$$c = \lambda f$$

In this equation, $c$ represents the speed of light, $\lambda$ is wavelength, and $f$ is frequency. Wavelength is the distance between a point in one wave and the same point in the following wave, measured in meters. Frequency is the number of wave cycles per unit time. The International System of Units (SI) unit of frequency is the hertz (Hz). This is a derived unit, equal to one cycle per second.

The different types of electromagnetic waves, in order of decreasing wavelength and increasing frequency, are radio waves, microwaves, infrared, optical or visible light, ultraviolet, x-rays, and gamma rays. All electromagnetic waves travel at the speed of light in a vacuum. Radio and optical waves are the easiest to study because they can pass readily through the atmosphere. Studying other forms of electromagnetism requires specially calibrated filters, cameras, and telescopes.

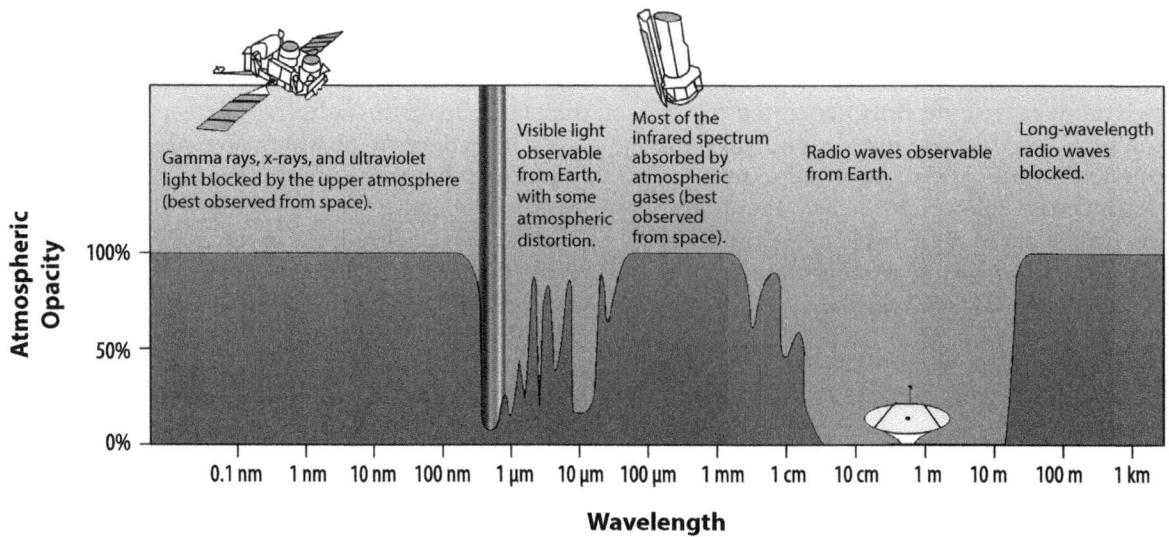

Percentage of atmospheric opacity for various electromagnetic wavelengths. Particular wavelengths cannot pass through the atmosphere while others can penetrate the atmosphere to varying degrees. Gamma rays, x-rays, and ultraviolet light are blocked by the upper atmosphere. Visible light is observable from Earth with some atmospheric distortion. Most of the infrared spectrum is absorbed by atmospheric gases. Radio waves are fully observable from Earth. Long-wavelength radio waves are blocked by the atmosphere. NASA (original); SVG by Mysid, public domain via Wikimedia Commons.

## Effects of Atmospheric Opacity

The opacity of an atmosphere affects the conditions on the surface below. Earth's atmosphere filters the amount of ultraviolet radiation that reaches the planet and helps keep Earth at the right temperature to sustain life. First it allows the warming rays of the sun to reach the surface and then it absorbs the heat energy as it rises, reflecting it back to space in what is known as the greenhouse effect.

Opacity also impacts radio-wave transmission. One of the highest levels of Earth's atmosphere, the ionosphere, is so named because it is made up of ions and free electrons. When atoms or molecules collide with ultraviolet and x-rays, they lose electrons, becoming positively charged ions. Together, the free electrons and positive ions form a plasma. A plasma is a state of matter that consists of unbound positively and negatively charged particles but is electrically neutral as a whole, because the positive and negative charges balance out. It was the ionosphere that bounced Marconi's radio waves across the ocean. The ionosphere also has sublayers that change seasonally and even daily. When there is no sunlight, one of the layers disappears and another two merge. This allows radio waves to travel farther at night than in the daylight.

## Importance of Atmospheric Opacity

The opacity of Earth's atmosphere is important to sustain human life, and humans may be just as important to atmospheric opacity. Some researchers think that the gases produced by certain human activities, such as farming, burning fossil fuels, and maintaining landfills, cause changes in the atmosphere's opacity that enhance the greenhouse effect. This could increase the surface temperature on all or part of Earth's surface. Scientists differ on how drastic these changes could be, their overall effects, and how fast they might occur.

Scientists also study the atmospheric opacity of other planets. NASA's Viking program, which sent two orbital probes to Mars in the 1970s, included studies of the depth and opacity of Mars's atmosphere. Further studies have been done on other planets in an effort to determine if their atmospheric opacity allows for the conditions necessary to sustain life.

—*Janine Ungvarsky*

**FURTHER READING**

"A Blanket around the Earth." *National Aeronautics and Space Administration.* California Inst. of Technology, n.d. Web. 27 Apr. 2015.

"Atmosphere." *National Geographic.* National Geographic Soc., n.d. Web. 27 Apr. 2015.

"Electromagnetic Radiation." *Cosmos: The SAO Encyclopedia of Astronomy.* Swinburne University of Technology, n.d. Web. 27 Apr. 2015.

"The Electromagnetic Spectrum. *National Aeronautics and Space Administration.* NASA, Mar. 2013. Web. 27 Apr. 2015.

"History of Discovery of the Atmosphere." *UCAR Center for Science Education.* U Corporation for Atmospheric Research, n.d. Web. 27 Apr. 2015.

"Karl Jansky and the Discovery of Cosmic Radio Waves." *National Radio Astronomy Observatory.* Assoc. U, 16 May 2008. Web. 27 Apr. 2015.

Randall, David A. *Atmosphere, Clouds, and Climate.* Princeton: Princeton University Press, 2012. Print.

# B

## BALL OF LIGHT PARTICLE MODEL

### FIELDS OF STUDY

Astrophysics

### SUMMARY

The ball of light particle model is a grand unification theory in the field of physics. John T. Nordberg developed the model and published material on it in 1995. Nordberg's model addresses many aspects of physics, including elementary particles, electric fields, magnetic fields, gravitational fields, standing waves, and moving waves. Nordberg claims that the ball of light particle model unifies the electric force, magnetic force, and gravitational force. In essence, then, it unifies all areas of physics. Nordberg also claims that his model can be used to explain different phenomena in astrophysics, including the big bang theory, black holes, quasars, and the cores of stars.

### PRINCIPAL TERMS

- **electric field:** a field around an object that is created by electric charges or magnetic fields.
- **grand unification theory:** a vision or goal in physics that involves unifying the fundamental forces such as strong, weak, and electromagnetic forces. This unification would allow for a better understanding of the way the universe is organized.
- **gravitational field:** a force field surrounding an object that has a gravitational influence.
- **harmonic standing spherical wave:** a wave that is created when moving waves crash into one another.
- **magnetic field:** an area around an object that has a magnetic influence.
- **moving wave:** a wave that travels along the medium; also known as a traveling wave.
- **polarization:** a property in which the oscillations in a wave are perpendicular to the direction of travel. It can cause waves to produce other waves.
- **standing wave:** a wave that stays in place as it vibrates up and down.

### BALLS OF LIGHT

The ball of light particle model is a grand unification theory that was developed in 1995 by theoretical physicist and industrial engineer John T. Nordberg (b. 1961) through the fusion research and development company Nordberg Fusion Technologies.

A basic knowledge of physics concepts is crucial to understanding the ball of light particle model. Physics is typically explained using forces called the strong nuclear force and the weak nuclear force. The strong nuclear force generally is a force that holds together matter in the universe. The weak nuclear force typically causes this matter to fall apart. Nordberg's ball of light particle model relates to these concepts in that it identifies three strong forces: electric force, magnetic force, and gravitational force.

Elementary particles also play an important role in understanding Nordberg's model. An elementary particle consists of an electric field, a magnetic field, and a gravitational field. Electric, magnetic, and gravitational fields come in the form of standing waves and moving waves (also called traveling waves). Moving waves may crash into one another. If certain conditions are present, including polarization, then this may cause the waves to combine and produce harmonic standing spherical waves. Harmonic standing spherical waves have electric, magnetic, and gravitational fields. This means that harmonic standing spherical waves can be thought of as elementary particles. Nordberg created a nickname for spherical electromagnetic and gravitational waves, which can be either standing or moving waves. The nickname he gave them was "balls of light."

### NORDBERG'S GRAND UNIFICATION EQUATION

Nordberg devised a grand unification equation that he used as the basis for the ball of light particle model. The equation, Nordberg believes, unifies all areas of physics in a way that is much like his view of the ball of light particle model. According to Nordberg, the gravitational field vector (G) is the cross product of the electric field (E) and the magnetic field (B) vectors, written as

$$G = E = B$$

In physics, a vector stands for a quantity with magnitude and direction. Nordberg explains the ball of light particle model using this equation with a sphere, demonstrating that the natural position of the fields makes the electric and magnetic fields point to the surface of the sphere. Therefore, the gravitational field, as the cross product, always points to the center of the sphere. Electric and magnetic fields can move around the sphere. Accelerating these fields creates greater magnitudes in the fields, which in turn creates a greater magnitude in the gravitational field. This higher gravitational field can make the particle more stable and also reduce its size. A simpler explanation of this is that the ball of light particle model unifies the electric, magnetic, and gravitational forces.

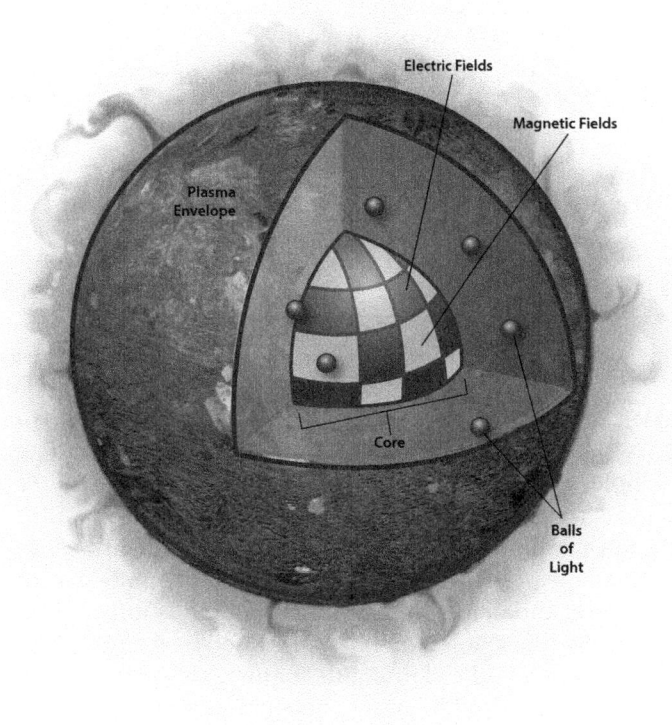

The sun described as predicted by the ball of light particle model. The sun's core is surrounded by a plasma envelope and has electric and magnetic fields flowing over its surface. The moving electric and magnetic fields can create smaller balls of light in the form of electrons, photons, and other particles.

## Scientific Significance

Physicists have long been working to find a grand unification theory in order to combine the various areas of physics. In the 1970s, physicists Steven Weinberg (b. 1933), Abdus Salam (1926–96), and Sheldon Glashow (b. 1932) developed a unification theory for the electromagnetic and weak forces. The theory eventually led to the grand unification theory and numerous variants. Such theories, however, did not address gravity. Nordberg's ball of light particle model does address gravity. It is believed to be the first working grand unification theory in the field of physics. This means that the model is the first theory that has unified all areas of physics, including the electric, magnetic, and gravitational forces.

Nordberg believes that his ball of light particle model can be used to explain various phenomena in the field of astrophysics. For example, he claims that the model can describe the big bang theory, which explains the creation of the universe. According to Nordberg, the ball of light particle model helps illustrate that the universe was created from energy that was contained in a single ball of light that was both spherical and harmonic. In other words, the model suggests that the big bang began with the decay of one enormous ball of light. Nordberg also claims that his model helps explain that black holes, quasars, the cores of stars, and other celestial objects are waves of energy that are spherical, standing, electromagnetic, and gravitational.

—*Michael Mazzei*

## Further Reading

Elert, Glenn. "Standing Waves." *Physics Hypertextbook*. Glenn Elert, n.d. Web. 6 May 2015.

"Forces and the Grand Unified Theory." *Particle Adventure*. Particle Data Group, n.d. Web. 6 May 2015.

Lucas, Jim. "What Is the Strong Force?" *Live Science*. Purch, 1 Nov. 2014. Web. 6 May 2015.

Lucas, Jim. "What Is the Weak Force?" *Live Science*. Purch, 24 Dec. 2014. Web. 6 May 2015.

Nordberg, John T. *Grand Unification*. John T. Nordberg, n.d. Web. 6 May 2015.

Odenwald, Sten. "What Is Grand Unification Theory?" *Astronomy Cafe*. Sten Odenwald, 1997. Web. 6 May 2015.

# BLACK HOLES

### FIELDS OF STUDY

Stellar Astronomy; Astronomy; Cosmology

### SUMMARY

Black holes are areas of space-time with gravitational fields so strong that nothing can escape. This phenomenon is caused by a mathematical singularity at the center of the black hole. The black hole is infinitely dense, and its nearby gravitational field is infinitely strong. Black holes were first proposed by scientist John Michell in the eighteenth century. However, they were popularized by Albert Einstein as a consequence of his general theory of relativity. While black holes cannot be directly observed, astronomers have found strong evidence for their existence.

### PRINCIPAL TERMS

- **event horizon:** the boundary of a black hole beyond which nothing can escape.
- **general relativity:** Albert Einstein's theory that gravity is the result of matter causing space-time to curve. Among other things, this implies that gravity can bend light.
- **supermassive black hole:** a black hole with a mass greater than one hundred thousand times that of the sun.

### DISCOVERY OF BLACK HOLES

The existence of black holes was first proposed by English clergyman and scientist John Michell (1724–93) in 1783. Michell published a letter stating that certain stars may grow so big that not even light can escape their gravitational pull. This was based on Isaac Newton's (1642–1727) "corpuscular" theory of light, which essentially stated that light is a particle rather than a wave. A particle could be affected by gravitational pull, while a wave could not. Soon after, a number of experiments seemed to disprove Newton's theory, and as a result Michell's proposal was largely ignored.

This all changed in 1915, when Albert Einstein published his famous theory of general relativity. When tested, Einstein's theories showed that light could be bent by powerful gravitational fields. Thus, if an area of space somehow acquired an infinitely powerful gravitational pull, light would not be able to escape it. The object would appear as a giant black sphere, effectively invisible in outer space.

Black holes are incredibly difficult to observe directly. However, astronomers can use other methods to determine their locations. First, black holes have a visible effect on the stars around them. When astronomers find a star that appears to be orbiting a large object that neither emits or reflects visible light, that object is probably a black hole. Second, the accretion disks of black holes create so much friction that they heat up and emit electromagnetic radiation, mostly in the x-ray range. Scientists can use x-ray telescopes to detect this radiation and trace it to its source.

### CLASSICAL BLACK HOLES

Black holes are formed by the collapse of incredibly massive stars. When such a star uses up all its fuel, the delicate balance that allowed it to exist is broken. The gravity from the star's dense core pulls its outer layers inward. As more matter is pulled into the core, the increase in the core's mass and density causes its gravitational pull to grow stronger, which in turn pulls in even more matter. This process increases in speed and power exponentially. Eventually, the area around the core becomes so dense and gravitationally powerful that not even light can escape it. As this area pulls everything around it inward, it continues to grow and increase its gravitational pull. This creates an infinitely dense region of space-time with an infinitely strong gravitational pull. Such a region is called a singularity.

Every singularity has a definitive boundary called an event horizon. An event horizon is the point at which the singularity's gravitational pull becomes infinitely powerful, making escape impossible. The larger the black hole, the larger the event horizon. In these cases, faraway objects can be captured permanently.

Black holes are typically surrounded by accretion disks. An accretion disk is a disk of gas and other matter that orbits the event horizon as it is slowly pulled toward the singularity. The gravity produced by the singularity compresses this matter as it rotates, creating intense amounts of friction. The friction generates so much heat that the accretion disk emits massive amount of radiation, mostly in the form of x-rays.

## Supermassive Black Holes

As their name suggests, supermassive black holes are many times larger than normal black holes. The singularities of supermassive black holes are at least one hundred thousand solar masses (one hundred thousand times the mass of the sun). Astronomers theorize that supermassive black holes as large as fifty billion solar masses could exist. The largest known supermassive black hole, SDSS J0100+2802, is twelve billion solar masses.

Many astronomers believe that a supermassive black hole lies at the center of every galaxy. The supermassive black hole at the center of the Milky Way is roughly 27,000 light-years from Earth and more than four million solar masses. Astronomers believe that supermassive black holes play a major role in the formation of galaxies. They have found a direct correlation between the mass of a galaxy's supermassive black hole and the size of the galactic bulge surrounding it.

A supermassive black hole's accretion disk is much brighter, hotter, and larger than that of traditional black holes. Occasionally, an extremely large object passes through a supermassive black hole's event horizon. The incredibly intense gravity near the event horizon crushes most objects, causing a reaction that converts roughly 10 percent of the object's mass into energy. The energy then dramatically shoots away from the black hole. Supermassive black holes sometimes capture objects so large that this reaction creates energy across multiple spectrums, including visible light and x-rays. These impressive events are called quasars.

Scientists are unsure how supermassive black holes grow to their tremendous sizes. While black holes can grow in mass by gradually absorbing more matter, it does not seem possible for one to grow quickly enough to reach such supermassive proportions during its lifetime. Therefore, some scientists theorize that supermassive black holes originated as giant clouds of gas soon after the big bang. This theory bypasses entirely the star phase of a black hole's existence. Proponents claim that some external force, such as large concentrations of dark matter in the early universe, could have compressed the gas quickly enough to form a large black hole, then continued to push more gas into the singularity at an incredibly fast rate. This would greatly increase the speed of the black hole's growth.

Another theory of the formation of supermassive black holes involves binary black hole systems. A binary black hole system is one in which two nearby black holes orbit a common

An artist's rendition of a black hole, Cygnus X-1, with an accretion disc. NASA/CXC/M.Weiss.

center of mass. This could happen when two galaxies collide, which would have happened more often early in the universe's history. The combined gravity from the two black holes would then pull their accretion disks out of alignment, which computer simulations show would cause the entirety of the disks to tear and collapse into their singularities. This would greatly increase the size of the black holes in a very short period of time. The two black holes would then merge, creating one black hole with the combined mass and gravitational pull of the entire binary system.

**INFORMATION PARADOX**

While black holes are viewed as necessary and mundane products of general relativity, their existence conflicts with a major tenet of quantum mechanics. General relativity deals primarily with the physics of large objects, while quantum mechanics deals with the physics of objects much too small to see. Scientists have been attempting to unify the two theories for many years.

One of the major principles of quantum mechanics is the idea that information can never be truly destroyed. If one has all the pieces of a current state of events, it should theoretically be possible to discover all past states. However, information pulled into the singularity of a black hole is effectively destroyed. It is compressed beyond any form of recognition and may never be able to escape.

British physicist Stephen Hawking (b. 1942) devised a way around this paradox. According to his calculations, due to quantum mechanics, all particles occasionally create temporary copies of themselves. Hawking proposed that black holes emit tiny amounts of radiation, slowly burning off all the matter they have collected over time. According to his models, black holes that fail to absorb new matter will eventually evaporate. Any information collected by a black hole will eventually be released, but not in any form humans can recognize or put back together.

An alternate solution for this paradox involves phenomena called white holes. According to this theory, black holes eventually burst, spewing all the information they have collected in a huge explosion of energy. Proponents of this theory say that this effect actually happens as soon as the black hole forms. However, due to the ways singularities bend time and space, the effect appears to take billions of years to anyone outside the singularity.

—*Tyler Biscontini*

**FURTHER READING**

Byrne, Michael. "The Black Holes We See in Space Might Already Be White Holes." *Motherboard*. Vice Media, 21 July 2015. Web. 30 Apr. 2015.

Cain, Fraser. "Never a Star: Did Supermassive Black Holes Form Directly?" *Universe Today*. Author, 7 Sept. 2007. 30 Apr. 2015.

GaBany, R. Jay. "A Singular Place." *Cosmotography*. Author, n.d. Web. 30 Apr. 2015.

Hall, Shannon. "How Do Black Holes Get Super Massive?" *Universe Today*. Fraser Cain, 13 Aug. 2013. Web. 30 Apr. 2015.

Marshall, John. "Introduction: Black Holes." *New Scientist*. Reed Business Information, 6 Jan. 2010. Web. 30 Apr. 2015.

Redd, Nola Taylor. "Black Holes: Facts, Theory, and Definition." *Space.com*. Purch, 9 Apr. 2015. Web. 30 Apr. 2015.

# BLAZARS

**FIELDS OF STUDY**

Astronomy; Extragalactic Astronomy

**SUMMARY**

Blazars are a type of active galactic nucleus (AGN). Like all AGNs, blazars are thought to be powered by supermassive black holes. Blazars are among the most energetic objects in the universe and emit large jets or bursts of matter at nearly the speed of light. Scientists study blazars to understand how they generate such fast and powerful jets. They may also reveal information about the early universe.

**PRINCIPAL TERMS**

- **active galactic nucleus (AGN):** a relatively compact region at the center of a galaxy that releases immense amounts of energy and electromagnetic

radiation, believed to be the result of mass accumulation by a supermassive black hole.
- **black hole:** a region of space with a gravitational pull so strong that even light cannot escape it.
- **gamma ray:** a form of electromagnetic radiation with high energy and a very short wavelength.
- **jet:** a plume of electrons and positrons that is emitted from a black hole when its magnetic field interacts with other space objects.

## ACTIVE GALACTIC NUCLEI

The universe is full of galaxies of many sizes and compositions. It is thought that most if not all massive galaxies have an immense black hole, known as a supermassive black hole, at their center. In some cases, the center of the galaxy is extremely luminous, brighter than all light sources in the rest of the galaxy combined and emitting a great deal of energy along the entire electromagnetic spectrum. Such galaxies are known as active galaxies, and their centers are known as active galactic nuclei (AGNs). Scientists believe that AGNs are the result of the sheer amount of matter accumulated by the supermassive black holes at the center of active galaxies. Research indicates that AGNs contain millions or even billions of times more material than Earth's sun.

The enormous gravitational force of a black hole constantly draws in matter. This creates bands of gas and space debris called accretion disks. Eventually, the black hole pulls all this matter in like a drain emptying a tub. Sometimes before gas disappears into the black hole, it is released in blasts or jets of atomic matter and debris. Fueled by energy from the supermassive black hole, some of these jets can shoot out at nearly the speed of light. Galactic jets can create more light than an entire galaxy of stars, making AGNs among the brightest and most luminous objects in space.

Several types of AGNs exist, including quasars, radio galaxies, Seyfert galaxies, and blazars. All of these AGNs blast out jets of gas, debris, and other matter. The distinguishing feature of blazars is that their jets are directed toward Earth. Scientists have theorized that this head-on view of the light explains why blazars appear much brighter than quasars and Seyfert galaxies to observers on Earth.

To qualify as a blazar, an AGN must meet certain criteria. It must have a flat radio spectrum with high radio brightness. It must be highly optically polarized. Finally, it must display major optical variation over a period of a few days.

## COMPOSITION OF BLAZARS

The National Aeronautics and Space Administration (NASA) has conducted several missions to study AGNs. NASA is particularly interested in the electromagnetic radiation that is released by blazars because they point toward Earth. Scientists are not sure how these jets are formed and what they are made of. One theory is that the magnetic field of the black hole interacts with the band of debris around it in a way that causes energy to be released in a certain direction.

Because of their great power and the amount of light they generate, blazar jets are believed to emit gamma rays along with other forms of electromagnetic radiation. Gamma rays are the brightest and highest-powered form of electromagnetic radiation known to exist.

Scientists use an array of powerful instruments and telescopes to study blazars. They are working to determine the percentages of electrons, protons, and positrons contained in the electromagnetic rays they emit. This will help scientists better understand

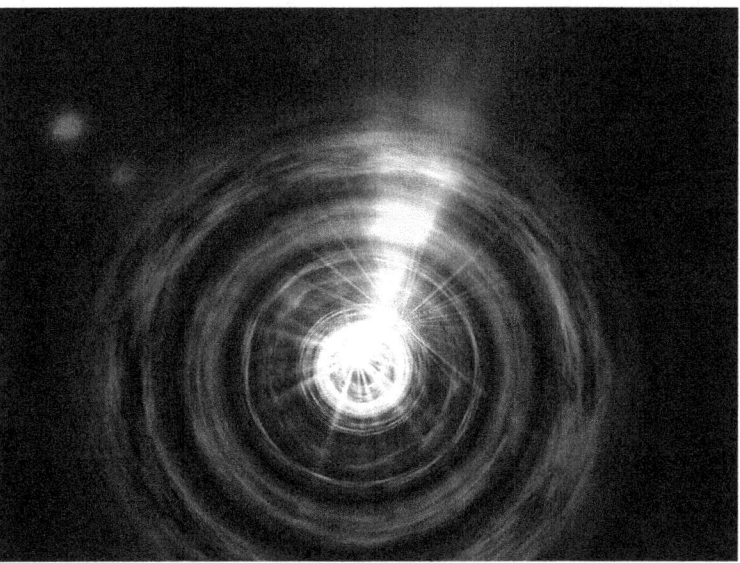

An artist's rendition of a blazar, with jets of particles emitting parallel to the viewer's line of sight, as matter falls into a supermassive black hole. NASA/Goddard Space Flight Center Conceptual Image Lab.

the power behind blazars and other AGNs, including how blazars generate jets with such force and speed.

### Detecting Blazars

Space-research organizations around the world have designed several missions to detect and study blazars. Between its launch in December 2009 and its temporary retirement in February 2011, NASA's Wide-Field Infrared Survey Explorer (WISE) identified more than two hundred blazars. Researchers believe that thousands more will be found.

WISE identified blazars by their infrared signatures in areas where large concentrations of gamma rays had been detected. Infrared light is outside the visible spectrum of light and requires special equipment to be seen. Blazars can be classified into two main subgroups based on their emission spectra. Flat-spectrum radio quasars (FSRQs) are blazars that produce emission spectra with strong, broad features. BL Lacertæ objects, or BL Lacs, are those with emission spectra that lack strong features. These differences are detected by variations in luminosity and redshift distributions.

NASA's Fermi Gamma-Ray Space Telescope discovered 709 suspected AGNs within a year of its activation in 2008. About three hundred are considered FSRQs, another three hundred are BL Lacs, and the rest are other types of AGNs.

### A Glimpse into Galactic History

All the AGNs studied by scientists are millions of miles away from Earth, so even the light from the extremely fast blazar jets takes a very long time to be seen. This means that AGNs provide information about the nature of black holes and the role they played in the formation of galaxies. It is possible that blazars will reveal how black holes affect the current state of the universe. In particular, the jets from blazars are clues to the differences between the supermassive black holes at the center of AGNs and the smaller black holes found in other parts of the universe.

—*Janine Ungvarsky*

### Further Reading

"Active Galactic Nuclei." *National Radio Astronomy Observatory.* Natl. Radio Astronomy Observatory / Assoc. Univs., n.d. Web. 13 Mar. 2015.

"Active Galaxies and Quasars." *Imagine the Universe!* NASA, 17 Nov. 2014. Web. 13 Mar. 2015.

"Black Holes." *NASA Science.* NASA, 21 Jan. 2015. Web. 13 Mar. 2015.

"Galactic Jets." *Cosmos: The SAO Encyclopedia of Astronomy.* Swinburne U of Technology, n.d. Web. 13 Mar. 2015.

"Galactic Nuclei." *Cosmos: The SAO Encyclopedia of Astronomy.* Swinburne U of Technology, n.d. Web. 13 Mar. 2015.

Harrington, J. D., and Whitney Clavin. "NASA's WISE Mission Sees Skies Ablaze with Blazars." *Wide-Field Infrared Survey Explorer.* NASA, 12 Apr. 2012. Web. 13 Mar. 2015.

"Inside Blazars." *Astronomy Magazine.* Kalmbach, 23 Apr. 2008. Web. 13 Mar. 2015.

Naeye, Robert. "Blazars and Active Galaxies." *Fermi Gamma-Ray Space Telescope.* NASA, 28 Aug. 2008. Web. 13 Mar. 2015.

# BOLIDE

### FIELDS OF STUDY

Astrometry; Astrophysics; Astronomy

### SUMMARY

A bolide is a meteor originating in outer space that explodes as it falls to Earth. Astronomers usually use the term to describe especially brightly burning objects or fireballs. Bolides are sometimes dangerous, but they usually land away from people.

### PRINCIPAL TERMS

- **asteroid:** a minor planet in the inner solar system that orbits the sun.
- **comet:** an icy small solar system body that heats up and releases gas when it passes near the sun.
- **fireball:** a meteor that is unusually bright.
- **meteor:** a small body of rock or metal that enters Earth's atmosphere and emits light.
- **meteoroid:** a small body of rock or metal that travels through space.

### Types of Bolides

Bolides start as one of three types of objects found in space: a comet, an asteroid, or a meteoroid. Comets are balls of ice and dust, asteroids are minor planets that are too small to be classified as actual planets, and meteoroids are small chunks of rock or metal. When these objects enter Earth's atmosphere, they heat up. If the object becomes hot enough to emit light, it is called a meteor. If the becomes brighter than the planet Venus in the night sky, it is called a fireball. If that fireball gets so hot that it explodes before reaching Earth's surface, the object is called a bolide. Especially bright bolides may be known as superbolides.

A bolide in the sky above Reno, Nevada, on April 22, 2012. NASA/Lisa Warren.

### Why Bolides Explode

On Earth, air slows down any moving object. An object's terminal velocity is the fastest that object can free-fall before the resistance of the atmosphere prevents it from accelerating further. However, there is no atmosphere in outer space. Because of this, many objects in outer space move much faster than objects on Earth.

Meteors that enter Earth's atmosphere are usually traveling much faster than their terminal velocity. Because of this, the atmosphere in front of the meteor is compressed, creating pressure and friction and causing the object to slow down. As the meteor slows, the air around it quickly heats up. This causes the meteor to burst into flames. When the heat and pressure become stronger than the meteor can bear, the meteor violently breaks apart.

### Bolides and Humans

When bolides explode near humans, the results can be catastrophic. On February 15, 2013, a bolide exploded near Chelyabinsk, Russia. According to the National Aeronautics and Space Administration (NASA), the bolide exploded with almost five hundred kilotons of force. For comparison, the atomic bomb dropped at Hiroshima exerted only sixteen kilotons of force. The explosion injured hundreds of people and damaged more than 297 buildings.

More than a century earlier, an even larger bolide exploded over Russia's Tunguska region. Scientists estimate that blast was one thousand times as powerful as the Hiroshima bomb. The Tunguska blast flattened trees for miles around the blast site and caused so much light and reflective dust to be thrown into the atmosphere that people as far away as western Europe reported glowing night skies.

—*Tyler Biscontini*

### Further Reading

Allain, Rhett. "Why Does a Meteor Explode in the Air?" *Wired*. Condé Nast, 18 Feb. 2013. Web. 11 Mar. 2015.

"Bolide." *Cosmos: The SAO Encyclopedia of Astronomy*. Swinburne U of Technology, n.d. Web. 27 Mar. 2015.

Bryner, Jeanna. "Meteor Explosion in Russia Hurts Hundreds of People: Reports." *LiveScience*. Purch, 15 Feb. 2013. Web. 11 Mar. 2015.

Popova, Olga, and Ivan Nemchinov. "Bolides in the Earth Atmosphere." *Catastrophic Events Caused by Cosmic Objects*. Ed. Vitaly Adushkin and Nemchinov. Dordrecht: Springer, 2008. 131–62. Print.

Reynolds, Mike D. *Falling Stars: A Guide to Meteors & Meteorites*. Mechanicsburg: Stackpole, 2010. Print.

Richardson, James, and James Bedient, comps. "Fireball FAQs." *American Meteor Society*. Amer. Meteor Soc., n.d. Web. 11 Mar. 2015.

# BREMSSTRAHLUNG RADIATION

## FIELDS OF STUDY

Astrophysics

## SUMMARY

Bremsstrahlung radiation is electromagnetic radiation that is produced when a highly charged particle, such as an electron, hits the nucleus of an atom or another electron and emits a photon. The bremsstrahlung process often produces x-rays. Bremsstrahlung occurs in ionized gas clouds and in cosmic rays.

## PRINCIPAL TERMS

- **continuous spectrum:** electromagnetic radiation that is emitted at all frequencies and wavelengths within a given range, with no apparent gaps.
- **electron:** a negatively charged subatomic particle.
- **free-free radiation:** radiation produced from electrons that are not associated with atoms or ions before or after bremsstrahlung.
- **x-rays:** electromagnetic radiation made of up of photons that can be produced through bremsstrahlung.

## BRAKING ELECTRONS

Bremsstrahlung radiation is a type of electromagnetic radiation that is produced when a charged particle, most often an electron, abruptly slows down because it hits an electric field, such as the nucleus of an atom. When the electron slows, its kinetic energy is transformed, and it emits a photon. Photons are the elementary particles that make up all forms of electromagnetic radiation. The process that creates bremsstrahlung radiation is called the bremsstrahlung process. Bremsstrahlung is one method of creating x-rays.

German physicist Wilhelm C. Röntgen (1845–1923) discovered x-rays in 1895. Four years later, another German physicist, Arnold Sommerfeld (1868–1951), discovered that x-rays come in two different types. The first type, known as characteristic x-rays, produces radiation at discrete frequencies, while the second type produces a continuous spectrum. Sommerfeld suggested that the second type be named *bremsstrahlung*, or "braking radiation," in reference to the sudden electron deceleration that causes it.

The first calculations for bremsstrahlung were devised in the 1930s. Decades later, in the 1960s, the first elementary measurements of bremsstrahlung were made. Scientists continue to study bremsstrahlung in hopes of learning more about cosmic rays, supernovas, and other phenomena in the universe.

Radiation is the emission of energy through electromagnetic waves. X-rays are a type of radiation that is made up of photons. They can be produced by bombarding metal x-ray tubes with high-energy electrons. When the energetic electrons hit the metal, they can produce x-rays in two main ways. Electrons that are associated with atoms produce x-rays through K-shell emissions, which occur when lower-energy electrons are knocked out of their orbits and then replaced by higher-energy electrons. The energy lost by the displaced electrons is emitted in the form of an x-ray photon.

Other electrons produce x-rays through bremsstrahlung. Electrons involved in bremsstrahlung deflect off the nucleus of an atom or off another electron. When the electrons are deflected, they lose some of their kinetic energy. As in K-shell emission, the energy lost is emitted as a photon. The type of photon emitted is determined by the change in kinetic energy. Electrons with more kinetic energy will release an x-ray photon. A less energetic electron may produce a photon with a longer wavelength.

## TYPES OF BREMSSTRAHLUNG

Bremsstrahlung happens in two forms: electron-nucleus or electron-electron. In electron-nucleus bremsstrahlung, the free electron collides with the nucleus of another atom. In electron-electron bremsstrahlung, the free electron collides with another electron. Electron-nucleus bremsstrahlung is the more common of the two types.

Some electrons that are involved in bremsstrahlung are not bound to an atom or an ion before or after the bremsstrahlung. Therefore, they are called "free" electrons. The free electrons create a type of continuous radiation. Radiation from these free electrons is sometimes called free-free radiation. Free-free

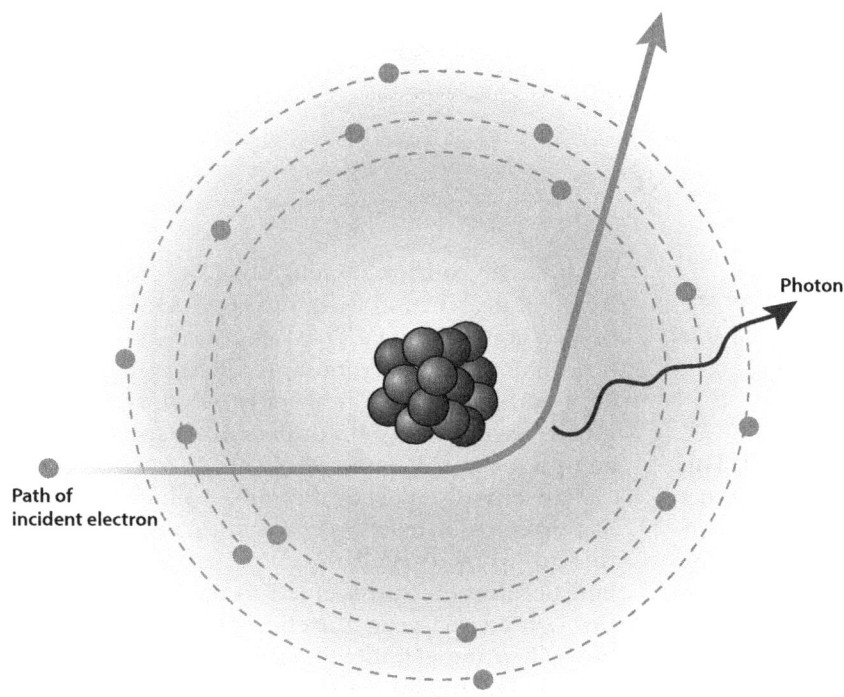

When an atom is bombarded with an electron, the incident electron is deflected by the nucleus of the atom and loses kinetic energy, which is then emitted as a photon.

radiation can come from hydrogen clouds, which can form around the exteriors of active comets.

### Studying Bremsstrahlung

Bremsstrahlung is an important concept in all areas of physics, including atomic, nuclear, solid-state, and elementary particle physics. However, it has been of particular importance to astrophysics, where it is used in experimental research. Astrophysicists have learned that bremsstrahlung affects which cosmic-ray particles reach Earth.

Electrons are more likely than other types of particles to be involved in bremsstrahlung because their smaller mass makes them easier to deflect in a collision. It is this deflection that causes the loss of kinetic energy in the form of a photon. Heavier particles, such as muons, have greater momentum due to their greater mass, so they lose less energy than electrons when they collide with other particles. Because of this, they are less likely to be involved in bremsstrahlung and can travel more deeply into a material. This is why muons that enter Earth's atmosphere in cosmic rays often travel all the way to Earth, while many electrons in cosmic rays do not.

—Elizabeth Mohn

### Further Reading

"Bremsstrahlung Radiation." *Cosmos: The SAO Encyclopedia of Astronomy.* Swinburne U of Technology, n.d. Web. 26 Mar. 2015.

Condon, J. J., and S. M. Ransom. "Free-Free Radio Emission from an HII Region." *Essential Radio Astronomy.* Natl. Radio Astronomy Observatory / Assoc. Univs., n.d. Web. 26 Mar. 2015.

Haug, Eberhard, and Werner Nakel. *The Elementary Process of Bremsstrahlung.* River Edge: World Scientific, 2004. Print.

Nave, Carl R. "Characteristic X-Rays." *HyperPhysics.* Georgia State U, n.d. Web. 24 Mar. 2015.

Somov, Boris V. *Plasma Astrophysics, Part II: Reconnection and Flares.* 2nd ed. New York: Springer, 2013. Print.

"What Is Bremsstrahlung?" *NDT Education Resource Center.* Iowa State U, n.d. Web. 26 Mar. 2015.

"X-Radiation." *NDT Education Resource Center.* Iowa State U, n.d. Web. 26 Mar. 2015.

# CANCER

## FIELDS OF STUDY

Astronomy; Observational Astronomy

## SUMMARY

Cancer is a constellation and part of the zodiac. This constellation represents a crab, and the word cancer means "crab" in Latin. The star pattern in the constellation resembles an upside-down Y. The constellation is located between the other zodiacal constellations Leo and Gemini. Cancer is the dimmest constellation in the zodiac. It includes a number of stars and other objects, including stars clusters and a planetary system. The crab in the constellation is linked to a crab from Greek mythology that was killed while fighting Hercules. Scientists study constellations such as Cancer to learn about their stars, deep sky objects, and exoplanets. Cancer contains a planetary system that scientists are studying because it has some similarities to Earth's solar system.

## PRINCIPAL TERMS

- **constellation:** an apparent pattern of stars, identified by humans, that can be seen in the night sky from Earth.
- **International Astronomical Union:** a recognized and authoritative professional organization for astronomers, founded in 1919 and based in Paris, France.

### Constellations and the Zodiac

Cancer is a constellation that is meant to be in the shape of a crab—hence its name, which means "crab" in Latin. It represents the crab that, according to ancient Greek mythology, Heracles (romanized as Hercules) fought during his twelve labors.

For thousands of years, people from many different cultures have identified different constellations in the night sky. Often, constellations are related to mythologies, stories, or religions. The International Astronomical Union (IAU) recognizes eighty-eight official constellations. The Cancer constellation has been recognized for thousands of years, though it is faint and hard to see, with only two stars above the fourth magnitude of brightness. Its shape resembles an upside-down Y.

Cancer is also part of the zodiac, which is a group of thirteen constellations. Ancient cultures believed this group of constellations was particularly important because the sun seemed to pass through them throughout the year, traveling along an imaginary line called the ecliptic. In fact, the sun does not actually move, but appears to move as the Earth orbits it. People remain interested in the constellations of the zodiac because they are well-known constellations and some of them are easily recognized.

The Cancer constellation is best seen in the Northern Hemisphere during March, though it remains visible between December and June. When viewed from Earth, it appears to be located between the zodiacal constellations Leo and Gemini. Other constellations located close to Cancer include Hydra, Lynx, and Canis Minor. People who want to view Cancer can first look for Regulus, the brightest star in the Leo constellation, and the stars Castor and Pollux from the Gemini constellation. If a line is drawn between Regulus and the twin stars, Cancer will be in the middle. However, Cancer is very dim and is best viewed with a telescope.

Cancer is made up of a number of different stars and objects. The brightest star in the constellation is beta Cancri, also known as Al Tarf, which is an orange star located approximately 290 light years from Earth. This star is in the leg of the crab and has a magnitude of 3.59. Another main star in the constellation is alpha Cancri, also known as Acubens. Even though alpha Cancri is only the fourth-brightest star in the constellation, with a magnitude of about 4.2, it was most likely named the alpha star because it represents one of the crab's claws. Iota Cancri is the star that represents Cancer's other claw.

The constellation Cancer also contains two Messier objects, Messier 44 (M44) and Messier 67 (M67). Messier objects are 110 deep-sky objects, such as galaxies, star clusters, and nebulae, that were catalogued by French astronomer Charles Messier (1730–1817) in the 1700s. Object M44 is cluster of stars sometimes called the Beehive Cluster or Praesepe, which means "manger" or "crib" in Latin. The Beehive Cluster is made up of approximately one thousand stars and somewhat resembles of a swarm of bees. It is the only part of Cancer that is easily visible with the naked eye or with binoculars. The Beehive is an open cluster, which means that its stars are gravitationally bound together and were formed from the same cloud or nebula. It is one of the closest open clusters to Earth, located about six hundred light-years away.

M67 is another open cluster, containing more than five hundred gravitationally bound stars. The stars in this cluster are approximately four billion years old, which is quite old for a star cluster, as clustered stars usually separate from one another over time. This star cluster is approximately 2,700 light-years from Earth and is roughly 12 light-years across.

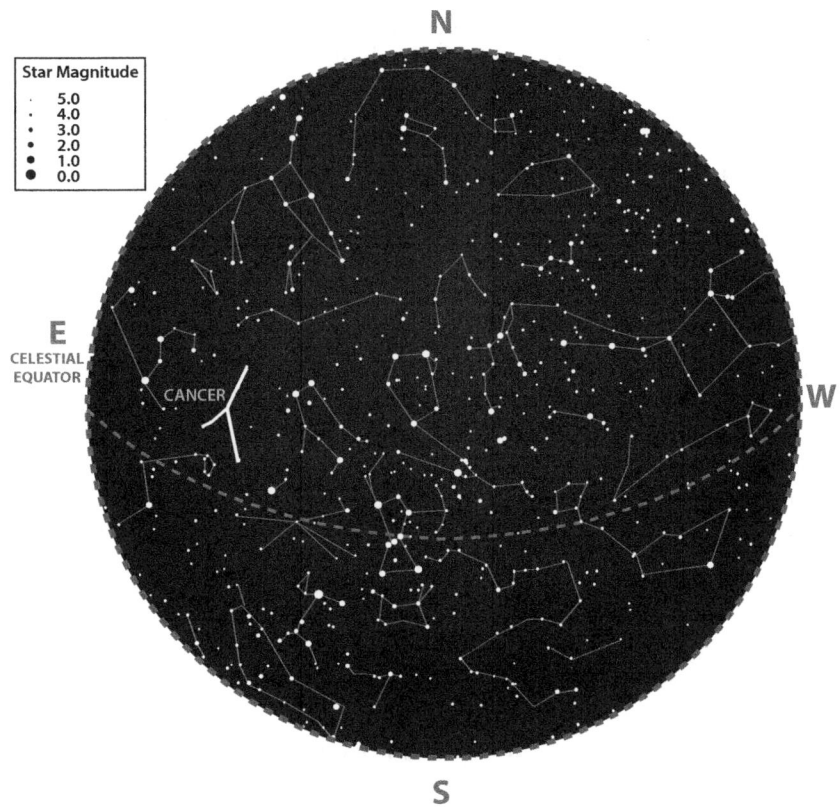

Star chart of the Northern and Southern Hemispheres in November. Lines connect the stars of the constellation Cancer.

## CANCER IN MYTHOLOGY

The Cancer constellation has its origins in ancient cultures. Most of the constellations recognized by the IAU were created by humans thousands of years ago. Some ancient cultures used constellations to tell stories about their religions. Others used them to identify seasons. Since the stars appear in the same part of the sky at the same time each year, farmers could observe the sky to find out when they should plant and harvest their crops.

Different stories are told about constellations over time. Many of the modern constellations are still related to Greek and Roman mythology. The Greek astronomer Ptolemy (ca. 100–170) wrote about the Cancer constellation, and the pattern was most likely identified before his lifetime.

In Greek mythology, Hercules encountered Cancer the crab during the twelve labors he had to undertake. One of his labors was to fight the sea serpent Hydra. While the two fought, Zeus's wife, Hera, sent a crab to help defeat Hercules. In one version of the legend, Hercules kicked the crab so hard that it flew into the sky. In another, Hercules crushed the crab, and Hera placed it among the stars as a reward for its efforts. The dimness of its stars was said to be because it failed to defeat Hercules.

## STUDYING CANCER AND OTHER CONSTELLATIONS

In the past, humans used the constellations to tell stories and to track the passage of seasons. Scientists use the constellations for other reasons. Since scientists can observe so many different stars and deep-sky objects, they use the constellations to identify and name some of the stars in the sky. For example, the stars beta Cancri and alpha Cancri received their names because they are part of the Cancer constellation.

Scientists also study Cancer and other constellations because they want to learn more about stars, deep-sky objects, and exoplanets. One binary star inside the Cancer constellation, called 55 Cancri, is orbited by five exoplanets. This star system is of interest to scientists because it is similar to Earth's solar system in a number of ways. The central star, 55 Cancri A, is about the same age and has about the same mass as Earth's sun. The planets farthest away from the star are giants, similar to the gas giants in Earth's solar system. However, these five exoplanets are larger than the planets in Earth's solar system. Furthermore, four of them are very close to 55 Cancri A, while the fifth is very far away. None of the planets are in the star's habitable zone, which is the distance at which a planet could contain liquid water, considered necessary to support life. The planets are therefore considered unlikely to hold life-forms of any kind.

—Elizabeth Mohn

### FURTHER READING

"Cancer, the Crab." *StarDate*. U of Texas McDonald Observatory, n.d. Web. 9 April 2015.

"Constellations: Frequently Asked Questions." *Tools for Science*. Coll. of St. Benedict & St. John's U, n.d. Web. 27 Mar. 2015.

Kaler, Jim. "Acubens (Alpha Cancri)." *STARS*. U of Illinois, n.d. Web. 9 Apr. 2015.

Kaler, Jim. "Al Tarf (Beta Cancri)." *STARS*. U of Illinois, n.d. Web. 9 Apr. 2015.

McClure, Bruce. "Cancer? Here's Your Constellation." *EarthSky*. Earthsky Communications, 7 Mar. 2014. Web. 9 Apr. 2015.

McClure, Bruce, and Deborah Byrd. "Beehive: 1,000 Stars in Cancer." *EarthSky*. Earthsky Communications, 23 Feb. 2015. Web. 9 Apr. 2015.

Nemiroff, Robert, and Jerry Bonnell. "Star Cluster Messier 67." *Astronomy Picture of the Day*. NASA, 9 Aug. 2007. Web. 9 Apr. 2015.

Newcomb, Simon. "Apparent Motion of the Sun." *Elements of Astronomy*. 1900. *Tools for Science*. Coll. of St. Benedict & St. John's U, n.d. Web. 27 Mar. 2015.

"Our Solar System's Cousin?" *NASA*. NASA, 11 June 2007. Web. 14 Apr. 2015.

Zimmermann, Kim Ann. "Cancer Constellation: Facts about the Crab." *Space.com*. Purch, 7 Aug. 2012. Web. 9 Apr. 2015.

# CANIS MAJOR

### FIELDS OF STUDY

Stellar Astronomy; Astronomy; Cosmology

### SUMMARY

Canis Major is a constellation that is visible from both the Northern and the Southern Hemispheres. In Western astronomy, it and nearby constellation Canis Minor are commonly interpreted as dogs following Orion, the hunter constellation. Canis Major has been studied by civilizations since ancient times. It is notable for containing Sirius, the brightest star in the night sky.

### PRINCIPAL TERMS

- **constellation:** a section of the night sky officially designated by the International Astronomical Union and recognized by patterns of stars.
- **International Astronomical Union:** the authoritative international organization of professional astronomers.

### THE GREAT DOG CONSTELLATION

Canis Major is a constellation, a section of the celestial sphere officially designated by the International Astronomical Union (IAU). Constellations are based on imagined patterns of stars called asterisms, which represent certain figures or shapes. Humans have noticed asterisms since at least 4000 BCE, when the earliest constellations were recorded in Mesopotamia. The stars of Canis Major have been part of the

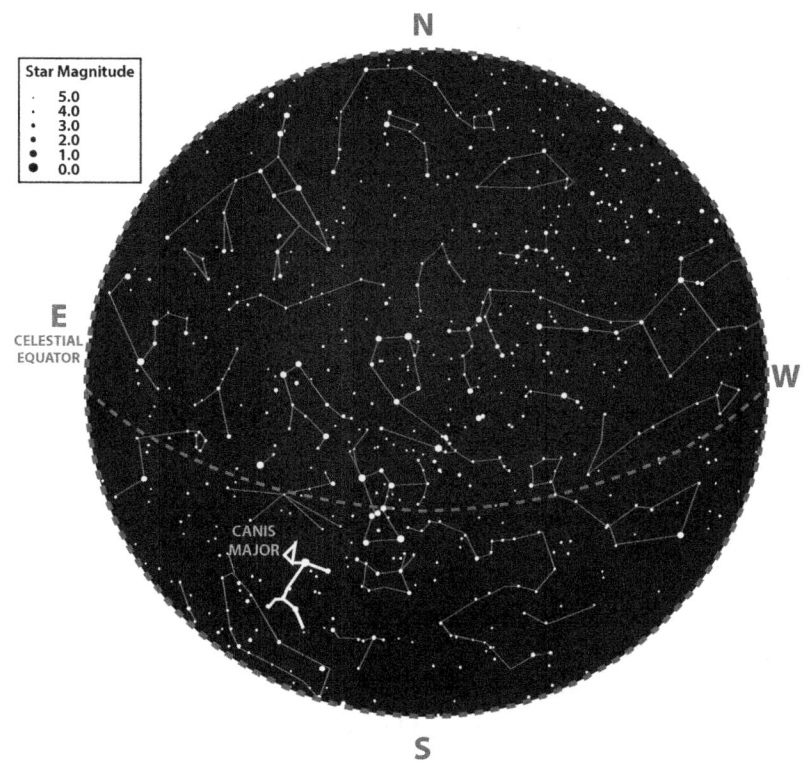

Star chart of the Northern and Southern Hemispheres in November. Lines connect the stars of the constellation Canis Major.

constellations and myths of civilizations around the world. Because Canis Major contains the brightest star in the night sky, it was observed by the ancient Egyptians, Chinese, Cherokees, Sumerians, and Greeks, among others.

The eighty-eight modern constellations recognized by the IAU are mostly derived from those of the ancient Greeks. Forty-eight of the constellations were collected by the astronomer Ptolemy (ca. 100–ca. 170 CE), who based them on historical sources and his own observations. While some of Ptolemy's work would later be proved incorrect, such as his assertion that Earth was the center of the solar system, his star charts were incredibly accurate for their time. His work *The Almagest* (ca. 150 CE) contained twenty-one northern constellations, twelve zodiacal constellations, and fifteen southern constellations. All the constellations found in *The Almagest* are used by the IAU. Because of Ptolemy and The Almagest, many people still use the Greek names for constellations.

Canis Major is one of Ptolemy's original constellations. In the ancient world, many stars and asterisms were given supernatural or religious meaning. Some people believed that the constellations were the gods' way of telling stories. Canis Major means "greater dog" in Latin. Though there are many different myths regarding the constellation, it is often seen as a hunting dog, along with its smaller neighbor, Canis Minor ("lesser dog"). Some Greeks associated the constellation with the mythical hunting dog Laelaps. Stories often claimed the dog was chasing Lepus, the nearby hare constellation, or battling Taurus, the bull constellation. The Romans associated Canis Major with Custos Europae, Europa's guard dog, or Janitor Lethaeus, the dog that guarded the entrance to the underworld. Many myths associated the dogs with the hunter Orion, another neighboring constellation. Because of this, Canis Major is included in the Orion family of constellations.

Constellations were commonly used for navigation. Stars move in consistent patterns, and star charts showed where each constellation could be found at any time on any given day. This allowed people to judge their position based on the stars. Canis Major was particularly important for navigators because it contains several very bright and easily identifiable stars.

### The Dog Star

The most noticeable star in the Canis Major constellation is Sirius, the brightest star in the night sky. Sirius is sometimes called the Dog Star or the Star in the Jaws of the Dog because of its position in Canis Major. It is more than twice as large as Earth's sun, twenty times brighter, and less than nine light-years away from Earth. Sirius is one of the three stars that form the Winter Triangle asterism, the other two being Procyon and Betelgeuse. Procyon is the largest star in Canis Minor, while Betelgeuse is a red giant found in Orion.

Civilizations have watched the Dog Star since ancient times. The Greeks and Romans believed that it was powerful enough to heat Earth, just like the sun. Because Sirius is sometimes visible during the day in the summer, they thought that the combined heat from Sirius and the sun was what made summer hotter than the other seasons. This is the source of the expression "the dog days of summer." Sirius was worshipped by or incorporated into the myths of many other cultures as well. Its brightness made it especially useful for navigation.

When most people look at Sirius in the night sky, they see a single star. In reality, however, Sirius is a binary star, or a system of two stars that orbit a common center of mass. The larger star in the binary system, called Sirius A, is the star most people see when they look at Canis Major. It is much larger and brighter than its companion star, Sirius B, which is a white dwarf star roughly the size of Earth's sun. While Sirius A is still in a stable period of its life cycle, Sirius B is slowly dying. The small star was once much larger than Sirius A. Over billions of years, however, Sirius B used all its internal fuel, shed its outer layers, and shrank to a fraction of its former size. It is no longer creating heat through chemical reactions and is slowly cooling down. Eventually, Sirius B will stop glowing entirely.

**OTHER FEATURES**
The red hypergiant star VY Canis Majoris is another important star in the Canis Major constellation. VY Canis Majoris is one of the largest known stars in the Milky Way galaxy. It is located more about 4,900 light-years from Earth, and its diameter is nearly two thousand times that of Earth's sun. For comparison, if the sun were replaced with VY Canis Majoris, then Mercury, Venus, Earth, Mars, and Jupiter would all be swallowed by the hypergiant.

Hypergiants burn through their internal fuel extremely quickly. For this reason, they tend to have much shorter life cycles than other types of stars. They live for only a few million years before undergoing a massive explosion called a supernova. Living hypergiants are extremely rare. After going supernova, some hypergiants become black holes.

Despite its size, VY Canis Majoris is not as bright as Sirius in the night sky because it is far away from Earth. In fact, it is one of the dimmest stars in Canis Major. The constellation's second-brightest star is epsilon Canis Majoris, also known as Adhara. It is the brightest known source of ultraviolet radiation in the night sky.

Another feature of Canis Major is the Messier 41 (M41) star cluster. This group of stars is around 2,300 light-years from Earth. It appears as a faint blotch of light to the naked eye, but binoculars reveal the individual stars.

**VIEWING CANIS MAJOR**
Canis Major can be seen from any location between sixty degrees north latitude and the South Pole. It is most visible during February. No special equipment is necessary to view it. Canis Major can be found by first locating Orion's Belt, three bright stars in the Orion constellation that form a straight line, and then following that line southeast. It will intersect with Sirius, which is much brighter than any other nearby star.

—*Tyler Biscontini*

**FURTHER READING**
"Canis Major." *Constellations of Words*. Constellation of Words, 2008. Web. 17 Apr. 2015.

"Canis Major, the Great Dog." *Star Date*. U of Texas McDonald Observatory, 2015. Web. 17 Apr. 2015.

Fuchs, Jim. "Constellation History." *Modern Constellations*. Author, n.d. Web. 17 Apr. 2015.

Holloway, April. "The Ancient Wonder and Veneration of the Dog Star Sirius." *Ancient Origins*. Ancient Origins, 1 Feb. 2014. Web. 17 Apr. 2015.

"Identify the Winter Circle and Winter's Brightest Stars." *EarthSky*. Earthsky Communications, 30 Jan. 2015. Web. 17 Apr. 2015.

Jones, Brian. "Skywatch Spring: Canis Major." *Cicerone Extra*. Cicerone, 10 Mar. 2015. Web. 17 Apr. 2015.

Plotner, Tammy. "Canis Major." *Universe Today*. Universe Today, 15 Oct. 2008. Web. 17 Apr. 2015.

Tate, Jean. "Sirius B." *Universe Today*. Universe Today, 15 Nov. 2009. Web. 17 Apr. 2015.

Theodossiou, Efstratios, et al. "Sirius in Ancient Greek and Roman Literature: From the Orphic *Argonautics* to the *Astronomical Tables* of Georgios Chrysococca." *Journal of Astronomical History and Heritage* 14.3 (2011): 180–89. *NARIT*. Web. 17 Apr. 2015.

Villaneuva, John Carl. "VY Canis Majoris." *Universe Today*. Universe Today, 8 Sept. 2009. Web. 17 Apr. 2015.

# CANIS MINOR

### FIELDS OF STUDY

Stellar Astronomy; Astronomy

### SUMMARY

Canis Minor is a small constellation made up primarily of two stars. In Western astronomy it is commonly interpreted as a small dog. The constellation's most notable star is Procyon, one of the brightest stars in the night sky. Canis Minor is a member of the Orion constellation family, which includes some of the most famous classical constellations. Canis Minor was known to a number of ancient peoples and has been featured in several cultural myths.

### PRINCIPAL TERMS

- **constellation:** a section of the night sky officially recognized by the International Astronomical Union and based on patterns of stars as seen from Earth.
- **International Astronomical Union:** the authoritative international organization of professional astronomers.

### The Small Dog Constellation

Canis Minor is a constellation, a segment of the night sky as designated by the International Astronomical Union (IAU), the main international organization of professional astronomers. There are eighty-eight official constellations, each based around asterisms, or imagined patterns of stars that are interpreted as images. Historians have found astronomical studies recognizing asterisms dating back at least as far as the ancient Babylonians, around 4000 BCE. Although virtually all cultures developed asterisms, it was the ancient Greeks who established many of the star patterns used by modern astronomers. The astronomer Ptolemy (ca. 100–ca. 170 CE) compiled forty-eight classical constellations in the second century, including Canis Minor.

Ptolemy studied astronomy, astrology, and mathematics. Some of his conclusions, such as his theory that Earth was the center of the solar system, were completely incorrect. However, his star charts were incredibly accurate for their time. His astronomical work *The Almagest* (ca. 150) contained twenty-one northern constellations, twelve zodiacal constellations, and fifteen southern constellations. He based them on historical records as well as his own observations. All of Ptolemy's constellations are officially recognized by the IAU. Ptolemy gave Canis Minor its Latin name, meaning "lesser dog" or "small dog."

Constellations were commonly used for navigation by many cultures. This is because stars always move in a consistent pattern. Once these patterns were mapped, people could use the stars to get a sense of direction, even at sea when land was not in sight. Bright, easily visible stars were most useful for navigation. Canis Minor's main star, Procyon, forms the highly visible Winter Triangle asterism along with fellow bright stars Sirius and Betelgeuse.

In the ancient world, people associated constellations with myths, gods, and religious ideals. Some civilizations, such as the ancient Greeks, believed that the gods placed the stars in the sky as reminders of important people or stories. For this reason, the Greeks named the constellations after their mythical heroes and monsters.

### Canis Minor in Myth and Legend

Canis Minor is an ancient constellation. A number of ancient cultures interpreted its main pair of visible stars in various different ways. Usually it is linked to the nearby Canis Major, or "greater dog." In fact, the ancient Greeks only recognized Canis Major, which was often considered to represent one of the dogs of the hunter Orion, another nearby constellation. The stars now known as Canis Minor were sometimes seen as a fox being chased by Canis Major. Alternatively, the star Procyon (Greek for "before the dog") was viewed separately, signaling the rise of Sirius, the Dog Star. By Roman times, Canis Minor was recognized as a second dog of Orion.

Canis Minor is the smallest member of the Orion family of constellations. This family includes Orion the hunter, which is known for containing the three bright stars that form the asterism Orion's Belt. Orion also contains the red supergiant alpha Orionis, better known as Betelgeuse. Canis Major is another member of the Orion family, known for the extremely bright star alpha Canis Majoris, more commonly called Sirius. The other constellations in this family are Lepus, the hare, and Monoceros, the unicorn.

According to Greek and Roman mythology, Orion was a great hunter. He boasted that he could defeat any animal on Earth. After his death, the gods placed Orion and his faithful dogs in the heavens as stars. Various versions of the myth depict Orion's dogs as chasing Lepus or attacking Taurus, the bull.

Another Greek legend associates Canis Minor with Maera, the dog of Icarius. The god Dionysus taught Icarius to make wine, and Icarius gave some of his wine to a group of local shepherds to taste. The shepherds had never tasted alcohol before; when they became drunk, they believed that Icarius had poisoned them, so they killed him. When Maera saw his dead owner, he ran to Icarius's daughter, Erigone, and led her back to her father's body. In response to Icarius's death, Maera and Erigone took their own lives. So the tragedy would not be forgotten, the Greek gods placed Maera, Erigone, and Icarius in the sky as the constellations Canis Minor, Virgo, and Boötes.

The stars of Canis Minor were referenced in the myths of other cultures as well. Procyon and Sirius are separated by the visible path of the Milky Way galaxy. In one Arabian myth, the Milky Way represents a great river, and Procyon and Sirius represent two sisters who tried to cross it. Sirius, the older sister, was strong enough to cross the river, but Procyon, the younger sister, was unable to make it across. This is why Sirius is on one side of the Milky Way and Procyon is on the other. It is also why Procyon is sometimes referred to as the "weeping one." Early Chinese astronomers also interpreted Procyon as part of a river, or alternately as a gate or guard.

**Notable Features of Canis Minor**
Procyon, also known as Antecanis or alpha Canis Minoris, is the brighter of the two visible stars in Canis Minor. Though not as bright as Sirius, Procyon is one of the ten brightest stars in the sky. This is because, at about 11.4 light-years away, it is relatively close to Earth. Though it looks like a single star to the naked eye, Procyon is actually a binary system. Binary stars are two stars in close proximity that orbit a common center of mass. The larger star in the system is known as Procyon A, and the smaller star is Procyon B.

Procyon A is a main-sequence star, meaning it is in the same stage of its life cycle as Earth's sun. It is 1.4 times the mass of Earth's sun and more than seven times as bright. Procyon B is a white dwarf, the degenerate core of a collapsed star that is no longer producing heat and is slowly cooling over billions of years. Procyon B is only 60 percent of the mass of the sun. It was first observed in 1896, but its existence was speculated in 1844 due to the wobbling of Procyon A.

Canis Minor's other main star is beta Canis Minoris, also known as Gomeisa. It is also a main-sequence star, classified as a gamma Cassiopeiae variable star. This means it has a fast rotation and its luminosity varies irregularly. This behavior is caused by the ejection of material from the star.

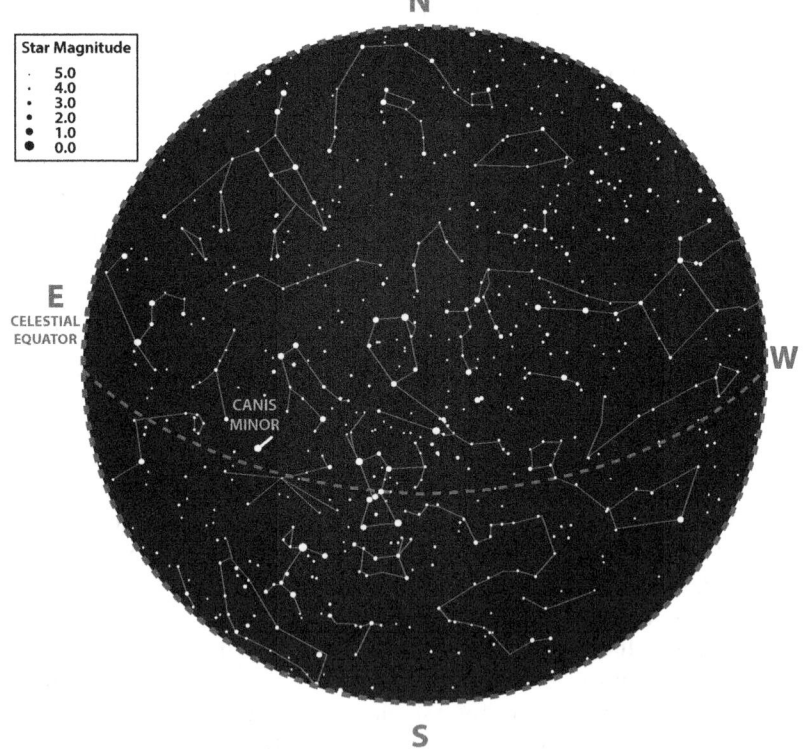

Star chart of the Northern and Southern Hemispheres in November. Lines connect the stars of the constellation Canis Minor.

Though there are few other easily visible stars in Canis Minor, the section of the Milky Way it contains is of great interest to astronomers. Many nebulae and star clusters can be seen in this region of the sky using binoculars or telescopes. The constellation is also linked to the Canis Minorids meteor shower.

**LOCATING CANIS MINOR**

The Orion family of constellations is visible from most parts of the world. Canis Minor can be seen anywhere from the North Pole to 75 degrees south latitude. It is most readily visible during March. While special equipment is not necessary for viewing Procyon and Gomeisa, Procyon B cannot be seen without a telescope.

Canis Minor is found in the second quadrant of the northern celestial hemisphere. It is bordered by the constellations Monoceros, Cancer, Gemini, and Hydra. It can commonly be located by finding the easily recognizable Orion's Belt; Canis Minor is on the same level as Orion's shoulders.

—*Tyler Biscontini*

**FURTHER READING**

"Canis Major." *Constellations of Words*. Constellation of Words, 2008. Web. 17 Apr. 2015.

"Canis Minor Constellation." *Constellation Guide*. Constellation-Guide.com, n.d. Web. 17 April, 2015

Fuchs, Jim. "Constellation History." *Modern Constellations*. Author, n.d. Web. 17 Apr. 2015.

Gaherty, Geoff. "Doggie Constellations Are a Skywatcher's Best Friends This Week." *Space.com*. Purch, 16 Feb. 2012. Web. 17 Apr. 2015.

Howell, Elizabeth. "Procyon: Bright Star with Hidden Companion." *Space.com*. Purch, 24 Sept. 2014. Web. 17 Apr. 2015.

Jones, Brian. "Skywatch Spring: Canis Major." *Cicerone Extra*. Cicerone, 10 Mar. 2015. Web. 17 Apr. 2015.

Plotner, Tammy. "Canis Minor." *Universe Today*. Universe Today, 15 Oct. 2008. Web. 17 Apr. 2015.

Ridpath, Ian. "Canis Minor: The Lesser Dog." *Ian Ridpath's Star Tales*. Author, n.d. Web. 17 Apr. 2015.

# CAPRICORNUS

**FIELDS OF STUDY**

Stellar Astronomy; Observational Astronomy

**SUMMARY**

Capricornus is a constellation in a roughly triangular shape that some people believe resembles a goat or a goat horn. Often portrayed as a goat with a fish tail, Capricornus can be found in the Southern Hemisphere and is most visible around September. Capricornus has a long history, as reflected in myths and tales of the ancient Greeks, Babylonians, and Sumerians. The Babylonians included Capricornus in the group of constellations known as the zodiac. Today, astronomers study Capricornus because it includes several interesting varieties of star as well as galaxies and clusters of stars.

**PRINCIPAL TERMS**

- **celestial equator:** the imaginary line above Earth's equator that halves the celestial sphere; it is equally distant from the celestial poles.
- **constellation:** a region of space defined by a pattern of stars that can be seen in the night sky from Earth.
- **declination:** the north-south position of a celestial body relative to the celestial equator expressed in degrees of arc.
- **International Astronomical Union:** an association of professional astronomers from all over the world who define astronomical constants while promoting research, education, and discussion on important astronomical topics.
- **right ascension:** the east-west position of a celestial body defined in relation to the celestial equator and expressed in hours and minutes, not degrees of arc.

## The Goat Horn Constellation

Capricornus is a constellation, an area of space defined by a grouping of stars that appear to some observers to form patterns or pictures in the night sky. People have studied constellations for thousands of years. The types and descriptions of these star patterns vary widely, but the International Astronomical Union acknowledges eighty-eight major constellations. One of the best-known and most historically important of these is Capricornus, a Latin name that means "goat horn." Capricornus, which is located in the Southern Hemisphere, contains about thirteen primary stars. These stars are arranged in a roughly triangular shape thought by some to resemble a goat or a goat horn.

Capricornus owes much of its worldwide recognition to its inclusion in the zodiac. The zodiac is a group of twelve or thirteen constellations through which the sun appears to move along a path called the ecliptic. The ecliptic is in fact an illusion resulting from Earth's orbit of the sun. However, the location of Capricornus and other constellations is measured relative to the celestial equator, not the ecliptic. Capricornus, for example, has a declination of about −20 degrees from the celestial equator and a right ascension of twenty-one hours.

## Attributes of Capricornus

Capricornus can be found in the Southern Hemisphere between the latitudes 60 degrees north and 90 degrees south. The constellation has a roughly triangular or arrowhead shape with an extra "spike" on the right top side. Although astronomers disagree on the exact number of stars in Capricornus, its main shape is composed of about thirteen primary stars. In addition, it includes many other stars and background objects. This constellation is just above average in size, being the fortieth largest of the officially recognized eighty-eight constellations. It is, however, unusually faint and can be challenging to locate.

This constellation is an important member of the zodiac and the star patterns of the night sky. It is located near many other constellations, including Aquila (the eagle), Microscopium (the microscope), Piscis Austrinus (the southern fish), and the zodiac constellations Aquarius (the water bearer) and Sagittarius (the archer).

Capricornus also contains many interesting stars. Greek letters such as alpha, beta, and so on were used to designate their relative brightness. Alpha Capricorni, also known as Algedi (the goat), is on the top right corner of the constellation and forms the tip of the horn. This star appears to be a binary star. However, that is actually an illusion based on two stars that coincidentally line up in the perspective of the Earth-based observer. The two stars involved are alpha 1, or Prima Giedi, and alpha 2, or Secunda Giedi. Just below and to the left of the Alpha stars is beta Capricorni (better known as Dabih), and farther left is Nu Capricorni (known as Alshat). These names, derived from Arabic, relate to the theme of goats and similar

Star chart of the Northern and Southern Hemispheres in June. Lines connect the stars of the constellation Capricornus.

animals. Alshat means "the sheep," and Dabih refers to "the butcher" who will prepare the sheep for feasts or sacrificial rituals.

Other names relate to parts of the goat's body, such as Deneb Algedi (goat's tail), an alternate name of delta Capricorni. Meanwhile, the giant red star omega Capricorni is referred to as Baten Algiedi (goat's belly) because it appears at the lower side of the constellation. Other important stars in the formation are Pi Capricorni, Rho Capricorni, and Omicron Capricorni, which form a triangle, and gamma Capricorni, better known as Nashira. The brightest of the stars in this constellation is delta Capricorni, which can often be seen without optical aids.

## THE GOAT AND THE SEA

Capricornus was one of the first constellations to be identified in recorded history. The ancient astronomer Ptolemy (ca. 100–170 CE) listed it among the forty-eight constellations he and his contemporaries recognized. It also became entwined in the study of the zodiac. Ancient Babylonian astronomers thought the constellations of the zodiac had special powers because they appeared to be touched by the path of the sun. Over the centuries, many cultures applied their own interpretations to the meaning of the constellation. Most of these interpretations related in some way to goats. Other themes, such as the sea, also worked their way into the complex and growing mythology.

Early Greek astronomers identified the shape of the constellation as representing a goat or a goat horn. The symbol of the goat was important to ancient Greeks. In Greek mythology, the chief god Zeus was raised by a goat. This goat, known as Amalthea, fed and protected Zeus through his vulnerable childhood. Later, the horn of Amalthea broke off and became a cornucopia, or a horn of plenty. Another Greek myth relating to goats is about Pan, the god of the wild, who had a half goat, half human form. In one tale, Pan leapt into a lake to avoid a monster, and his unexpected swim caused him to partially morph into a fish.

Other mythological combinations of humans, goats, and fish appear in Babylonian records as well. As early as 3000 BCE, some Babylonians drew fantastical images of goats with fish tails. Other Babylonian myths related to the god Ea, or Enki, who was half fish, half human and brought wisdom to humanity. Half-goat, half-fish creatures also figure into ancient Sumerian tales.

Although the exact reason is unclear, many astronomers began to imagine the figure of Capricornus as a goat with a fish tail—a "water goat." This depiction, while unusual, fits with the other constellations in that part of the sky, many of which have been assigned water-themed descriptions. The water goat joins the water-bearer Aquarius, the fish Pisces, and the river Eridanus. The water theme may have begun because, in some lands, the appearance of these constellations in the sky corresponded to the rainy season.

Capricornus may also have had a more spiritual meaning. Because in ancient times the sun reached winter solstice while in Capricornus, ancient people may have associated the constellation with the end of the year.

## ASTRONOMICAL IMPORTANCE TODAY

The people of the ancient world studied the stars and constellations for spiritual reasons as well as to help mark the seasons of the year. In the twenty-first century, many scientists turn to the constellations to help them study the various bodies of space. Like most other constellations, Capricornus includes many points of interest for modern astronomers and stargazers. It has also helped people study the path of the sun, which at one time reached its most southerly position in Capricornus at a spot still known today as the Tropic of Capricorn (though it is now in the constellation Sagittarius).

The stars of Capricornus are relatively faint and most are best viewed with binoculars or telescopes. Stargazers can find the constellation at its highest and brightest on September evenings around nine or ten p.m., though the stars are also readily spotted in October. Astronomers advise amateurs to locate Capricornus by finding three bright stars—alpha 2, beta, and omega Capricorni—that form an easy-to-spot triangle. Some of the most popular stars for stargazing include the optical double star alpha Capricorni, the binary star Prima Giedi, and the eclipsing binary star delta Capricorni.

Astronomers can study more than just stars in Capricornus. The constellation includes two stars with exoplanets as well as Messier 30, a globular cluster disk of many stars. Several notable galaxy groups, such as New General Catalog NGC 7103 and 6907, can be spotted with powerful telescopes in or near the constellation, and some of these have given astronomers valuable views of huge explosions called supernovas. Meteors are frequently spotted within Capricorn as well, particularly around July 30, when the Capricornid meteor shower passes through the constellation. Stargazers may spot about five to fifteen meteors each hour at the peak of the stream.

—*Mark Dziak*

**FURTHER READING**

"Capricornus (The Sea Goat)." *Chandra X-Ray Observatory*. NASA, n.d. Web. 7 Apr. 2015.

Dolan, Chris. "Capricornus." *The Constellations and Their Stars*. U of Wisconsin-Madison Department of Astronomy, n.d. Web. 7 Apr. 2015.

Kaler, James B. "Capricornus." *University of Illinois*. U of Illinois, n.d. Web. 7 Apr. 2015.

McClure, Bruce. "Capricornus? Here's Your Constellation." *EarthSky*. EarthSky, 4 Sept. 2014. Web. 7 Apr. 2015.

Nave, R. "Ecliptic Plane." *HyperPhysics*. Dept. of Physics and Astronomy, George State U, 2012. Web. 10 Apr. 2015.

Plotner, Tammy. "Capricornus." *Universe Today*. Universe Today, 15 Oct. 2008. Web. 7 Apr. 2015.

Zimmermann, Kim Ann. "Capricornus Constellation: Facts about the Sea Goat." *Space.com*. Space.com, 3 June 2013. Web. 7 Apr. 2015.

# CASSIOPEIA

**FIELDS OF STUDY**

Stellar Astronomy; Observational Astronomy

**SUMMARY**

Cassiopeia is a constellation with a zigzag shape that resembles the letter M or W, depending on the constellation's position in the sky. Cassiopeia is a Northern Hemisphere constellation with a long and rich mythological history. It is named after a boastful queen from Greek mythology. Astronomers and amateur stargazers look to Cassiopeia for examples of double and triple stars, variable stars, nebulae (clouds of dust and gas), and star clusters.

**PRINCIPAL TERMS**

- **celestial equator:** the imaginary line above Earth's equator that halves the celestial sphere; it is equally distant from the celestial poles.
- **circumpolar constellation:** a constellation that is always visible in the night sky; the Northern and Southern Hemispheres have different sets of circumpolar stars.
- **constellation:** a region of space defined by a pattern of stars that can be seen in the night sky from Earth.
- **declination:** the north-south position of a celestial body relative to the celestial equator expressed in degrees of arc.
- **International Astronomical Union:** an association of professional astronomers from all over the world who define astronomical constants while promoting research, education, and discussion on important astronomical topics.
- **right ascension:** the east-west position of a celestial body relative to the celestial equator and expressed in hours and minutes, not degrees of arc.

**A QUEEN OF CONSTELLATIONS**

Cassiopeia is a constellation, or a region of space defined by a group of stars that appear to form a pattern or picture. Since ancient times, astronomers have studied the constellations. Cassiopeia was first identified in antiquity and was included in star charts as early as the second century CE. It is one of the eighty-eight constellations officially recognized by the International Astronomical Union. Although many cultures have observed and told tales about this constellation, the name Cassiopeia is derived from an arrogant queen from Greek mythology.

A star formation of the Northern Hemisphere, Cassiopeia is notable for being circumpolar, which means it appears to move around the pole star. Due to this type of motion, Cassiopeia often appears reversed in the night sky. Cassiopeia has a declination

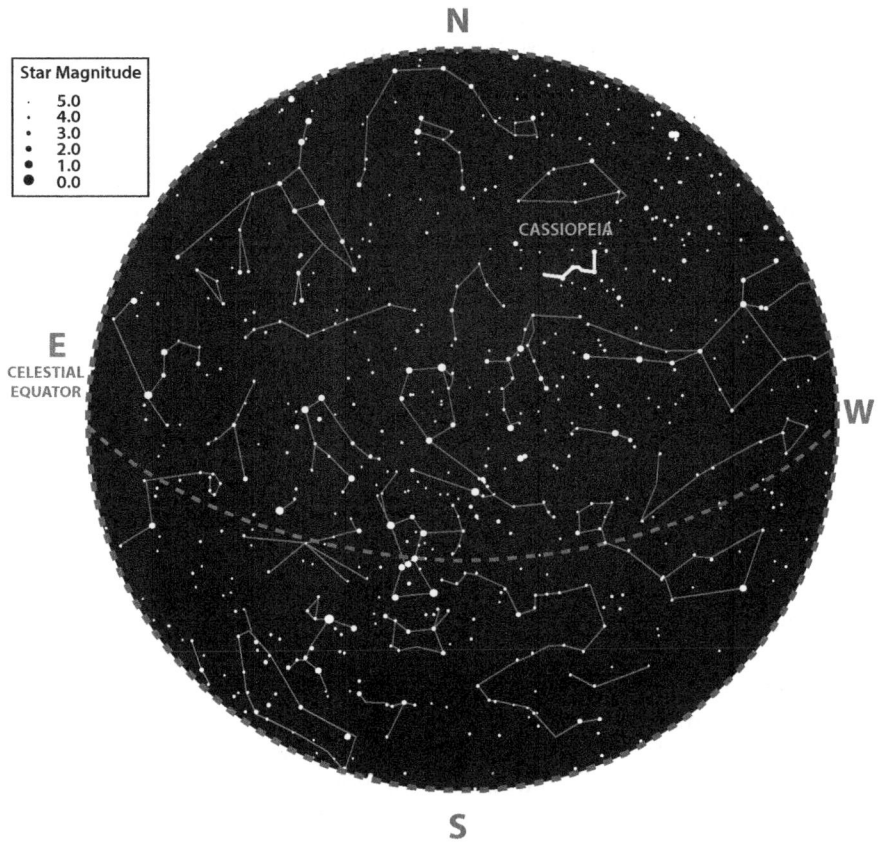

Star chart of the Northern and Southern Hemispheres in November. Lines connect the stars of the constellation Cassiopeia.

actually pairs or groups of stars. Eta Cassiopeiae is a double star, and iota Cassiopeiae is a triple star.

The constellation's leading star, alpha Cassiopeiae (also known as Shedir), is an irregularly variable star. That means the star occasionally changes in brightness, although astronomers have not observed any major changes on Shedir since the nineteenth century. Gamma and rho Cassiopeiae are also types of variable stars. The gamma star, in the center of the constellation, sometimes dramatically increases in brightness due to unpredictable emissions of gas. The rho star, which is about a half-million times brighter than the sun, has shown signs of increasing instability and astronomers suspect it will eventually explode into a supernova. Two known supernovas have occurred in Cassiopeia in the past several centuries.

of approximately + 60 degrees from the celestial equator and a right ascension of one hour.

Cassiopeia is most visible in the Northern Hemisphere between the latitudes of 90 degrees north and 20 degrees south. Evenings in late autumn, particularly November, are the best times to view the constellation. During those times, the constellation appears in the northeastern sky. Cassiopeia looks like a flattened zigzag form or the letter *W* or *M* (depending on the direction the constellation is facing). It is set against the backdrop of the Milky Way, the galaxy that includes Earth.

The constellation of Cassiopeia contains many diverse space objects. Among these objects are five primary stars as well as fifty-three other stars designated by letters of the Greek alphabet, roughly in descending order of brightness. Astronomers have discovered a great variety of star types in the constellation. For example, some of the stars in Cassiopeia are

Despite the great brightness of the variable stars, beta Cassiopeiae (also known as Caph) is the consistently brightest star in the constellation. People can often see Caph without optical aids. The name Caph is one of several star names in Cassiopeia that refer to parts of the body. Caph comes from an Arabic word meaning "palm" or "hand." Shedir means "breast," and Ruchbah (delta Cassiopeiae) means "knee."

Cassiopeia also contains several other deep-sky objects. There are several nebulae, vast clouds of dust and gas in space in which new stars may form, in or near the constellation. There are also clusters of stars, many of which are relatively new, having formed in the nearby nebulae. Unfortunately, many of these new stars are obscured from sight by the thick clouds of dust in that region of space.

## A Tragic Boast

As with many other constellations, Cassiopeia has inspired myths and tales from cultures around the world. In China, people identified the star pattern as a representation of a bold charioteer. Among the Welsh, the constellation may have been considered the court of the mother goddess Don. Of all these diverse interpretations, the most famous one, and the one from which the name Cassiopeia is derived, originated in ancient Greece.

In Greek mythology, Cassiopeia was a queen in the legendary realm of Aethiopia. A beautiful but arrogant woman, she ruled the kingdom with her husband, King Cepheus. One day Cassiopeia boasted that she and her daughter, Andromeda, were more beautiful than the sea nymphs. This claim upset the sea god Poseidon, who decided to take revenge on the entire kingdom for its queen's imprudent words. Poseidon cast a disastrous flood upon Aethiopia. Worse, he also unleashed a sea monster called Cetus to terrorize its people.

Cassiopeia and Cepheus were helpless against the wrath of the sea god. They consulted an oracle, an all-knowing source of wisdom, which advised them that only a human sacrifice could appease Poseidon. The king and queen agreed and had their daughter Andromeda chained to a rock by the sea as an offering to Cetus. Andromeda's doom seemed assured until the hero Perseus discovered and rescued her, thus confounding the plans of the gods and monarchs.

Poseidon felt that he had been denied his vengeance against the boastful queen, and so devised a new, more personal punishment. He banished both Cassiopeia and Cepheus into the sky to live forever among the stars. Just as Andromeda had been chained to a rock, Cassiopeia was chained to her royal throne. To humiliate the arrogant queen, Poseidon arranged for the throne to move in the sky so that it frequently turned upside-down, leaving the queen helplessly hanging in an undignified stance. Meanwhile, Cepheus also became a constellation, but the king and queen were positioned so they could never see or talk to one another again.

Long ago, astronomers referred to the constellation as Cassiopeia's Chair. In the 1930s, a group of scientists at the International Astronomical Union decided to change the name to Cassiopeia the Queen. This name change, however small, deemphasized the ancient myth of Poseidon's vengeance.

The myth of Cassiopeia fits into the larger scheme of constellation mythology. The constellation of the queen is part of the Perseus constellation family, a group of constellations sharing thematic connections. This family includes constellations representing the hero Perseus, Queen Cassiopeia, King Cepheus, and Andromeda. It also includes the sea monster Cetus, Perseus's flying horse (Pegasus), a charioteer (Auriga), and a lizard (Lacerta).

## Cassiopeia's Importance Today

In ancient times when astronomers and storytellers first imagined the constellations, the stars were of great importance to everyday life. People looked to the night sky to find answers to their questions about spirituality and nature. Stars and constellations also served a practical purpose of helping farmers mark the changing of the seasons and decide when crops should be planted and harvested. In modern times, most people enjoy stargazing just for the beauty of it, and scientists use the constellations to help organize the stars and learn more about the always-changing universe.

Cassiopeia has many features of interest to amateur and professional astronomers. It is generally a fairly bright constellation, and some of its features may be seen with the naked eye. Once stargazers find Cassiopeia, they have numerous features to examine. The stars of the constellation include many types such as double stars, triple stars, and variable stars. Other sky objects are also of interest to astronomers, such as nebulae and star clusters. The astronauts of the *Apollo* missions used gamma Cassiopeiae as one of the thirty-six stars by which they navigated. Astronaut Gus Grissom (1926–67) nicknamed the star Navi, a backward spelling of his own middle name, Ivan.

For the earthbound observer, Cassiopeia is also a valuable reference point for locating other constellations in the sky. Some of the neighboring constellations include Perseus, Andromeda, Cepheus, and Lacerta, as well as Camelopardalis (the giraffe).

—*Mark Dziak*

## Further Reading

Beatty, Kelly. "Tour November's Sky: Cassiopeia." *Sky & Telescope*. F+W Media, 31 Oct. 2014. Web. 22 Apr. 2015.

"Cassiopeia." *Chandra X-Ray Observatory*. NASA, 2 Dec. 2013. Web. 9 Apr. 2015.

"Cassiopeia, the Queen." *StarDate*. U of Texas McDonald Observatory, n.d. Web. 9 Apr. 2015.

"Greek Mythology and the Constellations." *Skywise Unlimited*. Western Washington U, n.d. Web. 9 Apr. 2015.

Kronberg, Christine. "Cassiopeia." *Munich Astro Archive*. Munich Astro Archive, 10 Mar. 1997. Web. 9 Apr. 2015.

Plotner, Tammy. "Cassiopeia." *Universe Today*. Universe Today, 16 Oct. 2008. Web. 9 Apr. 2015.

Sen, Nina. "Queen of the Night: Classy Cassiopeia Constellation Reigns over Meteor." *Space.com*. Purch, 9 Oct. 2012. Web. 9 Apr. 2015.

# CELESTIAL MECHANICS

## FIELDS OF STUDY

Astronomy; Astrophysics; Orbital Mechanics

## SUMMARY

Celestial mechanics is a subfield of classical mechanics and a branch of astronomy that studies the motion of celestial objects and the forces affecting this movement. Several types of forces can affect the orbit of bodies in space. Gravity is the main cause of orbital motion. Other forces such as atmospheric interference, radiation pressure, and electromagnetic fields affect the movement of celestial bodies.

## PRINCIPAL TERMS

- **classical mechanics:** the study of the motion of bodies, rooted in Isaac Newton's physical and mathematical principles; also called Newtonian mechanics.
- **eccentricity:** the extent to which a celestial body's orbit deviates from a perfect circle.
- **ellipse:** a shape that resembles an elongated circle; mathematically speaking, a closed conic section.
- **n-body problem:** a mathematical model used to determine how gravity affects the motions and interactions of a group of celestial bodies.
- **Newton's laws of motion:** the three laws that describe how bodies respond to the application of force.
- **perturbation:** a change in the orbit of a celestial object caused by the gravitational force of another object.
- **tidal evolution:** the change in the rise and fall of an ocean caused by the gravitational force of a nearby celestial object.

## EXPLAINING THE MOTIONS OF THE HEAVENS

Ancient peoples viewed the stars, moon, and planets as objects of worship. They made up stories about them and believed in their power to affect conditions on Earth. Early observers were aware of the consistent motions of these objects; Mesopotamian, Egyptian, and Indus Valley civilizations understood them well enough to predict eclipses. However, few astronomical observers questioned the cause of this motion, although several Greek philosophers tried to calculate precise movements of the sun, moon, and planets.

The Greek Egyptian astronomer and mathematician Ptolemy (ca. 100– ca. 170) proposed that Earth was the center of the universe and all the other planets and stars orbited around it. This theory was known as "geocentrism." Ptolemy used mathematics to calculate and predict the movement of celestial bodies. His calculations told him that planets move in epicycles, or small circles, while simultaneously orbiting Earth. Ptolemy was not certain if this was true, but he simply could not calculate a better model of planetary motion.

Ptolemy's theories were widely accepted for many centuries before Polish mathematician Nicolaus Copernicus (1473–1543) disproved them. Copernicus held that Earth orbited the sun; therefore, the sun was the center of the solar system. This theory was called "heliocentrism." Many scholars were slow to accept heliocentric views of the universe. However, Copernicus's mathematical calculations proved more accurate than Ptolemy's. This breakthrough led to what is sometimes called the "Copernican revolution."

Several other key astronomical breakthroughs occurred in the years following Copernicus's death. The celestial observations of Danish astronomer Tycho

Brahe (1546–1601) greatly contributed to the accurate measurement of planet positions. Brahe's work influenced seventeenth-century German astronomer Johannes Kepler (1571–1630), considered by many to be the father of celestial mechanics. Kepler's laws of planetary motion state, among other things, that a planet's orbit is shaped like an ellipse.

English physicist Isaac Newton (1643–1727) refined Kepler's laws, working out the mathematics behind the movements of the planets. His *Philosophiæ Naturalis Principia Mathematica* (published 1687 and translated as *The Mathematical Principles of Natural Philosophy* in 1729) laid the foundations for classical mechanics. In it, Newton put forth three fundamental principles, now commonly referred to as Newton's laws of motion, to explain how different forces affect the movement of any physical body. He also defined the law of universal gravitation, which played a key role in the development of celestial mechanics.

## THE NATURE OF CELESTIAL MECHANICS

Classical mechanics laid the groundwork for celestial mechanics. Further progress was made by astronomers such as Félix Tisserand (1845–96), who compiled all known studies in the field into the compendium *Traité de mécanique céleste* (Treatise on celestial mechanics, 1889–96), and physicist Albert Einstein (1879–1955), whose theory of general relativity improved on Newton's description of gravity to allow celestial movement to be calculated with greater accuracy. Celestial mechanics, also referred to as dynamic astronomy, calculates the motion and orbit of celestial bodies by measuring the effects of gravity and other forces.

Several basic problems make up the bulk of celestial mechanics equations. All of these problems rely on Newton's laws of classical mechanics. Most are variations on the $n$-body problem, which attempts to account for the overall motion of a group of $n$ bodies that are each affected by the others' gravitational forces. When the gravitational forces of more than one body act on a single massive body, the result is a complex series of motions called perturbation. Perturbation also happens when air resistance and atmospheric pressure disturb a celestial body. Scientists must determine the orbital properties of a group of celestial bodies and how their forces will affect one another to predict their future orbital motions. Astronomers and physicists have only solved the $n$-body problem for situations in which $n = 2$ (two-body problem) and $n = 3$ (three-body problem). One well-known version of the three-body problem is the orbital relationship between Earth, the moon, and the sun.

## CALCULATING THE N-BODY PROBLEM

Early mathematics did not account for the ways in which gravity affects the orbits of celestial objects. With the advent of Newton's law of universal gravitation, mathematicians could more accurately calculate the motions of two- and three-body systems. However, the unpredictability of perturbations makes solving the $n$-body problem difficult in cases involving three or more bodies.

The $n$-body problem involves a series of equations. The first calculates the number of celestial bodies being measured. Once this is established, the next

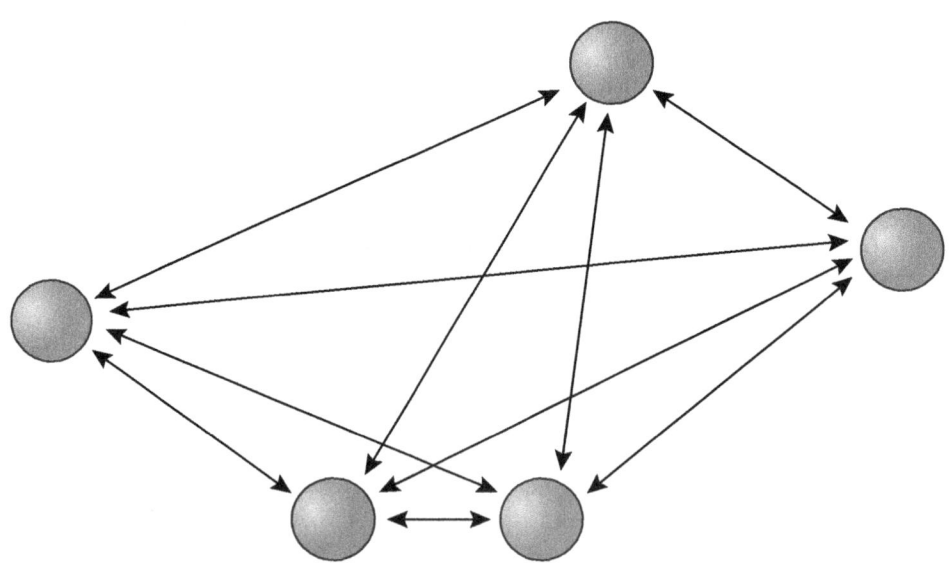

Objects often interact with one another in space. The more celestial bodies involved, the more complicated the mechanics of studying and identifying their interactions.

step is to calculate each body's initial velocity, relative position and time, and mass. The motion of the bodies can be determined from the size and eccentricity of their orbits and their interactions according Newton's law of gravitation. However, numerical calculations of celestial motion offer only indefinite predictions of future orbits, given the chaotic nature of the universe. The motion of the eight planets of the solar system is an *n*-body problem that incites much debate. Scientists continually try to determine whether the current movement of the solar system is ultimately stable or will eventually change motion due to altered external forces.

## Impact of Celestial Mechanics

Understanding the motion of celestial bodies has allowed scientists to predict astronomical events such as eclipses. It has also contributed to the subfields of astrodynamics and lunar theory and led to a greater understanding of Earth's ocean tides. The moon's gravitational force is primarily responsible for Earth's tidal evolution. As the moon's orbit expanded over time, its effect on the tides dissipated, causing the Earth's rotation to slow and thus lengthening Earth's days.

Technological innovations have given scientists easier ways to calculate celestial motions. Dutch American astronomer Dirk Brouwer (1902–66) pioneered the use of digital computers to solve orbital problems. The digital calculations proved incredibly accurate, and Brouwer's methods were soon adopted worldwide. Computer-based problem solving enabled more accurate calculations of the orbits of artificial satellites, thus facilitating the development of satellite communications.

—*Cait Caffrey*

## Further Reading

Coolman, Robert. "What Is Classical Mechanics?" *LiveScience*. Purch, 12 Sept. 2014. Web. 6 May 2015.

Hockey, Thomas, et al., eds. *Biographical Encyclopedia of Astronomers*. 2nd ed. New York: Springer, 2014. Print.

Klioner, Sergei A. "Lecture Notes on Basic Celestial Mechanics." *Department of Astronomy / Lohrmann Observatory*. Technical U Dresden, 2011. Web. 14 May 2015.

Matzner, Richard A., ed. *Dictionary of Geophysics, Astrophysics, and Astronomy*. Boca Raton: CRC, 2001. Print.

Stern, David P. "How Orbital Motion Is Calculated." *From Stargazers to Starships*. Author, 6 Apr. 2014. Web. 14 May 2015.

# CENTAUR

## FIELDS OF STUDY

Astronomy; Astrophysics; Sub-planet Astronomy

## SUMMARY

Centaurs are small objects in solar orbit between Jupiter and Neptune. Their orbits are unstable due to interactions with larger planets. Centaurs have characteristics of both comets and asteroids. Determining the true nature of centaurs and their origin will provide important information about the development of the early solar system.

## PRINCIPAL TERMS

- **asteroid:** small, rocky space objects that orbit other larger celestial bodies such as the sun and planets. Asteroids are categorized in groups defined by their orbits.
- **scattered disk objects:** space bodies made of ice and rock with very eccentric, or flattened oval, orbits. They are found in the area beyond Neptune in the outer area of the solar system.

## Kuiper Belt Objects and Centaurs

The Kuiper Belt is a section of the solar system beyond Neptune's orbit. It is populated with small icy objects called Kuiper Belt Objects (KBOs). The Kuiper Belt, also known as the Edgeworth-Kuiper Belt, was named after the men who first theorized its existence, Irish astronomer Kenneth Essex Edgeworth (1880–1972) and Dutch American astronomer Gerard Kuiper (1905–73).

KBOs are divided into classes based on their orbit. There are classical KBOs, resonance KBOs, and

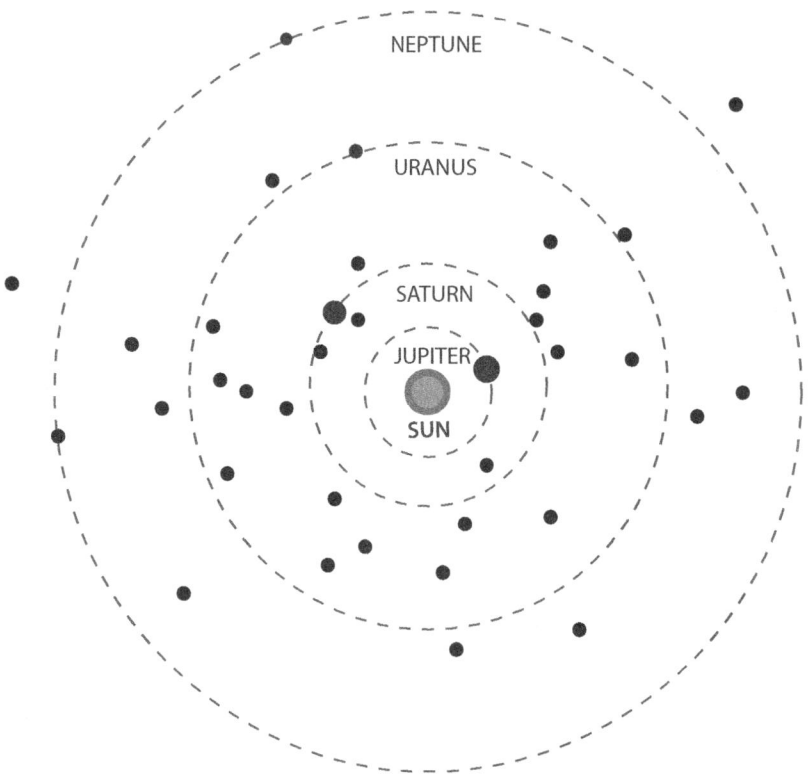

Centaurs orbit in the region of the orbits of the outer planets. Most are found between the orbits of Saturn and Neptune, but some can extend within Jupiter's orbit and beyond Neptune's orbit.

scattered KBOs or scattered disk objects (SDOs). When SDOs are affected by the forces of other celestial bodies, they can move into the area between Neptune and Jupiter. They are then called centaurs. Chiron, the first centaur object to be discovered, was detected in 1977. As of 2014, there are about 250 known centaurs and SDOs.

**CHARACTERISTICS OF CENTAURS**

Centaurs were classified as asteroids. However, since Chiron was first seen, it has developed a coma. A coma is a cloud of debris. Therefore, it is now both a comet and an asteroid. This dual nature led scientists to dub the objects centaurs, after the half-horse, half-man creature in Greek mythology.

Scientists believe that centaurs were once caught by Neptune's strong gravitational forces when they were knocked loose from their place in the Kuiper Belt. Their orbits are unstable because they cross the orbits of the larger planets. The unstable nature of their positions has led to the theory that their status as centaurs is temporary. Scientists think they will either be ejected from the solar system or become comets.

**CENTAUR RESEARCH**

In 2013, the National Aeronautics and Space Administration's (NASA) Wide-field Infrared Survey Explorer (WISE) mission found information that led astronomers to believe centaurs may be more like comets than asteroids. Astronmers used WISE's infrared data to examine centaurs. They also used knowledge about albedos. (The albedo is an object's surface reflectivity.) They already knew that many of the objects were either blue-gray or red. The infrared data let them know whether a centaur was light and reflective or dark and matte. Asteroids tend to be brighter and more reflective, like the moon, while comets tend to be darker. The WISE study showed that nearly two-thirds of the centaurs studied are more likely to be comets. Additional observations conducted by WISE showed that at least some centaurs may be comets from beyond the solar system. These comets were pulled into orbit by the strong gravitational pull of the bigger planets.

On April 5, 2014, astronomers found two rings circling a centaur known as 10199 Chariklo. It was the first time that astronomers saw rings around a small celestial object. Scientists are not sure what caused the rings.

**CENTAURS AND COMETS**

The word *comet* brings to mind a bright object with a long glowing tail streaking across the sky. People may wonder how a comet can possibly be confused with a rocky asteroid. But a comet starts out as an irregularly shaped dark object. It only develops a tail as it nears the sun and its coma warms up and starts to blow away. Sunlight reflected off the coma forms the comet's tail. The tail disappears as the comet begins to move away from the sun and the sun no longer shines on the debris. Without seeing the coma, it can be hard for scientists who are millions of kilometers away to tell if an object is a comet or an asteroid.

NASA's WISE mission is an attempt to reveal the true nature of these and other objects from the far reaches of the solar system. The discovery that many or most centaurs may be comets does not change the importance of these dual-natured space objects. Scientists think that both comets and asteroids have original materials from the development of the solar system 4.6 billion years ago locked in their cores. Whether they are comets or asteroids, centaurs hold key information about the origins of the solar system.

—*Janine Ungvarsky*

**Further Reading**

"Asteroids and Comets." *National Geographic.* Natl. Geographic Soc., 1996–2015. Web. 21 Feb. 2015.

"Centaurs." *Cosmos: The SAO Encyclopedia of Astronomy.* Swinburne U of Technology, 2014. Web. 21 Feb. 2015.

"Comet Facts for Students." *Jet Propulsion Lab Education.* California Inst. of Technology, 2 Nov. 2010. 21 Feb. 2015.

Crockett, Christopher. "Surprising Rings Circle Comet-Asteroid Hybrid." *Science News for Students.* Soc. for Science & the Public, 30 Mar. 2014. Web. 3 Mar. 2015.

Ferron, Karri. "What Are We Learning about the True Identity of Centaurs?" *Astronomy* Jan. 2014: 19. *Science Reference Center.* Web. 3 Mar. 2015.

"Kuiper Belt Objects." *Cosmos: The SAO Encyclopedia of Astronomy.* Swinburne U of Technology, 2014. Web. 21 Feb. 2015.

"NASA's WISE Finds Mysterious Centaurs May Be Comets." *NASA.* NASA, 25 July 2013. 21 Feb. 2015.

"Pluto and the Developing Landscape of Our Solar System." *International Astronomical Union.* IAU, n.d. Web. 21 Feb. 2015.

# CERES: 1 CERES

**FIELDS OF STUDY**

Astronomy; Cosmology; Observational Astronomy

**SUMMARY**

1 Ceres was the first asteroid to be discovered in the asteroid belt. It was considered the largest asteroid in the solar system before it was reclassified as a dwarf planet in 2006. Research using the Hubble Space Telescope reveals that Ceres is more similar in composition to terrestrial planets such as Earth and Mars than to other asteroids. NASA's *Dawn* space probe moved into orbit around Ceres in March 2015, after spending sixteen months orbiting Vesta, another large planetary object in the asteroid belt. Scientists hope that learning more about Ceres's composition and history will provide insight into the early solar system.

**PRINCIPAL TERMS**

- **asteroid belt:** an oval-shaped ring located between the orbits of Mars and Jupiter, containing millions of asteroids in orbit around the sun.
- **Dawn mission:** a NASA-funded mission launched in 2007 to study the two largest objects in the asteroid belt, the asteroid Vesta and the dwarf planet Ceres.
- **dwarf planet:** a celestial body that orbits the sun and has enough mass to attain a nearly round shape but lacks the gravity necessary to keep its neighborhood free of other space objects.

**First and Largest Asteroid**

1 Ceres is the largest space object orbiting in the asteroid belt, a doughnut-shaped ring of objects located between the orbits of Mars and Jupiter. The asteroid belt is about 2 to 4 astronomical units (AU), or 186 to 370 million miles, from the sun. Ceres accounts for about one-third of the total mass of all objects in the asteroid belt.

Ceres was the first asteroid to be discovered and, with an average radius of about 476 kilometers (296 miles), is the largest known asteroid in the solar system. However, due to Ceres's size and other characteristics that make it more planet-like than asteroid-like, it was classified as a dwarf planet in 2006. Some sources still consider it to be an asteroid as well as a dwarf planet.

The International Astronomical Union (IAU) created the category of dwarf planet to differentiate small planetary bodies from the regular planets of the solar system. In 2006, the IAU released Resolution 5A, which defined the criteria for both regular and dwarf planets. Dwarf planets are defined as space objects that orbit the sun and have enough mass and gravity to attain a spherical (round) shape. Unlike regular planets, however, dwarf planets lack the gravity required to clear their orbit of other space objects. Additionally, they do not act as satellites (moons) to other space objects.

After establishing this classification system, the IAU categorized Ceres and two larger space objects, Pluto and Eris, as dwarf planets. Pluto had been considered a regular planet from its discovery in 1930 until its reclassification in 2006. Eris, a celestial body that is nearly 30 percent larger than Pluto, was discovered in 2005. The IAU has since added additional dwarf planets to this category and expects that many more will be discovered in the future.

## Identification of Asteroids

Italian priest and astronomer Giuseppe Piazzi (1746–1826) discovered Ceres on January 1, 1801, while searching for a planet that astronomers believed existed between Mars and Jupiter. Piazzi tracked Ceres through the night sky over the course of several weeks. At first he believed Ceres was a comet. However, by June 1801, Piazzi and most astronomical experts had concluded that Ceres was a new planet due to its circular shape and relatively rapid speed of orbit.

In 1802, astronomers discovered a second small space body, Pallas, in the asteroid belt. British astronomer Sir William Herschel (1738–1822), already famed for his discovery of the planet Uranus, studied the observable characteristics of both Ceres and Pallas. After closely comparing them to planets and comets, Herschel realized they formed an entirely new category of space objects, which he dubbed "asteroids" (from the Greek for "starlike"). However, the scientific community did not immediately accept this new categorization of Ceres, and it remained classified as a planet for about fifty years.

## Characteristics of Ceres

Ceres is the only known dwarf planet in the inner solar system. It is located in the asteroid belt, about 2.77 AU (414 million kilometers, or 257 million miles) from the sun. It orbits the sun in a mostly circular path that takes about 4.6 Earth years to complete. It rotates quite quickly, completing one full rotation in about nine hours.

Ceres is quite different from its rocky asteroid neighbors, which are composed mainly of rock and metals. Scientists using the Hubble Space Telescope have found evidence of water vapor on Ceres's surface. They believe this vapor may be produced by ice near the surface as it transforms from a solid into a gas. It may also come from cryovolcanoes, or ice volcanoes that spew volatile compounds such as water and methane rather than hot lava. The presence of water vapor supports the idea that the surface of Ceres is covered in ice.

Computer models show that round space objects such as Ceres most likely have what is called a differentiated interior. This means that Ceres is composed of a dense material at its core and lighter material near its surface. All terrestrial planets, such as Earth and Mars, have a differentiated interior. Asteroids do not have a differentiated interior.

## NASA *Dawn* Mission

NASA launched the *Dawn* spacecraft from Cape Canaveral, Florida, in 2007. It is the first spacecraft to have the goal of visiting two widely spaced planetary bodies: the asteroid Vesta and the dwarf planet Ceres. The objective of the *Dawn* mission is to learn as much as possible about the composition and evolution of

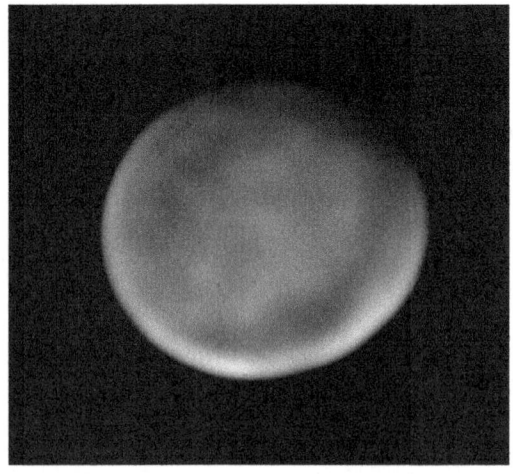

1 Ceres, dwarf planet. Dumas, C., et al., NASA-JPL.

these two planetary bodies, which are the two largest space objects in the asteroid belt.

One of the most important questions the mission hopes to address is how size and the presence of water affect planetary evolution. Vesta and Ceres are the two most massive planetary objects left over from the early days of the solar system. Scientists believe that their evolution to full planetary status was interrupted by the formation of the gas giant Jupiter. According to this theory, Jupiter's massive gravitational force disrupted the process of accretion, or the gradual buildup of space objects to form planets. Jupiter's gravity also caused smaller space bodies to crash into each other, creating the millions of asteroids that form the asteroid belt.

*Dawn* reached Vesta in July 2011 and orbited the asteroid until September 2012. During this leg of its mission, *Dawn* sent more than thirty thousand images back to Earth. It also recorded a variety of measurements that scientists will use to better understand Vesta's composition and history.

Afterward, *Dawn* traveled across the asteroid belt to its next destination. It entered Ceres's orbit on March 6, 2015. NASA planned for *Dawn* to remain in orbit for about sixteen months, getting as close as 375 kilometers (233 miles) away from Ceres's surface.

One question *Dawn* was sent to investigate is the nature of two bright spots spotted on Ceres's surface. They were first seen with the Hubble Space Telescope and later confirmed by images from *Dawn*. Scientists believe these spots may be evidence of ice or volcanic activity.

—Marie Keenan, MS

**FURTHER READING**

"Asteroids: Overview." *Solar System Exploration*. NASA, 6 Jan. 2015. Web. 22 Apr. 2015.

"Ceres: Overview." *Solar System Exploration*. NASA, 6 Jan. 2015. Web. 22 Apr. 2015.

"*Dawn* Mission: Science." *Jet Propulsion Laboratory*. California Inst. of Technology, n.d. Web. 22 Apr. 2015.

Landau, Elizabeth. "'Bright Spot' on Ceres Has Dimmer Companion." *Jet Propulsion Laboratory*. California Inst. of Technology, 25 Feb. 2015. Web. 22 Apr. 2015.

"Largest Asteroid May be 'Mini Planet' with Water Ice." *HubbleSite*. NASA, 7 Sept. 2005. Web. 22 Apr. 2015.

"NASA Spacecraft Becomes First to Orbit a Dwarf Planet." *NASA*. NASA, 6 Mar. 2015. Web. 22 Apr. 2015.

"Pluto and the Developing Landscape of Our Solar System." *International Astronomical Union*. IAU, n.d. Web. 22 Apr. 2015.

Wayman, Erin. "Cryovolcano." *Science News*. Soc. for Science and the Public, 2 Dec. 2013. Web. 22 Apr. 2015.

Weintraub, David. *Is Pluto a Planet? A Historical Journey through the Solar System*. Princeton: Princeton UP, 2007. Print.

# CETO: 65489 CETO

## FIELDS OF STUDY

Astrometry; Astrophysics; Cosmology

## SUMMARY

65489 Ceto is a small trans-Neptunian object. Its similarly sized moon, Phorcys, qualifies Ceto as a binary system. Because of Ceto's location and orbit, it is designated a centaur. Ceto is the second binary centaur ever discovered.

## PRINCIPAL TERMS

- **binary centaur:** two small solar system bodies that orbit around a common center of mass while also following a larger, nonresonant, unstable orbit around the sun that crosses the path of one or more outer planets.
- **binary trans-Neptunian object:** two minor planets that orbit around a common center of mass while also following a larger orbit around the sun at a greater average distance than Neptune.

## Binary Systems

In astronomy, most objects orbit another object. A binary system is a system in which two objects are close enough together that they orbit one another. More precisely, they orbit a common center of mass that lies between them.

65489 Ceto is a binary trans-Neptunian object. Its companion in the binary system is its moon, Phorcys. Ceto was discovered on March 22, 2003, but it was not identified as binary until Phorcys was discovered on April 11, 2006. Although Ceto is larger than Phorcys, the two are fairly close in size. Ceto has an estimated diameter of 174 kilometers (108.1 miles). Phorcys is thought to be about 132 kilometers (82 miles) across.

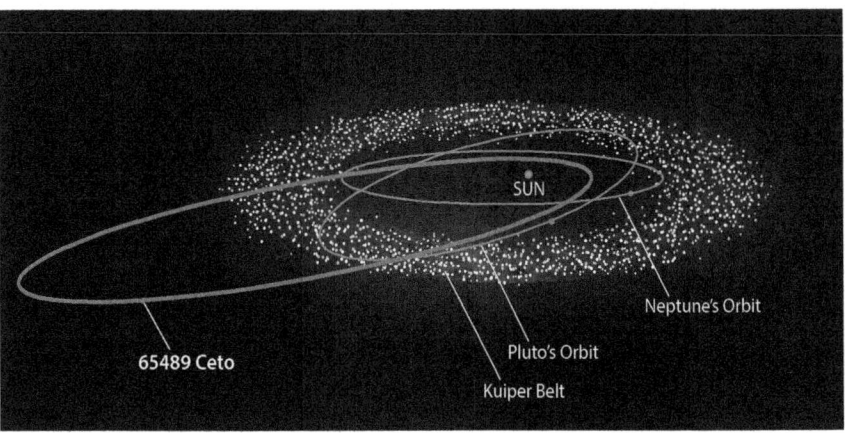

Orbit of 65489 Ceto, asteroid.

Ceto classified as a trans-Neptunian object (TNO) because its average distance from the sun is greater than that of Neptune. However, at times its orbit brings it closer to the sun than Neptune, crossing the paths of both Uranus and Neptune. Because of this, it is also classified as a binary centaur. Ceto is only the second binary centaur to be discovered. The first one was (42355) Typhon, whose binary companion is its moon, Echidna. Typhon itself was discovered on February 5, 2002. Echidna was discovered on January 20, 2006.

A centaur is a small solar system body (SSSB) whose orbit crosses that of one or more of the outer planets. (A TNO is a type of SSSB.) Centaurs have very unstable orbits. An unstable orbit is one that will eventually decay. A centaur's orbit lasts only a few million years. After that the centaur will either escape its orbit or be destroyed in an impact.

Most centaurs exhibit attributes of both asteroids and comets. For this reason, they were named after the mythical centaur, which was said to have the torso and upper body of a human and the lower body of a horse. NASA's Wide-Field Infrared Survey Explorer mission found that two-thirds of centaurs reflect less light, like comets, though some are brighter like asteroids.

## Ceto and Phorcys

Like many space objects, Ceto and Phorcys were named after ancient Greek gods. Ceto was the goddess of the dangers of the sea, specifically sharks, whales, and sea monsters. Phorcys was her brother and husband, another powerful and dangerous sea god. Ceto and Phorcys were said to be the parents of many mythical Greek monsters, including Scylla and the Gorgons.

—*Tyler Biscontini*

## Further Reading

Barber, Elizabeth. "What Is a Space 'Centaur'? Scientists Now Know the Answer." *Christian Science Monitor*. Christian Science Monitor, 26 July 2013. Web. 20 Mar. 2015.

Grundy, W. M., et al. "The Orbit, Mass, Size, Albedo, and Density of (65489) Ceto/Phorcys: A Tidally-Evolved Binary Centaur." *Icarus* 191.1 (2007): 286–97. Print.

"Keto." *Theoi Project: Greek Mythology*. Aaron J. Atsma, n.d. Web. 20 Mar. 2015.

Lacey, Sheridan, and Sarah Maddison. "Binary Asteroids." *SAO Astro News*. Swinburne Astronomy Online, 13 Apr. 2014. Web. 20 Mar. 2015.

"NASA's WISE Finds Mysterious Centaurs May Be Comets." *Jet Propulsion Laboratory*. Caltech, 25 July 2013. Web. 25 Mar. 2015.

Noll, Keith S., et al. "Binaries in the Kuiper Belt." *The Solar System beyond Neptune*. Ed. M. A. Barucci et al. Tucson: U of Arizona P, 2008. 345–63. Print.

"Phorkys." *Theoi Project: Greek Mythology.* Aaron J. Atsma, n.d. Web. 20 Mar. 2015.

# CNO CYCLE

## FIELDS OF STUDY

Astrochemistry; Astrometry; Astrophysics

## SUMMARY

The CNO cycle is the dominant energy production mechanism in main sequence stars with masses greater than 1.1 times the solar mass. It involves the stepwise addition of protons to $^{12}C$ nuclei to produce nitrogen, and oxygen nuclei which finally fission to regenerate the $^{12}C$ and release a $^{4}He$ nucleus. Because the proton in the first step must approach a nucleus with six positive charges already present a very high thermal velocity and thus a high temperature is required.

## PRINCIPAL TERMS

- **CNO chain:** The energy producing mechanism in main sequence stars. It requires the presence of 12C nuclei in the core of the star, and a high temperature.
- **main star:** In the Hertzsprung-Russell diagram, a star whose spectral type and luminosity are consistent with hydrogen fusion as the principal energy source.
- **nuclear fusion:** The combining of two low atomic number nuclides to produce a nucleus of higher atomic number and the release of mass-energy.
- **proton-proton chain:** The predominant mechanism by which hydrogen nuclei are fused to form helium nuclei in lower mass main sequence stars.
- **reaction mechanism:** In chemistry or nuclear physics, the precise sequence of reactions by which the product is formed.

## FINDING THE POWER SOURCE OF STARS

Prior to the twentieth century, the power source of the stars had been a long-standing mystery. Discarding supernatural explanations, the best that nineteenth century, physics could provide was that the stars were self-gravitating systems with the gravitational potential energy that was released as the star collapsed. Lord Kelvin, leading physicist of his age did the calculation and found that the sun could have lasted only a few thousand years. This contradicted the evidence of geology and the new theory of biological evolution.

The atomic nucleus was discovered by Ernest Rutherford by about 1908, and the mass-energy relationship, $E=mc2$, by Albert Einstein few years earlier. As it became apparent that nuclear reactions might be responsible for the energy released as the stars began to shine, astrophysicists began to theorize about the exact sequence of reactions, the reaction mechanism, by which the energy could be released.

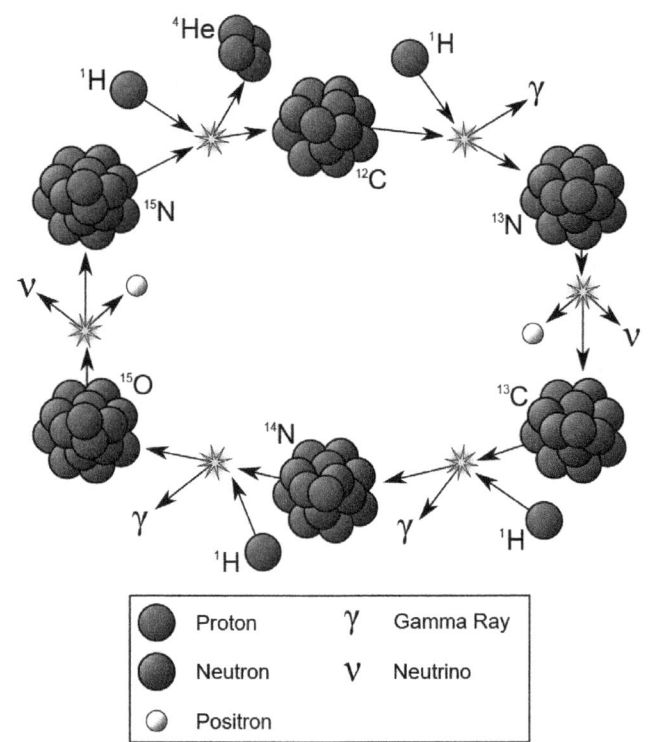

Overview of the CNO-I Cycle

The result was that for stars of mass less than 1.1 solar masses energy was released by the proton-proton chain while for stars greater in mass the carbon nucleus might serve a catalytic role.

### THE CNO CYCLE

The CNO cycle was proposed in 1939 by Hans Albrecht Bethe of Cornell University, and was one of the reasons cited for the award of the Nobel Prize in Physics for 1967.

The CNO cycle requires six steps to fuse 4 nuclei of hydrogen into one of 4He:

$$12C + 1H \rightarrow 13N + g$$
$$13N \rightarrow 13C + e+ + n$$
$$13C + 1H \rightarrow 14N + g$$
$$14N + 1H \rightarrow 15O + g$$
$$15O \rightarrow 15N + e+ + n$$
$$15N + 1H \rightarrow 4He + 12C$$

Here g is a gamma ray photon and n is an electron neutrino. Note that for the CNO chain to be operative some concentration of 13C must be established in the stellar core. The first, third, fourth and sixth steps are energetically unfavorable as they require letting a proton get close enough to a heavy nucleus. The net result of the cycle is turning four protons into one nucleus of 4He, two positrons, three gamma ray photons, and two neutrinos. The mass difference between the reactants and the product is accounted for in the conversion of the mass loss into the kinetic energy of the nuclei, the energy of the gamma rays produced, and the energy of the neutrinos produced.

### CONFIRMING THE CNO CYCLE MECHANISM

Confirmation of the mechanism required detecting the neutrinos produced. Raymond Davis, Jr. designed a detector, a neutrino telescope which was simply a large tank of cleaning fluid (tetrachloroethylyene) which could be circulated through a detector once a month to determine the number of Argon atoms created by the reaction 17Cl + n -> 18Ar + e-. The Davis detector found only one-third the number of neutrinos from the sun predicted by theory. This caused some controversy, originally but seems now to be resolved by the hypothesis of neutrino oscillation for which Kajita and Macdonald received the Nobel Prize in 2015. According to the Kajita-Macdonald theory, the electron neutrinos created the reactions like those in the CNO chain are only approximately in energy eigenstates. They oscillate among the three neutrino flavors, electron, muon, and tauon on their journey to the earth. By the time they arrive they are pretty much evenly distributed among the three flavors. Since the detector is only capable of detecting electron neutrinos, it is reasonable that two-thirds will be missed. The Davis telescope has detected solar neutrinos and the supernovas of 1987 which occurred in the Large Magellanic Cloud, a near neighbor to the Milky Way Galaxy.

—Donald R. Franceschetti, PhD

### FURTHER READING

Burchfield, Joe D. *Lord Kelvin and the Age of the Earth.* Chicago: University of Chicago Press: U of Chicago, 1990. Print.

Kaufmann, William J. *Universe.* 4th ed. New York: W.H. Freeman, 1994. Print.

Pais, Abraham. *Inward Bound: Of Matter and Forces in the Physical World.* New York: Oxford UPon, 1986. Print.

Seeds, Michael A., and Dana E. Backman. *Foundations of Astronomy.* 12th ed. Boston, MA: Brooks/Cole, Cengage Learning, 2013. Print.

# COMA (COMET ANATOMY)

### FIELDS OF STUDY
Astronomy; Observational Astronomy

### SUMMARY

A coma is the cloud of gas and dust around a comet's nucleus. When it is near the sun, a comet has three parts: a nucleus, a coma, and a tail. The coma is formed by the sublimation of water and other compounds from the nucleus's surface and creates an atmosphere around the comet. When the comet is very close to the sun, solar winds blow the coma away from the sun, creating the tail. Scientists continue to research comets and their comas because of the significant impact comets could have on Earth.

### PRINCIPAL TERMS

- **comet:** celestial bodies mostly made up of carbon dust and ice.
- **dust coma:** the cloud of dust surrounding a comet's nucleus.
- **gas coma:** the cloud of gas surrounding a comet's nucleus.

### COMA FORMATION

A comet is a celestial body made of ice, rock, and dust. When a comet is far away from the sun, it is frozen and solid. However, when its orbit brings it within 2 to 3 astronomical units (AU) of the sun, a comet undergoes a transformation.

As its surface warms, the comet undergoes sublimation. In this process, materials such as ice change directly from a solid to a gas, without going through the intermediate liquid stage. The gas forms the coma, an atmosphere of gas and dust around the comet. A coma grows depending on conditions, and can become extremely large, larger than planets even. The shapes of comas vary. The coma of a particular comet will also change according to how far it is from the sun.

### COMAS TYPES

Each comet actually has two comas: the dust coma and the gas coma. The gas coma is made up of the gas that is released when heat from the sun causes the comet to sublimate. Hydrogen molecules also form an invisible gas halo. The dust coma is made up of dust particles that are pulled from the comet when gases are expelled. Solar wind and radiation from the sun eventually push both comas to form the comet's gas and dust tails.

### HUMANS AND COMETS

People have always been fascinated by comets, but before the 1600s they perceived them mainly as omens. After this, people identified what comets were and how they worked using science. Ever since, scientists have tried to learn as much as they can about comets. They have undertaken a number of space missions to try to learn more about comets' anatomy, including the coma. In 2014, scientists landed the first spacecraft on a comet to study its composition.

Potential Impact of Comets

Scientists are interested in comets and their comas because they carry hydrogen, carbon, and other materials that are vital for life. Some scientists believe that much of the water on Earth was brought there by comets. Scientists study comets because they could potentially impact life on Earth or elsewhere in the solar system.

—*Elizabeth Mohn*

### FURTHER READING

Amos, Jonathan. "Controllers Now Banking on Philae Wake-Up Call." *BBC*. BBC, 30 Jan. 2015. Web. 5 Mar. 2015.

"Anatomy of a Comet." *Rosetta*. Jet Propulsion Laboratory, ESA, n.d. Web. 5 Mar. 2015.

"Comet Facts for Students." *Jet Propulsion Laboratory*. NASA, 2 Nov. 2010. Web. 5 Mar. 2015.

"Cometary Coma." *Cosmos*. Swinburne U of Technology, n.d. Web. 5 Mar. 2015.

"Exploring Comets and Asteroids: Time Capsules of the Solar System." *NASA*. NASA, 13 Nov. 2014. Web. 5 Mar. 2015.

"History of Cometary Missions." *ESA*. ESA, 1 Aug. 2014. Web. 5 Mar. 2015.

"What Is a Comet?" *Qualitative Reasoning Group*. Northwestern U, n.d. Web. 5 Mar. 2015.

# COMBUSTION IN SPACE

## FIELDS OF STUDY

Aerospace Engineering; Spacecraft Propulsion; Space Technology

## SUMMARY

Combustion, or the burning of fuel, is essential for generating the thrust needed to move rockets through space. However, it creates risk of explosion and fire. Scientists have learned that combustion in space is different from combustion on Earth. Studying combustion in space helps researchers learn how to make technology safer and more efficient.

## PRINCIPAL TERMS

- **Combustion Integrated Rack (CIR):** a combustion chamber on the International Space Station (ISS) that allows scientists to remotely conduct combustion experiments in the extreme low-gravity conditions of space.
- **Flame Extinguishment Experiments (FLEX):** a series of experiments conducted in the CIR from 2009 to 2013 that studied the burning of fuel droplets in the microgravity conditions of space and their response to various gases intended to suppress burning.
- **gravity:** a fundamental force that attracts two or more bodies to each other.
- **staged combustion:** a chemical process in which fuel is ignited, reacts with oxygen, and begins burning; the resulting heat is forced into a second stage chamber, where it ignites additional fuel and completes the combustion process.
- **vacuum:** an empty space devoid of air or any other gas; also sometimes defined as the absence of matter, although scientists now believe that even the vacuum of space, where no atomic matter is found, is occupied by dark energy and dark matter.

## IMPORTANCE OF COMBUSTION

Airplanes, jets, rockets, and even cars all burn fuel to create the power that allows them to move. This fuel-burning process is called combustion. During the combustion process, a fuel source is combined with an oxidizer, producing a chemical reaction that creates heat. It also generates a substance called exhaust. Exhaust is often in gaseous form, but some combustion products, such as the soot produced by burning wood or candles, can be solid. Because the exhaust is hot, it provides the heat necessary to keep the combustion process going.

Some propulsion systems use staged combustion, where the heat of the first combustion is forced into a second chamber containing more fuel. The exhaust heat from the first combustion ignites this fuel, completing the staged combustion process.

While combustion is crucial to space flight, it also poses the danger of explosion and fire. To find ways to protect astronauts and the vehicles used for space exploration, scientists study combustion both on Earth and in space.

## COMBUSTION STUDIES

Fire burns differently in space than on Earth. This is due to the vacuum of space and the extremely low gravity, referred to as microgravity. On Earth, the hot exhaust gases from combustion rise, pulling in more oxygen to aid the burning process. In space, these gases do not rise, and there is no oxygen to feed the fire. Flames burn more slowly and at lower temperatures.

To study these differences in detail, researchers have equipped the International Space Station (ISS) with the Combustion Integrated Rack (CIR). This hundred-liter combustion chamber contains the hardware and diagnostic equipment needed to perform experiments in space. With it, scientists have learned much about combustion in space, including how flames behave in microgravity, how combustion produces power and exhaust, and what happens to the waste products.

The CIR includes a fluid-oxidizer system to condition the gases and a chromatograph to measure them. Lights, cameras, and other diagnostic equipment are used for recording and analysis. The system is designed for remote operation, either from within the Payload Operations Center on the ISS or from Earth.

The chamber offers a variety of components for different experiments. Scientists can control the gas level, temperature, and position of the chamber. Eight removable optics windows allow them to position the diagnostic equipment for the best results for each experiment. An environmental system removes heat generated during the tests. The CIR also includes a component called the Multi-User Droplet Combustion Apparatus (MDCA), which allows researchers to conduct experiments with single droplets of fuel.

**CIR EXPERIMENTS**
Scientists can customize the environment in the chamber for each experiment. One key experiment that used the MDCA insert in the CIR was the Flame Extinguishment Experiment (FLEX). From 2009 to 2013, tests were conducted to determine the effectiveness of various gas-based fire suppressants in space. FLEX researchers studied how fuel vaporizes, how heat is created and dispelled in microgravity, and how different flammable chemicals interact in space. The results of these tests will help researchers develop new ways to control fires in future spacecraft.

The CIR was also the site of the 2013–14 Italian Combustion Experiment for Green Air (ICE-GA). This experiment focused on the efficiency and rate of evaporation of several types of biofuel. In 2014–15, the Burning and Suppression of Solids II (BASS-II) experiments used the CIR to test the combustion of plastics, fabrics, and other solid matter.

Additional experiments in the CIR include a second round of FLEX experiments designed to study the behavior of flame when fuel is sprayed (2015) and an MDCA/Cool Flames experiment to look for ways to decrease exhaust emissions in the second stage of staged combustion (2016). Starting in 2017, new equipment installed in the CIR will allow for studies aimed at developing high-efficiency, low-emission fuels. These fuels will affect space travel as well as all devices powered by combustion.

**APPLICATIONS OF CIR COMBUSTION EXPERIMENTS**
The results of the CIR experiments have far-reaching implications not just in space but also on Earth. Scientists believe that the microgravity of space enables them to study properties of fire that may be obscured by the effects of gravity, such as buoyancy. For instance, on Earth, hot gases rise, causing flames to flicker and present an elongated appearance. In space, the minimal gravitational effects and altered atmosphere makes flames burn in a rounder shape, with less flicker and less interference from buoyancy.

Through the FLEX studies, scientists have discovered that even after a flame has been extinguished, small droplets of a relatively unstable, highly flammable, colorless liquid hydrocarbon called heptane sometimes continue to burn. The cause of this phenomenon remains under investigation. Further research could provide insight into why some fires are more difficult to put out.

The results of the many CIR combustion tests provide insights that will lead to safer spacecraft, improved space suits for astronauts, and new ways to control fire during in-space emergencies. They also have applications on Earth, allowing scientists to develop better ways to prevent, detect, and extinguish fires. Researchers also hope to develop new and greener fuels as well as more efficient

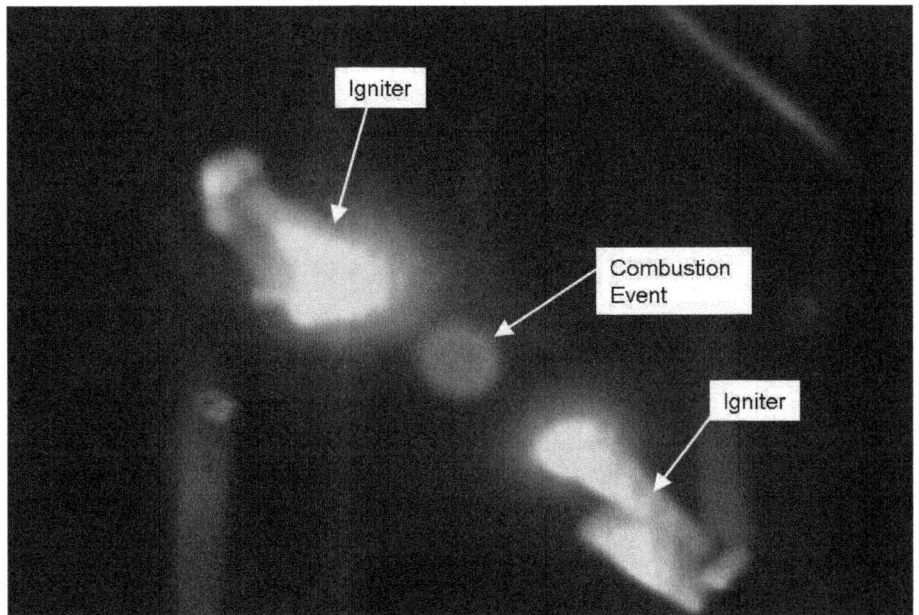

Multi-User Droplet Combustion Apparatus (MDCA) for space combustion experimentation, at the beginning of a combustion event. Glow of powered igniters and the combustion event are labeled. NASA.

combustion systems for everything from jet engines to cars to home furnaces.

—Janine Ungvarsky

**FURTHER READING**

Benson, Tom, ed. "Combustion." *Glenn Research Center*. NASA, 12 June 2014. Web. 10 Apr. 2015.

"CIR." *Space Flight Systems*. NASA, 29 Dec. 2014. Web. 10 Apr. 2015.

Giannone, Mike. "Combustion Continues to Draw Researchers to Space Station." *International Space Station*. NASA, 17 Jan. 2014. Web. 10 Apr. 2015.

Giannone, Mike. "FLEX-ible Insight into Flame Behavior." *International Space Station*. NASA, 28 Nov. 2011. Web. 10 Apr. 2015.

Nave, Carl R. "Gravity." *HyperPhysics*. Georgia State U, n.d. Web. 10 Apr. 2015.

Stiennon, Patrick J. G., and David M. Hoerr. *The Rocket Company*. Reston: AIAA. 2005. Print.

"Vacuum." *Cosmos: The SAO Encyclopedia of Astronomy*. Swinburne U of Technology, n.d. Web. 10 April 2015.

Williams, Forman A. "Flame Extinguishment Experiment (FLEX)." *International Space Station*. NASA, 21 Oct. 2014. Web. 10 Apr. 2015.

# COMET ANATOMY

### FIELDS OF STUDY

Astronomy; Cosmology; Astrochemistry

### SUMMARY

A comet is an icy celestial body comprised mainly of frozen gases and dust that has an elliptical orbit around the sun. The main part of a comet is the nucleus. When a comet's orbit takes it near the sun, the nucleus begins to sublimate, and gas and dust form a cloud around it called a coma. When the comet comes even closer to the sun, solar wind and heat push the cloud behind the nucleus, creating tails.

### PRINCIPAL TERMS

- **comet:** an icy body made primarily of frozen gases and dust that orbits the sun.
- **dust coma:** a cloud of dust that surrounds the nucleus of a comet when it approaches the sun.
- **gas coma:** a cloud of gas that surrounds the nucleus of a comet when it approaches the sun.
- **hydrogen cloud:** an invisible layer of hydrogen that surrounds a comet's coma.
- **nucleus:** the frozen body of a comet.
- **tail:** ionized gas and dust that is pushed away from the head of a comet.

### A Dirty Snowball

A comet is a celestial body made of ice from various gases and dust that orbits the sun. Because of its composition, a comet is often called a "dirty snowball." Scientists believe that comets were created during the formation of our solar system about 4.6 billion years ago in either the Kuiper Belt or the Oort Cloud regions. A comet has three parts: a nucleus, a coma, and a tail. The frozen body of a comet is called the nucleus.

A comet has an elliptical orbit, so at times it is close to the sun. When a comet approaches the sun, its nucleus begins to sublimate (go directly from solid to gas) and release gas and dust, forming a coma, or cloud, around it. When the comet comes very close to the sun, solar winds blow the dust and gas in the coma away from the sun, which creates the coma's tail. Comets that have a bright tail can be seen from Earth.

The size of comet's orbit determines how often it can be seen from Earth. A comet with a smaller orbit (short period) is visible more often. The famous Comet Halley (or Halley's Comet) travels around the sun approximately every seventy-six years. People on Earth will see Halley again in 2061. The comet Hale-Bopp has a very large orbit.

Hale-Bopp was visible to the naked eye for eighteen months in 1995 and 1996, but will not be seen again for about 2,400 years.

### A Comet's Anatomy

The main part of a comet is the nucleus, which is a large chunk of ice. This ice may be frozen gas, such as carbon dioxide, nitrogen, ammonia, and methane, along with rock, iron, and dust. The composition of a

Anatomy of a comet; the nucleus is surrounded by the coma; a dust tail and an ion tail both trail from the coma away from the sun. By E. Kolmhofer and H. Raab, Johannes-Kepler-Observatory, Linz, Austria, via Wikimedia Commons.

comet's nucleus varies. When a comet is far from the sun, it consists only of a nucleus. A comet's nucleus may range from a few hundred feet to many miles wide. Hale-Bopp, a very large comet, has a nucleus that is 30 to 40 kilometers (19 to 25 miles) wide.

The ice in a comet's nucleus begins to melt when the comet is about 3 to 4 astronomical units (AU) from the sun. Gas, water, and dust "boil off" or are expelled from the nucleus and form a cloud around it. This process is referred to as *sublimation*. The gas portion of the cloud that forms around the nucleus is called the gas coma. A dust coma is the dust portion of the cloud. An invisible layer of hydrogen surrounds the coma. It is called the hydrogen cloud or hydrogen envelope. A comet's coma grows and can become very large—occasionally even bigger than the sun. The nucleus and the coma are referred to as the "head" of a comet.

As the comet comes closer to the sun, solar wind and heat push the coma behind the nucleus. This forms tails that reflect sunlight, making the comet a beautiful sight in the night sky. Dust and gas each form their own tail. A comet's tail is sometimes hundreds of millions of kilometers long.

When a comet orbits the sun, it leaves behind some of its material. In time, it will not have enough material to form a coma and a tail. When this happens, the comet becomes extinct and may be classified as an asteroid.

## The Study of Comets

Throughout history people were both fascinated and terrified by the comets they saw in the night sky. They did not understand the existence and workings of the solar system. Comets seemed to appear and disappear randomly. People worried that a comet might be a sign of bad luck to come, such as a storm or an earthquake. They tried to use comets to predict the future.

Scientists today study comets to learn about the past. Because comets formed so long ago, they can help scientists learn about the formation and evolution of our solar system. They hope that spacecraft and probes will help them learn more about the chemistry of comets. They theorize that learning more about comets may help them find life in other places within the universe.

When their orbit takes them close to Earth, some comets can be seen at night with the naked eye. However, comet chasers often use binoculars and telescopes to view them more closely.

—*Adrienne A. Kennedy*

## Further Reading

"Anatomy of a Comet." *Rosetta*. California Inst. of Technology, n.d. Web. 5 Mar. 2015.

Choi, Charles Q. "Comets: Facts about the 'Dirty Snowballs' of Space." *Space.com*. Purch, 15 Nov. 2014. Web. 5 Mar. 2015.

"Comet Facts for Students." *Jet Propulsion Laboratory*. NASA, 2 Nov. 2010. Web. 5 Mar. 2015.

"Cometary Coma." *Cosmos: The SAO Encyclopedia of Astronomy*. Swinburne U of Technology, n.d. Web. 5 Mar. 2015.

"How Does a Comet Work?" *Rosetta*. California Inst. of Technology, n.d. Web. 5 Mar. 2014.

Peterson, Carolyn Collins. *Astronomy 101: From the Sun and Moon to Wormholes and Warp Drive, Key Theories, Discoveries, and Facts about the Universe*. Avon: Adams, 2013. Print.

"What's in the Heart of a Comet?" *SpacePlace*. NASA, 4 Mar. 2015. Web. 5 Mar. 2015.

# COMPTON SCATTERING

### FIELDS OF STUDY

Astrophysics; Theoretical Astrophysics

### SUMMARY

Compton scattering is a phenomenon in which energy and momentum are transferred between electromagnetic radiation and solid matter during a collision between the two. Arthur H. Compton discovered this phenomenon in the 1920s. His discovery supported a key element of Einstein's quantum theory and proved that electromagnetic radiation can behave as both waves of light and streams of particles.

### PRINCIPAL TERMS

- **Bragg spectrometer:** an instrument that uses x-rays to examine and measure variations in the scattering angles of crystals; originally invented by William Henry Bragg.
- **conservation of energy:** a fundamental law of physics that states that the amount of energy in a domain remains constant over time. Although the energy in the domain can be converted from one form to another, it cannot be created or destroyed.
- **conservation of momentum:** a fundamental law of physics that states that the amount of momentum in a domain remains constant over time. Although momentum can be changed through the action of forces, it cannot be created or destroyed.
- **photoelectric effect:** a process in which electrons are freed from a solid after the solid is exposed to electromagnetic radiation.
- **Planck relationship:** an equation that relates the energy of a moving particle to its frequency; first proposed by physicist Max Planck to explain the properties of radiation.
- **relativistic energy expression:** a physics expression that shows that in high-speed collisions, particles with mass can be created at the expense of kinetic energy.
- **x-ray:** a type of electromagnetic radiation with very high energy and very short wavelengths.

### THE NATURE OF LIGHT

Washington University professor Arthur H. Compton (1892–1962) discovered Compton scattering while working with electromagnetic radiation in 1922. Compton conducted a series of experiments involving collisions between light in the form of high-energy electromagnetic radiation and electrons in carbon atoms. Compton aimed x-rays and gamma rays at solid targets from different angles. He then used an instrument known as a Bragg spectrometer to examine and measure the results.

Through these experiments, Compton demonstrated that some of the energy and momentum from the charged photons (the x-rays and gamma rays) could be transferred to the electrons in the solid object (the carbon) during the collision. The electrons gained energy and momentum, and the photons lost energy and momentum. Imagine a pool player

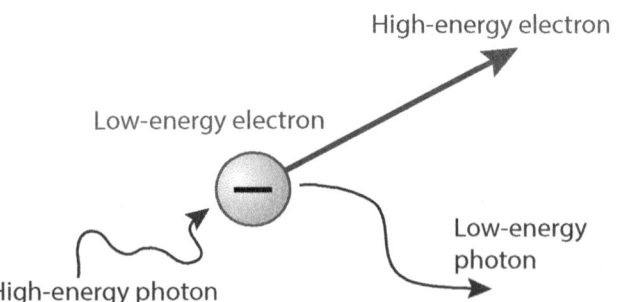

In the Compton scattering process, photons lose energy. High-energy photons are bombarded with low-energy electrons, resulting in high-energy electrons and low-energy photons.

shooting a cue ball at a stationary ball on a pool table. The cue ball loses energy and momentum, like the photon. The stationary ball gains energy and momentum, like the electron.

Compton's work was instrumental to the acceptance of Albert Einstein's particle theory of light because it offered supporting evidence that light was made of photons that possessed particle-like momentum and energy. Einstein's theory was the basis for the new field of quantum mechanics.

**PHOTONS VERSUS WAVES**
Compton's work was important because it provided evidence for a hotly debated issue on the nature of light. Many scientists of that time believed that light behaved as smooth, continuous waves. They thought these waves traveled smoothly through space the way water ripples over the surface of the ocean.

Others disagreed. In a seminal 1905 paper, "Concerning an Heuristic Point of View toward the Emission and Transformation of Light," German physicist Albert Einstein (1879–1955) theorized that light comes in packets called photons. Photons are small particles with no mass and a fixed amount of energy. According to Einstein, the total energy in a beam of light is the sum total of the individual energies of each of the particles, which he referred to as "light quanta," or photons.

Einstein's paper explained the photoelectric effect, a phenomenon first documented by German physicist Heinrich Hertz (1857–94). In 1886 Hertz accidentally discovered that some metals produced visible sparks when they were subjected to certain frequencies of light. These sparks were later identified as photoelectrons, or light-excited electrons, leaving the surface of the metal. German physicist Philipp Lenard (1862–1947) continued experiments into this phenomenon. His work attracted Einstein's attention.

Einstein explained the photoelectric effect using the quantum hypothesis of German physicist Max Planck (1858–1947). The Planck relationship (sometimes called the Planck-Einstein relation) is a mathematical formula that describes the energy of moving particles. According to the Planck relationship, energy can be quantified. Einstein used the formula to show that photons contain a specific amount of energy that varies by electromagnetic frequency. Einstein's theory stated that in high-speed collisions, relativistic energy expression could occur without breaking two fundamental laws of physics: the conservation of energy and the conservation of momentum.

**COMPTON SCATTERING IN MODERN TIMES**
Compton scattering is a well-accepted phenomenon in physics. It is also a key component in fields that rely on the use of x-ray technologies. In astronomy, Compton scattering is vital to the detection and study of objects far off in space. Very hot objects in space emit x-rays and gamma rays. Astronomers use x-ray telescopes and gamma-ray detectors in space to study these rays. The telescopes rely on Compton scattering to collect and measure photons. These measurements offer insights into the composition, temperature, and density of space objects.

X-ray technologies are also a key part of many of NASA's robotic missions. For example, rovers such as *Spirit* and *Curiosity* rely on x-ray technologies to detect the composition of rocks on Mars.

Understanding how energy and momentum are exchanged can yield important insights into the smallest elements of the universe, as well as some of the most powerful.

—*Marie Keenan, MS*

**FURTHER READING**
"Arthur H. Compton and Compton Scattering." *OSTI: DOE R&D Accomplishments*. US Dept. of Energy, 28 Jan. 2015. Web. 15 Apr. 2015.

"Astronomer's Toolbox: Gamma-Ray Detectors." Oct. 2013. *Imagine the Universe!*. NASA Goddard Space Flight Center, 20 Nov. 2014. Web. 15 Apr. 2015.

"Astronomer's Toolbox: X-Ray Astronomy." Sept. 2013. *Imagine the Universe!*. NASA Goddard Space Flight Center, 17 Nov. 2014. Web. 15 Apr. 2015.

Benson, Tom. "Conservation of Energy." *NASA Glenn Research Center*. NASA, 12 June 2014. Web. 15 Apr. 2015.

Benson, Tom. "Conservation of Mass." *NASA Glenn Research Center*. NASA, 12 June 2014. Web. 15 Apr. 2015.

Cassidy, David C. "Focus: *Landmarks*: Photons Are Real." *Physics Rev. Focus* 13.8 (2004): n.pag. *Physics*. Web. 15 Apr. 2015.

Dunbar, Brian. "Mission to Mars: Mars Science Laboratory." *NASA.* NASA, 28 July 2013. Web. 15 Apr. 2015.

Fowler, Michael. "The Photoelectric Effect." *Modern Physics.* U of Virginia, n.d. Web. 15 Apr. 2015.

"Nobel Prize in Physics 1927: Arthur H. Compton—Biographical." *Nobel Prize.* Nobel Media, 2015. Web. 15 Apr. 2015.

# CONSTELLATIONS

## FIELDS OF STUDY

Astronomy; Observational Astronomy

## SUMMARY

Constellations are patterns of stars in the night sky that can be seen from Earth. People have seen and told stories about these patterns for thousands of years. Constellations may look like everyday objects, animals, mythical creatures, or gods. In the past, people used these patterns to tell the time of year.

## PRINCIPAL TERMS

- **International Astronomical Union:** a recognized and authoritative professional organization for astronomers, founded in 1919 and based in Paris, France.
- **zodiac:** an imaginary band across the sky that follows the ecliptic (the path along which the sun appears to travel over one year when viewed from Earth) and is divided into twelve equal sections, each occupied by one of twelve constellations.

## GROUPS OF STARS

Throughout human history, when people have looked up at groups of stars in the night sky, the stars have seemed to form familiar patterns. The patterns that people see in the stars are called constellations. They may take the form of people, animals, gods, or everyday objects. The stars themselves are only abstract representations. In order to see the full figure represented by a constellation, it is necessary to imagine lines between and around the stars that show its rough shape and outline. The stars that make up constellations reside in many different places in the Milky Way Galaxy.

The appearance of the constellations has not changed much since they were first identified thousands of years ago. Nevertheless, all stars are constantly moving. This means that over tens of thousands of years, the constellations will eventually change. However, most people consider the constellations fixed in the sky because the changes in their locations will not be noticeable to humans for hundreds of lifetimes.

## COMMON CONSTELLATIONS

Constellations can be seen in the night sky with the naked eye. The best time to view them is on cloudless

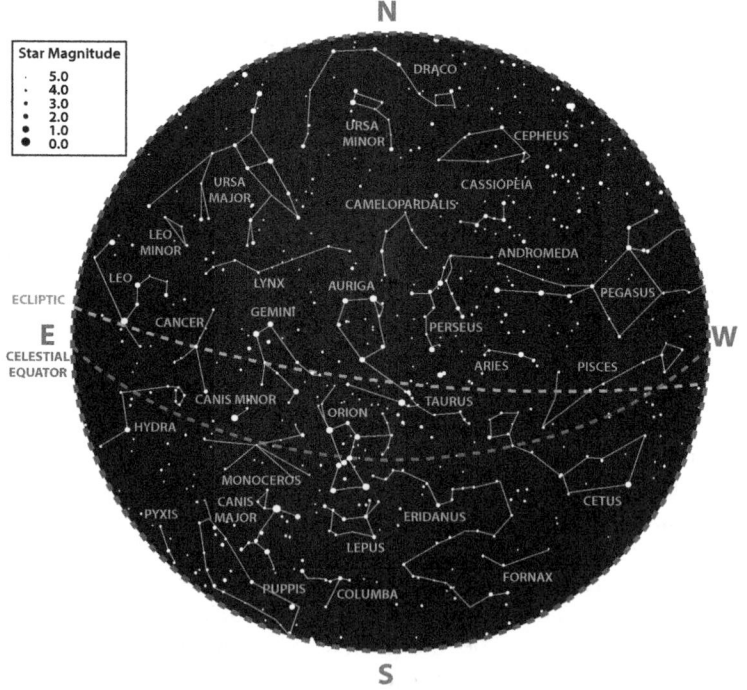

Star chart of the Northern and Southern Hemispheres in November. Lines connect the stars of common constellations.

nights. Different constellations are visible at different times of the year in the Northern and Southern Hemispheres. In the Northern Hemisphere, even beginning stargazers can easily see and identify Ursa Major and Ursa Minor without a telescope. Ursa Minor (the small bear) helps people locate Polaris (the North Star), which was historically used for navigation. A well-known constellation that is only visible from the Southern Hemisphere is Crux, or the Southern Cross. The Southern Cross is so distinctive that it appears on the official flag of several nations in the Southern Hemisphere, including Australia, New Zealand, and Brazil. Other commonly viewed constellations are Orion (the hunter), Draco (the dragon), and Taurus (the bull). Taurus is one of the twelve constellations of the zodiac, on which Western astrology is based.

Even ancient cave-dwelling humans looked up at the sky and wondered how and why the stars and planets worked. The first constellations were created by humans thousands of years ago, perhaps even as early as 17,300 years ago.

Some scholars believe that the first constellations were most likely religious, and the people who created them saw images of their god or gods in the sky in the form of stars. The constellations helped these people tell stories about their gods and their religions. Over time, many different cultures identified constellations.

The constellations recognized today came mostly from the ancient Greeks. Many stars seen close to the South Pole are missing from their records, however. The Greeks may have borrowed some or all of these from other peoples, such as the ancient Babylonians and Sumerians, who lived farther south, though still not close to the equator. Other constellations were added by ancient Egyptians and early modern Europeans as well. Other cultures, including those of Asia and the Americas, also developed their own constellations to tell stories about their culture and religion.

In 1930, the International Astronomical Union officially recognized eighty-eight constellation regions. These regions include the space beyond the stars

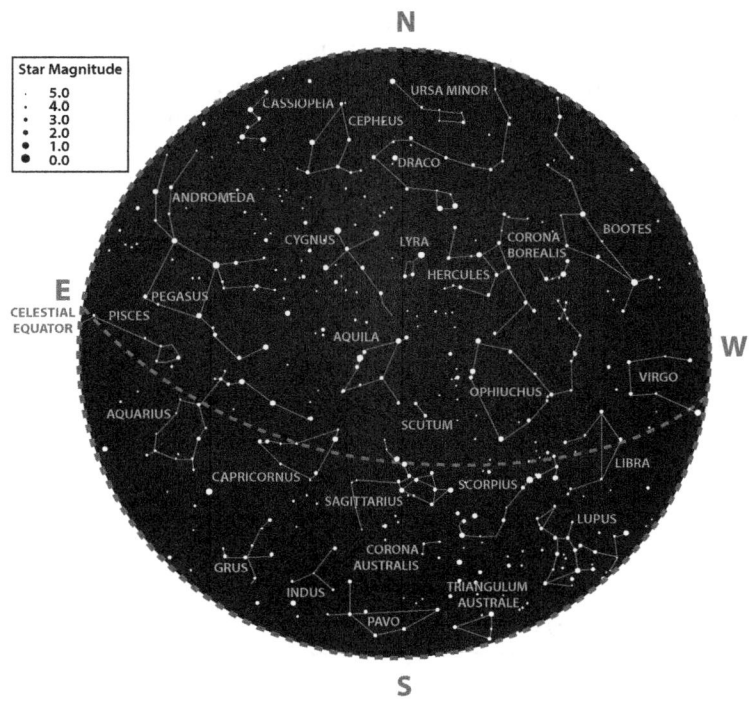

Star chart of the Northern and Southern Hemispheres in June. Lines connect the stars of common constellations.

that make up the constellations. Since then, the scientific community has not added new constellations. Instead, when new stars are found, they are added to existing constellation regions.

## USES OF CONSTELLATIONS

The various uses of constellations have changed over time. Farmers used to use the appearance of different constellations to help them know when to plant and harvest crops. Thousands of years ago, before calendars were available, people determined the time of year by tracking the locations of different constellations. Constellations are seen in different parts of the sky at the different times of the year, so they act as a type of calendar.

Constellations were also used to help navigate. Sailors and other travelers looked for the constellation now known as Ursa Minor, which in turn pointed them to Polaris. Because Polaris is almost directly above the North Pole, it serves as a compass, indicating which direction is north. It can also be used to determine one's latitude, which helped people travel across seas to different continents.

One modern use for constellations is in naming stars. When new stars are discovered, they are named based on how bright they are and which constellation they appear in. For example, the star α (alpha) Tauri is the brightest star in the Taurus constellation. This naming structure allows astronomers to understand which star is being discussed.

—Elizabeth Mohn

**FURTHER READING**

"About Constellations." *SkyTellers*. Lunar and Planetary Inst., 21 May 2007. Web. 20 Mar. 2015.

"The Constellations." *International Astronomical Union*. IAU, n.d. Web. 20 Mar. 2015.

Falkner, David E. *The Mythology of the Night Sky: An Amateur Astronomer's Guide to the Ancient Greek and Roman Legends*. New York: Springer, 2011. Print.

Kirkman, Thomas W. "Constellations: Frequently Asked Questions." *Tools for Science*. Coll. of St. Benedict & St. John's U, 1 May 2006. Web. 24 Mar. 2015.

Krco, Marko. "What Are Constellations Used For?" *Ask an Astronomer*. Curious Team, Oct. 2002. Web. 24 Mar. 2015.

Sparrow, Giles. *Constellations: A Field Guide to the Night Sky*. London: Quercus, 2013. Print.

WhiteHorne, Mary Lou. "How Far Are the Constellations?" *Space.com*. Purch, 24 Jan. 2007. Web. 20 Mar. 2015.

# CORE NAVIGATION SYSTEM

**FIELDS OF STUDY**

Space Technology; Astronautics

**SUMMARY**

Core navigation systems are used to direct and track vehicles and people as they move from one location to another. A range of instruments and tools have been developed for this purpose, including systems that use information relayed from satellites in space to determine position both in space and on Earth. These systems are important because they make locating and directing vehicles and people in real time easy and accurate.

**PRINCIPAL TERMS**

- **GPS:** the Global Positioning System (GPS), a network of satellites and Earth-based control stations used to determine the location, direction, and speed of objects. Established and maintained by the US government, it can be accessed freely with any GPS receiver device.
- **satellite navigation:** a system of navigation in which location is determined based on information transmitted from satellites. A receiver records the time of regular transmissions from multiple satellites equipped with atomic clocks, allowing it to determine its longitude, latitude, altitude, and time.

**NAVIGATION IN THE TWENTY-FIRST CENTURY**

The concept of navigation can be viewed as both a science and an art. In the scientific view, navigation is a means of determining an object's position and speed in relation to a set point in space. The art of navigation involves determining how to get from one point to another while avoiding as many obstacles as possible.

In the past, travelers used the positions of the sun, moon, and stars to determine their geographic location as well as the time of day. Eventually, instruments such as astrolabes and compasses were created to help refine the navigation process, making it easier to accurately determine and plot a course. Accurate clocks became invaluable for establishing speed and location.

As technology improved, navigational equipment also became more advanced. Navigation based on triangulation of radio signals became popular in the early twentieth century. However, land-based systems had limited accuracy and did not work everywhere. With the advent of space travel in the mid-twentieth century, researchers began work on a new generation of tools. Satellites made it possible to transmit radio signals anywhere. A web of satellites encircling the Earth could send information to a specially equipped receiver, enabling it to precisely determine its location. Multiple readings could establish how fast and in what direction the receiver was traveling. This also made it possible to look ahead of the receiver's projected path and plot a course that would fulfill any of a range of variables. Computers made these calculations far more quickly and accurately than a human could. Scientists sought to create a core navigation system that would more completely automate the navigation process.

## Navigation from Space

In the mid-1990s, the United States and Russia established the first satellite navigation systems that provided complete coverage of any location on the planet. Subsequently, China and the European Union also placed satellites in orbit with the intent of covering the entire globe, while several other countries pursued regional satellite navigation systems to cover certain areas.

The United States initiated its Navigation Signal Timing and Ranging, or Navstar, Global Positioning System (GPS) in the 1970s. It was intended to consolidate the growing number of navigation systems used by the US military. The first satellite for this system was launched in 1974. By the mid-1980s, enough satellites were in orbit to make the system functional for some uses. The full array of twenty-four satellites was completed in 1993, and fully operational worldwide coverage was achieved by 1995. GPS was also made available to the public as well as the military. By 2015, a total of thirty-two satellites were available for use as part of the GPS. The full array of satellites is called a constellation.

At least twenty-four satellites are operational at any given time. Twice a day, these satellites circle Earth in overlapping orbits at a distance of about 20,200 kilometers (12,550 miles) above the planet's surface. The satellites are programmed to send signals at exactly the same time. Receivers can tell their distance, latitude, longitude, and time based on how long it takes a signal to travel from the satellite to the receiver. The satellites are calibrated and synchronized so precisely that a GPS-enabled device can triangulate its position to within a few meters or even centimeters of its actual location, depending on the sophistication of the device. The Federal Aviation Administration (FAA) has reported that even their most basic receivers have an accuracy within 7.8 meters (about 25.6 feet).

The satellites constitute the space segment of the GPS, one of three components that make up the overall system. Another portion, the command segment, keeps track of the satellites and their information. From numerous bases on Earth, US Air Force personnel coordinate, analyze, and synchronize the data flowing to and from the GPS satellites and monitor their maintenance needs. These tasks are controlled by the Master Control Station in Colorado. There are also a number of monitoring stations around the world that track the satellites as they pass overhead, gathering and sending data to ensure the system remains operational. In addition, information is gathered from a number of ground antennas that are also part of the monitoring and command system.

The space and command segments of the system are controlled and monitored by the United States government. The final component of the system, the user segment, is in the hands of individuals and companies. The GPS is available free for use by all, and this has led to the development of thousands of devices that allow for the tracking of everything from vehicles and shipping containers to cell phones. In most cases, users do not need to know anything about where the satellites are or how they work. They merely need to turn on their receivers to benefit from the automated navigation system.

Though GPS is the most widely used core navigation system, others are also in use or development. Russia maintains the GLONASS system that functions similarly to GPS.

The European Union's Galileo global satellite navigation system and China's Compass system are expected to be fully operational by 2020.

## Applications of Core Navigation Systems

Satellite navigation systems were originally conceived and designed for military use. In the twenty-first century, satellite navigation is used to assist with weapons firing, supply and troop deployment, search-and-rescue maneuvers, and other military operations. Some US military vehicles are equipped to operate without human direction, relying totally on GPS navigation for guidance.

Because the US military has become so dependent on GPS for so many operations, it carefully protects its core system. The GPS used in civilian applications is not as precise as the military system. Military GPS transmissions are also encoded differently to prevent people with civilian equipment from jamming or taking over the system.

As GPS and its capabilities grew, commercial applications for it increased. In the 1980s, scientists began using handheld units to provide precise locations for archaeological and geological research sites. In 1989, the first handheld units for hikers, boaters, and other outdoor enthusiasts became available. As awareness of the technology grew, so did the demand. By the 2000s, small, portable units were available for use by motorists. Within a few years, GPS technology was being built into everything from automobiles and farming equipment to electronic tablets and cell phones.

While some applications of a core navigation system are more of a convenience, others increase safety and save lives. First responders such as fire, police, ambulance, and rescue workers can use GPS to more precisely locate given addresses or vehicles with onboard navigation systems. This enables them to arrive more quickly when help is needed, with fewer delays caused by misdirection, saving lives in the process.

GPS has also greatly improved the accuracy of spacecraft navigation. Early space probes that used other navigation systems were hard to track, making precise missions difficult. GPS allowed better calculation of position and ability to set a specific course. Between the 1960s and 2012, the accuracy with which a spacecraft could be tracked increased by a factor of one hundred thousand.

—*Janine Ungvarsky*

**FURTHER READING**

"Challenges of Space Navigation." *Time and Navigation*. Smithsonian Inst., n.d. Web. 22 Apr. 2015.

*Europe's Satellite Navigation Systems*. European Space Agency, 2000–2010. Web. 22 Apr. 2015.

"The Global Positioning System." *GPS.gov*. Natl. Coordination Office for Space-Based Positioning, Navigation, and Timing, 11 Feb. 2014. Web. 22 Apr. 2015.

Groves, Paul D. *Principles of GNSS, Inertial, and Multisensor Integrated Navigation Systems*. 2nd ed. Boston: Artech, 2013. Print.

Nave, Carl R. "Global Positioning Satellites." *HyperPhysics*. Georgia State U, 2014. Web. 22 Apr. 2015.

"Satellite Navigation: GPS: How It Works." *Federal Aviation Administration*. FAA, 4 Feb. 2015. Web. 22 Apr. 2015.

# CORONA AUSTRALIS

**FIELDS OF STUDY**

Stellar Astronomy; Observational Astronomy

**SUMMARY**

Corona Australis is a constellation that resembles a crown or wreath. It is located in the Southern Hemisphere and is most visible in August. This constellation has been interpreted in many ways over thousands of years and has a rich history of myths and symbolism. Many stargazers and astronomers study Corona Australis to see examples of binary and variable stars as well as nearby nebulae.

**PRINCIPAL TERMS**

- **celestial equator:** the imaginary line above Earth's equator that halves the celestial sphere; it is equally distant from the celestial poles.
- **circumpolar constellation:** a constellation that is always visible in the night sky near a celestial pole; the Northern and Southern Hemispheres have different sets of circumpolar stars.
- **constellation:** a region of space defined by a pattern of stars that can be seen in the night sky from Earth.
- **declination:** a space object's angular distance north or south of the celestial equator, expressed in degrees of arc.
- **International Astronomical Union:** an association of professional astronomers from all over the world who define astronomical constants while

- **right ascension:** a space object's longitudinal arc along the celestial equator, measured eastward from the vernal equinox and expressed in hours.

## CROWN IN THE SOUTHERN SKY

For tens of thousands of years, people have studied the stars and searched for constellations. Corona Australis was one of the earliest constellations to be documented by ancient astronomers. It was given the name Corona Australis, Latin for "southern crown," because its stars form a semicircular shape that resembles a crown or wreath. A similar crown-shaped constellation in the Northern Hemisphere—Corona Borealis, the "northern crown"—is considered by astronomers to be a counterpart to Australis. Corona Borealis is by far the larger and brighter of the two. Both circumpolar constellations are among the eighty-eight constellations that the International Astronomical Union recognizes today.

Corona Australis is visible throughout the Southern Hemisphere and in some parts of the Northern Hemisphere, particularly below the latitude of 40 degrees north. Some more northern latitudes can see it during its highest point in August. However, it cannot be seen at any time above 53 degrees north. Corona Australis can be found using its declination (about –41.5 degrees from the celestial equator) and right ascension (18.64 hours).

## ATTRIBUTES OF CORONA AUSTRALIS

Corona Australis is a relatively small and compact constellation. It consists of six primary stars and fourteen other space objects. Many other stars, planets, and other bodies are located near the constellation in the night sky, some of which are of significant interest to astronomers.

The main star in Corona Australis formation is alpha Coronae Australis, also known as Alfecca Meridiana. It is the only star in the constellation with a common name. Alfecca Meridiana is about 125 to 140 light-years from Earth. While not particularly bright, it is noteworthy for the high speed at which it rotates—approximately 180 kilometers per second at its equator—and for its excess infrared radiation. Such infrared excess is a sign that a star may be surrounded by a disk of dust. This, in turn, suggests that it may host its own planetary system.

Another notable star in the constellation is epsilon Coronae Australis. It is an eclipsing binary variable star, consisting of two stars that regularly eclipse one another as they orbit a common center of mass. In addition, this star is a contact binary, meaning that its two stars are close enough to share an atmosphere. Gamma Coronae Australis is also a binary star, while kappa Coronae Australis is an optical double star. This means that while its two stars appear close together when viewed from Earth, they are in fact very far away from one another. R Coronae Australis is an

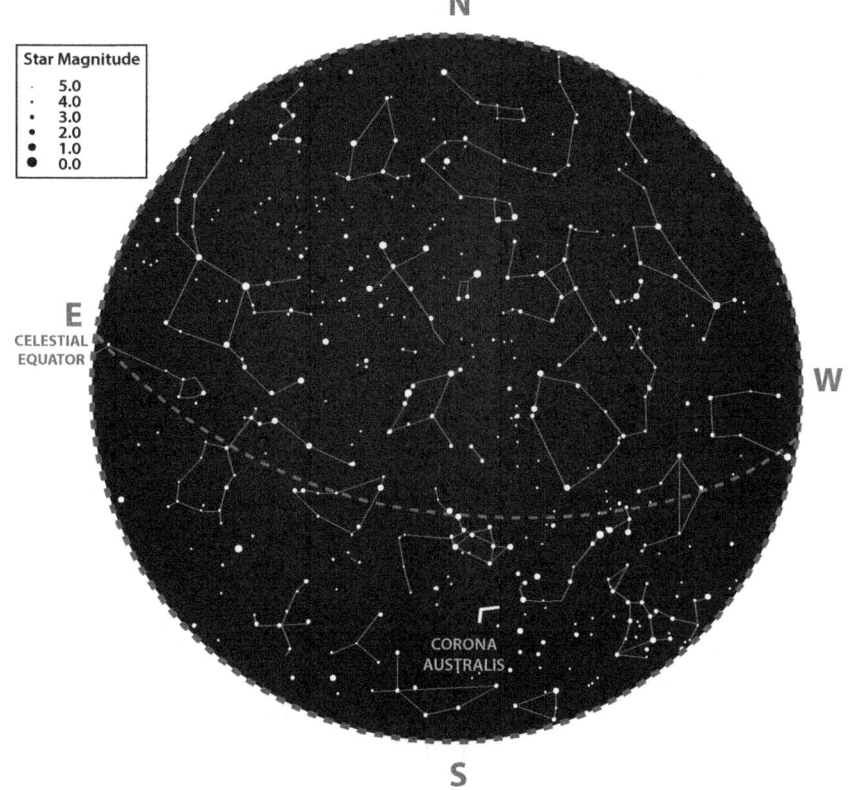

Star chart highlighting the Corona Australis constellation as seen in the Northern and Southern Hemispheres in June.

irregular variable star that undergoes great changes in brightness.

Despite this assortment of stars, Corona Australis is best known for the other astronomical objects it contains. One of the most famous of these is the Corona Australis molecular cloud complex, an enormous group of nebulae. A nebula is a vast cloud of dust and gas. Nebulae are often the site of new star and planet formation; over millions of years, drifting matter clumps together to form increasingly large bodies. Astronomers have identified many newly created stars in the Corona Australis complex, and more are constantly forming. At a distance of about 420 light-years, the complex is one of the closest star-forming regions to Earth's solar system. Corona Australis also contains a number of globular clusters, or huge, spherical groupings of orbiting stars. One such cluster is New General Catalog 6541 (NGC 6541). It is estimated to be about fourteen billion years old—as old as the universe itself.

## Monsters, Gods, Heroes, and Poets

Although Corona Australis appears mainly in the Southern Hemisphere, the ancient Greeks noticed and took great interest in it. The Greek Egyptian astronomer Ptolemy (ca. 100–170 CE) included it among the forty-eight constellations he charted during the second century CE. Corona Australis achieved its lasting fame thanks largely to the mythical stories of the ancient Greeks, although records of its underlying mythology are uncertain and varied.

The earliest Greek astronomers may not have considered the constellation a crown at all. According to some sources, it was originally rendered as a quiver of arrows in the hand of the nearby constellation Sagittarius, a centaur. In Greek myth, centaurs were generally portrayed as warlike half-human, half-horse creatures that favored bows and arrows during their attacks. Other Greek astronomers interpreted Corona Australis as a crown belonging to Sagittarius, while still others thought it represented a laurel wreath, an ancient Greek symbol of great honor. According to some myths, the god Dionysus honored his dead mother with such a wreath, then hung the wreath in the sky as an eternal monument. Some historians suggest that ancient people may have considered the wreath to be in honor of the great Greek poet Corinna, who won a contest against her rival Pindar.

Although the exact mythological meaning of Corona Australis is unclear, its importance is demonstrated by its inclusion in a "family" of constellations related to the mythical hero Hercules. In addition to the Hercules constellation, other members of this constellation family include Aquila, the eagle; Sagitta, Hercules's arrow; and Scutum, the shield. Within the same family are several of Hercules's foes, including Hydra, the sea serpent; Serpens, the serpent; Centaurus, the centaur; and Lupus, the wolf.

## Modern Astronomical Importance

Corona Australis is most visible during its highest point in August. Even at its brightest, however, it is a relatively faint constellation, and most of its components are best viewed with binoculars or a telescope. Observers can study alpha, beta, and epsilon Coronae Australis with binoculars, but telescopes of varying power are necessary to see deep-sky objects such as NGC 6541 and many of the more distant clusters and nebulae. The faintness of the constellation is largely due to the amount of cosmic dust in the region. In many cases, though, this dust reflects light and creates stunning patterns of bright colors.

Despite the faintness of the constellation, astronomers often look to Corona Australis to study its binary stars, variable stars, and other types of stellar objects. In mid-March, stargazers frequently monitor this constellation to see the Corona Australids, a meteor shower that can yield about five to seven falls per hour.

Corona Australis is an important piece of the southern night sky, easily found and studied in conjunction with other nearby constellations. Its nearest neighbors in the night sky include Sagittarius, Scorpius, Ara, and Telescopium. Observers can view the layout of these stars and constellations as well as study their rich histories of myth and symbolism.

*—Mark Dziak*

## Further Reading

"Corona Australis." *Peoria Astronomical Society*. Peoria Astronomical Soc., n.d. Web. 10 Apr. 2015.>

Kaler, James B. "Alfecca Meridiana." *Stars*. U of Illinois, n.d. Web. 10 Apr. 2015.

Kaler, James B. "Corona Australis." *Stars*. U of Illinois, n.d. Web. 10 Apr. 2015.

Kronberg, Christine. "Constellation Families." *Munich Astro Archive*. U Observatory Munich/Leibniz Supercomputing Centre, 5 Aug. 1996. Web. 10 Apr. 2015.

Kronberg, Christine. "Corona Australis." *Munich Astro Archive*. U Observatory Munich/ Leibniz Supercomputing Centre, 29 Sept. 1997. Web. 10 Apr. 2015.

Malin, David. "The Tail of the Corona Australis Reflection Nebula." *Australian Astronomical Observatory*. Commonwealth of Australia, 25 July 2010. Web. 10 Apr. 2015.

Nemiroff, Robert, Jerry Bonnell, and Ignacio Diaz Bobillo. "Stars and Dust across Corona Australis." *Astronomy Picture of the Day*. NASA, 12 Sept. 2013. Web. 10 Apr. 2015.

O'Meara, Stephen James. *Southern Gems*. New York: Cambridge UP, 2013. Print.

# CORONA BOREALIS

### FIELDS OF STUDY

Stellar Astronomy; Observational Astronomy

### SUMMARY

Corona Borealis (also called the Northern Crown) is a constellation, or a group of stars, in a semicircular shape that resembles a crown. This constellation is found in the Northern Hemisphere and is most visible around July. Corona Borealis has a rich history. Cultures around the world identified it and linked it to a wide variety of myths, tales, and beliefs. Astronomers are interested in the constellation because it contains an assortment of various types of stars including binary stars, variable stars, and recurrent novas.

### PRINCIPAL TERMS

- **celestial equator:** the imaginary line above Earth's equator that halves the celestial sphere; it is equally distant from the celestial poles.
- **circumpolar constellation:** a pattern of stars that appear in the night sky near a celestial pole and can be seen all night long, year-round.
- **constellation:** a section of the night sky identified by patterns of stars seen from Earth.
- **declination:** the north-south position of a celestial body relative to the celestial equator expressed in degrees of arc.
- **International Astronomical Union:** a worldwide association of professional astronomers that sets the rules for naming celestial bodies and features on them and defines scientific constants of importance to astronomy.
- **right ascension:** the east-west position of a celestial body defined in relation to the celestial equator and expressed in hours and minutes, not degrees of arc.

### THE CROWN CONSTELLATION

Corona Borealis is a constellation, an officially designated section of the night sky. Since ancient times, people have imagined patterns of stars, or asterisms, to form various images. Different cultures around the world have identified their own sets of patterns in the stars. Since 1930, the International Astronomical Union has recognized eighty-eight official constellations.

The Latin name Corona Borealis, meaning Northern Crown, comes from the semicircular shape of the circumpolar constellation. The ancient Greeks saw it as a crown. It was one of the original forty-eight constellations noted by the early astronomer Ptolemy (ca. 100–170 CE). People in many different cultures have identified Corona Borealis in different ways. Some have seen a wreath, dish, boomerang, castle, or even a group of dancers.

Corona Borealis can be found in the Northern Hemisphere between the latitudes of 90 degrees north and 50 degrees south. It is far from the celestial equator, with a declination of +30 degrees, and has a right ascension of sixteen hours. It is most visible during its highest point in July. The form of Corona Borealis is made of many stars, which are designated with Greek letters alpha, beta, gamma, and so on. Some of the more notable stars have separate names, such as Alphecca and Nusakan. The stars of

Corona Borealis vary in brightness, but the major ones are visible to the naked eye.

Stargazers can find Corona Borealis between several other constellations. To its north and west is the Boötes constellation, which has been interpreted as a farmer or shepherd. To its south is Serpens Caput, or Serpent's Head. To its east is the Hercules constellation, named after the mythical hero. Astronomers consider Corona Borealis to be a counterpart to a similar constellation in the Southern Hemisphere known as Corona Australis, or the Southern Crown.

Overall, among the many constellations, Corona Borealis is not particularly large. However, it is still highly significant to astronomers. Perhaps the most notable object is the central star of the semicircle. This star is known as Alphecca, Gemma, Gnosia, Asteroth, or alpha Coronae Borealis. The name Gemma means "gem" or "jewel," a name the star received due to its resemblance to a jewel on a crown. This

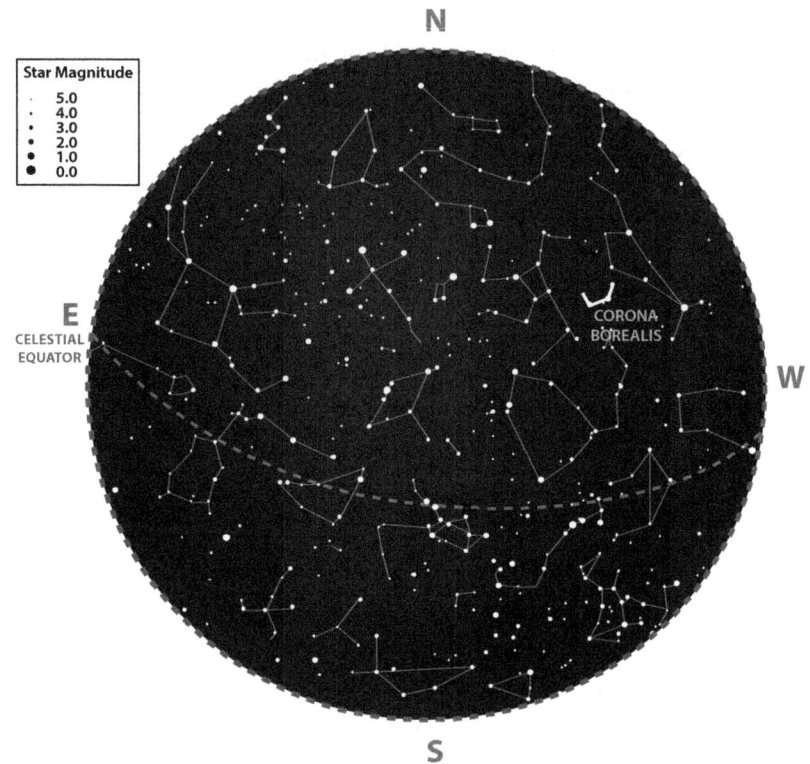

Star chart of the Northern and Southern Hemispheres in June. Lines connect the stars of the constellation Corona Borealis.

star, the brightest star in Corona Borealis, is actually not one star but two. It is an eclipsing binary star, a system in which two nearby stars revolve around a common center of mass. About every 17.4 days the two stars eclipse each other. This creates a small drop in brightness.

Another important star in the Corona Borealis is known as R Coronae Borealis (or R CrB). This star is known to some astronomers as the Fadeout Star or the Reverse Nova due to its unusual behavior. Although normally somewhat bright, it suddenly and unpredictably dims before returning to normal. Astronomers believe this odd change occurs when the star emits clouds of carbon that fill its atmosphere and block light from escaping.

A third notable star from the constellation is T Coronae Borealis, known to many astronomers as the Blaze Star. This star appears to have opposite behavior from R Coronae Borealis. T Coronae Borealis is a recurrent nova, a star that occasionally and briefly increases its brightness greatly due to explosions within the star. Normally, the Blaze Star is not very bright, but it can become as bright as Alphecca during its eruptions.

### MYTHS AND TALES OF THE ANCIENT WORLD

Since ancient times, people around the world have been fascinated by the night sky. These people looked for constellations and created stories about the patterns (asterisms) they saw. Their stories often reflected religious beliefs or myths and folktales that helped explain the mysteries of nature. Stars and constellations were also important for practical reasons. People used these heavenly bodies as a sort of calendar because the apparent movement of the stars followed similar paths each year. Ancient farmers, for instance, often studied the positions of constellations to decide when it was time to plant or harvest crops.

Corona Borealis is thought to be one of the earliest constellations identified, and different groups around the world made their own interpretations of the semicircle shape. The most famous story, and the one from which the official name derives, originated in ancient Greece. Ancient Greek astronomers felt the semicircle looked like a crown. This observation linked with myths of the princess Ariadne of Crete, who helped the hero Theseus defeat a monster called the Minotaur. Afterward, according to some versions of the myth, Ariadne married Dionysus, the Greek god of wine and fertility. The god presented his wife with a splendid bejeweled crown, which he later placed in the heavens as an eternal memorial to Ariadne.

In Western Asia, astronomers believed the semicircular shape represented a cracked dish or a broken pot. In Australia, some native people felt the star shape resembled a boomerang, a curved wooden stick used for hunting. The Cheyennes and Shawnees of North America believed the constellation represented a circle of eternally dancing maidens in the sky. The circle became a semicircle when one maiden left to marry. In ancient Celtic lands, some people envisioned the stars as a celestial castle called Caer Arianrhod.

Although the Greek interpretation is the most famous and influential due to the work of Ptolemy, astronomers and historians preserve the other tales for their cultural importance.

**Astronomical Importance Today**

Constellations like Corona Borealis are still a favorite of stargazers as well as a helpful tool for astronomers. Astronomers can use constellations as maps to quickly locate stars and determine whether stars have changed location, appearance, or patterns of movement.

Corona Borealis is most important to astronomers and stargazers today only for the wide variety of space objects that can be located within the constellation. There are several binary stars in the grouping, including Alphecca, beta (or Nusakan), and gamma. These stars are interesting to astronomers, who monitor the revolutions and eclipses of the star pairs.

The activity of R Coronae Borealis makes it an astronomical curiosity that has led space researchers to study the different factors that can affect a star's brightness. Astronomers began to use this star as a prototype star, or an example. Officially, all stars that exhibit similar changes in luminosity are classified as R Coronae Borealis (RCB) Type Variables. Similarly, astronomers have taken great interest in T Coronae Borealis for its qualities as a recurrent nova, another unusual and interesting type of star.

Another interesting feature of the constellation is the Corona Borealis galaxy cluster, also called Abell 2065. This extremely distant cluster of about four hundred galaxies is more than 1,200 million light-years from Earth. Due to its extreme distance, this formation can only be seen with very powerful telescopes.

—*Mark Dziak*

**Further Reading**

"Corona Borealis." *Chandra X-Ray Observatory*. NASA, 2 Dec. 2013. Web. 1 Apr. 2015.

"Corona Borealis." *Royal Astronomical Society of New Zealand*. Royal Astronomical Soc. of New Zealand, 2015. Web. 1 Apr. 2015.

Kaler, James B. "Corona Borealis." *STARS*. U of Illinois, n.d. Web. 1 Apr. 2015.

Koupelis, Theo. *In Quest of the Stars and Galaxies*. Sudbury: Jones, 2011. Print.

Kronberg, Christine. "Corona Borealis." *The Munich Astro Archive*. C/ Kronberg, 14 June 2003. Web. 1 Apr. 2015.

Olcott, William Tyler. *Star Lore of All Ages*. New York: Knickerbocker, 1911. *AAVSO*. Web. 1 Apr. 2015.

Plotner, Tammy. "Corona Borealis." *Universe Today*. Universe Today, 31 Oct. 2008. Web. 1 Apr. 2015.

# COSMIC INFLATION

### FIELDS OF STUDY

Astrophysics; Theoretical Astronomy; Cosmology

### SUMMARY

The big bang theory states that the universe developed slowly after the initial burst that created it. In contrast, cosmic inflation is a theory says that there was a short, intense period of expansion immediately afterward. This explains the flatness, horizon, and magnetic monopoles problems with big bang theory.

### PRINCIPAL TERMS

- **big bang theory:** a widely accepted explanation of the origin of the universe that says the universe was formed by the sudden expansion of a very hot, dense primordial fireball about 13.7 billion years ago.
- **blackbody spectrum:** the range of wavelengths of thermal radiation given off by a blackbody, an object that neither reflects nor transmits received light but gives off some light at all wavelengths, with shorter wavelengths for hotter objects.
- **cosmic microwave background radiation:** thermal radiation that is present in nearly uniform quantities throughout the observable universe and produces a blackbody spectrum characteristic of a temperature of approximately 2.7 kelvins. It believed to be left over from an early stage of universe expansion..
- **cosmology:** the astrophysical science that seeks to understand why the universe is the way it is by looking at how it formed, how it has changed, how it is structured now, and how it may change in future.
- **flatness:** a point of near-critical density, or the point at which the universe stops expanding.
- **horizon:** the limit of the observable universe, or the farthest region of the universe from which Earth can receive information (i.e., radiation), located about forty-five billion light-years away.
- **light element:** one of the first three elements of the periodic table—hydrogen (in the form of the isotope deuterium), helium, and lithium—which are believed to have been formed within the first few minutes of the big bang.
- **magnetic monopole:** a hypothetical magnetic particle that has only a north pole, unlike ordinary magnets, which have both a north and a south pole.

### THE BIG BANG RA

The idea that the universe had a specific beginning in a sudden burst has its origins in the thoughts of ancient astronomers. Advancements in technology helped refine those ideas into the big bang theory. However, as researchers began to study the universe in more detail, they discovered some problems with the theory. Some phenomena that should have been associated with the big bang were absent, while others were present that the theory did not explain.

In 1980, physicist Alan Guth (b. 1947) proposed a theory to address some of those issues. Guth's theory had a few errors, but fellow physicists Andreas Albrecht (b. 1957), Andrei Linde (b. 1948), and Paul Steinhardt (b. 1952) soon added further research and information. These four men are considered to be the architects of the cosmic inflation theory.

### BUILDING ON THE BIG BANG

Scientists in the field of cosmology and astrophysics study the universe for clues to its origin. They have discovered several clues that are compatible with the predictions of the big bang theory. For example, researchers predicted in 1948 that the big bang should have left behind heat in the form of cosmic microwave background radiation (CMB), which would produce a blackbody spectrum. In 1965, this was found to be true. The theory also predicted the early formation of light elements, which researchers were also able to prove.

However, there were several problems with the big bang theory. Three of these are the flatness problem, the horizon problem, and the magnetic monopole problem. The flatness problem is so called because, according to the theory, the universe should have continued to curve as it expanded, but cosmologists determined that it is nearly flat. The horizon problem refers to the near-uniformity of the CMB throughout the universe. All CMB produces a blackbody spectrum characteristic of a specific temperature—about 2.7 kelvins—which suggests that when it was first emitted, the universe was a constant temperature throughout. However, the horizons of the observable universe would have been too distant from one another to reach thermal equilibrium. The time required for information to travel that far was greater than the age of the universe. Finally, the theory predicted the existence of a large number of magnetic monopoles, while no such particle has ever been found.

The theory of cosmic inflation addresses all these issues. While the standard big bang theory proposed that the universe expanded at a steady rate, cosmic inflation theory says that there was first a short period of rapid inflation. During this period, the universe grew exponentially, expanding faster than the speed of light, before losing energy and slowing its growth. This accounted for the flatness problem: the initial expansion altered the universe's shape from the expected sphere into something that appears flatter. It also meant that when the CMB was produced, the now-distant reaches of the universe would have been much closer together than previously thought—close enough to reach thermal equilibrium before being separated by inflation. Cosmic inflation also suggests that magnetic monopoles were formed shortly after the big bang but were then spread so far and wide by inflation that they have yet to be detected.

Cosmic inflation theory also helps explain the formation of stars and galaxies. It says that the building blocks of these stellar features were microscopic until inflation expanded them exponentially. The areas that maintained the highest densities of the original matter of the universe formed stars, galaxies, and galaxy clusters.

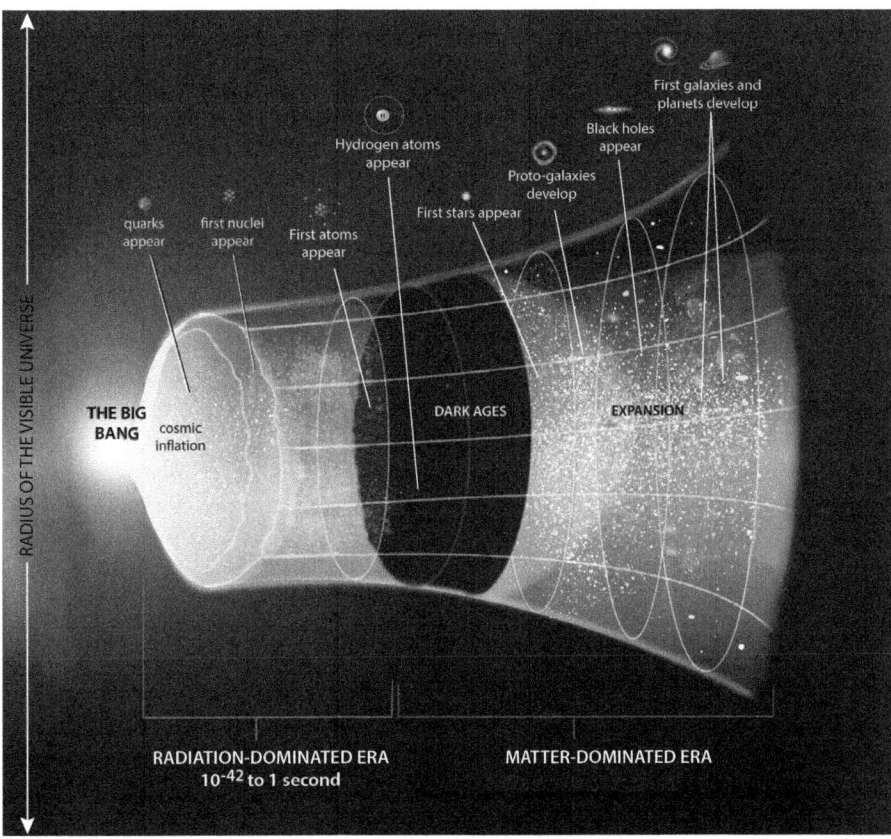

After the Big Bang occurred, the universe expanded rapidly in l ess than $1 \times 10^{32}$ seconds. During cosmic inflation quarks and the first nuclei appeared. Following cosmic inflation, the universe continues to expand.

### EVIDENCE OF COSMIC INFLATION

In March 2014, scientists using the Background Imaging of Cosmic Extragalactic Polarization (BICEP2) telescope in Antarctica said they had detected gravitational waves in the CMB. The BICEP2 team believed these waves showed that inflation had occurred, validating the theory. However, in January 2015, information from the European Space Agency's (ESA) *Planck* satellite reported a large dust cloud in the Milky Way galaxy that could have caused some or all of the gravitational waves. The ESA team did not completely discount the BICEP2 scientists' potentially groundbreaking discovery, and the search for information about the origins of the universe continues.

### VISUALIZING COSMIC INFLATION

The effects of cosmic inflation can be broadly likened to the inflation of a balloon. The first puff of air causes a rapid change in the size and shape of the balloon, after which it grows more gradually as it inflates. This accounts for why the universe remains flat and has not grown rounder: it is maintaining the shape it gained during the initial inflation period. The theory explains how parts of the universe can have such similar properties, since they were once much closer together, just as the ink of a design drawn on the balloon would have consistent properties even after the balloon was inflated. The increasing size of the universe would also allow for any monopoles to be scattered widely, just as dots drawn on the balloon would spread farther apart as the balloon expanded.

While scientists continue to study the universe for clues to its formation, cosmic inflation provides an important way to understand how the universe formed as it did.

—*Janine Ungvarsky*

**FURTHER READING**

"Beyond Big Bang Cosmology." *Wilkinson Microwave Anisotropy Probe.* NASA, 16 Apr. 2010. Web. 22 Apr. 2015.

Connolly, Amy R. "Scientists: Evidence of Big Bang Theory Fails to Space Dust." *UPI.com.* United P Intl., 31 Jan. 2015. Web. 22 Apr. 2015.

"Cosmic Microwave Background." *Cosmos: The SAO Encyclopedia of Astronomy.* Swinburne U of Technology, n.d. Web. 22 Apr. 2015.

"Cosmology: The Study of the Universe." *Wilkinson Microwave Anisotropy Probe.* NASA, 21 Dec. 2012. Web. 22 Apr. 2015.

"The Origins of the Universe: The Big Bang." *The Stephen Hawking Centre for Theoretical Cosmology.* U of Cambridge, n.d. Web. 22 Apr. 2015.

Palma, Christopher. "Blackbody Radiation." *Astronomy 801: Planets, Stars, Galaxies, and the Universe.* Penn State U, 2014. Web. 22 Apr. 2015.

# CRUX

**FIELDS OF STUDY**

Stellar Astronomy; Observational Astronomy

**SUMMARY**

Crux is a cross-shaped constellation that appears in the Southern Hemisphere. Because of its location, Crux was not identified by many cultures in the Northern Hemisphere, but it was identified by various Aboriginal Australian and Pacific Islander peoples, among others. Europeans first saw the constellation in the sixteenth century. Scientists study Crux and other constellations to learn more about stars, the galaxy, and the universe.

**PRINCIPAL TERMS**

- **celestial equator:** the imaginary line above Earth's equator that halves the celestial sphere; it is equally distant from the celestial poles.
- **circumpolar constellation:** a constellation that is always visible above the horizon of the night sky.
- **constellation:** a pattern of stars identified by humans that can be seen in the night sky from Earth.
- **declination:** a space object's angular distance north or south of the celestial equator, expressed in degrees of arc.
- **International Astronomical Union:** an association of professional astronomers from all over the world who define astronomical constants while promoting research, education, and discussion on important astronomical topics.
- **right ascension:** a space object's longitudinal arc along the celestial equator, measured eastward from the vernal equinox and expressed in hours.

**THE SOUTHERN SKY**

Crux, also known as the Southern Cross, is a constellation of five stars that form the shape of a cross. Even though Crux is probably the most famous constellation in the Southern Hemisphere, its apparent area is also the smallest of all eighty-eight constellations recognized by the International Astronomical Union. This popular constellation pattern appears on flags and postage stamps of a number of countries in the Southern Hemisphere. Crux can be seen from anywhere in the Southern Hemisphere at almost any time of year. It appears to change positions in the sky during the year; to viewers on Earth, it can look as if it standing straight up, lying on its side, or standing

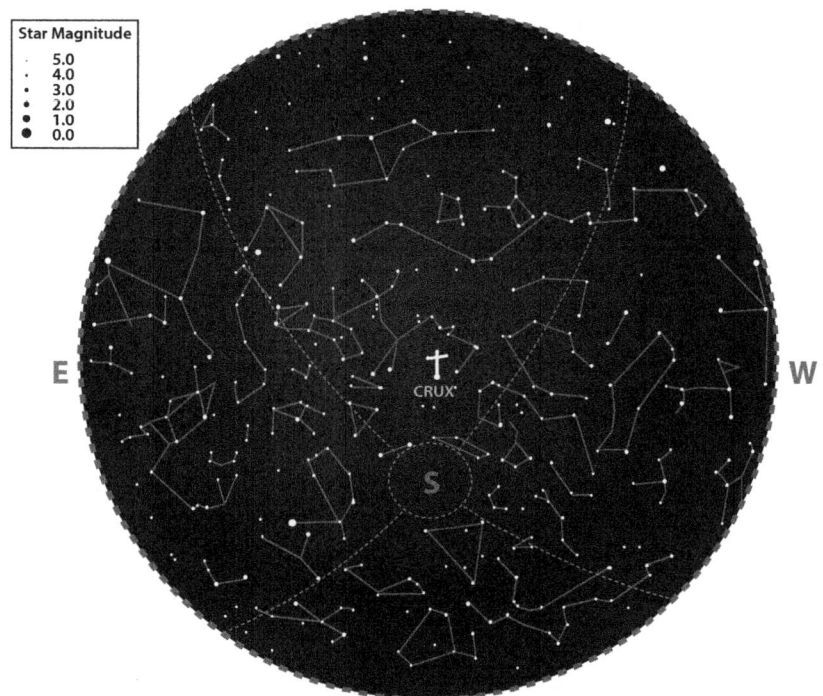

Star chart of the Southern Hemisphere in March. Lines connect the stars of the constellation Crux.

upside down. At latitudes between 23.5 degrees south and the South Pole, Crux is a circumpolar constellation, meaning that it can be seen all year long at any time of night. However, it is most visible in May and June.

Centuries ago, Crux was visible in parts of the Northern Hemisphere, including ancient Greece and what is now the United States. However, the subtle shifting of Earth's axis over time caused it to appear mostly in the Southern Hemisphere. During spring in the Northern Hemisphere, Crux can be seen at some lower latitudes, starting at about 25 degrees north. Potential viewing locations include Hawaii and some parts of southern Texas and the Florida Keys.

Some novice stargazers looking for the Crux may be fooled by a pattern that has become known as the False Cross. The False Cross is made up of four different stars, two from the constellation Vela and two from the constellation Carina. Because these four stars appear to make an X shape, they are easy to mistake for the Crux constellation. One way to tell the difference is to remember that the stars of the real Crux always point toward the celestial South Pole.

The brightest star in the Crux constellation is alpha Crucis, also known as Acrux. Beta Crucis, or Becrux, is the constellation's second-brightest star, sometimes called Mimosa. The third-brightest star is gamma Crucis, or Gacrux. Both Becrux and Gacrux are binary stars. Binary stars are star systems in which two stars orbit a common center of mass. Acrux, Becrux, Gacrux, and delta Crucis are the four main stars that form the cross shape. Crux also contains several Cepheid variables, including R Crucis, S Crucis, and T Crucis. A Cepheid variable is a star can be used to calculate the distances between Earth and other galaxies.

Crux contains some deep-sky objects as well. One such object is the Coalsack, a large dark nebula in the Milky Way. The Coalsack is a significant object in Australian Aboriginal astronomy. For a number of indigenous Australian peoples, it forms part of the quasi-constellation known as the "emu in the sky." Another deep-sky object in Crux is the star cluster New General Catalog 4755 (NGC 4755), commonly called the Jewel Box for its bright, colorful stars.

### HISTORICAL OBSERVATION

Although many of the classic constellations were named by ancient Greek or Roman astronomers, Crux was not one of them, as these civilizations were too far north to see it. However, Crux and its stars were visible to the ancient cultures of the Southern Hemisphere, including Aboriginal Australians and Pacific Islanders. Different cultures have seen different patterns in the constellation. While some saw a cross, others saw a bird or a fish.

For centuries, sailors have used the stars to help them navigate. When European sailors began exploring the Northern Hemisphere, they learned that they could use Polaris, also called the North Star, and the Big Dipper, a famous pattern of stars that makes up part of the constellation Ursa Major, to navigate while sailing. When Europeans first entered the

Southern Hemisphere, they realized that Polaris and the Big Dipper vanished from view. They learned, however, that the formation that would later be called the Southern Cross could serve the same purpose. In the early sixteenth century, during his third voyage to South America, Italian explorer Amerigo Vespucci (1454–1512) noticed Crux in the sky. Vespucci called the pattern "the four stars." Other European sailors learned to use the constellation to find the celestial South Pole.

### STUDYING CRUX AND OTHER CONSTELLATIONS

Modern astronomers use the various constellations to identify and name the stars found inside them. Because there are so many stars in the sky, this makes it easier to study and track them. Modern astronomers use constellations to help locate stars and other space objects. For instance, the stars of Crux can be found using its declination (about −60 degrees) and right ascension (12.45 hours) relative to the celestial equator, the projection of Earth's equator onto the night sky. Astronomers study the stars and other objects inside the constellations to learn more about the galaxy and the universe.

—Elizabeth Mohn

### FURTHER READING

Gendler, Robert, Lars Lindberg Christensen, and David Malin. *Treasures of the Southern Sky*. New York: Springer, 2011. Print.

Jansen, Albert. *Star Maps for Southern Africa: An Easy Guide to the Night Skies*. Cape Town: Struik, 2004. Print.

McClure, Bruce. "Southern Cross: Signpost of Southern Skies." *EarthSky*. Earthsky Communications, 15 Apr. 2014. Web. 7 Apr. 2015.

Privett, Grant, and Kevin Jones. *The Constellation Observing Atlas*. New York: Springer, 2013. Print

Sessions, Larry. "Acrux, Brightest Star in Southern Cross." *EarthSky*. Earthsky Communications, 24 Mar. 2015. Web. 7 Apr. 2015.

Sessions, Larry. "Star of the Week: Mimosa Second-Brightest in Southern Cross." *EarthSky*. Earthsky Communications, 1 Apr. 2015. Web. 7 Apr. 2015.

# CRYSTAL SPECTROSCOPY

### FIELDS OF STUDY

Astrophysics; Space Technology

### SUMMARY

Spectroscopy breaks down the light emitted by an object into separate wavelengths so that the object can be studied. This study helps determine properties of the object. These include its luminescence, light absorption and reflection, and reaction to external forces, such as electricity and radiation. In crystal spectroscopy, a spectrometer passes the object's light through a crystal with a known makeup to measure its wavelengths and analyze its characteristics. Astronomers use this data to understand the makeup of celestial bodies, especially those that cannot be studied directly.

### PRINCIPAL TERMS

- **absorption:** the process by which the energy of an electromagnetic wave passing through matter is retained and transferred to a molecule or atom.
- **luminescence:** the light given off by matter that does not get its energy from the creation of heat.
- **photon:** an elementary particle of light that moves and has energy but lacks mass and electrical charge.
- **reflection:** the return of a light or sound wave to its source. The wave bounces back after interacting with a surface.
- **spectroscopy:** the study of the spectrum of light wavelengths emitted by atoms and molecules. This tool helps researchers determine the wavelengths of an object's light. It can also verify the object's velocity, temperature, and makeup.
- **spectrum:** the range of wavelengths that make up light, such as visible, infrared, and ultraviolet light.
- **x-ray:** a type of electromagnetic radiation with very short wavelengths and very high energy. X-rays

are produced in space by a number of sources, including supernovas, neutron stars, and quasars.

**HOW SPECTROSCOPY WORKS**
Scientists in many fields use spectroscopy to gather information about an object by studying the energy it gives off in the form of light. In astronomy, spectroscopy provides a way for scientists to learn about objects that are too far away to be studied directly.

Spectroscopy is a useful tool because scientists already know a great deal about light. Light is a form of energy that also has properties of a particle. It travels in waves and is affected by things around it. Obstacles can interrupt the travel of light. This is similar to the way water flowing along a river is affected by an encounter with a bridge support. A fast-moving river will make waves with great rippling energy. A slow river will create larger but lower energy waves. Likewise, light traveling at a higher speed has greater energy than slower-moving light. Observing the type of light waves given off by an object can help scientists understand much about the object. This includes how fast it is moving, the conditions of the atmosphere where it is located, and even its composition.

In research using crystal spectroscopy, scientists can break down the light emitted by an object into its electromagnetic spectrum of colors and wavelengths. This includes the visible colors (the rainbow seen when sunlight passes through a prism) as well as rays outside the visible spectrum. Studying the wavelengths that are invisible without special equipment (such as gamma, ultraviolet, infrared, and x-rays) allows scientists to see known objects in a new way and gather crucial information about unknown ones. It can also reveal objects that cannot be observed by sight alone. By examining an object's light spectrum, scientists can learn about properties such as its absorption, reflection, and luminescence. Thus, with the help of a crystal spectrometer, they can determine characteristics of light emitted from a far-off celestial body.

**APPLICATION OF CRYSTAL SPECTROSCOPY**
A spectrometer is an instrument calibrated to provide standard and meaningful data from the light being analyzed. In crystal spectroscopy, this instrument involves using one or more crystals of known makeup (concave, bent, or sometimes flat) to disperse the light toward a detector for study. Scientists then look at the different aspects of the light spectrum separately and record how much of each type there is. With this data they can plot a graph of the light spectrum of that particular object. Using light spectrum data from known crystals, scientists can identify

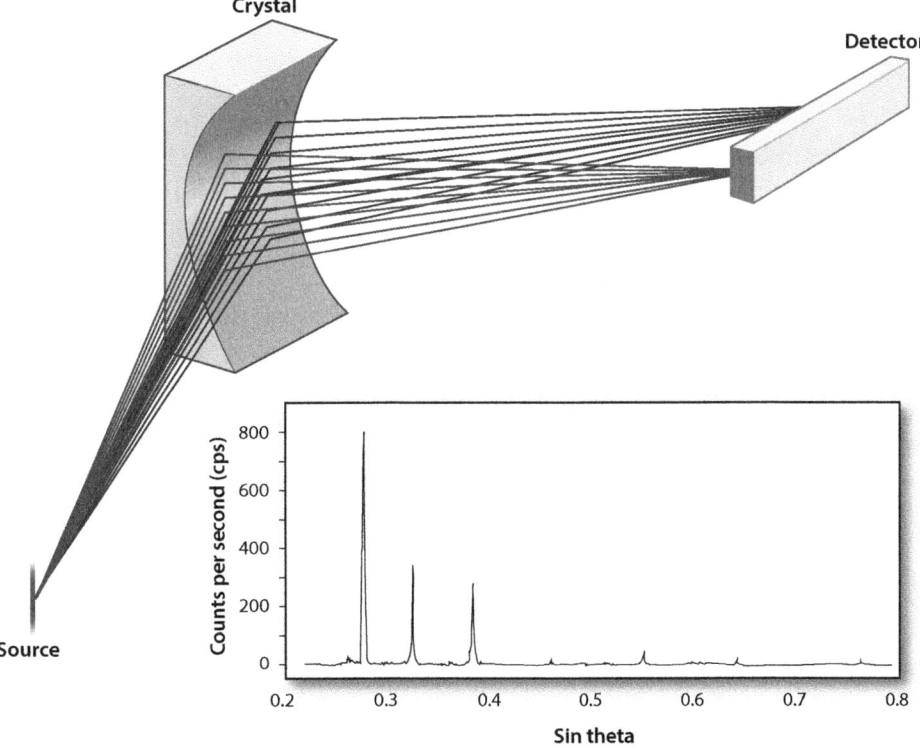

The setup for a crystal spectrometer includes a concave crystal and a detector positioned so that the x-rays projected from the source and refracted (bounced) off the crystal will hit the detector. Each wavelength of the source x-rays will refract off the crystal at different angles, which determine where they hit the detector. The spectrometer provides an output of the frequency (counts per second) at which different wavelengths (sin theta) hit the detector.

the unknown chemical makeup of the light source being studied.

However, the spectra of objects can be quite complicated. This is especially true when studying distant celestial objects. A star might be giving off light from more than one source, for instance. It may be made up of a number of different chemicals or have other objects interfering with the emitted light. Scientists plot the light onto a graph to help make sense of the various types of energy given off by the object. These graphs are very useful in making sense of x-ray emissions often found coming from accretion disks around black holes. These emissions are the result of a continuous release of x-ray photons of all levels of energy.

Bragg's law was identified in 1913 when English physicist Sir W. H. Bragg (1862–1942) and his son, Sir W. L. Bragg (1890–1971), searched for an answer to why crystal faces seemed to reflect x-rays at certain angles. In their equation, $\lambda$ is the wavelength of x-rays hitting the crystal, $d$ is the spacing of the crystal lattice or framework, and $\theta$ is the angle in degrees. Scientists apply Bragg's law with the following equation:

$$n\lambda = 2d \sin\theta$$

Application of the Bragg's law in crystal spectroscopy lets scientists angle the crystal to study a particular wavelength of the source object's light spectrum.

## Astronomic Spectroscopy Findings

Spectroscopic study of the universe has led to many major discoveries. The fact that elements such as hydrogen and helium that are found on Earth are also found in the sun and many other celestial bodies was the result of spectroscopic study. Scientists have uncovered the makeup of other planets using this technique.

Orbiters launched by the National Aeronautics and Space Administration used spectroscopy to confirm suspected water jets on Saturn's moon Enceladus in 2007. It also helped identify ice-filled craters on Mars. Similar studies revealed the presence of light wavelengths suggesting the existence of hydrogen and oxygen—the makings of water— on Earth's moon in 2009.

Spectroscopic equipment on the Hubble Telescope and other orbiters seeks to reveal the secrets of quasars as well as some of the oldest parts of the galaxy.

## Importance of Crystal Spectroscopy

Due to the vastness of the universe and limitations of space travel, very little of what is known about the universe has come from direct, hands-on study. Humans have spent a few days on the moon, landed exploratory craft on Mars, and circled other planets and moons with orbiters equipped to make observations. Thus, the information that can be gathered through spectroscopy is key to understanding the solar system and the universe beyond.

As scientists continue to refine this technique, they will be able to examine even the most distant celestial objects in greater detail. The tools planned for use in space will also have applications on Earth. For instance, scientists have been experimenting with using orbiting spectroscopic equipment to study the changes in volcanoes before, during, and after an eruption. This could lead to a better understanding of how and why eruptions happen and to more precise predictions of when they may occur.

—*Janine Ungvarsky*

## Further Reading

Cartwright, Jon. "Reading between the Lines: Space Spectroscopy." *Chemistry World* Dec. 2009: 46–49. Print.

Hearnshaw, J. B. *The Analysis of Starlight: Two Centuries of Astronomical Spectroscopy*. New York: Cambridge UP, 2014. Print.

"Introduction to Spectroscopy." *National Aeronautics and Space Administration*. NASA, n.d. Web. 30 Apr. 2015.

Knowles, J. W. "Crystal Diffraction Spectroscopy of Nuclear γ-Rays." *Alpha-, Beta- and Gamma-Ray Spectroscopy*. Vol. 1. Ed. Kai Siegbahn. Rpt. Burlington: Elsevier, 2012. Digital file.

Lawrence, Andy. "Spectroscopy." *Astronomical Measurement: A Concise Guide*. Heidelberg: Springer, 2014. 121–44. Print.

Nave, Carl R. "Bragg's Law." *HyperPhysics*. Dept. of Physics and Astronomy, Georgia State U, n.d. Web. 30 Apr. 2015.

"Spectroscopy." *Cosmos: The SAO Encyclopedia of Astronomy*. Swinburne U of Technology, n.d. Web. 30 Apr. 2015.

"Wavelength Dispersive Spectrometry (WDS)." *Northern Arizona University*. Northern Arizona U, n.d. Web. 30 Apr. 2015.

# CURIOSITY

### FIELDS OF STUDY

Space Technology; Remote Sensing; Observational Astronomy

### SUMMARY

*Curiosity* is an unmanned mobile unit about the size of a large car that can move independently on the surface of Mars. It was launched on November 26, 2011. Its mission includes taking photographs, gathering scientific data and samples, and transmitting information to Earth-based scientists. The information provided by the Curiosity rover about Mars in the present and the past helps scientists understand more about that planet and the solar system as a whole.

### PRINCIPAL TERMS

- **Aeolis Palus:** a plain that stretches between the northern rim of Gale Crater and the mountain Aeolis Mons on the planet Mars.
- **Deep Space Network:** a collection of large radio antennae positioned in three locations around the world that provide a communications system that can reach into space.
- **National Aeronautics and Space Administration:** the government agency established in 1958 by an act of Congress to run and oversee the United States' space program and research.
- *Opportunity*: a rover that, as of 2015, has spent the longest amount of time and traveled the greatest distance on the surface of Mars to date. It landed in 2004 and has provided crucial data about the planet.
- **rover:** an unmanned, self-propelled mobile research unit that can travel independently to gather information about a location, especially a planet, moon, or other celestial body.
- *Sojourner*: the first robotic rover to land on Mars. It landed in 1997 and began sending information from Mars's surface back to Earth.
- *Spirit*: a rover that landed on Mars in 2004. It sent information about the planet back to Earth until it ceased transmission in 2011.

### HISTORY OF MARS EXPLORATION

Aside from Earth, Mars has been discovered to have the most hospitable climate of all of the planets in the solar system. Therefore, scientists have made efforts to explore and learn more about this planet and its potential for life since early in the world's various space programs. In 1964, less than a decade after the establishment of the National Aeronautics and Space Administration (NASA), the United States launched its first exploratory missions toward Mars. The first attempt, the spacecraft *Mariner 3*, experienced technical difficulties and never reached Mars. However, its identical twin spacecraft, *Mariner 4*, reached Mars in July 1965. It took the first photographs of a planet other than Earth.

In 1976, *Viking 1* and *Viking 2* became the first spacecraft to touch down on the surface of Mars. They were equipped with a number of instruments for measuring the soil and atmosphere of the red planet. The *Viking* craft allowed scientists to conduct the first tests seeking life on another planet.

In 1992, the *Mars Observer* orbiting spacecraft failed just before deployment. However, the 1996–7 *Mars Global Surveyor* spacecraft was more successful. Among the information it recorded was data showing that Mars has repeating weather patterns.

The Mars Pathfinder mission launched in 1996 with *Sojourner*, the first robotic rover to land on Mars. Several other exploratory missions, including others with rovers, were launched during the first decade of the twenty-first century. The *Opportunity* and *Spirit* rovers both landed on Mars's surface in January 2004. After *Spirit* helped scientists learn more about the presence of water on the planet in the past, it became lodged in soft soil. NASA lost contact and ended its mission in 2011. Opportunity was still transmitting information more than ten years later. It has investigated the makeup of craters such as Endurance and Victoria. By late 2011, NASA was ready to launch the Mars Science *Laboratory* mission with *Curiosity*, its most ambitious and advanced rover to date. Curiosity would look for any habitable places and signs of microbial life.

**ABOUT THE ROVER**

The spacecraft that included the *Curiosity* rover was launched with an Atlas rocket from the Cape Canaveral Air Force Station in Florida on November 26, 2011. The payload, including the rover, landed on Mars on August 6, 2012.

*Curiosity* measures 3 meters (about 10 feet) long, 2.7 meters (about 9 feet) wide, and 2.2 meters (7 feet) tall, with an arm that can reach an additional 2.2 meters (7 feet). It is roughly the size of a small sport utility vehicle and weighs about 899 kilograms (1,982 pounds). The rover is equipped with a geology lab, a rock-vaporizing laser, and a variety of cameras to help it gather data.

The rover has six tires equipped with cleats along with a special suspension to enable it to move across the rocky surface of Mars. Unlike its predecessors, *Curiosity* is powered by a radioisotope thermoelectric generator (RTG). The generator's thermocouples use heat produced by the decay of plutonium-238 to create an electric current that powers the rover. Residual heat from the process is also used to heat other parts of the rover's operating equipment without losing any of its electrical power. This makes the RTG very efficient. The RTG has a lifespan of at least a full Martian year, or 687 Earth days. This allows *Curiosity* to move farther and faster than previous rovers.

The RTG provides power for four individual steering motors on Curiosity's two front and two rear wheels. This four-wheel steering allows the rover to turn in a complete circle in place. A special suspension system called a rocker-bogie keeps *Curiosity* stable and balanced on Mars's rocky, hilly terrain. While its programming is set to prevent the rover from attempting any inclines greater than 30 degrees, the rocker-bogie system is designed to keep it from tipping over at any angle up to 45 degrees. *Curiosity* averages a speed of 30 meters (98 feet) per hour, depending on the terrain it is covering.

The rover's robotic arm has three main joints, just like a human arm. It can hold instruments, pick up objects, and maneuver a camera to take photos. Other instruments onboard the rover include a spectrometer, special chemistry equipment such as an x-ray diffraction and fluorescence instrument, and a radiation assessment detector. An array of environmental monitoring sensors, sample-collecting instruments such as a drill, and specialized cameras for taking pictures in different conditions are also included.

Since the mission is unmanned, the rover needs a communications system to send information back

Mars rover, composite self-portrait. NASA/JPL.

to scientists on Earth. *Curiosity* has three antennae that have different functions. One is an ultrahigh-frequency antenna that sends messages to Earth by way of the *Odyssey* and *Reconnaissance* orbiters put in place in previous missions. Also included are high-gain antennae that receive signals from the Mars Science Laboratory mission team on Earth. Communication is helped by NASA's Deep Space Network, an array of antennae in three locations on Earth that provides continuous communication with planetary missions.

### *Curiosity's* Accomplishments

The *Curiosity* rover took its first drive on August 22, 2012, after NASA scientists completed tests on its systems and determined that the ground around it was safe. It had landed on Aeolis Palus, a plain in Gale Crater that scientists hoped would be rich with geological deposits to study. During its first trip, *Curiosity* traveled about 4.5 meters (15 feet) forward and then 2.5 meters (8 feet) back. It has since logged more than 10 kilometers (about 6 miles) through the early part of 2015. During its two-year planned mission and the time it has remained functional beyond that, *Curiosity* has benefited the space program and future exploration in many ways.

Scientists pioneered a new landing method to bring *Curiosity* safely to Mars's surface. Instead of airbags, a soft-landing technique known as a sky crane was designed and used to lower the rover from the vehicle. NASA engineers believe this technique could be useful for future large-payload missions, including manned missions to Mars and other celestial bodies.

Once *Curiosity* began roving, it determined that radiation levels on Mars are not high enough to prevent manned missions to the red planet. Data collected during the trip to Mars indicated that the levels of radiation in space are greater than those *Curiosity* detected on the planet's surface. In addition, the rover drilled beneath the surface of the planet to collect samples, marking the first time this has been done on any planet other than Earth. Continued investigation revealed the location of an ancient streambed that once held water several feet deep. This suggests that there could once have been life on Mars. The rover has identified sulfur, nitrogen, hydrogen, oxygen, phosphorus, and carbon as well as clay minerals in collected samples. These are all indicators that the planet once had an environment with enough water to sustain microbial life.

### Significance and Future Mars Study

*Curiosity* has explored many regions on Mars, gathering large amounts of data that will help NASA teams plan future missions to Mars and to other planets. The mission's success has encouraged NASA to plan additional missions with rovers similar in design to Curiosity.

*Curiosity's* exploration of Mars is important for many reasons. While still risky, Mars is the best bet for a manned landing on another planet. The ability to land and live for a time on another planet would provide the opportunity to develop new techniques and technologies that could be applicable on Earth. The lunar missions of the 1960s and 1970s provided innovations that have since become part of everyday life. Mars could also become a base for colonies to launch exploration further into the universe. Having the tools, technology, and know-how to successfully live on a temporary or permanent basis on Mars could also provide the opportunity for humans to live somewhere other than Earth for the first time.

—*Janine Ungvarsky*

### Further Reading

Bilger, Burkhard. "The Martian Chroniclers." *New Yorker*. Condé Nast, 22 Apr. 2013. Web. 5 June 2015.

"Curiosity Mission Overview." *Mars Science Laboratory: Curiosity Rover*. California Inst. of Technology, n.d. Web. 5 June 2015.

Kaufman, Marc. "Mars Rover Finds Ancient Streambed—Proof of Flowing Water." *National Geographic*. Natl. Geographic Soc., 28 Sept. 2012. Web. 5 June 2015.

Lamb, Robert. "Why Explore Mars?" *Discovery News*. Discovery Communications, 4 May 2010. Web. 5 June 2015.

"NASA's Curiosity Rover Making Tracks and Observations." *Mars Science Laboratory: Curiosity Rover*. California Inst. of Technology, 16 Apr. 2015. Web. 5 June 2015.

Ouellette, Jennifer. "The Brilliance behind the Plan to Land Curiosity on Mars." *Smithsonian.com*. Smithsonian Inst., Dec. 2013. Web. 5 June 2015.

Palmer, Brian. "Mars Rover Gets Instructions Daily from NASA via a Network of Antennae." *Washington Post*. Washington Post, 29 Oct. 2012. Web. 5 June 2015.

# D

## DACTYL: (243) IDA I DACTYL

### FIELDS OF STUDY

Sub-planet Astronomy; Astrometry; Cosmology

### SUMMARY

(243) Ida I Dactyl (Dactyl) is the only moon of the asteroid 243 Ida. It was discovered in February 1994 when astronomers analyzed photographs taken by the *Galileo* spacecraft. This was the first known example of a natural satellite orbiting an asteroid.

### PRINCIPAL TERMS

- **asteroid:** a irregularly shaped space object that orbits the sun and is larger than one meter across but smaller than a planet.
- **Ida:** a heavily cratered asteroid that orbits the sun in the asteroid belt. It is likely at least a billion years old.

### 243 Ida and Dactyl

243 Ida is an asteroid found in the region between Mars and Jupiter known as the asteroid belt. An asteroid is a rocky space object that is smaller than a planet and orbits the sun. Ida is a silicate (S-type) asteroid. S-type asteroids are fairly bright and are usually made of stone, nickel, and iron. First observed in 1884, Ida was the 243rd asteroid discovered. It was named after a mythical Greek nymph.

Ida is shaped irregularly. Instead of being spherical, the asteroid has two wide ends connected by a narrow waist. At its largest points, Ida's diameter is roughly 31 kilometers (19.3 miles). Its surface is covered in many craters, suggesting huge numbers of collisions with smaller asteroids. From this, astronomers estimate that Ida is more than one billion years old. Ida has one small moon, Dactyl.

### Dactyl's History

Upon examination of photos of Ida, astronomers discovered that the asteroid has a small moon. It was the first time an asteroid was found with a natural satellite. The moon was named Dactyl, after the mythical Greek creatures said to be found on Mount Ida. It is composed of materials similar to Ida and has a diameter of about 1.6 kilometer (1 mile).

There are several theories about how Dactyl was formed. Ida's gravitational field is too weak to have captured a separate object. Ida and Dactyl may have broken away from a larger object at the same time. Another possibility is that Dactyl was formed by an impact on Ida.

### Galileo Mission

The *Galileo* spacecraft flew by Ida in 1993, providing the main data on the asteroid and its moon. Scientists continue to study this data to better understand asteroids. The age of such bodies gives insight into the origins of the solar system. Also, other asteroids may affect humans, both through the danger of

(243) Ida I Dactyl, moon orbiting asteroid Ida. NASA/JPL.

an impact with Earth and as a potential source of mineral resources.

—*Tyler Biscontini*

**FURTHER READING**

"Asteroids: The Rocky Debris of Space." *Astronomy Today*. Astronomy Today, n.d. Web. 16 Mar. 2015.

Belfiore, Michael. "How to Mine an Asteroid." *Popular Mechanics*. Hearst Digital Media, 27 Oct. 2014. Web. 20 Mar. 2015.

Bond, Peter. "Galileo Spacecraft Crashes into Jupiter." *Spaceflight Now*. Spaceflight Now, 21 Sept. 2003. Web. 16 Mar. 2015.

"Fast Facts: Asteroid Ida." *Amazing Space*. Space Telescope Science Inst., n.d. Web. 16 Mar. 2015.

Francis, Matthew. "Moonday: A Bite-Sized Moon." *Galileo's Pendulum*. Galileo's Pendulum, 5 Mar. 2012. Web. 16 Mar. 2015.

"Images of Asteroids Ida & Dactyl." *Near Earth Object Program*. NASA, 22 Aug. 2005. Web. 16 Mar. 2015.

Johnston, Wm. Robert. "(243) Ida and Dactyl." *Asteroids with Satellites Database: Johnstons's Archive*. Wm. Robert Johnston, 21 Sept. 2014. Web. 16 Mar. 2014.

"243 Ida: Overview." *Solar System Exploration*. NASA, 13 May 2014. Web. 20 Mar. 2015.

# DARK ENERGY

**FIELDS OF STUDY**

Cosmology; Astrophysics; Stellar Astronomy

**SUMMARY**

It was long thought that the effects of gravity would slow the expansion of the galaxy. However, evidence has shown that it is instead speeding up. Scientists believe the explanation for this is dark energy, a force that counters gravity and allows the universe to continue expanding at a faster rate. Dark energy thus has a significant impact on the size and future of the universe.

**PRINCIPAL TERMS**

- **cosmological constant:** developed by Einstein, the concept that empty space has sufficient density and pressure to prevent the universe, once believed to be static, from collapsing under the pressure of its own gravity.
- **gravity:** a force of mutual attraction between two masses that increases with increased proximity.
- **quantum physics:** the subfield of physics that studies the often contradictory and counterintuitive behavior of the smallest units of matter and energy, such as subatomic particles, atoms, and molecules.
- **theory:** an explanation of some aspect of the natural universe that has been developed and thoroughly tested by means of the scientific process of observation, development of a hypothesis or reasoned explanation, experimentation, and conclusion.

**A FORCE AGAINST GRAVITY**

In the 1990s, researchers discovered that the universe was neither static nor expanding more and more slowly, as many had believed for decades. Instead, the pace of expansion was speeding up. This seemed to contradict gravity, a force that affects all normal matter and causes objects to be attracted to each other. For the universe to be expanding at an increasing rate, something would have to be working against gravity. Scientists differed on how this "something" came to be a force, but they agreed on one thing. Whatever was exerting the force to counter gravity and allow the universe to expand was something previously unknown. Researchers called it "dark energy."

**A CHANGING UNDERSTANDING OF THE UNIVERSE**

In his theory of relativity, German-born physicist Albert Einstein (1879–1955) proposed the theory that time, space, duration, and distances are not absolutes but relative and depend upon the observer. This general theory of relativity also found that time and space are affected by gravity and motion. It was the first significant change to gravitational theory since English physicist Isaac Newton (1642–1727) articulated his law of gravity. Einstein's equations involving space and time relativity in a static universe had indicated that the universe would collapse in on itself. He resolved this by adding an extra term he called the cosmological constant to these equations. Represented by the Greek letter lambda, the cosmological constant posited that unoccupied areas of space are filled with a constant force that is dense enough and exerts enough pressure to prevent gravity from collapsing the universe.

Then American astronomer Edwin Powell Hubble (1889–1953) discovered that the rate at which a distant galaxy moves away from another increases the farther apart they are. This indicated that the universe was not static but was expanding. Thinking an expanding universe made the concept irrelevant, Einstein discarded the idea of a cosmological constant from his theory of relativity.

For many years after Hubble's discovery, scientists believed that gravity would slow the rate at which the universe expands. All objects made of normal matter—objects made of protons, neutrons, and electrons—are pulled together by gravity, and this pull would limit and might even eventually stop the universe from growing, scientists believed. Then in 1998, a team of California-based astronomers using the Hubble Space Telescope and an Australian team at Mount Stromlo Observatory were studying a distant supernova. They made the remarkable discovery that the universe was in fact expanding faster than it had when the supernova had occurred. Additional research showed that the universe began expanding at a faster rate about 4 to 7 billion years ago.

This left researchers in a quandary. If the universe were expanding as fast as the evidence showed, then Einstein's theory of general relativity was wrong. Research indicated that the universe expanded at a steady rate for billions of years and then experienced a growth spurt beginning 4 to 7 billion years ago that kept increasing. This was unexpected and contrary to all known ways the universe worked.

Scientists eventually came up with three possible answers to this dilemma. One theory was that Einstein's cosmological constant concept was in fact in play and that something could not only counter the pressure of gravity but get faster at doing it over time. Other scientists suggested that space is filled with an unknown but very fluid energy about to suppress the effects of gravity. Still others thought Einstein's theory of relativity might need to be revised to include calculations accounting for

The gravity of dark matter tries to pull the universe together, while dark energy tries to push it apart. Dark matter dominated the early universe, but dark energy began to dominate about five billion years ago. As the universe gets larger, dark energy's domination increases. A timeline comparing relative quantities of dark energy and dark matter in the universe at 9 billion years ago, 5 billion years ago, and the present state of the universe show the progression of an expanding universe due to increased dark energy over time.

the acceleration. While the theories of what was wrong differed, all three pointed to dark energy as the solution.

**THE MYSTERY OF DARK ENERGY**
Scientists know little about dark energy. No one has ever seen it, and what is known about it is inferred from its effects on the things around it. Based on the degree to which it counters gravity as the universe expands, scientists estimate that about 68 to 74 percent of the entire universe is made of dark energy.

This becomes even more surprising when the estimated percentage of the universe that is made of dark matter is added. Dark matter is made up of unseen particles that have mass but emit no light. It was discovered when researchers determined that the visible portions of galaxies did not contain enough mass to prevent the galaxies from spinning apart. Like dark energy, dark matter cannot be studied directly, but researchers know how much there is based on its effect on galaxies. An estimated 22 to 27 percent of the universe is made of dark matter. When this is added to the 68 to 74 percent thought to be made of dark energy, it becomes clear that, for as large as stars and planets can be, all of them together with all other celestial bodies make up only a tiny fraction (just 4 to 5 percent) of the material of the universe.

Scientists have advanced several explanations for dark energy. Building on Einstein's observation that empty space is not necessarily a void, some scientists suggest that dark energy is a property of space. Humankind's ability to understand space is limited by available technology and by the fact that most observations are not made firsthand. Thus, it is likely there are many things about even empty space yet to be discovered. For example, Einstein theorized that space could have its own energy and that this energy could itself increase and grow. If this were the case, this higher energy production could explain why the growth of the universe was accelerating.

Quantum physics plays a role in another explanation of dark energy. Under quantum theories, the tiniest particles of matter and energy do not behave the same way that larger particles do. If the seemingly empty portions of space are in fact filled with tiny particles acting in unexpected ways, this could account for the effects of dark energy. However, when scientists attempted to determine how much energy could be produced this way, their answers seemed mathematically impossible.

Some researchers are investigating the possibility that the effects of dark energy mean that Einstein's theory of gravity is wrong. Scientists are studying how galaxies are pulled together to form clusters as a way of testing this hypothesis. To prove that Einstein was wrong would require the development of an alternate theory of gravity that explains the accelerated growth of the universe while still adequately accounting for the way the solar system and galaxy clusters hold together, as Einstein's long-standing theory does.

**THE DARK ENERGY SURVEY**
On August 31, 2013, more than one hundred American, Spanish, German, Brazilian, and British researchers began work on the Dark Energy Survey (DES). Using the most powerful camera available, the team launched a five-year effort to survey a wide expanse of the southern sky. They sought to measure just how fast the universe has been growing throughout each part of its 13.7-billion-year history.

The camera used for this project is called DECam, short for Dark Energy Camera. It is a 570-megapixel camera mounted on a 4-meter (13.1-foot) telescope at the Cerro Tololo Inter-American Observatory in the Chilean Andes Mountains. With lenses up to 1 meter (3.3 feet) across and a mass over four tons, the camera was the most powerful then known. Even so, researchers would not be able to see dark energy. Instead, they planned to use these devices to photograph the far-distant galaxies moving apart. This data can help researchers determine how fast the galaxies are moving and how the expansion rate has changed over time.

The work of the DES team and other ongoing research will help unravel the mysteries of cosmological acceleration and provide important glimpses into aspects of the universe that have remained hidden. By learning how fast the growth of the universe has been increasing and observing the effects dark energy is having on faraway galaxies, scientists hope to better understand the mysterious force that makes up more than two-thirds of the universe.

—*Janine Ungvarsky*

**Further Reading**

"Albert Einstein." *History.com.* A&E Network, n.d. Web. 5 June 2015.

"Cosmological Constant, Also Known as Lambda." *Astronomy Department.* Cornell U, n.d. Web. 5 June 2015.

"Dark Energy." *Hubble Discoveries.* Hubble Site–NASA, 2015. Web. 5 June 2015.

"Dark Energy, Dark Matter." *NASA Science: Astrophysics.* NASA, June 2015. Web. 5 June 2015.

"Edwin Powell Hubble." *Hubble Telescope.* NASA Goddard Space Flight Center, 6 Feb. 2012. Web. 5 June 2015.

Moskowitz, Clara. "Right Again, Einstein! New Study Supports Cosmological Constant." *Scientific American.* Scientific Amer., 17 Jan. 2013. Web. 5 June 2015.

Rousseaux, Charles. "Dark Energy Cam: Fermilab Expands Understanding of Expanding Universe." *Energy.gov.* US Dept. of Energy, 12 Mar. 2012. Web. 5 June 2015.

"What Is Causing Cosmic Acceleration?" *Dark Energy Survey.* Dark Energy Survey, n.d. Web. 5 June 2015.

"What Is a Cosmological Constant?" *Universe 101: Our Universe.* NASA, 2015. Web. 5 June 2015.

# DARK MATTER

**FIELDS OF STUDY**

Cosmology; Stellar Astronomy; Theoretical Astronomy

**SUMMARY**

Dark matter is an unseen, theoretical part of the universe. The visible bodies of normal matter that make up the universe are not massive enough to generate the amount of gravity it takes to hold the universe together. Thus, the nature of this dark matter that has mass but no light is a key to understanding the past and future of the universe.

**PRINCIPAL TERMS**

- **axion:** an as-yet undetected type of low-mass WIMP particle that could explain the absence of a measure of polarity for neutrons and that may have been made in large quantities during the big bang. Axions are thought to be one possible component of dark matter.
- **black hole:** a section of space with such strong gravitational forces that nothing can escape, not even light.
- **brown dwarf:** a celestial body that is between 15 and 75 times the mass of Jupiter and cannot maintain the hydrogen fusion needed to become a star.
- **gravitational lens:** a phenomenon that occurs when space is distorted by a large object in the foreground of the field of vision, such as a black hole, a sun, or a galaxy. The light from distant objects behind these larger objects appears brighter and larger as it is warped by the gravity of the larger object, much as a drop of water magnifies an image under it.
- **massive compact halo objects (MACHOs):** a category of objects considered likely possibilities to be dark matter. MACHOs are objects made of ordinary matter— electrons, neutrons, and protons—and include neutron stars, brown dwarfs, and black holes, all of which emit no light and are large enough to cause gravitational lensing.
- **neutralino:** an as-yet undetected type of very large, light WIMP particle that may be a kind of dark matter and that is part of supersymmetry, a physics theory that endeavors to combine all the known forces of physics.
- **neutrino:** a type of fundamental particle created in large amounts by nuclear reactions in stars, and one possible type of WIMP that could contribute to dark matter. Neutrinos pass right through Earth without causing any effect, making them difficult to detect.
- **weakly interacting massive particle (WIMP):** a category of objects considered likely possibilities to be dark matter. WIMPs are not made up of ordinary matter but do have mass, can be light or heavy, and can pass through ordinary matter without effect. Possible types include axions, neutrinos, and neutralinos.

## The Case for the Unseen

In 1933, the Swiss astronomer Fritz Zwicky (1898–1974), then working at the California Institute of Technology, was studying the speed with which the Coma galaxy cluster revolved. He observed that the galaxy cluster was not spinning as fast as would be expected based on the mass of its objects. Zwicky proposed a theory to explain this. He said that there was something else in the galaxy that could not be seen with the available instruments, and that it had enough mass to affect the speed at which the cluster spun. This substance was named "dark matter," because it had mass but gave off no detectable light.

In 1978, a team of researchers led by Vera Rubin (b. 1928) both confirmed and expanded on Zwicky's theory. They determined that not only did clusters of galaxies appear to have more substance to them than could be accounted for by the visible light generated by their objects, but single galaxies did as well. Researchers now had confirmation that dark matter existed. There was still much speculation, however, about what it was.

Some scientists have theorized that celestial bodies that do not emit light might comprise dark matter. This would include white or brown dwarfs, stars that never formed enough mass to produce the hydrogen fusion that fuels starlight. It would also include the remnants of exploded stars such as neutron stars or black holes, which form after a massive star experiences an explosion known as a supernova. One problem with this theory is that these objects are uncommon. It is therefore unlikely there are enough of them to account for all the dark matter scientists say exists—about 22 to 27 percent of the entire mass of the universe.

This improbability led scientists to speculate that dark matter may be a different kind of matter than normal matter, which is made up of electrons, neutrons, and protons. They coined the term massive compact halo objects (MACHOs) for the dark objects such as black holes and neutron stars that they could not see. They also began looking for particles they dubbed weakly interacting massive particles (WIMPs). The supersymmetry theory, which seeks to combine all the known theories of physics into one, predicts these non-light-emitting particles with mass exist. There could be several different types of WIMPs, each with unique characteristics. Axions, neutrinos, and neutralinos have all been proposed as dark matter objects. The existence of neutrinos has been confirmed, but all three are elusive. Scientists must turn to indirect methods of study.

## Search for the Missing Puzzle Piece

Someone building a puzzle and searching for a particular piece knows things about that piece even before he or she sees it. The shape of the gap where it fits and the properties of the pieces next to it provide clues to the unseen missing piece. Likewise, scientists searching for dark matter know how much must exist in a particular galaxy by looking at the speed with which the galaxy spins to determine how much total mass the galaxy has. Subtracting the mass of the visible objects provides the mass of the unseen dark matter. The challenge for scientists is to detect something that cannot be seen. The gravitational lens effect is one way scientists can learn about far-off galaxies. The distortion of light from a faraway object by something closer to the observer magnifies the distant object. This phenomenon, predicted by Albert Einstein (1879–1955), enables researchers using the Hubble Telescope to see far-off objects in greater detail than they otherwise would.

The Fermi Gamma-Ray Telescope, launched on June 11, 2008, provided another way for scientists to apply indirect means of locating the leading dark matter candidate, the neutralino. Although proof of their existence remains elusive, supersymmetry predicts that colliding neutralinos destroy each other, generating gamma rays. Applying supersymmetry, scientists say that these gamma rays behave in predictable ways. For instance, they should have a specific wavelength that differs from those emitted by other celestial objects. Neutralinos are thought to only release gamma rays, allowing scientists to ignore gamma rays accompanied by other forms of radiation. Their gamma rays should appear as large patches of radiation, not small streaks. They should also come continuously, not in short bursts. By searching for gamma rays that meet these criteria, scientists seek to pinpoint WIMPs and close in on locating dark matter.

## Why Dark Matter Matters

Identifying and understanding dark matter is important to understanding the past and future of the universe. Objects sitting on a spinning platform will slip and eventually fall off if something does not keep them in place. The same is true of spinning galaxies. Many have been spinning for millions of years, but with very little slipping because of gravity. Scientists have determined, however, that there is not enough mass in the known, visible objects in the universe to generate the amount of gravity that is holding everything together. There must be more, as-yet invisible matter out there to account for the forces of gravity.

Finding and identifying this missing dark matter will help scientists understand the full size and shape of the universe. Once dark matter is identified and better understood, scientists should gain insight into whether the universe's growth has plateaued and remains a constant size or if it is continuing to grow. If it is growing, identifying dark matter should help scientists determine whether the universe will reach a finite point and then begin collapsing.

Whatever dark matter turns out to be, it has been around since the universe was formed and is as much a part of it as any other matter, organic or inorganic. Understanding how dark matter affected the formation of the universe and the role it plays in maintaining the universe's gravitational forces would help scientists anticipate the future.

—*Janine Ungvarsky*

**FURTHER READING**

"Dark Energy, Dark Matter." *NASA Science: Astrophysics.* NASA, June 2015. Web. 5 June 2015.

"Hubble Zooms in on Magnified Galaxy." *Hubble Space Telescope.* NASA, 2 Feb. 2012. Web. 5 June 2015.

"Neutron Stars: Incomprehensible Density." *Science and Space.* Natl. Geographic Soc., 2015. Web. 5 June 2015.

"Possibilities of Dark Matter." *Educators' Corner: Imagine the Universe.* NASA, 2015. Web. 5 June 2015.

"What Is Dark Matter?" *NASA Education.* NASA, 2012. Web. 5 June 2015.

"What Is Fermi?" *Fermi Gamma-Ray Telescope.* NASA Goddard Space Flight Center, 2007. Web. 5 June 2015.

Woo, Marcus. "Dark Matter." *Fermi Gamma-Ray Space Telescope.* NASA, 23 Aug. 2007. Web. 5 June 2015.

# *DAWN* MISSION

**FIELDS OF STUDY**

Sub-planetary Astronomy; Space Technology

**SUMMARY**

The Dawn Mission is a space mission organized by NASA to study the celestial bodies Vesta and Ceres. The mission began in 2007 and was scheduled to conclude in 2016. The Dawn Mission included two historic firsts. The mission's spacecraft, *Dawn*, became the first spacecraft in history to orbit an object in the main asteroid belt between Mars and Jupiter. *Dawn* also became the first probe ever to orbit a dwarf planet.

**PRINCIPAL TERMS**

- **asteroid:** a rocky, irregularly shaped space body that is thought to be a remnant of the formation of the solar system.
- **Ceres:** a dwarf planet in the main asteroid belt, an oval-shaped ring of objects between the orbits of Mars and Jupiter. Ceres is being studied on the Dawn Mission.
- **ion propulsion:** a system found on certain spacecraft that gives xenon atoms a negative electrical charge, accelerates them, and then emits them to propel the vehicle through space.
- **NASA:** abbreviation for the National Aeronautics and Space Administration; the US government agency responsible for developing aviation and space technologies and conducting both manned and unmanned space missions to research Earth, its solar system and galaxy, and the universe.
- **Vesta:** the largest asteroid in the main asteroid belt, an oval-shaped ring of objects between the orbits of Mars and Jupiter. Vesta was studied on the Dawn Mission.

**MISSION HISTORY AND GOALS**

The Dawn Mission was initially proposed by the National Aeronautics and Space Administration's Discovery Program in 2001. NASA is a US agency focused on space exploration, space technology, and aeronautics. The agency was established by President Dwight D. Eisenhower in 1958.

The Dawn Mission is led by Discovery Program principal investigators (PIs). The PIs also lead the scientific team, consisting of twenty-one members from the United States, Germany, and Italy. The Dawn Science Center, which is located at the University of California, Los Angeles, organizes science and instrument operations. The center is directed by the PIs. The Jet Propulsion Laboratory (JPL) became responsible for project management. Scientific instruments for the mission were donated by Germany and Italy. Both countries were also involved in design, fabrication, and testing.

The mission was meant to study celestial bodies in the main asteroid belt, a field of asteroids, or large rocky objects, between the planets Mars and Jupiter. Specifically, its purpose was to study the asteroid Vesta and the dwarf planet Ceres. The mission would first take the mission's spacecraft, named *Dawn*, to Vesta. After remaining in orbit around Vesta for fourteen months, performing various tasks to study the asteroid, *Dawn* would then head for Ceres. The spacecraft would orbit Ceres and study the dwarf planet until June 2016.

An artist's rendition of the *Dawn* spacecraft orbiting the asteroid 4 Vesta. NASA/JPL-Caltech.

By studying Vesta and Ceres, NASA expected to gain an understanding of the evolution of the two bodies. NASA also expected to be able to compare their evolutionary paths. Furthermore, because Vesta and Ceres are believed to be remnants of the formation of the solar system, NASA hoped a study of the celestial bodies would provide breakthroughs in understanding the solar system's origins.

The Dawn Mission was cancelled on March 2, 2006, after a congressional budget hearing revealed cost and technical issues. By March 16, NASA made the decision to review the mission's termination after the JPL presented new evidence to support the mission. Initially, NASA refused to discuss the evidence or provide any commentary to the press. The Dawn Mission was later reinstated after the JPL resolved all technical issues and volunteered to restaff the project. The JPL also became responsible for all costs and resources, as well as the completion of the project.

## THE DAWN SPACECRAFT

The *Dawn* spacecraft is categorized as an orbiter. It uses many modern technologies and consists of both new components and spare components from previous space missions. One of the unique characteristics of *Dawn* is its ion propulsion system. The system accelerates xenon ions and produces very small amounts of thrust. Although this means that *Dawn* accelerates very slowly, the spacecraft can ultimately reach incredible speeds because its ion drive allows the engines to keep firing for extended amounts of time—even as long as several years. To complete the entire mission, the spacecraft requires about 2,100 days of thrust time. Three ion propulsion engines are therefore needed to provide enough thrust time.

*Dawn* is also equipped with backups for most of its systems in the event of a problem with a system. Certain software onboard can sense the problem and then try to replace the faulty system with a backup.

*Dawn* measures 2.36 meters (7 feet, 9 inches) in length when its solar array is in the retracted position. When the solar array is extended, the spacecraft measures 19.7 meters (65 feet) in length.

To complete the Dawn Mission, the *Dawn* spacecraft was equipped with instruments that would be able to characterize Vesta and Ceres and also provide gravity field data and data regarding bulk properties and the internal structure of the bodies. The instruments included a camera, a visible mapping spectrometer, an infrared mapping spectrometer, a gamma-ray spectrometer, and a neutron spectrometer.

### Vesta and Ceres

The targets of the Dawn Mission, Vesta and Ceres, are the two largest objects in the main asteroid belt between Mars and Jupiter. Vesta was discovered by German astronomer Heinrich Wilhelm Olbers (1758–1840) in 1807. It has a diameter of about 530 kilometers (330 miles). The asteroid rotates every five hours and twenty minutes. Vesta was the fourth asteroid to have its orbit calculated, so its official name is "4 Vesta."

Ceres was discovered by Italian astronomer Giuseppe Piazzi (1746–1826) in 1801. The dwarf planet has a diameter of about 950 kilometers (590 miles). Ceres rotates every nine hours, four minutes, and thirty seconds. It was the first asteroid or dwarf planet ever discovered and its orbit computed; therefore, its official name is 1 Ceres. Ceres has a long history of changing categorization. It was first considered a planet. As more objects in the asteroid belt were discovered, Ceres became regarded as an asteroid before finally being categorized as a dwarf planet.

### Mission Details

The $473 million Dawn Mission began when NASA launched *Dawn* from the Cape Canaveral Air Force Station in Florida on September 27, 2007. On February 17, 2009, *Dawn* passed Mars and continued on to its first target, Vesta. The spacecraft entered orbit around Vesta on July 15, 2011. *Dawn* continued to orbit it for the next fourteen months. During this time, *Dawn* took pictures of Vesta, imaged its surface, established its interior structure, performed various measurements of its composition and temperature, and searched for possible moons. The spacecraft was also able to drop down to an altitude of 210 kilometers (130 miles), which was made possible by its ion propulsion system. On September 4, 2012, *Dawn* left Vesta and set out for Ceres. On March 13, 2015, *Dawn* entered orbit around Ceres and began investigations.

### Significance of the Mission

Although the Dawn Mission was the ninth project in NASA's Discovery Program, it was the first mission to orbit a main belt asteroid and the first to orbit a dwarf planet. The Dawn Mission was also the first mission to orbit two different extraterrestrial bodies that were neither the earth nor the moon.

The Dawn Mission has made several important findings regarding Vesta. The mission found that Vesta has a dense core that is enveloped by a mantle and crust. Vesta also resembles a terrestrial planet more than it resembles an asteroid. The mission further revealed that Vesta has produced more meteorites on Earth than the moon or Mars has. Another finding was that Vesta has a large crater with a mountain that is twice as tall as Earth's Mount Everest. Lastly, the Dawn Mission found that Vesta has a network of about ninety chasms, some with dimensions similar to Earth's Grand Canyon.

—*Michael Mazzei*

### Further Reading

"About the Mission." *Jet Propulsion Laboratory, California Institute of Technology*. NASA, n.d. Web. 7 May 2015.

"Asteroids and Comets." *National Geographic*. Natl. Geographic Soc., n.d. Web. 7 May 2015.

"Dawn Mission Overview." *NASA*. NASA, 30 Apr. 2015. Web. 7 May 2015.

"Dawn Spacecraft and Instruments." *NASA*. NASA, 30 Apr. 2015. Web. 7 May 2015.

"Dawn—Ceres and Vesta." *NASA*. NASA, 30 Apr. 2015. Web. 7 May 2015.

Geveden, Rex D. "DAWN Cancellation Reclama." Letter to director, Jet Propulsion Laboratory. 27 Mar. 2006. PDF file.

"NASA Reviewing Canceled Mission." *CNN*. Cable News Network, 16 Mar. 2006. Web. 7 May 2015.

"NASA Spacecraft Becomes First to Orbit a Dwarf Planet." *NASA News*. NASA, 6 Mar. 2015. Web. 7 May 2015.

Rayman, Marc D., et al. "Dawn: A Mission in Development for Exploration of Main Belt Asteroids Vesta and Ceres." *Acta Astronautica* 58 (2006): 605–16. Print.

"What Does NASA Do?" *NASA*. NASA, Nov. 2014. Web. 7 May 2015.

# DEEP-SPACE NAVIGATION

### FIELDS OF STUDY

Space Technology; Planetary Astronomy; Extragalactic Astronomy

### SUMMARY

Deep-space navigation is an advanced space technology that enables scientists to target and to explore faraway solar-system bodies and specific deep-space sites. Scientists have developed many kinds of spacecraft to accomplish deep-space navigation. Deep-space navigation technology is often unmanned and controlled from a ground unit on Earth. Deep-space missions have allowed scientists to analyze the atmosphere of distant planets and to retrieve land samples from their surfaces.

### PRINCIPAL TERMS

- **astronomical unit:** a unit of measure used to estimate long distances in space; equal to equal to about 149.6 million kilometers (about 93 million miles).
- **ground-based navigation:** a grounded communication unit that controls the navigation of a spacecraft.
- **inertia:** the principle that states that an object remains at rest or in motion unless an outside force acts upon it.
- **optical navigation system:** a navigation system that uses optical physics to measure the degree of the relative motion (both speed and magnitude) between a navigation device and the surface being navigated. ONS processes the reflection and scattering of light to aid in the navigation of surfaces.
- **trajectory:** the curved path that an object, either natural or artificial, takes through space.

### THE SCIENCE OF DEEP-SPACE NAVIGATION

Deep-space navigation is an area of astronomical study that investigates distant space and its components. Space navigation systems exist all over the world. They aid ground-based navigation with exploring and mapping the outermost reaches of space. Directing a spacecraft to far-flung locations in the solar system involves collaboration between scientists and engineers. They use advanced radio systems, large antennas, computers, and accurate timing equipment. Space-navigation software uses radio waves to communicate with navigation technology in space. Instruments used in deep-space navigation include satellites, probes, rovers, flybys, manned orbiters, landers, and sample-collecting spacecraft. Navigation techniques and equipment, such as range projection, Doppler radar, networks of large dish antennas, and optical navigation systems, help a spacecraft maneuver through space with the utmost accuracy. These techniques also work to ensure a deep-space mission runs smoothly. Exact timing is important to space navigation. Calculating the precise time it takes for a spacecraft to receive a radio signal ensures the vessel stays on course. If a spacecraft veers off course, scientists can transmit signals to make the vehicle fix its trajectory. Scientists must account for all possible obstacles when organizing the navigation of a deep-space mission.

Deep-space navigation requires a number of mathematical calculations to figure out potential conditions that could affect the course of a spacecraft as it travels to a distance as far as several astronomical units away in outer space. The navigator attempts to match the predicted data with the observed data so that the differences between the two are essentially nonexistent. Mathematical models account for interferences, such as charged particles in the ionized layers of the upper atmosphere, the force of gravity acting upon a vessel, and solar radiation pressure. Even the smallest of forces can greatly affect a spacecraft's mapped trajectory. The spacecraft's inertia is another important factor. Mathematical model accuracy is crucial and must account for every possibility or risk navigational failure.

Scientists use laboratories and research centers to simulate deep-space missions before the actual launch of a spacecraft. Using software and instruments that mimic the conditions of deep space, mission control attempts to predict the various effects celestial forces will have on a vehicle. Mission designers then analyze this data and base their navigation on the findings. Sometimes an outside force acts to guide a spacecraft along its way. Deep-space navigators often rely on the assistance of gravity to propel their vehicles in the right direction. The National Aeronautics and Space Administration (NASA) and the Jet Propulsion Laboratory have developed an autonomous space-probe navigation software called AutoNav. The software enables a spacecraft to navigate itself through space and transmit data such as images and trajectory information back to Earth. This eliminates the need for ground-based navigation.

## A NEW HORIZON

During and after World War II, the fields of science and engineering experienced a surge of innovation. Compelled to develop greater communication and transport technologies, research facilities across the globe began delving into the areas of space, microwaves, and lasers. During the war, scientists established the Jet Propulsion Laboratory (JPL) to build and test missiles. The research involved studying radar and rocket science. JPL quickly became the breeding ground for future space-navigation work.

The decades-long Cold War between the United States and Soviet Russia that followed World War II further propelled the development of space navigation. The Soviets' 1957 launch of Sputnik 1, the first artificial satellite sent into space, triggered the space race. American laboratories were soon developing their own Earth-orbiting satellites to aid in the range and tracking of missiles. The first successful US satellite, Explorer 1, launched into orbit in January 1958. NASA was established in October of that year. It merged all in-development space-exploration programs of the US Army, Navy, and Air Force. NASA took control of JPL two months later. Soon a number of space-exploration initiatives were underway as the United States entered the space race with Russia. The two countries competed to develop the most advanced space-exploration technology as quickly as possible.

Also established in 1958, the Advanced Research Projects Agency (ARPA; now known as Defense Advanced Research Projects Agency, or DARPA) was responsible for all US space projects. ARPA was charged with advancing space-flight technology and developing the technology that would allow humans to observe the moon closely. After experiencing several design failures, ARPA eventually navigated a successful lunar probe into space in 1959. ARPA's lunar probes carried instrumentation that measured the magnetic fields of the earth and moon as well as charged particles in near-Earth space. ARPA's probes were not the first to complete a close flyby of the moon, however. Soviet probes had already achieved this feat a few weeks before. Soviet crafts also took the first photographs of the far side of the moon. This knowledge, coupled with the desire to land on the moon before the Soviets, encouraged NASA to focus on the development of superior space-flight and navigation technology. NASA formed several space-navigation organizations over the following decades. These included the Space Network, the Near Earth Network, and the Deep Space Network (DSN). In 2006, these organizations consolidated under the umbrella organization known as the Space Communications and Navigation Program (SCaN).

## THE IMPACT OF DEEP SPACE NAVIGATION

NASA's space shuttle program executed successful manned missions to the moon. This gave humans their first physical experience with an extraterrestrial surface. Deep-space navigation developed further as scientists set their eyes on the farthest reaches of space. In 1977, NASA launched the twin spacecraft *Voyager 1* and *2*. These were space probes designed to explore the planets Jupiter, Saturn, Uranus, and Neptune. The vessels returned helpful data about the planets before moving even deeper into space. In August 2012, *Voyager 1* crossed into interstellar space, the area outside the Milky Way galaxy's star system. Both spacecraft continued to send scientific data about their surroundings through the DSN.

Space navigation also made possible the exploration of the planet Mars. Space rovers with advanced autonomous navigational instruments landed on Mars in August 2012. The most famous of the rovers, *Mars Curiosity Rover*, collected and analyzed surface sediment. It sent this information back to Earth using the signals from the DSN antennas. Navigators used Doppler radar data and ranging methods to track the vehicle's progress accurately. NASA's Mars Exploration Program included other long-term robotic navigation efforts to explore the surface of Mars further.

—*Cait Caffrey*

**FURTHER READING**

Bhaskaran, Shyam. "Autonomous Navigation for Deep Space Missions." *Twelfth International Conference on Space Operations, June 11–15, 2012.* Pasadena: Jet Propulsion Laboratory, 2012. PDF file.

"Deep Space Navigation." *Jet Propulsion Laboratory at the California Institute of Technology.* NASA, n.d. Web. 7 May 2015.

"Deep Space Navigation." *Princeton Satellite Systems.* Princeton Satellite Systems, n.d. Web. 7 May 2015.

Garg, Rahul. "Optical Navigation Systems: The Foundation of Modern Pointing Devices." *EDN Network.* UBM Canon, 12 Aug. 2013. Web. 7 May 2015.

Hall, R. Cargill. *Lunar Impact: The NASA History of Project Ranger.* Chelmsford: Courier, 2013. 7–12. Print.

Jones, Jeremy. "How Do Space Probes Navigate Large Distances with Such Accuracy and How Do the Mission Controllers Know When They've Reached Their Target?" *Scientific American.* Scientific American, 27 Mar. 2006. Web. 7 May 2015.

"Mars Curiosity Navigation." *Jet Propulsion Laboratory at the California Institute of Technology.* NASA, n.d. Web. 7 May 2015.

"Navigating in Deep Space." *Navigating in Space.* Smithsonian Inst., n.d. Web. 7 May 2015.

# DYSNOMIA: (136199) ERIS I DYSNOMIA

**FIELDS OF STUDY**

Astronomy; Observational Astronomy; Space Technology

**SUMMARY**

(136199) Eris I Dysnomia is the only moon of Eris, a dwarf planet on the edge of our solar system. Astronomers confirmed Eris's existence in 2005 and at first considered Eris the tenth planet because it was larger than Pluto. The discovery of Eris sparked a debate as to what constitutes a planet. This controversy led to Pluto being demoted from planet to dwarf planet. Later discoveries revealed that Eris and Pluto are nearly identical in size and composition.

**PRINCIPAL TERMS**

- **adaptive optics:** a system installed on astronomical telescopes that allows astronomers to see celestial bodies that are very bright.
- **dwarf planet:** a celestial body that orbits the sun, has a nearly round shape, and has not cleared its orbital path of other celestial bodies.
- **Eris:** a dwarf planet on the edge of our solar system.

**A FRIGID WORLD**

According to the International Astronomical Union (IAU), a dwarf planet is a celestial body that (1) orbits the sun, (2) has enough mass to be round in shape, and (3) has not cleared the neighborhood around its orbit. A non-dwarf planet has a cleared orbit because its gravity is strong enough to either attract or push away smaller bodies in its path. Dwarf planets lack such gravity, so other objects may cross into their path. Most dwarf planets are in the Kuiper Belt, a region of the solar system beyond Neptune's orbit.

Eris is a dwarf planet on the outer edge of our solar system, extremely far from the sun. It has a maximum distance from the sun of 97.56 astronomical units (AU), or 97.56 times the distance between Earth and the sun. (136199) Eris I Dysnomia, or simply Dysnomia, is its only moon. Dysnomia is estimated to be between 100 kilometers (62 miles) and 250 kilometers (155 miles) in diameter. Its orbit around Eris takes approximately sixteen days.

**THE GREAT PLANET DEBATE**

Eris was first sighted in 2003 by a team of researchers led by Michael E. Brown, a professor of planetary science at the California Institute of Technology (CalTech). The team conducted additional observations and confirmed and announced the discovery in

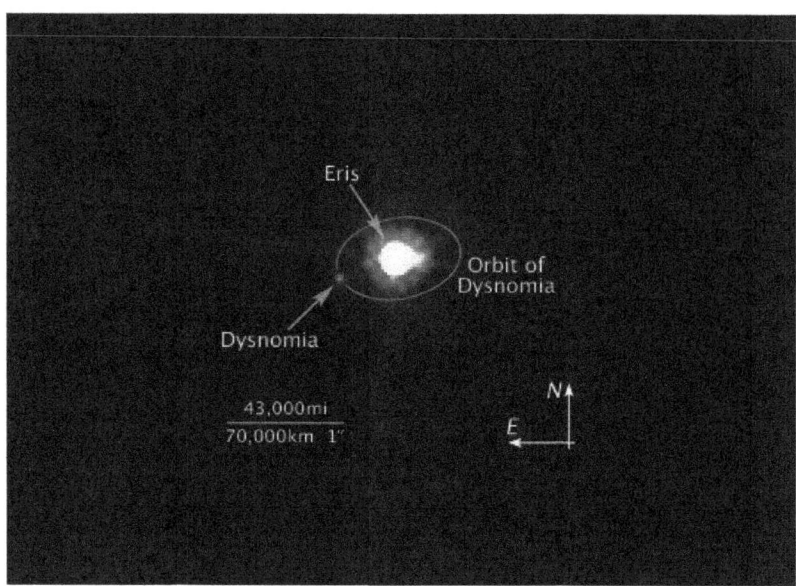

Eris I Dysnomia, moon of dwarf planet Eris. M. Brown, public domain via Wikimedia Commons.

2005. That year, Brown used a laser-guided adaptive optics system to discover Eris's moon.

By observing the interactions between Eris and Dysnomia, astronomers were able to calculate the likely mass of Eris. They determined that Eris's mass is greater than that of Pluto. This discovery initiated a debate about whether Eris should be considered the tenth planet of the solar system or whether Pluto should be considered a dwarf planet instead. In the end, Pluto was reclassified as a dwarf planet.

—*Adrienne A. Kennedy*

**FURTHER READING**

Brown, Michael E., Al Conrad, and Linda Copman. "Planetary Astronomy: The New Solar System." *Cosmic Matters.* W. M. Keck Observatory, 7 Sept. 2007. Web. 9 Mar. 2015.

Brown, Michael E., and Emily L. Schaller. "The Mass of Dwarf Planet Eris." *Science* 316.5831 (2007): 1585. Print.

"Eris: Overview." *Solar System Exploration.* Natl. Aeronautics and Space Administration, 15 Jan. 2015. Web. 9 Mar. 2015.

Peterson, Carolyn Collins. *Astronomy 101.* Avon: Adams, 2013. Print.

"Pluto and the Developing Landscape of Our Solar System." *International Astronomical Union.* IAU, n.d. Web. 9 Mar. 2015.

Redd, Nola Taylor. "Eris: The Dwarf Planet That Is Pluto's Twin." *Space.com.* Purch, 27 Jan. 2015. Web. 9 Mar. 2015.

# EARTH

## FIELDS OF STUDY

Planetary Science; Planetary Geology; Orbital Mechanics

## SUMMARY

The planet Earth is the third terrestrial planet, and the only planet in the solar system to have water in solid, liquid and vapor phases in dynamic equilibrium. It is a dynamic planet, with a solid iron or nickel-iron core, and a nitrogen-oxygen atmosphere extending to 150 kilometers above the surface of the planet. Earth is the only planet known to harbor life. There is one natural satellite, and hundreds of artificial satellites in orbit due to human activities. As a human activity, Earth is also the starting point of the science of astronomy.

## PRINCIPLE TERMS

- **celestial equator:** the imaginary projection of Earth's equator onto the night sky.
- **common center of mass:** balance point of a system of orbiting celestial bodies; also known as the barycenter.
- **declination:** a space object's angular distance north or south of the celestial equator, expressed in degrees of arc.
- **orbital mechanics:** the study of the movements of artificial satellites and spacecraft; also called flight mechanics or astrodynamics.
- **orbital robotics:** field that focuses on using robotic technology such as satellites, space probes, and service spacecraft and equipment in space.
- **satellite:** technically, any small object, either natural or artificial, that orbits around a larger object. However, the term is most often applied to man-made objects that are placed in orbit around Earth or another celestial body for research, monitoring, or communication purposes.
- **spin-orbit coupling:** the rotation rate of a celestial object in relation to the rate at which it orbits another body; also known as orbital resonance.

## EARTH AND ITS PLACE

The planet Earth is the third of the four terrestrial planets, along with Mercury, Venus and Mars. It orbits the Sun at an average distance of $149.6 \times 10^6$ kilometers (km) in an elliptical orbit having an eccentricity of 0.0167. The orbit of Venus, in comparison, has an eccentricity of just 0.0068, and a perfect circle has an eccentricity of 0.

Earth has a mass of $5.97 \times 10^{24}$ kilograms (kg), which is designated as 1 astronomical unit (AU) of distance, and a diameter of 12,756 kilometers at the equator. The overall density of Earth is 5.52 grams per cubic centimeter ($g.cm^{-3}$).

The composition of the planet is highly complex, and the planet contains all of the known natural chemical elements in various quantities. The atmosphere consists primarily of nitrogen and oxygen in a 77-to-21 ratio, with the remaining 2 percent consisting of small quantities of other gases such as ozone, carbon dioxide, water, argon, and many other compounds. Clouds composed of water are normally seen in the atmosphere. The relative quantities of these gases are extremely small, yet some of them are vitally important for the status of the planet as the only known haven of life in the solar system.

The planet rotates on its axis once in 23.934 hours, and completes one orbit around the Sun in 365.26 days. The axis of rotation of the planet is tilted at 23.44° to the plane of Earth's orbit about the Sun. Earth is an active planet and has a magnetic field that aligns approximately to the axis of rotation and that is known to shift location over time.

It is the only known planet to have liquid water on its surface, in a dynamic equilibrium with both water ice and water vapor. This gives the planet the appearance of a "blue marble" when viewed from space. A single moon, itself the size of a small planet, orbits

Earth at an average distance of 384,400 km. The Moon has no atmosphere and its surface is marked with a great many impact craters, far more than can be identified on Earth's surface.

## QUALITIES

Planet Earth is the most diverse planet in the solar system, although the other planets also have their own unique features of interest. Earth is the only planet in which solid, liquid and gaseous water exist in equilibrium in an open atmosphere. Other planets have solid and liquid water, but these are always either encased beneath a solid crust or sequestered underground, and never as all three phases in a dynamic equilibrium system. The proximity of Earth to the Sun permits an open atmosphere constrained by the gravitational force of the planet. Earth's atmosphere consists of 77 percent nitrogen, 21 percent oxygen, and small amounts of carbon dioxide, argon, water vapor, ozone and several other gases. Of these, carbon dioxide accounts for about 0.035 percent of the atmospheric content, and ozone accounts for even less. Yet without the presence of these two gases in the atmosphere, Earth would be a lifeless ball of ice floating in space.

The carbon dioxide and water vapor are known as "greenhouse gases." Sunlight, arriving from the Sun, enters the atmosphere and is absorbed by the planet's surface materials. The energy is subsequently re-emitted from those materials as infrared energy directed towards space. Molecules of carbon dioxide, water and trace quantities of methane absorb the infrared energy that they encounter and are thereby raised to an energetically higher state. The energy is given off again when those molecules revert back to their normal, lower energy state. Half of those emissions are directed back toward the surface, and so are able to be reabsorbed by other molecules. The process repeats an untold number of times, effectively trapping enough of the heat energy within the atmosphere to maintain elevated temperatures around the planet.

Calculations of the amount of heat transferred in this way have shown that the average temperature of the planet would be at least 35°C (95°F) colder than they are presently, and Earth would be frozen in a permanent ice age. This same effect on Venus, where the atmosphere is 95 percent carbon dioxide, has produced a consistent surface temperature of at least 460°C, which is well above the melting point of some metals and minerals. Earth also benefits from the presence of ozone ($O_3$) in the atmosphere, where it forms a barrier against destructive ultraviolet (UV) radiation. Without the "ozone layer" UV radiation would have wreaked havoc with the chemical processes that produce and maintain life on Earth. Even with the ozone layer, however, some UV radiation reaches Earth's surface and is responsible for causing skin cancer and other genetic abnormalities.

The interior of Earth consists of a central core of solid iron or iron and nickel. This massive amount of metal, having a density of approximately 7.5 g.cm-3, accounts for the relatively high overall density of the planet. It also produces Earth's magnetic field by means of the "dynamo effect" as it turns. Surrounding this solid core is a broad layer of molten metal, and over that is the "mantle," a layer of molten material called magma. Magma consists of heavy silicate materials, in a very viscous liquid state.

Surrounding the mantle is Earth's crust, which varies in thickness, ranging up to 60 kilometers. The crust is cracked into several different tectonic plates that essentially float upon the magma below and fit together like a jigsaw puzzle. The dynamic motion of the mantle drives the movement of the tectonic plates, causing them to move relative to each other, in turn spawning the formation of volcanoes and earthquakes. In some places, where two plates are moving away from each other, magma is able to reach the surface of the planet, where it is known as lava. In other places, the motion of one plate forces it to go beneath another plate, and earthquakes or volcanoes in such areas can be particularly hard.

The surface of Earth is characterized by the presence of large land masses and oceans of liquid water atop the crust of the planet, all of which is surrounded by a gaseous atmosphere extending up to 150 kilomters above the surface.

Earth's magnetic field encompasses the region of space all around the planet and provides a protective layer against the influx of charge particles that emanate from the Sun and travel through space. The atmosphere provides significant protection for the planet against meteors that would impact the surface. The vast majority of such objects vaporize due to the heat of friction generated when they enter the atmosphere, and therefore, they never reach the surface. There have been numerous impacts from meteorites

(*meteorites* is the term used for meteors that actually strike the surface of the planet) in Earth's history, as evidenced by the craters and crater remnants that can still be seen.

The Moon, in orbit about Earth, also acts to shield the planet from many impacts, as evidenced by the high number of craters that can be seen on its surface. The Moon is Earth's only natural satellite, and due to spin-orbit coupling, it always has the same hemisphere facing the planet as the planet and its natural satellite orbit each other about their common center of mass.

### Earth in History and Mythology

In mythology, the personification of Earth is known as Gaea to the Greeks, and as Tellus to the Romans. Gaea is described as the first being that sprang from Chaos, who gave birth to Uranus (Heaven) and Pontus (Sea). She is also said to have mated with her son, Uranus, and to have given birth to the Titans, the most powerful of whom was Cronus (Saturn). Cronus defeated Uranus and became the ruling Titan.

Cronus's surviving children—Zeus, Poseidon and Hades Jupiter, Neptune and Pluto—defeated Saturn and became the principal gods, known as the Olympians, of the Roman pantheon (their Roman counterparts were known as Jupiter, Neptune and Pluto respectively). The Greek and Roman pantheons now provide the conventional names for astronomical objects.

The history of Earth comprises vast body of knowledge that is stored within the nature of its rocky structure, able to tell us the whole story from the formation of the planet some 4.5 billion years ago to the present day. It is the only known abode of "life" (at least as humans define it) in the solar system, although the potential for life to exist on some of the moons of other planets is believed to be real. As the home of the human race, Earth is also the birthplace of astronomy, a science that developed over time and in tandem with the development of the planet itself in reference to Earth itself.

Astronomers divide the sky is divided by astronomers along an imaginary plane called the celestial equator, which extends outward from Earth's own equator. The positions of stars and other objects are described by their declination relative to the celestial equator. The study of their motions, as well as those of satellites—both natural and artificial—are described by the mathematics of orbital mechanics. Even though men have made and continues to conduct first-hand exploration of outer space, far more astronomical exploration is carried out with the technology of orbital robotics, especially as we seek to explore space objects ever more distant from Earth's orbit.

The Blue Marble by NASA/JPL-Caltech/MSSS

### Measuring Earth

The development of geometry was an essential factor in understanding

Earth and its place within the solar system. The very size of the planet relative to humans makes it easy to understand how men might see the planet as flat, since the great disparity in size masks the planet's nearly spherical shape. The word *geometry* comes from two Greek words that translate literally as "Earth" and " measurement." The first accurate estimate of Earth's circumference obtained using basic geometric principles was made by the Eratosthenes, around 200 B.C.E. He knew that at midday on a specific day of the year in the town of Syene in Egypt, the Sun shone directly down a certain well when it reached its zenith, Eratosthenes measured the length of a shadow cast by the Sun at Alexandria at that same time, and used the angular distance of the shadow and the distance from Alexandria to Syene to calculate the circumference of Earth. His estimate, though too large by perhaps as much as 25 percent, was nevertheless essentially correct, according to the method he used. The same method used today, with standardized distances, produced a measurement that is much closer reality.

Given that the angular distance of the Sun at Alexandria is 7° and the distance from Alexandria to Syene is 800 kilometers, one can calculate the diameter of Earth using Eratosthenes' method as follows.

Assume that the triangle formed by the Sun, Syene and Alexandria is a right triangle, with the opposite side having a length of 800 km, and the opposite angle of 7°. The solution is then to relate the distance spanned by 7° to the distance that would be spanned by a full circle of 360°. Let the total circumference = $x$.

$x$ / 800 km = 360 / 7
$x$ / 800 km = 51.43
$x$ = (800 km) (51.43)
$x$ = 41,144 km

—*Richard Renneboog, MSc*

**FURTHER READING**

Bjornerud, Marcia. *Reading the Rocks: The Autobiography of the Earth.* New York, NY: Basic Books, 2005. Print.

Cobb, Allan B. *Earth Chemistry.* New York, NY: Chelsea House, 2009. Print.

Cockell, Charles, ed. *An Introduction to the Earth-Life System.* New York, NY: Cambridge University Press, 2007. Print.

Martin, Ronald. *Earth's Evolving Systems:. The History of Planet Earth.* Burlington, MA: Jones & Bartlett, 2013. Print.

Stott, Carole, Robert Dinwiddie, David Hughes, and Giles Sparrow. *Space from Earth to the Edge of the Universe.* New York, NY: DK Publishing, 2010. Print.

Ward, Peter D., and Donald Brownlee. *Rare Earth: Why Complex Life is Uncommon in the Universe.* New York, NY: Springer-Verlag, 2000. Print.

# EARTH-IMAGING SATELLITES

## FIELDS OF STUDY

Aerospace Engineering; Orbital Mechanics; Remote Sensing

## SUMMARY

An Earth-imaging satellite orbits Earth while gathering images and information. Scientists use this information to learn about Earth's surface, oceans, and atmosphere as well as to monitor weather patterns and natural disasters. Earth-imaging satellites provide vital information to geologists, meteorologists, astronomers, and other scientists.

## PRINCIPAL TERMS

- **geospatial technology:** the equipment used to see, measure, and analyze Earth and its geography, weather, and other features. Examples include global positioning systems (GPS), geographical information systems (GIS), and remote sensing (RS).
- **radiometric resolution:** the ability of an imaging system or sensor to capture small increments of energy in an image.
- **spatial resolution:** the number of pixels, or individual units, that make up a digital image. A greater number of pixels provides a higher spatial resolution and therefore greater clarity and detail.

- **spectral resolution:** the ability of an astronomical device called a spectrograph to differentiate various wavelengths of light.
- **temporal resolution:** the amount of time that it takes a remote sensing device to make a return trip to an exact location over Earth and acquire information. A shorter delay between revisits means the temporal resolution is high.
- **US Explorer 6:** the sixth in a series of Earth-imaging satellites launched in the late 1950s and the first to send back a picture of Earth from orbit in 1959.

## EARTH OBSERVATION AND SATELLITES

Since the late 1950s, artificial satellites have been launched into orbit around Earth. The first such satellite was the Soviet-built *Sputnik 1* in 1957. The United States National Aeronautics and Space Administration (NASA) launched the third satellite, *Explorer 1*, in 1958. In 1959, NASA launched *US Explorer 6*, a small rounded satellite. Scientists used it to study radiation, cosmic rays, and Earth's magnetism. *Explorer 6* sent back the first picture of Earth from orbit.

The success of these early Earth-orbiting satellites paved the way for the thousands of artificial satellites that followed. Twenty-first-century satellites carry an array of geospatial technology that provides vast amounts of data about what is occurring on and around Earth.

## SATELLITE REMOTE SENSING IMAGERY

Satellites in Earth's orbit gather information on active and passive remote sensors. Active sensors send out electromagnetic radiation. They also measure the strength of the signal returned to the sensor. Passive sensors detect and measure the amount of electromagnetic radiation reflected or emitted by Earth. The sensitivity of these sensors determines their radiometric resolution. This in turn affects the clarity and detail of the resulting image.

Spatial resolution is another factor in the clarity of satellite images. Images captured with greater numbers of pixels produce images with higher spatial resolution. These images are sharper, clearer, and more detailed. The ability of satellite equipment to capture and record the wavelength and intensity of light is known as spectral resolution. This can also affect the quality of the images revealed at ground stations on Earth when the satellites transmit data for processing.

In addition, the temporal resolution of a satellite sensor array helps determine the usefulness of the information and images received from the satellite. Temporal resolution describes the time it takes for a satellite to return to an exact location to record more data. For example, scientists monitoring a natural disaster, such as a flood, can tell much more about the situation on the ground if the satellite can revisit the area quickly and provide updated information. Temporal resolution is measured in days or hours.

Satellite images can be highly detailed and enable scientists to examine relatively small areas or objects. However, the image from a satellite does not look like a photograph taken with a camera. The image is returned in different color values based on

Map of the United States compiled from multiple image strips collected by the *Landsat* satellites. Courtesy of the U.S. Geological Survey.

123

the type of data being processed. For example, an image of an ocean may show colder areas in blue and warmer areas in red. This image is helpful to scientists but does not resemble a photograph of the ocean.

Uses for Satellite Data and Geospatial Technology

The information gathered by Earth-imaging satellites can be used in many different ways. These include cartography and the creation of specialized maps of specific geographic features such as mountains. Satellite data are also essential to monitoring and planning water resources, soil surveys and crop assessments, and natural-disaster monitoring and response coordination.

In addition to their research uses, the geospatial equipment on Earth-imaging satellites is at the heart of things people use daily. All global navigation systems, such as the global positioning system (GPS), rely on data from a network of satellites. Every day both scientists and average citizens use information shared between places on Earth that are linked by satellites orbiting high above the planet.

—*Janine Ungvarsky*

### FURTHER READING

"About Earth Observation and Satellite Imagery." *Geoscience Australia*. Australian Govt., n.d. Web. 24 Feb. 2015.

"Explorer 6." *NASA Space Science Data Center*. NASA, 26 Aug. 2014.Web. 24 Feb. 2015.

Parry-Hill, Matthew J., et al. "Spatial Resolution in Digital Images." *Molecular Expressions Optical Microscopy Primer*. Florida State U, 26 Mar. 2014. Web. 24 Feb. 2015.

"Radiometric Resolution." *Natural Resources Canada*. Govt. of Canada, 29 May 2013. Web. 24 Feb. 2015.

Smith, Nick. "Geospatial Technology: Mapping the Future." *E & T*. Inst. of Engineering and Technology, 20 Aug. 2012. Web. 24 Feb. 2015.

"Temporal Resolution." *Natural Resources Canada*. Govt. of Canada, 22 Apr. 2014. Web. 24 Feb. 2015.

"What Is a Satellite?" *NASA Education*. NASA, 12 Feb. 2014. Web. 24 Feb. 2015.

# 136199 ERIS

### FIELDS OF STUDY

Sub-planet Astronomy; Cosmology; Observational Astronomy

### SUMMARY

136199 Eris is a Kuiper Belt object (KBO) that orbits far beyond Neptune. It was found in 2003 and officially classified as a dwarf planet in 2006. Its similarity to Pluto forced the scientific community to determine the characteristics that qualify a space object as a planet. The ensuing debate created a new category of space objects called dwarf planets, and Pluto was demoted from its planetary status.

### PRINCIPAL TERMS

- **dwarf planet:** a celestial body that has enough mass to attain a nearly round shape, lacks the gravity to keep its neighborhood free of other space objects, orbits a star, and does not act as a satellite to another space object.
- **Kuiper Belt:** a disk-shaped region of space beyond the orbit of Neptune, about thirty to fifty-five astronomical units (AU) from the sun. The Kuiper Belt contains a collection of more than a thousand icy space objects that may be untouched material from the formation of the solar system.

### AN IMPORTANT DISCOVERY

A team of researchers at the Palomar Observatory, a research center operated by the California Institute of Technology (Caltech) outside San Diego, California, imaged 136199 Eris in 2003. Astronomer Mike Brown and his team conducted nightly surveys of the Kuiper Belt, a disk-shaped region of space beyond the orbit of Neptune. This area of space is filled with at least one thousand icy space objects of varying sizes. The team was looking for large space objects that had not yet been identified due to their distance from Earth.

The team first photographed Eris, then called 2003 UB313, on October 21, 2003. They confirmed

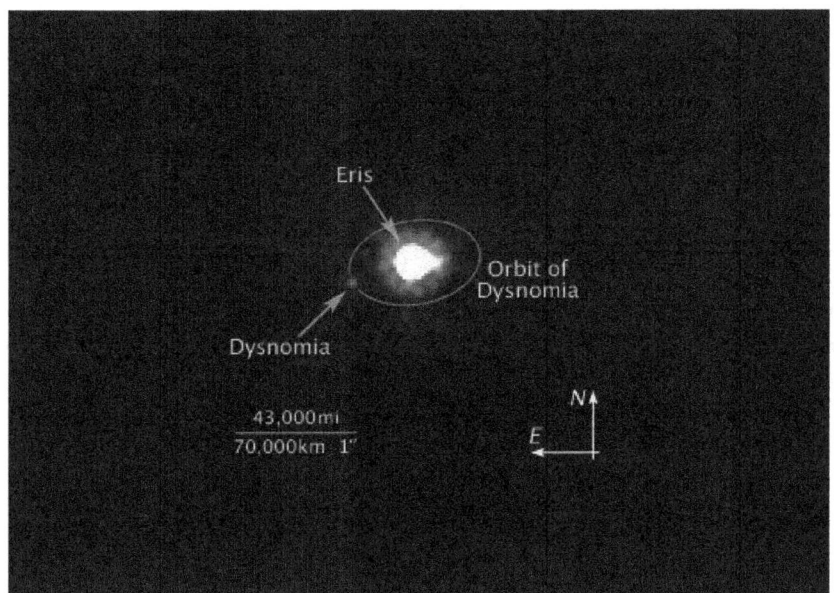

136199 Eris, dwarf planet, with its moon Dysnomia orbiting. M. Brown, public domain via Wikimedia Commons.

the discovery in January 2005 and submitted the object to the International Astronomical Union (IAU) for consideration as the tenth planet of the solar system.

The discovery was to have a serious impact on astronomical science. It forced the scientific community to consider seriously the question, "What is a planet?" Astronomers had long known that Pluto, discovered in 1930, differed greatly from the other eight planets of the solar system. At only about 2,390 kilometers (1,485 miles) in diameter, Pluto is smaller than Earth's moon. Pluto's location beyond the orbit of Neptune places it billions of miles from the sun, making it a cold, icy place. Additionally, the shape and angle of its orbital path vary from the elliptical orbital path traveled by the other eight planets. The discovery of 2003 UB313, which was then estimated to be about 3,000 kilometers (1,864 miles) in diameter, forced the question. If Pluto were a planet, it seemed logical that 2003 UB313 should qualify, too.

## A Dwarf Planet

After intense debate, the International Astronomical Union (IAU) resolved the issue by creating a new class of space objects, dwarf planets, in 2006. This category was meant to distinguish between small, round planetary bodies and the eight regular planets of the solar system. This difference was outlined in IAU Resolution B5.

IAU defined dwarf planets as space objects that orbit the sun, have enough mass and gravity to attain a round shape, and lack the gravitational power to clear other space objects out of their orbit. Additionally, dwarf planets are not satellites, or moons, to other space objects.

Once the dwarf planet category was established, IAU placed 2003 UB313 in it. The IAU named the object Eris after the Greek goddess of conflict, a nod to the debate that followed its discovery. At the same time, IAU demoted Pluto from a full planet to a dwarf planet. (This was a controversial decision for many in the public who did not understand the necessity for the change.) The IAU also designated Ceres, a unique object found in the asteroid belt, as a dwarf planet. Eventually, the IAU added more objects to the dwarf planet category. The group expects the number of known dwarf planets to grow in future as many more of these objects are discovered in the Kuiper Belt.

## Characteristics of Eris

Eris travels in the outermost part of the solar system, 97.56 astronomical units (AU) from the sun at its farthest point. Eris is therefore more than ninety-seven times farther from the sun than Earth is. However, Eris also has an unusual, tilted orbit, so its distance from the sun can vary quite a bit. At its closest, Eris is 38.5 AU from the sun. This takes it closer to Earth than Pluto.

Eris's orbit follows a long path that takes it far beyond Neptune and the Kuiper Belt into a zone of icy debris. In fact, Eris's orbital path is 561.37 Earth years long. Due to the distance Eris travels, its surface temperatures are extreme. It is very cold, with surface temperatures estimated to be between −217 and −243 degrees Celsius (−359 and −405 degrees Fahrenheit). At its farthest point from the sun, the atmosphere collapses and freezes. When Eris moves closer to the sun, its atmosphere expands.

Eris's size has been open to debate since its discovery. While early estimates placed the object at about 3,000 kilometers (1,864 miles) in diameter, later research indicated that its diameter is 2,326 kilometers (1,445.3 miles). Thus, Eris is about the same size as Pluto, not larger, as first believed.

Despite the controversy over size, scientists agree that calculations from Eris's satellite, Dysnomia, prove that Eris has 27 percent more mass than Pluto does. This extra mass makes it the denser of the two dwarf planets. Scientists believe this density is evidence that Eris is composed mainly of a rocky material covered with a thin surface of ice. (Pluto, on the other hand, is less dense and therefore more likely to be made up of ice and rock.) Eris is very bright and reflects nearly all light. Because of this and its color, scientists believe its surface is made of nitrogen ice and solid methane.

## Eris's Satellite

Eris has one known satellite, Dysnomia. It was discovered on September 10, 2005, at the Keck Observatory in Hawaii. This moon is named for the daughter of the goddess Eris who was known for celebrating lawlessness. The moon Dysnomia may be composed of frozen water, but scientists are unsure of this.

Dysnomia is quite small. Astronomers estimate Dysnomia's diameter at about 150 to 250 kilometers (93 to 155 miles). Dysnomia orbits close to Eris. It is located only about 37,370 kilometers (23,200 miles) from Eris. (By contrast, Earth's moon is 384,400 kilometers, or 238,855 miles, from Earth.) Its orbit takes sixteen days to complete.

Scientists believe that Dysnomia formed much like Earth's moon. They hypothesize that Eris collided with a large space object that ejected material into space. That material eventually came together to form Dysnomia.

## Vastness of the Solar System

Like Eris and Pluto, Dysnomia is considered a Kuiper Belt object (KBO), which is simply a designation for objects that formed in the region of space beyond Neptune. More than 1,300 KBOs were discovered between 1992, when the belt first was identified, and 2014. These objects are important for several reasons. KBOs are thought to be untouched fragments from the formation of the solar system about 4.6 billion years ago. As such, they may teach scientists much about how the solar system formed and evolved. In addition, KBOs have proven that the solar system is far more vast than previously imagined.

—Marie Keenan, MS

**Further Reading**

Brown, Mike. "Dysnomia, the Moon of Eris." *Division of Geological and Planetary Sciences.* Caltech, n.d. Web. 27 Mar. 2015.

Brown, Mike, Al Conrad, and Linda Copman. "Planetary Astronomy: The New Solar System." *Cosmic Matters.* W. M. Keck Observatory, Fall 2007. Web. 27 Mar. 2015.

"Dwarf Planets: Overview." *Solar System Exploration.* NASA, 2 Feb. 2015. Web. 24 Mar. 2015.

"Eris." *Theoi Project.* Aaron J. Atsma, 2000–2011. Web. 27 Mar. 2015.

"Eris: Overview." *Solar System Exploration.* NASA, 2 Feb. 2015. Web. 24 Mar. 2015.

"Faraway Eris Is Pluto's Twin." *ESO.* European Southern Observatory, 26 Oct. 2011. Web. 27 Mar. 2015.

Harbison, Rebecca. "Is There Really a 10th Plant?" *Ask an Astronomer.* Curious Team, 1997–2015. Web. 27 Mar. 2015.

"Kuiper Belt Objects: Dwarf Planets." *Laboratory for Atmospheric and Space Physics.* U of Colorado at Boulder, Aug. 2007. Web. 24 Mar. 2015.

"Pluto and the Developing Landscape of Our Solar System." *IAU.* International Astronomical Union, n.d. PDF file.

Redd, Nola Taylor. "Eris: The Dwarf Planet That Is Pluto's Twin." *Space.com.* Purch, 27 Jan. 2015. Web. 24 Mar. 2015.

# 433 EROS

## FIELDS OF STUDY

Observational Astronomy; Cosmology; Sub-planet Astronomy

## SUMMARY

433 Eros is a large near-Earth asteroid (NEA). Eros was the first NEA to be discovered and the second-largest known. It was also the object of study of the NEAR Shoemaker mission, which sent a spacecraft to orbit the asteroid for one year beginning in 2000. Planetary scientists were unexpectedly able to land the spacecraft on the asteroid before it lost power.

## PRINCIPAL TERMS

- **asteroid:** a small, irregularly shaped celestial body made of rock, silicates, and metals.
- **near-Earth asteroid (NEA):** a small, irregularly shaped celestial body with an orbital path that brings it close to Earth. NEAs travel within 0.3 astronomical units (AU) of Earth's orbit and within 1.3 AU of the sun. An astronomical unit is equal to about 149.6 million kilometers (about 93 million miles).
- **S-type asteroid:** one of three broad classes of asteroids. S-type ("silicaceous") asteroids are mainly composed of silicate (stone) and metals such as iron and nickel.

## Near-Earth Asteroids

433 Eros is an asteroid, or a relatively small, irregularly shaped space object that orbits the sun. Asteroids are believed to be matter left from the formation of the solar system about 4.6 billion years ago.

There are millions of asteroids in the solar system. Most of these are located in the asteroid belt, a doughnut-shaped ring of space between the orbits of Mars and Jupiter. The asteroid belt is about 2 to 4 astronomical units (AU), or 186 to 370 million miles, from the sun. Researchers theorize that after the gas giant Jupiter formed, the massive gravity surrounding it caused many small, nearby planetary bodies to crash into each other, creating asteroids.

Not all asteroids stay in the asteroid belt, however. Some asteroids are ejected from the belt when they collide with each other or get too close to the gravity surrounding Jupiter. The asteroids can be flung into the larger solar system. Near-Earth asteroids (NEAs) are asteroids that orbit within 0.3 AU (about 44.8 million kilometers, or about 31 million miles) of Earth. Eros is a NEA that has passed Earth at a distance of 0.178 AU.

NEAs are classified as one of three types (Amor, Aten, or Apollo) based on their average distance from the sun and the size and shape of their orbit. Eros is considered an Amor asteroid. Amor is the type of asteroid that approaches Earth but does not cross its orbit. However, like most other Amor asteroids, Eros does cross the orbit of the nearby planet Mars.

Asteroids vary greatly in size. The largest known asteroid, Ceres, measures about 966 kilometers (600 miles) in diameter. However, most asteroids are much smaller. Most asteroids are the size of small rocks. Eros is the second-largest NEA, at forty kilometers (twenty-five miles) in diameter.

## Characteristics of Eros

Eros was the first NEA to be discovered. Two different observers spotted Eros on the same date in August 1898: Gustav Witt of Berlin, Germany, and Auguste Charlois of Nice, France. Eros attracted their attention because it was observed orbiting near Mars rather than in the asteroid belt, which defied astronomical knowledge of the time.

Eros is an S-type asteroid, or one that is mainly composed of silicates (stone) and metals such as iron and magnesium. S-class asteroids are among the brightest asteroids in the solar system due to their ability to reflect the light of the sun. They are also the most common type of asteroid found in the inner asteroid belt.

Eros is shaped like a giant peanut or a horse's saddle. Two thicker, more rounded ends flank its thinner midsection. Craters and pits mark its surface. Scientists believe that these types of marks, which are seen on most asteroids, are the result of collisions with other objects in space.

Eros circles the sun once every 1.76 Earth years and spins on its axis once every 5.27 hours. Its gravitational force is relatively weak but is strong

433 Eros, asteroid. NEAR Project, NLR, JHUAPL, Goddard SVS, NASA, public domain via Wikimedia Commons.

enough to hold a spacecraft, as shown by the NEAR Shoemaker mission.

### NEAR Shoemaker Mission

Eros was the first asteroid ever to be orbited by a space probe and to have the probe touch down on its surface. The Near-Earth Asteroid Rendezvous (NEAR) mission, later named the NEAR Shoemaker mission, sent the *NEAR Shoemaker* spacecraft to orbit Eros for one year beginning in 2000.

The goal of the mission was to orbit the asteroid and send back data on its characteristics, including properties such as surface features, chemical composition, internal mass distribution, and magnetic field. The mission also sought to clarify the relationship among asteroids, comets, and meteorites, as well as to increase general understanding of the solar system. Although the spacecraft was not designed with landing gear, the mission ended with a touchdown of the spacecraft in the saddle region of Eros in February 2001. The spacecraft lost power shortly thereafter.

—*Marie Keenan, MS*

### Further Reading

"Amor Asteroids." *Cosmos: The SAO Encyclopedia of Astronomy*. Swinburne U of Technology, 2014. Web. 9 Mar. 2015.

"Asteroids: Overview." *NASA*. NASA, 13 Jan. 2015. Web. 9 Mar. 2015.

"Basics of Space Flight. Chapter 1: The Solar System." *Jet Propulsion Laboratory*. California Inst. of Technology, 2013. Web. 9 Mar. 2015.

"Near Earth Asteroid Rendezvous." *Discovery is NEAR*. JHUAPL, 2001. Web. 10 Mar. 2015.

"NEAR Shoemaker." *National Space Science Data Center*. NASA, 26 Aug. 2014. Web. 13 Mar. 2015.

"Small Worlds: The Neighborhood." *Marshall Space Flight Center Discovery Program*. NASA, n.d. Web. 9 Mar. 2015.

# FIREBALL

**FIELDS OF STUDY**

Astronomy; Observational Astronomy

**SUMMARY**

A fireball is one of a small number of meteors that appear brighter in the sky than the planet Venus and may have a visible trail as they approach Earth. Fireballs are significant because they may impact on Earth or explode in the atmosphere and cause a sonic boom.

**PRINCIPAL TERMS**

- **bolide:** a fireball that explodes in Earth's atmosphere and creates a sonic boom.
- **meteor:** a piece of rock or metal that enters Earth's atmosphere and becomes visible as a burning, glowing streak in the sky. The streak of light is also called a meteor.

## Meteors and Fireballs

Meteors are small pieces of asteroids. Every day, thousands of meteors enter Earth's thermosphere. The thermosphere is between 80 and 120 kilometers (50 to 75 miles) above Earth. About one in every thousand meteors is a fireball. Fireballs are generally defined as meteors that appear brighter than Venus when viewed from Earth. Venus has an apparent magnitude, or degree of brightness, of between −4.9 (brightest) and −3.8 (least bright). Fireballs are also commonly called "shooting stars." They may have a visible tail of light or smoke.

The meteors that cause fireballs can be larger than one meter, or slightly more than three feet, across. As they enter the atmosphere at speeds between 11 and 72 kilometers per second (25,000 to 160,000 miles per hour), their speed puts pressure on the air. This causes the meteor to heat up to 2000 kelvins (1727 degrees Celsius or 3140 degrees Fahrenheit).

Fireballs are grouped into different classes. The classes are based on what the fireball is made of. The first two types of fireballs are chondritic meteoroids. These meteoroids are stony and come from smaller asteroids. They are created by collisions. Type 1 fireballs are ordinary chondrites. Type 2 fireballs are carbonaceous chondrites. They are rich in carbon.

The other two types of fireballs are made by cometary meteoroids. Cometary meteoroids come from comets. Type 3a fireballs come from high-density comets. Type 3b fireballs come from low-density comets.

## Seeing a Fireball

Thousands of meteors with potential to be fireballs enter Earth's atmosphere daily, but only a few are actually seen. There are different reasons for this. Meteors may fall over oceans or other areas of the planet where few people live; they may fall during daytime hours and thus be obscured by sunlight; or they may fall late at night, when most people are asleep. Experts estimate that only two to twelve fireballs per day have the potential to actually be seen by people. However, several times a year Earth passes through a meteor stream. A meteor stream is a cloud of debris left behind by a comet. When Earth passes through it, the stream creates a shower of meteors that appear to come from the same spot in the sky. The showers are named after the constellation in which that spot appears to occur. These meteor showers increase the likelihood of seeing a fireball. The Perseid meteor shower in mid-August is among the best-known showers.

The Leonid meteor shower in mid-November is also well known. The first recorded sighting of the shower was in China in 902 CE. Every thirty-three years, a peak Leonid shower phase occurs. Peak phases offer better chances of seeing a fireball. The last peak Leonid shower phase was between 1998 and 2002.

Time-lapse image of the glowing track of a fireball, a small interplanetary rock heating up as it enters the atmosphere and passes through the sky behind the ALMA array in the Chilean Andes. By ESO/C. Malin via Wikimedia Commons.

**METEORITES AND BOLIDES**

While some fireballs fall all the way through Earth's atmosphere to the surface, most are not large enough and burn up before they reach the ground. Some are large enough to reach Earth's surface in fragments that are called meteorites. Experts say that these pieces of debris cool as they complete their fall to Earth and are not generally hot once they land. Usually only fireballs with a magnitude of −8 or brighter will reach Earth's surface as meteorites.

Some fireballs explode as they enter the atmosphere. They create a brilliant flash of light followed some time later by a sonic boom, a loud explosion-like noise that objects make when they travel faster than the speed of sound. These loud, explosive fireballs are called bolides. They may cause damage on the ground. The meteor that shattered windows and rattled buildings in Cheylabinsk, Russia, in February 2013, injuring more than one thousand people and causing more than $30 million in damages, exhibited the fiery appearance and sonic boom characteristic of a bolide.

The National Aeronautics and Space Administration (NASA) operates a near-Earth object (NEO) detection system and monitors space objects such as asteroids that could threaten Earth. However, some objects, such as the meteors that become fireballs and bolides, are small and therefore difficult to observe and track.

—*Janine Ungvarsky*

**FURTHER READING**

"Fireball." *Cosmos: The SAO Encyclopedia of Astronomy.* Swinburne U of Technology, n.d. Web. 9 Mar. 2015.

"Fireball and Bolide Reports." *Near Earth Object Program.* NASA, 9 Mar. 2015. Web. 9 Mar. 2015.

"Fireball FAQs." *American Meteor Society.* Amer. Meteor Soc., 2013–15. Web. 9 Mar. 2015.

"Fireball Observations." *International Meteor Organization.* Intl. Meteor Org., 1997– 2013. Web. 9 Mar. 2015.

Kuzmin, Andrey. "Meteorite Explodes over Russia, More Than 1,000 Injured." *Reuters.* Thomson, 15 Feb. 2013. Web. 9 Mar. 2015.

*NASA's All Sky Fireball Network.* NASA, 8 Mar. 2015. Web. 9 Mar. 2015.

"Near Earth Asteroids (NEAs): A Chronology of Milestones, 1800–2200." *International Astronomical Union.* IAU, 7 Oct. 2013. Web. 9 Mar. 2015.

# FUEL SYSTEMS

## FIELDS OF STUDY

Aerospace Engineering; Orbital Mechanics; Space Technology; Spacecraft Propulsion

## SUMMARY

A fuel system is the part of a vehicle's engine that combines fuel and air in amounts that allow for combustion. This process of burning fuel generates the vehicle's thrust, or motion. Without fuel systems, rockets could never reach space, and the development of new and more efficient types of fuels and fuel systems will further space exploration.

## PRINCIPAL TERMS

- **combustion:** a chemical process in which a fuel source combined with an oxidizer reacts to create heat.
- **fuel injection:** a computer-controlled process in a vehicle's fuel system in which the fuel and oxygen are mixed in the ratio needed to power the vehicle's engine

## FUEL SYSTEM BASICS

Since the Wright brothers' first flight in 1903, every engine-propelled vehicle of flight has required a fuel system to generate thrust, the force that propels forward movement. While systems may vary in design and type of fuel burned, thrust is always generated by the application of Isaac Newton's (1642–1727) third law of motion. This law states that for every action, there is an equal and opposite reaction. In the case of rockets, the hot gases produced by fuel combustion flow backward and out of the rocket. This causes the rocket to move forward with an equal force. The speed of a rocket can be altered by changing the amount of fuel used and the way the resulting gases are concentrated through the nozzle (shaped opening) as they exit the fuel system.

All combustion systems require both a fuel source and an oxidizer to produce the heat that creates thrust. However, space vehicles must compensate for the fact that there is no oxygen in space. Therefore, their systems must be built to contain both the fuel and the oxidizer.

### TYPES OF ROCKET FUEL SYSTEMS

There are three main types of rocket engines: those that use liquid fuel, those that use solid fuel, and those that use a combination of the two. In a liquid-fuel rocket, the fuel source (usually liquid hydrogen) and the oxidizer are stored separately as liquids. Then they are combined in a process known as fuel injection. The fuel injectors combine the two in computer-controlled proportions in the combustion chamber, where the fuel burns and provides the force for thrust. The speed of the rocket can be controlled by regulating the flow of fuel into the system. As a result, liquid-fuel systems are more versatile; however, the additional mechanics needed to control the injection process make them more complicated and heavier than their solid-fuel counterparts.

In a solid-fuel rocket, the fuel source and oxidizer are combined and loaded into a cylinder. In the case of the booster rockets that launch space shuttles into orbit, the solid, dry fuel mix contains ammonium perchlorate for an oxidizer and aluminum for the fuel, plus a binder. Solid-fuel rockets are the simplest of rocket designs and generally the most stable. They will not burn until exposed to an igniter. However, once combustion begins, it continues until all the fuel is used up. The only way it can be stopped is by destroying the cylinder containing the fuel. This makes solid-fuel rockets less versatile, but they are both lighter and less complicated in design.

Hybrid rockets use a combination of liquid and solid propellants, usually a solid fuel and a liquid or gaseous oxidizer. These combine the advantages of both solid-fuel and liquid-fuel rockets: the fuel is more stable and requires fewer components for the fuel-injection process, but the use of fuel can still be regulated. These rockets are considered safer, cheaper to operate, and more environmentally friendly than the other types. Their fuel sources are not combustible on their own, and they have fewer working parts to break down and allow accidental explosions. They are cheaper because the oxidizer can be oxygen, hydrogen peroxide, or nitrous oxide—all substances that are readily available and

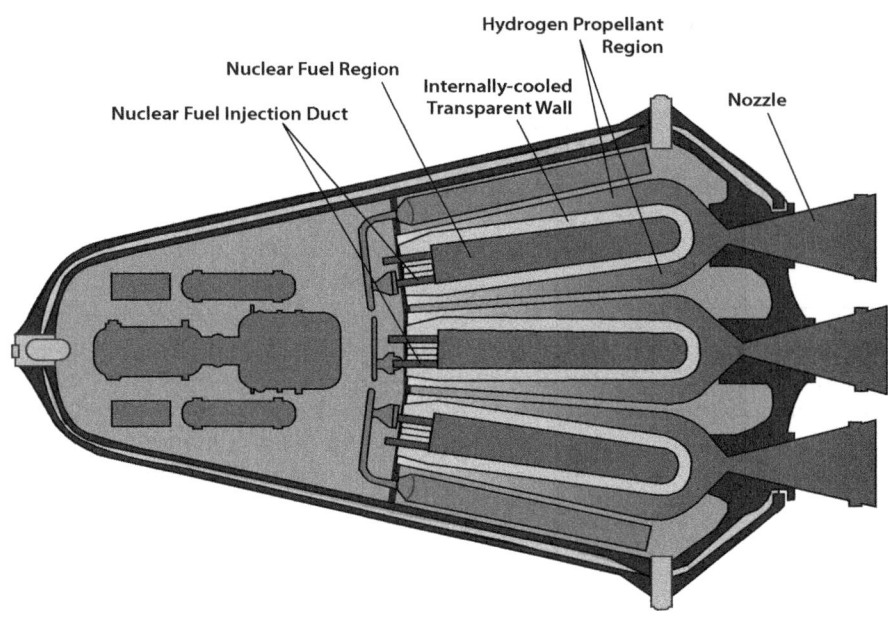

Nuclear gas-core engine designed using a closed cycle would keep the gaseous uranium fuel contained within an internally-cooled wall surrounded by hydrogen.

years. Even before the first successful moon landing in 1969, scientists were working on the idea of a nuclear-powered spacecraft.

While no such spacecraft have yet been launched, scientists have created designs for liquid-, solid-, and gas-core fuel systems with variations including closed- and open-core systems. Some solid-core prototypes were created, and NASA established programs for long-duration missions that would use nuclear-powered spacecraft, but they were canceled due to safety concerns and budget cuts.

less expensive than custom-made solid fuel mixes. And because the fuels used in hybrids produce fewer hazardous by-products, they are considered to be a greener alternative.

Some vehicles, such as the space shuttle, use several types of fuel systems. Solid-fuel booster rockets are used to launch the vehicle, which is then propelled through space by a cryogenic, or very low temperature, liquid-fuel system.

### NUCLEAR ENGINES

Solid, liquid, and hybrid fuel systems all share an inherent problem for space exploration: their weight and slow speed make reaching objects farther out in space impractical. The increased efficiency of nuclear power might mean lighter, faster spacecraft that could allow astronauts to reach Mars in as few as four months. Nuclear-powered rockets could trim the flight to Saturn from seven years to as few as three

—*Janine Ungvarsky*

### FURTHER READING

"Combustion." *Glenn Research Center*. NASA, 12 June 2014. Web. 21 Apr. 2015.

"Engine Fuel System." *Glenn Research Center*. NASA, 12 June 2014. Web. 21 Apr. 2015.

Fishbine, Brian, et al. "Nuclear Rockets: To Mars and Beyond." *National Security Science*. Los Alamos Natl. Laboratory, 2011. Web. 21 Apr. 2015.

"Nuclear Thermal Rocket Propulsion." *Space Propulsion and Mission Analysis Office*. NASA, 9 July 2008. Web. 21 Apr. 2015.

"Rocket Propulsion." *Glenn Research Center*. NASA, 12 June 2014. Web. 21 Apr. 2015.

Shimada, Toru. "Hybrid Rocket Will Go for the Next Generation of Space Transportation." *Institute of Space and Astronautical Science*. Japanese Aerospace Exploration Agency, 2012. Web. 21 Apr. 2015.

# G

## GALAXY TYPES

### FIELDS OF STUDY

Astronomy; Cosmology; Extragalactic Astronomy

### SUMMARY

Galaxies are collections of dust, gas, and stars. They are usually found in clusters. Galaxies come in many shapes and sizes and are classified into four major types: elliptical, spiral, irregular, and lenticular. Other subtypes also exist. Elliptical galaxies are oval in shape, while spiral galaxies look like a pinwheel. Galaxies that do not look like either of these shapes are classified as irregular. Lenticular galaxies share qualities of both ellipticals and spirals. Edwin Powell Hubble is the American astronomer who discovered the existence of galaxies outside the Milky Way and devised a method of classifying them. The Hubble Space Telescope was named in honor of his achievements.

### PRINCIPAL TERMS

- **bulge:** a dense, spheroidal group of stars found in the center of many spiral and lenticular galaxies.
- **disk:** the roughly flat, circular component of a spiral or lenticular galaxy that contains the majority of the galaxy's stars, most of which orbit in the same direction; also spelled "disc."
- **elliptical galaxy:** a roughly spherical or ellipsoid galaxy with no bulge or disk, consisting mainly of old stars that follow random orbits.
- **globular cluster:** a dense, spheroidal collection of old stars of the same age and origin, bound together by gravity and orbiting within the halo of a large galaxy.
- **halo:** a roughly spherical region that extends beyond the visible reaches of a galaxy, containing old, dim stars, dark matter, gas, and globular clusters.
- **irregular galaxy:** a galaxy without a distinct shape or structure.
- **lenticular galaxy:** a disk galaxy with a prominent central bulge, no arms, and little interstellar medium; so named because its shape is similar to that of a lens.
- **spiral galaxy:** a disk galaxy with spiral arms that extend into the disk from either a central bulge or a bar that passes through the bulge.

### THE KNOWN UNIVERSE EXPANDS

In 1923, American astronomer Edwin Hubble (1889–1953) was working at the Mount Wilson Observatory in California when he spotted a new Cepheid variable in Messier 31 (M31), then known as the Andromeda Nebula. A Cepheid is a star that can be used to calculate distance in space. At the time, most astronomers, led by Harvard College Observatory director Harlow Shapley (1885–1972), believed that the Milky Way galaxy was the extent of the universe. However, when Hubble used Shapley's methods to calculate the distance to the Cepheid, the result placed it about a million light-years away—far beyond the Milky Way, which Shapley himself had determined to be only three hundred thousand light-years across. This meant that other galaxies outside of the Milky Way did exist, including M31, now called the Andromeda galaxy.

Hubble's discovery expanded the universe significantly and was a key finding in the astronomical world. Many other objects once thought to be nebulae within the Milky Way were soon found to be separate galaxies as well. Hubble began to devise a classification scheme for galaxies. This scheme, now called the Hubble sequence or the Hubble tuning-fork diagram (due to its shape), forms the basis of the modern galaxy-classification scheme.

Hubble initially divided galaxies into three main categories: spiral, elliptical, and irregular. The earliest version of his tuning-fork diagram, published in 1936, deals primarily with spirals and ellipticals. The left side of the diagram shows three types of elliptical galaxies (E0, E3, and E7), ordered from the

133

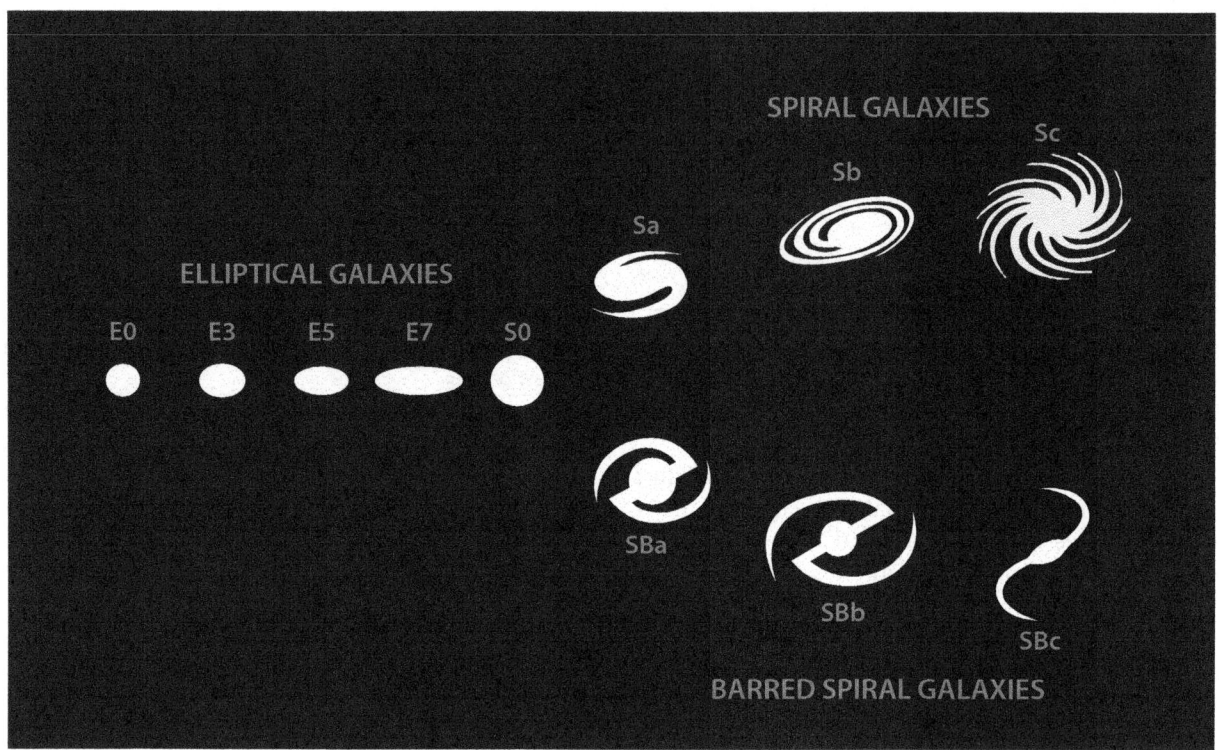

Edwin Hubble's classification scheme for galaxies includes elliptical, spiral, and barred spiral galaxies. Elliptical galaxies range from more spheroid galaxies (E0) to extremely elliptical galaxies (E7). Spiral galaxies range from tightly wound with a prominent bulge (Sa) to loosely wound with a faint bulge (Sc). Barred spiral galaxies follow the same progression as unbarred spiral galaxies (SBa through SBc). Lenticular galaxies are represented by type S0.

most circular (E0) to the most elongated (E7). The diagram then splits into two branches—the "tines" of the tuning fork—representing the two categories of spiral galaxies, barred (SBa, SBb, and SBc) and "normal," or unbarred (Sa, Sb, and Sc). Irregular galaxies were any that did not fall into one of these two categories.

In the center of the diagram, at the point where the two spiral branches meet the elliptical branch, Hubble included type S0, representing a hypothetical intermediate form. This type later came to represent lenticular galaxies.

### FOUNDER OF THE GALAXIES

Edwin Powell Hubble was born on November 20, 1889, in Marshfield, Missouri. His family moved to Wheaton, Illinois, less than a year later. Hubble attended the University of Chicago on a partial scholarship and graduated in 1910 with a bachelor of science degree in mathematics and astronomy. He then won a Rhodes Scholarship to attend the University of Oxford, where he studied law.

After returning to the United States, Hubble practiced law and taught high school for a year before deciding to pursue his first love. He returned to the University of Chicago in 1914 to complete his doctorate in astronomy. Upon graduation, Hubble was offered a job at the Mount Wilson Observatory near Pasadena, California, but he enlisted in the army instead.

After World War I ended, Hubble returned and accepted the position at Mount Wilson. There, he worked with the largest and newest telescope in the world: the hundred-inch Hooker Telescope, completed in 1917. It was with this telescope that Hubble first observed the Cepheid variable that would ultimately lead him to revolutionize humanity's view of the universe.

**ELLIPTICAL GALAXIES**

Elliptical galaxies are among the oldest in the universe. They have a three-dimensional ellipsoid shape, ranging from nearly spherical to an elongated oval. Due to their age, they contain very little dust or gas, as they contain mostly old stars and generally do not form new ones. This gives elliptical galaxies a reddish color and makes them appear fainter than other galaxies. In the sky, they look like a blurry, rounded patch of light. While the majority of galaxies observed thus far have been spiral galaxies, scientists suspect that elliptical galaxies are in fact more common.

Of all galaxy types, ellipticals have the largest range in terms of size and mass. The smallest ones are called dwarf elliptical galaxies. These can be as small as ten thousand light-years in diameter. The largest elliptical galaxies can be a million or more light-years across and contain trillions of stars. They often contain unusually large numbers of globular clusters. Messier 87 (M87), the largest known elliptical galaxy in the universe, is believed to have as many as fifteen thousand globular clusters orbiting in its halo.

Elliptical galaxies are believed to be formed by collisions between smaller galaxies. When two spiral galaxies collide, they lose their spiral arms and merge into one large, disordered elliptical galaxy. A large number of elliptical galaxies are found in the Coma Cluster, a cluster of more than one thousand galaxies. At the center sit two massive elliptical galaxies, New General Catalog 4874 (NGC 4874) and New General Catalog 4889 (NGC 4889), surrounded by numerous other ellipticals, both dwarf and giant.

**SPIRAL GALAXIES**

More than three-quarters of all known galaxies are spiral galaxies. They are one of two types of disk galaxies, so called because they are shaped like a flattened, rotating disk. In a spiral galaxy, the disk has a distinct spiral structure, typically with arms extending from the center. Approximately two-thirds are classified as "barred," meaning that a bar-shaped structure extends through the middle, with arms emerging from either end. About 60 percent of spiral galaxies have multiple arms, and 10 percent have only two arms. It is believed that the remaining 30 percent used to have arms but that they have since faded. The Milky Way is an example of a barred spiral galaxy with multiple arms.

Spiral galaxies are younger than elliptical galaxies. They are made up of dust, gas, and mostly young stars. Because of the abundance of gas and dust, they are able to form new stars and appear brighter than ellipticals. The majority have a central bulge, a dense collection of older, dimmer stars that protrudes above and below the disk. Scientists believe that such bulges contain a supermassive black hole at their center.

The disk surrounding the bulge and is made up of both old and new stars. The spiral arms contain large numbers of young, blue stars, causing them to shine more brightly than the rest of the disk. Scientists are uncertain how these arms form. Some believe that they are formed by regions of increased density causing stars to briefly slow their orbit, creating a kind of stellar traffic jam. More recent evidence suggests that they result from the gravitational effects of giant molecular clouds.

Spiral galaxies range in size from about one billion to one trillion times the mass of the sun. Their disks are typically between ten thousand and three hundred thousand light-years across. The largest known spiral galaxy is New General Catalog 6872 (NGC 6872), which is about five times larger than the Milky Way. The oldest one is Q2343-BX442, which is thought to be more than ten billion years old.

As spiral galaxies spin, they burn up their gas, which slows down new star formation. Scientists believe that as this happens, they lose their spirals and evolve into elliptical galaxies.

**LENTICULAR GALAXIES**

Lenticular galaxies are the second type of disk galaxy. They have properties of both elliptical and spiral galaxies. Some scientists believe that this class is a transitional form between the two. The term "lenticular" means "lens-like." Lenticular galaxies are so named because when viewed from the side, the shape of their disks resembles a convex lens.

While lenticular galaxies appear visually similar to fading spiral galaxies, their compositions are closer to those of elliptical galaxies. Like spiral galaxies, lenticulars have a disk and bulge, but unlike spirals, they do not have arms. Like ellipticals, lenticulars contain little gas and mostly consist of old, reddish stars, with few bluish new stars. Lenticulars, however, contain a large amount of dust. Because of these characteristics, some scientists believe that lenticulars are very old, faded spiral galaxies. Others believe that they are formed by the merging of two old spiral galaxies.

**IRREGULAR GALAXIES**

Irregular galaxies are ones that cannot be categorized by shape. Their age, shape, and structure vary.

They do not have spiral arms or bright centers. Some may have been spiral or elliptical at one time, but a collision or merger caused them to take on an irregular shape. Irregular galaxies contain large amounts of gas and dust and a mix of old and new stars. They are also smaller than both elliptical and spiral galaxies.

Two classifications of irregular galaxies exist. Type I irregular galaxies have some sort of structure but are not considered elliptical, spiral, or lenticular galaxies. Type II galaxies do not have any structure.

—*Angela Harmon*

**FURTHER READING**

Buta, Ronald J., Harold G. Corwin Jr., and Stephen C. Odewahn. *The de Vaucouleurs Atlas of Galaxies.* New York: Cambridge UP, 2007. Print.

Cain, Fraser. "Irregular Galaxy." *Universe Today.* Author, 4 May 2009. Web. 11 May 2015.

Crockett, Christopher. "What Are Elliptical Galaxies?" *EarthSky.* Earthsky Communications, 5 Dec. 2012. Web. 11 May 2015.

Dick, Steven J. *Discovery and Classification in Astronomy: Controversy and Consensus.* New York: Cambridge UP, 2013. Print.

"Edwin Powell Hubble: The Man Who Discovered the Cosmos." *Hubble Space Telescope.* European Space Agency, n.d. Web. 11 May 2015.

"New Insights on How Spiral Galaxies Get Their Arms." *Harvard-Smithsonian Center for Astrophysics.* Harvard U, 2 Apr. 2013. Web. 15 May 2015.

Sandage, Allan. "Classification and Stellar Content of Galaxies Obtained from Direct Photography." *Galaxies and the Universe.* Ed. Sandage, Mary Sandage, and Jerome Kristian. Chicago: U of Chicago P, 1975. 1–35. *Level 5: A Knowledgebase for Extragalactic Astronomy and Cosmology.* Web. 12 May 2015.

"The Third Way of Galaxies." *Hubble Space Telescope.* European Space Agency, 12 Jan. 2015. Web. 11 May 2015.

"Violent Origins of Disc Galaxies: Why Milky Way–Like Galaxies Are So Common in the Universe." *ScienceDaily.* ScienceDaily, 17 Sept. 2014. Web. 11 May 2015

# GALILEO (ESA)

**FIELDS OF STUDY**

Astronomy; Space Technology

**SUMMARY**

Galileo is a satellite navigation system developed by the European Union (EU) and the European Space Agency (ESA). It is expected to be fully operational by about 2020. Using data from space-based satellites orbiting Earth, Galileo can pinpoint the location of satellite-enabled devices to within a meter of their location. Galileo is designed to give the EU access to satellite navigation data that is independent of systems controlled by countries, such as GPS (the United States) and GLONASS (Russia).

**PRINCIPAL TERMS**

- **Global Positioning System (GPS):** a satellite navigation system developed in the United States; uses satellite data to determine the position, navigation, and timing of objects on Earth.
- **satellite:** an object that orbits another object; most commonly refers to a machine that is launched into space to orbit Earth.

**SATELLITE NAVIGATION SYSTEMS**

A satellite navigation system uses a series of space-bound satellites traveling precise orbits around Earth to pinpoint the latitude and longitude of Earth-bound receivers with a great deal of accuracy. Satellites are highly sophisticated machines with exposure to large areas of Earth at once. This allows them to collect large amounts of data very quickly. Because satellites orbit high above Earth, beyond the planet's atmosphere, their radio signals are not blocked by objects in the atmosphere or on the surface.

The United States and the Soviet Union developed the first satellite navigation systems. The United States began establishing its system, the Navigation Signal Timing and Ranging Global Positioning System (NAVSTAR GPS)—now usually just called Global Positioning System (GPS)—in 1978. By 1993, GPS had twenty-four satellites in orbit, giving the system worldwide coverage. Eventually, the system grew to include a total of around thirty satellites, with at least twenty-four in operation at any one time. GPS is operated and maintained by the US Department of Defense, with a subsystem made available for commercial purposes and use.

The Soviet Union (now Russia) began developing its system, the Global Navigation Satellite System (GLONASS), in 1976. Despite setbacks caused by the breakup of the Soviet Union, GLONASS was briefly operational in the mid-1990s, with twenty-four satellites in orbit. However, the system soon fell into disrepair. It was brought back online in October 2011, with twenty-four satellites again giving the system global coverage. GLONASS is operated and maintained by the Russian Federal Space Agency.

Galileo, named for famed Italian astronomer Galileo Galilei (1564–1642), is the satellite navigation system of the European Union (EU). Galileo is meant to be interoperable with both GPS and GLONASS while also being completely independent of both systems.

The European Space Agency (ESA) manages the technological development of Galileo in close cooperation with EU ministers and civil aviation experts. Galileo's first two satellites were launched into orbit in 2011. Additional satellites were launched in 2012, 2014, and 2015. The system will eventually include a total of thirty satellites, which will provide worldwide coverage. It is expected to be fully operational around 2020.

**Why Navigation Systems Were Developed**
Satellite navigation systems were originally conceived and designed by governments for military uses. They can be used to assist in the accurate firing of weapons, the deployment of supplies and troops, and other military operations. These systems are invaluable for search-and-rescue maneuvers, especially in terrain that is difficult to otherwise reach or view. Satellite navigation is also used to propel and guide unmanned military vehicles. The Russian and US governments can limit or reserve the strength and precision of signals from their navigational systems at their discretion.

As navigation systems grew and became more sophisticated, commercial applications became more popular. Civilian scientists used handheld units in the 1980s to provide precise locations for archaeological and geological sites. By the early 2000s, portable units were available to motorists, who used them for travel directions from one location to another. Within a few years, satellite navigation technology was being built into everything from automobiles and airplanes to farming equipment and cell phones.

Since that time, satellite navigation has been

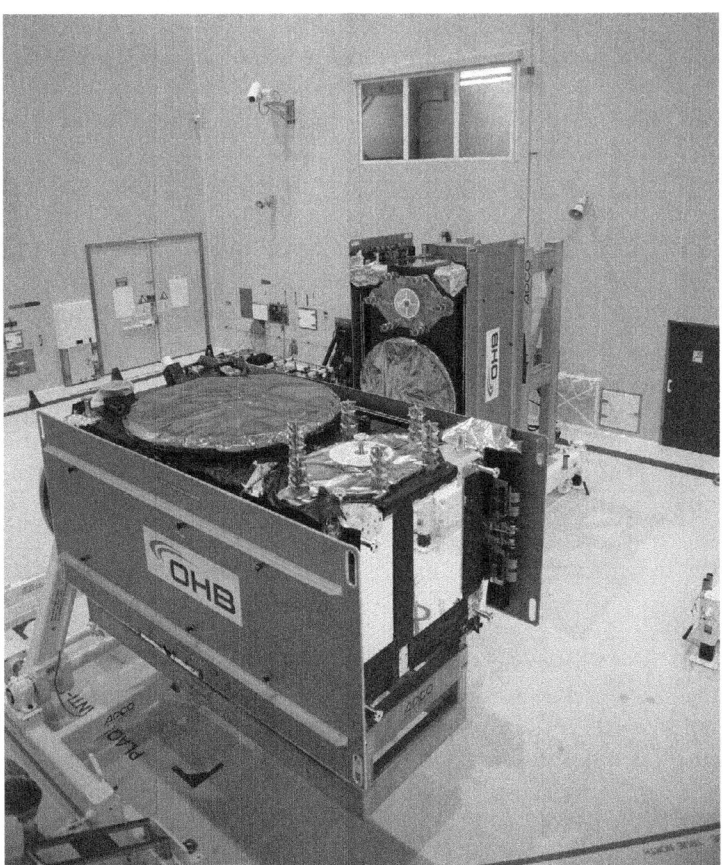

Galileo Full Operational Capability satellites in preparation prior to launch. ESA–S. Corvaja, 2014.

incorporated into ships, trains, planes, cars, shipping containers, and smartphones. In most cases, users do not need to know anything about where the satellites are or how they or the receivers work. They merely need to turn on their device to benefit from the automated navigation system.

While some applications of satellite navigation systems are more of a convenience, others save lives. The ability to pinpoint addresses to specific coordinates enables first responders such as firefighters, police officers, ambulance drivers, and rescue workers to more precisely identify the location where help is needed. As a result, first responders reach sites more quickly, with fewer delays caused by misdirection.

**How Satellite Navigation Systems Work**
Roughly twice a day, the satellites circle Earth in their unique but overlapping orbits at varying distances from Earth. Galileo's satellites are located in circular medium-Earth orbit (MEO) about 23,222 kilometers (14,429 miles) above Earth. The satellites send radio signals to receivers based on Earth. Each satellite system is carefully calibrated so that its satellites send their signals at exactly the same time. Earth-bound receivers then calculate latitude, longitude, altitude, and time based on the amount of time it takes the signal to travel from the satellite to the receiver. Satellites are programmed and synchronized with precision so that they can locate and track an enabled receiver, such as a dedicated unit or smartphone, to within a few meters of its actual location, depending on the device's sophistication level.

**Galileo's Rollout**
The Galileo rollout was originally divided into two distinct phases: In-Orbit Validation (IOV) and Full Operational Capability (FOC). The IOV phase, which began in 2011, was designed to validate the system, to make sure that it worked as expected and successfully transmitted accurate information. IOV involved assessing performance through a variety of tests using four satellites and ground infrastructure. Galileo was validated through this phase.

The FOC phase began in 2014 and will continue until a total of thirty satellites are in orbit. This is expected to be completed by 2020. Additionally, during the FOC phase, supporting ground infrastructure will be built out as necessary. Eventually this will include control centers throughout Europe and a network of sensor stations and uplink stations installed around the globe. Interested parties will be allowed to utilize the signals from Galileo in their products and services.

Galileo works in conjunction with the European Geostationary Navigation Overlay Service (EGNOS). This system improves the accuracy and reliability of Galileo using signals from mostly ground-based stations.

**Galileo's Services**
Galileo differs from previous navigational systems in that it is designed specifically for civilian rather than military use. Unlike other countries' systems, Galileo's full signal strength will be available for civilian and commercial purposes at all times, except in extreme circumstances such as armed conflict.

Galileo uses dual frequencies in order to deliver real-time data without delay. This allows Galileo to transmit a location to within one meter. Additionally, Galileo can inform users immediately if a satellite goes offline, helping users make more informed decisions. This feature also allows Galileo to be used for safety-critical applications, such as train automation and aircraft landing.

In addition to its free and open public service, Galileo will provide encrypted, secure service mainly for government operations, and encrypted and highly accurate commercial services will available for a fee. A search-and-rescue service will be used to help locate people, vessels, and aircraft in distress. This service will transfer distress signals from user transmitters to regional rescue centers. An important advance in this part of the system is that it will let the user know that the signal was received and help is on the way.

—*Marie Keenan, MS*

**Further Reading**
"European Satellite Navigation." *European Space Agency*. ESA, n.d. Web. 26 June 2015.
"Frequently Asked Questions—Galileo, the EU's Satellite Navigation Programme." *European Commission*. European Commission, 28 Mar. 2015. Web. 26 June 2015.
"Galileo Fact Sheet." *European Space Agency*. ESA, 15 Feb. 2013. Web. 26 June 2015.

"Global Positioning Satellites." *HyperPhysics*. Georgia State U, n.d. Web. 26 June 2015.

"The Global Positioning System." *GPS.gov*. Natl. Coordination Office for Space-Based Positioning, Navigation, and Timing, 11 Feb. 2014. Web. 26 June 2015.

"Launching Galileo." *European Space Agency*. ESA, 9 Mar. 2015. Web. 26 June 2015.

"What Is a Satellite?" *NASA*. NASA, 12 Feb. 2014. Web. 26 June 2015.

"What Is Galileo?" *European Space Agency*. ESA, 27 June 2014. Web. 26 June 2015.

# GAS-GRAIN MODELS

### FIELDS OF STUDY

Astrochemistry; Stellar Astronomy

### SUMMARY

Information gathered by specially equipped telescopes and in laboratories has shown that the relatively small amount of fine dust in molecular clouds has a key role in interacting with gas molecules. Such interactions initiate the chemical reactions that lead to the formation of stars. This gas-grain model of star formation is important because it advances the understanding of the formation of the universe and possibly the development of life.

### PRINCIPAL TERMS

- **astrochemistry:** the scientific study of chemical elements in space and how they interact with each other and with radiation.

### STAR FORMATION MODELS

The area between stars is filled with a blend of material that scientists call interstellar medium (ISM). This mixture of gas and dust combines in areas with thicker or denser collections of ISM known as molecular clouds. The bulk of the gases within them are sharing electrons in a molecular state.

These clouds are often called "stellar nurseries," because the conditions within them foster the birth of most stars. Star formation occurs when areas of the gas in the cloud become dense, causing the cloud to collapse in on itself to form the core or seed of a new star. This happens when something changes or reacts with the gas to increase its density. Astrochemistry focuses on the study of these and other kinds of chemical reactions in space. When all the aspects of the reaction that creates the star are gaseous, astrochemists refer to it as a gas-phase model.

Scientists have discovered that the fine dust grains in a molecular cloud can also play a role in star formation. They are important even though they make up a very small percentage of the cloud's matter. Sometimes these tiny grains of dust interact with the gas to change their temperature and start the reaction that makes the gas become denser. This is known as the gas-grain model of star formation.

Molecular clouds are estimated to be about 99 percent gas. The other 1 percent consists of solid particles called interstellar dust or dust grains. Most of the gas is hydrogen gas created at the same time as the universe. These clouds are quite cold. Temperatures stay at about 10 to 20 kelvins, or just above absolute zero.

This extreme cold makes chemical reaction between the gases impossible. Instead, most of the chemistry in the gas phase occurs by ion-molecule reactions. Ultraviolet light emitted by young stars provides the heat and energy needed to make molecular hydrogen. This compound is formed when two hydrogen atoms share their electrons. This molecular hydrogen is crucial to the formation of new stars.

Researchers have also recognized the importance of dust grains to the formation of stars. Dust grains make up a much smaller percentage of the stellar nurseries than the gaseous elements. However, scientists determined that many of the species or types of molecules they could identify in the molecular clouds could not be formed just by gas-phase elements. It is now believed that many stars are formed by the molecular gas interacting with dust grains. In the cold state of the cloud, the gas-phase molecules freeze into an icy mantle around the dust. This gas-grain model of star formation helps explain the richness of the chemistry in molecular clouds. This richness

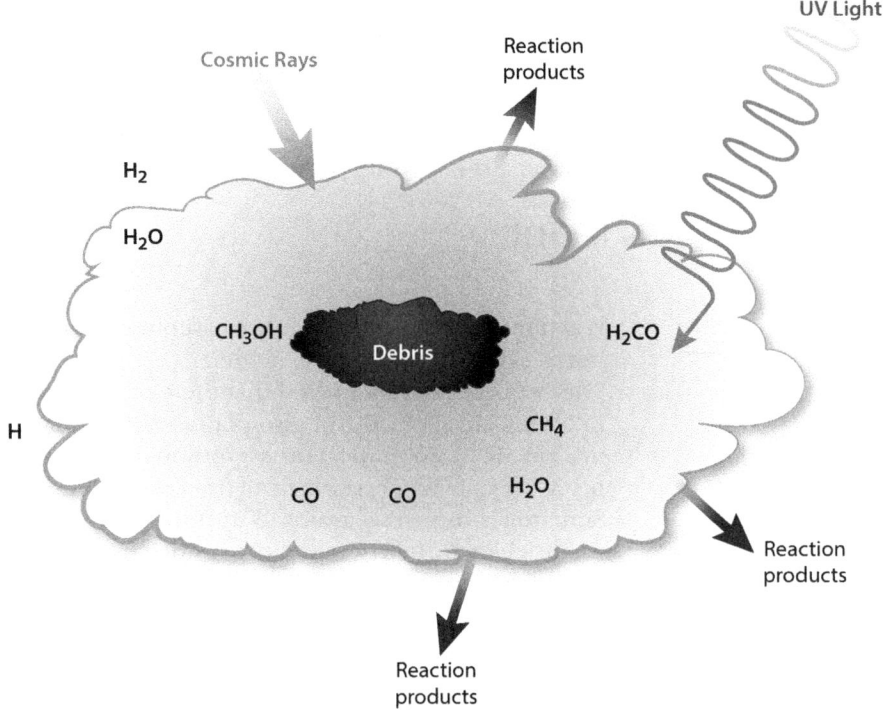

The gas-grain astrochemistry model describes chemical reactions among gases that occur around some grain of debris. The grain provides a surface upon which molecules can interact, resulting in more complex molecules that may either evaporate into the surrounding interstellar gas cloud or remain on the grain surface.

cannot be accounted for in models that only allow for gas-phase interactions.

## Dust and Gas Interactions

The dust in molecular clouds is very fine and is the result of material cast off from nearby stars as they die. It is largely made up of carbon and silicate and picks up atoms of other elements, such as hydrogen, oxygen, and nitrogen. These interactions then result in an outer coating of ice. This ice mantle can also include methane, ammonia, and carbon monoxide. As this mantle reacts with ultraviolet light, the outer molecules can form more complex organic compounds.

This icy dust plays key roles in the formation of stars. The "sticky" outer surface of the grains attracts atoms and holds them. This fosters the formation of molecules that might not take place otherwise in the conditions of the cloud. For instance, two hydrogen atoms colliding with the dust are likely to share electrons and react because of the dust. The dust also absorbs ultraviolet radiation, protecting molecules from the radiation's effects while reducing the ionization level of the cloud. This is important because excessive ionization would limit the ability of gravity to collapse the gas into a star core. Later, this radiation will be released as infrared energy, a cooler form of energy.

The interaction of the dust and gas causes the gas to freeze into a new state and come in contact with elements in ways that it could not in a gaseous state. The grain of dust becomes the nucleus for the new star. It collects hydrogen molecules and other atoms until the clump is dense enough for its gravity to make it collapse into the core of a new star.

This gas-grain model is also thought to explain the development of many of the more than one hundred different element species scientists have identified in molecular clouds. Many of these cannot be accounted for in a pure gas-phase model.

## Research Continues

Scientists continue to probe deep into molecular clouds, looking for more details in the star formation process. Using infrared and radio telescopes, researchers conduct ongoing studies of far-away stellar nurseries to learn more about the stars and other elements found there. Information from the Herschel Space Observatory and Planck satellite telescopes has been crucial in advancing knowledge about star formation. This research has allowed scientists to observe and catalog more of the species of elements found in distant stellar nurseries. The Hubble Telescope's infrared capabilities have given scientists many insights into the birth of stars, even though they are covered in clouds of dust. Infrared light can be detected through dust via the use of these special telescopes that can reveal the light's electromagnetic spectrum for study.

The discoveries continue in labs on Earth as well. For example, scientists at the Harvard-Smithsonian Center for Astrophysics in Cambridge, Massachusetts, performed experiments in a specially designed ultra-high vacuum surface lab. These experiments were designed to test the effects of ultraviolet light, x-rays, and electrons on icy surfaces similar to those found in star formation areas. This provides scientists with ways to understand and even predict how those icy mantles behave during star formation.

Some of these research efforts, such as the Spitzer Space Telescope mission from the National Aeronautics and Space Administration (NASA), have identified chemical combinations on distant stars that resemble the amino acids that are vital to life on earth. These discoveries stress the importance of continued study of the gas-grain model of star formation and its potential implications for the formation of the universe and the development of life on Earth.

—*Janine Ungvarsky*

**FURTHER READING**

"Chemical Soups around Cool Stars." *Jet Propulsion Laboratory*. California Inst. of Technology, 7 Apr. 2009. Web. 18 May 2015.

Choi, Charles Q. "Simple Recipe for Star Formation Revealed." *Space.com*. Purch, 10 Apr. 2014. Web. 18 May 2015.

"Dust Grain." *Cosmos: The SAO Encyclopedia of Astronomy*. Swinburne U of Technology, n.d. Web. 18 May 2015.

Kemsley, Jyllian. "Space Dust Science." *Chemical and Engineering News* 19 Apr. 2010: 36–38. Print.

"Molecular Cloud." *Cosmos: The SAO Encyclopedia of Astronomy*. Swinburne U of Technology, n.d. Web. 18 May 2015.

"Molecular Hydrogen." *Cosmos: The SAO Encyclopedia of Astronomy*. Swinburne U of Technology, n.d. Web. 18 May 2015.

"Star Formation & Molecular Astrophysics." *Blake Research Group*. California Inst. of Technology, n.d. Web. 18 May 2015.

"Stars." *Herschel Space Observatory*. NASA, n.d. Web. 18 May 2015.

"Stars." *NASA Science Astrophysics*. NASA, n.d. Web. 18 May 2015.

"The Story of Star Formation." *Cool Cosmos: Star Formation*. California Inst. of Technology, n.d. Web. 18 May 2015.

# GAS-PHASE MODELS

**FIELDS OF STUDY**

Astrochemistry; Stellar Astronomy

**SUMMARY**

Elements in a gaseous state make up the vast majority of the dense molecular clouds where stars are formed. Scientists have identified several ways to calculate or model the chemical reactions between these gases and other items around them that are responsible for the formation of stars. The gas-phase model is important to understanding the origins of stars and the role that they played in the formation of the universe.

**PRINCIPAL TERMS**

- **astrochemistry:** the study of chemical elements in the universe and their interactions with each other and with radiation.
- **gas-grain model:** the theory that grains of space dust interact with gas to form stars.
- **interstellar medium:** a mix of gas, dust, and cosmic rays that fills the space between star systems in a galaxy.
- **protostars:** masses of gas formed from the collapse of a giant molecular cloud that will eventually become stars.

**MODELS OF STAR FORMATION**

Space is often envisioned as groups of stars or planets surrounded by vast stretches of empty space. However, the area between star systems in a galaxy is filled with matter called interstellar medium (ISM).

This mixture of gas and dust combines to form interstellar clouds of varying density, temperature, and composition. Areas with denser collections of ISM are known as molecular clouds. This is because most of the gases within them are sharing electrons in a molecular state.

Molecular clouds are sometimes called "stellar nurseries" because most, if not all, stars are born within such clouds. Star formation occurs when something reacts with the gas in a molecular cloud, causing it to become denser and collapse in on itself. This gaseous mass then breaks away from the cloud to form a protostar, the core or seed of a new star. The study of these and other chemical reactions in space is called astrochemistry. When all of the components of the reaction that creates the star are gaseous, astrochemists refer to it as a "gas-phase model."

Scientists believe that the very fine dust in a molecular cloud can also play a role in star formation, even though it makes up a very small portion of the cloud's matter. Sometimes these tiny grains of dust interact with the gas in a way that changes its temperature

## THE CHEMISTRY OF STAR FORMATION

Molecular clouds are estimated to be about 99 percent gas and 1 percent dust. The majority of the gas is hydrogen gas, though some helium and other elements are also present. These clouds are very cold, with temperatures of about 10 to 20 kelvins, or just above absolute zero.

In this extreme cold, it is generally not possible for a chemical reaction to take place among the various gaseous elements. Instead, most of the chemistry in the gas phase takes place via ion-molecule reactions. Ultraviolet light, often emitted by young stars nearby, can provide the heat and energy needed to make molecules. Molecular hydrogen is formed when two hydrogen atoms share their electrons. This molecular hydrogen is essential to the formation of new stars.

The cold temperature causes the gas molecules to form clumps. These clumps grow denser until they collapse under their own gravity. The collapse causes pressure to build, increasing the temperature in the newly formed core until fusion becomes possible and the star is created.

In many cases, clusters of young stars are found close to other clusters of young stars. This is often caused by a supernova explosion of a massive star. The explosion creates shock waves that cause the molecules of hydrogen to compress and form new stars.

Researchers have also come to appreciate the importance of dust in star formation. Scientists came to understand that many of the types of molecules they could identify in the molecular clouds could not be created solely by gas-phase elements. It is now believed that some stars are formed by the molecular gas interacting with dust particles. In the cold conditions in the cloud, the gas-phase molecules freeze into an icy mantle around the dust. The hydrogen molecules later return to the gas phase. This gas-grain model of star formation helps account for the

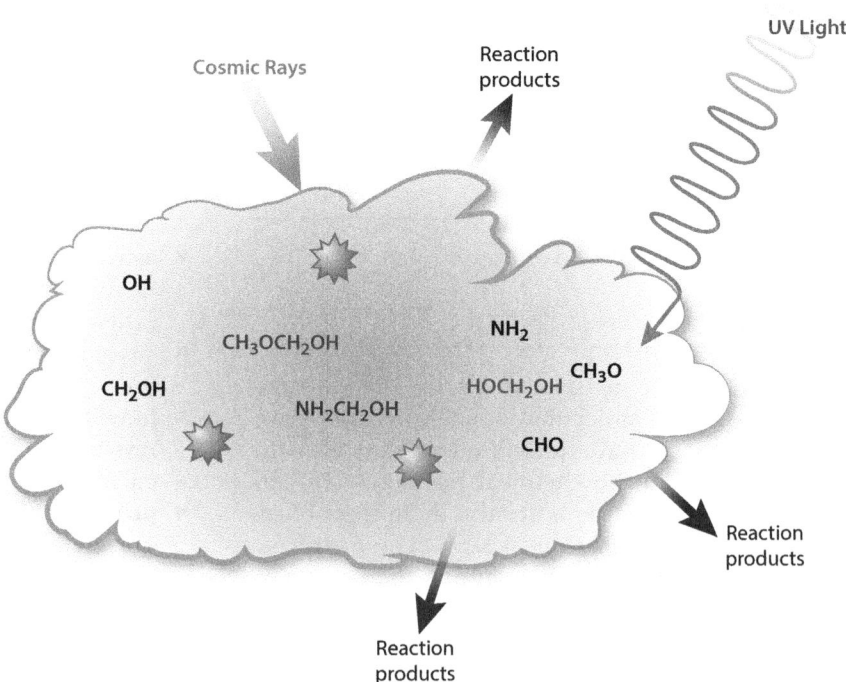

The gas-phase astrochemistry model describes gas-phase chemical reactions among molecules released from ices into the clouds surrounding protostars. Molecules evaporate from the surface of interstellar dust grains and react to form more-complicated molecules in star-forming clouds.

richness of the chemistry in molecular clouds, which cannot be accounted for in models that only allow for gas-phase interactions.

## Challenges of Studying Star-Formation Models

Astrochemists encounter several difficulties in trying to study any model of star formation. All of the processes that lead to the birth of stars occur within the molecular cloud, obscured from the view of even the most powerful telescopes. In addition, molecular hydrogen is extremely difficult to detect. These processes may also take tens of thousands of years to move from stage to stage. Therefore, it is virtually impossible to follow the development of any particular star.

Embryonic and newborn stars do not give off their own visible light. However, they and the processes that create them do emit light in the nonvisible spectrum, such as infrared. Specially equipped telescopes can detect this light, allowing researchers to observe the process. They can find the areas of molecular hydrogen ripe for star formation by looking for carbon monoxide. Scientists have determined that carbon monoxide is found in space at a ratio of one carbon monoxide molecule for every ten thousand hydrogen molecules. Carbon monoxide is easier to detect in space, so searching for those molecules allows researchers to find areas of hydrogen molecules as well. In addition, scientists can determine the age of a star based on the light it emits and the elements it is burning to produce it, among other factors. So although studying one star through its entire lifetime is not possible, scientists can form theories about the future of young stars by comparing them to other, older stars.

## Ongoing Efforts to Understand the Gas-Phase Model

Astrochemists study the gas-phase reactions of star formation using satellites and specially equipped telescopes. They also seek to recreate some of the reactions and processes that are part of star formation in laboratories on Earth. Understanding the processes that lead to the formation of stars is important to understanding how the universe was created, how it is changing, and what the future might have in store.

Studying the origin of stars is important for another reason. The same matter that makes up stars makes up everything on Earth, including people. In laboratory conditions, scientists have been able to observe organic molecules that are the precursors of life in ice that has been treated to replicate the conditions in star-formation areas. Through continued study of stars, scientists in a range of disciplines from astrophysics and astrochemistry to biology hope to gain a greater understanding of how the formation of the universe led to life on Earth.

—*Janine Ungvarsky*

## Further Reading

Choi, Charles Q. "Simple Recipe for Star Formation Revealed." *Space.com*. Purch, 10 Apr. 2014. Web. 23 Apr. 2015.

Kemsley, Jyllian. "Space Dust Science." *Chemical and Engineering News* 19 Apr. 2010: 36–38. Print.

"Molecular Cloud." *Cosmos: The SAO Encyclopedia of Astronomy*. Swinburne U of Technology, n.d. Web. 23 Apr. 2015.

"Molecular Hydrogen." *Cosmos: The SAO Encyclopedia of Astronomy*. Swinburne U of Technology, n.d. Web. 23 Apr. 2015.

"Star Formation & Molecular Astrophysics." *Blake Research Group*. California Inst. of Technology, n.d. Web. 23 Apr. 2015.

"Stars." *Herschel Space Observatory*. NASA, n.d. Web. 23 Apr. 2015.

"Stars." *NASA Science Astrophysics*. NASA, 2015. Web. 7 May 2015.

"The Story of Star Formation." *Cool Cosmos: Star Formation*. California Inst. of Technology, n.d. Web. 23 Apr. 2015.

Ward-Thompson, Derek, Helen Fraser, and Jonathan Rawlings. "The Chemistry of Star Formation." *News and Reviews in Astronomy and Geophysics* 43.4 (2002): 26–27. Print.

# GEMINI

## FIELDS OF STUDY

Stellar Astronomy; Observational Astronomy

## SUMMARY

A part of the zodiac, the constellation Gemini is located between fellow zodiac constellations Taurus and Cancer. The stars Castor, Pollux, and Alhena are the brightest and most easily identified in the constellation. The Gemini constellation has been identified by many different cultures, but it is most often associated with the Greek myth about twins named Castor and Pollux. The constellation Gemini is important to modern astronomers for its pulsating Cepheid variable stars, which have helped determine the relationship between a star's period and its luminosity.

## PRINCIPAL TERMS

- **celestial equator:** the imaginary line above Earth's equator that halves the celestial sphere; it is equally distant from the celestial poles.
- **constellation:** a section of the night sky identified by patterns of stars seen from Earth.
- **declination:** the north-south position of a celestial body relative to the celestial equator expressed in degrees of arc.
- **International Astronomical Union:** a worldwide association of professional astronomers that sets the rules for naming celestial bodies and features on them and defines scientific constants of importance to astronomy.
- **right ascension:** the east-west position of a celestial body defined in relation to the celestial equator and expressed in hours and minutes, not degrees of arc.

## Gemini and the Zodiac

Gemini, which means "twin" in Latin, is a constellation, or area of the night sky defined by pattern of stars, has been seen as representing twin boys from Greek mythology. Gemini, like other constellations, has been identified by cultures around the world since ancient times. The International Astronomical Union recognizes Gemini among its eighty-eight official constellations.

Gemini is part of the zodiac. The sun appears to pass through Gemini and the other twelve constellations of the zodiac. This path, called the ecliptic, is an illusion caused by Earth's orbit of the sun. However, the constellations, including Gemini, are noted by their distance relative not to the ecliptic but to the celestial equator. Gemini, for instance, has a declination of +20 degrees from the celestial equator and a right ascension of about seven hours.

When drawn, the constellation Gemini looks like twin boys standing or sitting side by side. The stars in the constellation look similar to two stick figures. Two bright stars represent the heads; faint stars represent the bodies and arms; and a few bright stars represent the twins' feet. Gemini is one of the few constellations that has an outline that looks similar to what the constellation represents.

Gemini is located between Taurus to the west and Cancer to the east. It is northeast of Orion. The Milky Way is visible at the southwestern tip of the constellation. Stargazers in the Northern Hemisphere have their best view of Gemini during February when it is located high in the winter sky. By April and May, the constellation is far to the west and can be seen after sunset.

The Gemini constellation is made up of a number of stars and other objects. Castor and Pollux are two of the main stars in the constellation. They are named after figures from Greek mythology and are the "heads" of the two twins.

The star Pollux, also known as Beta Geminorum, is the seventeenth brightest star in the sky and the brightest star in the Gemini constellation. Pollux is an orange giant located about thirty-five light-years away from Earth. Castor, also known as Alpha Geminorum, is actually a sextuplet star system, meaning it is made up of six stars bound together by gravity. Castor is about fifty light-years from Earth. Castor appears less luminous than Pollux in the night sky. Another bright star in the constellation is Alhena, also known as Gamma Geminorum. It is the foot of one of the twins. In places where pollution is an issue, Castor, Pollux, and Alhena are often the only stars of the constellation visible to the naked eye.

The Eskimo Nebula and the Medusa Nebula are also located inside the Gemini constellation.

## The Myth of Gemini

For tens of thousands of years, humans have looked to the sky and found objects in the patterns of the stars. Scholars believe that humans used constellations to represent religious stories. They also used them to tell the time of year in order to plant and harvest at the right times. Since constellations appear in about the same place in the sky at the same time year every year, farmers could track the time of year by the position of the constellations.

Gemini has been identified by humans for thousands of years. The constellation is one of the forty-eight discussed by the early astronomer Ptolemy (ca. 100–170 CE). A number of different stories have been told about the constellation. The most enduring story is of the twins Castor and Pollux.

In Greek mythology, Castor and Pollux were twin brothers. They had the same mother—Leda, the queen of Sparta—but different fathers. According to the story, the god Zeus disguised himself as a swan and seduced Leda. Leda had four children, two of whom were Zeus's, while two were those of her husband, Tyndarus. Castor's father was Tyndarus, but Pollux's father was Zeus. Because his father was a god, Pollux was immortal. When Castor eventually died, Pollux became distraught. He asked Zeus to allow him to die so he could be with his brother. Zeus then put Castor and Pollux in the sky so they could always be together.

The Gemini constellation has been an important symbol in other cultures as well. Ancient sailors believed that the Gemini twins protected those who sailed the ocean. People in Elizabethan England considered the twins protectors of the entire sea.

## Studying Gemini and Other Constellations

Scientists study constellations to learn more about the universe. Scientists sometimes use constellations to the name stars, so that the stars can be easily categorized and located. For example, the stars Castor and Pollux were named by astronomers many years ago; however, as scientists identified more stars in the constellation, they wanted to give all the stars easily recognizable names. So scientists gave the stars in the Gemini constellation names such as alpha Geminorum and beta Geminorum. The Greek letters alpha, beta, and so forth indicate the luminosity of each star in the constellation.

One star of interest in Gemini is Mekbuda, also known as zeta Geminorum. This star is a supergiant with a radius of about 220,000 times the size of the sun. This star is important to scientific research because it is a Cepheid variable. Cepheid variables help scientists better measure the distances between the stars and Earth. These large stars have used up their hydrogen fuel supplies, so they begin to pulsate. Scientists have found the periods of the stars are related to their luminosity. Scientists can determine how luminous a Cepheid variable is based on the period. Then, they can see how luminous the stars appear in the sky. With these two pieces of information,

Star chart of the Northern and Southern Hemispheres in November. Lines connect the stars of the constellation Gemini.

scientists can determine how far away stars are from Earth. This measurement method has been refined over the years and has become integral to stellar research.

—Elizabeth Mohn

**FURTHER READING**

"The Constellations." *IAU.* Internatl. Astronomical Union, n.d. Web. 10 Apr. 2015.

"Constellations in the Zodiac." *NASA.* Natl. Aeronautics and Space Administration, 1 Apr. 2015. Web. 6 Apr. 2015.

Kambic, Bojan. *Viewing the Constellations with Binoculars: 250+ Wonderful Sky Objects to See and Explore.* New York: Springer, 2010. Print.

Nave, R. "Ecliptic Plane." *HyperPhysics.* Dept. of Physics and Astronomy, George State U, 2012. Web. 10 Apr. 2015.

Rao, Joe. "How to Observe Gemini, the Heavenly Twin Constellation." *Discovery.com.* Discovery Communications, 12 Feb. 2012. Web. 30 Mar. 2015.

Sen, Nina. "Interesting Facts about the Gemini Constellation." *LiveScience.org.* Purch, 8 Sept. 2012. Web. 30 Mar. 2015.

Soper, Davison E. "Cepheid Variable Stars as Distance Indicators." *Institute of Theoretical Science.* U of Oregon, n.d. Web. 30 March 2015.

Zimmermann, Kim Ann. "Gemini Constellation: Facts about the Twins." *Space.com.* Purch, 30 July 2012. Web. 30 Mar. 2015.

# GLONASS

**FIELDS OF STUDY**

Geodetics; Space Technology

**SUMMARY**

GLONASS is a Russian space-based satellite navigation system. GLONASS is short for Global Navigation Satellite System. It is an alternative to the United States' Global Positioning System (GPS), but can be used alongside GPS to provide position information more speedily. GLONASS is used to provide air, marine, and other users with positioning, velocity measuring, and timing data.

**PRINCIPAL TERMS**

- **Global Positioning System (GPS):** satellite navigation system developed by the United States. GPS is used to determine the position, navigation, and timing of objects through data from satellites.
- **satellite:** an object that orbits another object; most commonly, a machine that is launched into space to orbit Earth.

**MAN-MADE CONSTELLATIONS**

A satellite navigation system is a network of artificial satellites used to map and position precise locations across the planet. Scientists place satellites into orbit around Earth by attaching them to rockets. The rockets launch and then release the satellites at a specific altitude. Once in orbit, satellites can send their time and positioning signals by radio waves to electronic receivers on the ground. Computers can then interpret these signals to determine the receiver's position.

Scientists use satellite systems to calculate factors such as longitude, latitude, and altitude. Satellites determine great distances by measuring the amount of time it takes for the satellite signal to reach the signal receiver on the ground. Satellites also make time synchronization possible by coordinating their signals to transmit at exactly the same time. Most positioning devices need to be in contact with at least four satellites to get an accurate reading. The more satellites a receiver connects to, the more accurately it will locate a position. Errors may occur in satellite navigation, however. Atmospheric conditions can interfere with a satellite's signal. Inaccuracies also occur when signals bounce off surfaces such as tall buildings before reaching the ground receiver.

Satellites travel around the earth in what is termed medium Earth orbit (MEO). MEO is from about 10,000 to 20,000 kilometers (6,300 to 12,500 miles) above sea level. In MEO, the satellites travel slowly enough for observation and can be arranged to achieve maximum signal transmission. Ground systems must constantly monitor satellite positions to ensure the highest possible positioning accuracy. Advanced systems with optimal conditions may provide positional accuracy to within a few centimeters.

GLONASS is operated by the Russian government and is one of several government-funded satellite navigation systems orbiting Earth. It followed the Global Positioning System (GPS) of the United States as the second such system available. GPS and GLONASS were also the first to provide coverage of the entire Earth, earning designation as global navigation satellite systems (GNSS). Other GNSSs under development include Europe's Galileo satellite network and China's Compass system. Regional systems include the BeiDou Satellite Navigation System (BDS) of China, and the Indian Regional Navigation Satellite System (IRNSS) of India. Simultaneous use of multiple satellite systems give the greatest amount of positioning data for a receiver.

### HISTORY AND DEVELOPMENT

Development of GLONASS began in 1976 as a Soviet military initiative. It was intended to provide ballistic missile navigation and targeting capabilities, similar to the origins of GPS in the United States. The Soviet Union's goal was to have a fully functional comprehensive satellite system in operation by 1991. Beginning in 1982, various rocket launches released forty-three satellites into space to create the constellation of devices needed to achieve coverage.

However, by 1991 the Soviet Union had collapsed, and only twelve working satellites remained. The last satellites were added in 1995 to make GLONASS operational, with twenty-four total satellites. However, the Russian economy experienced a massive downturn, and GLONASS operations suffered as a result. In 1996 the system was opened to civilian use, but was too expensive to maintain. Satellites were left to deteriorate in space and the program was essentially abandoned for nearly a decade.

Russia began restoring GLONASS satellites in 2001 under the government of Vladimir Putin, who made the project a national priority. Improved satellite models were developed and launched, beginning with the GLONASS-M models. All twenty-four satellites in the constellation were fully reestablished by 2011, giving Russia full global coverage. Over the next few years, Russia continued to release new satellites with improved navigational systems and longer lifespans. The next series, GLONASS-K, was first launched in February 2011. The second launch was delayed but took place in November 2014. GLONASS-K featured a lighter design and better accuracy. A second series of the GLONASS-K, GLONASS-K2, would follow soon after.

### GLONASS CONSTELLATION CONFIGURATION

A set of twenty-four operational satellites must be in orbit around Earth for the GLONASS system to be fully functional. Only twenty-one of the satellites transmit signals. The remaining three are spares. The satellites move in three orbital planes containing eight satellites each. The three planes follow a nearly circular orbit and are displaced from each other by 15 degrees of latitude. Each satellite takes just over eleven hours and fifteen minutes to completely orbit the earth.

When the constellation is fully populated, at least five satellites are visible from any given point at any given time, ensuring constant availability of GLONASS's global navigation. Due to its roughly circular orbit, a satellite only goes by the exact same spot approximately every eight days. This helps reduce the resonance effect that results from continuous usage of one frequency.

Russia's GLONASS differs from the American GPS in several ways. Each GLONASS satellite transmits its own frequency, unlike GPS satellites. Each GLONASS satellite uses frequency division multiple access (FDMA) in the L1 and L2 radio frequency bands. The L1 band covers 1602.5625 megahertz (MHz) to 1615.5 MHz, while L2 is 1240 to 1260 MHz. FDMA allows multiple users to access satellite frequencies simultaneously.

Early GLONASS satellites operated between the radio frequencies 1610.6 and 1613.8 MHz. This caused interference for radio astronomy research, however. In the early 1990s, the GLONASS administration and the Russian Federation came to an agreement with other countries to clear the band of this interference. All satellites launched after 2005 were designated specific frequency channels and had filters incorporated into their design to block out any transmissions other than those within the band.

### COMMERCIAL USE OF GLONASS

Twenty-first-century GLONASS developers intended their satellites for more than military usage. Russia wanted a navigation system on par with the United

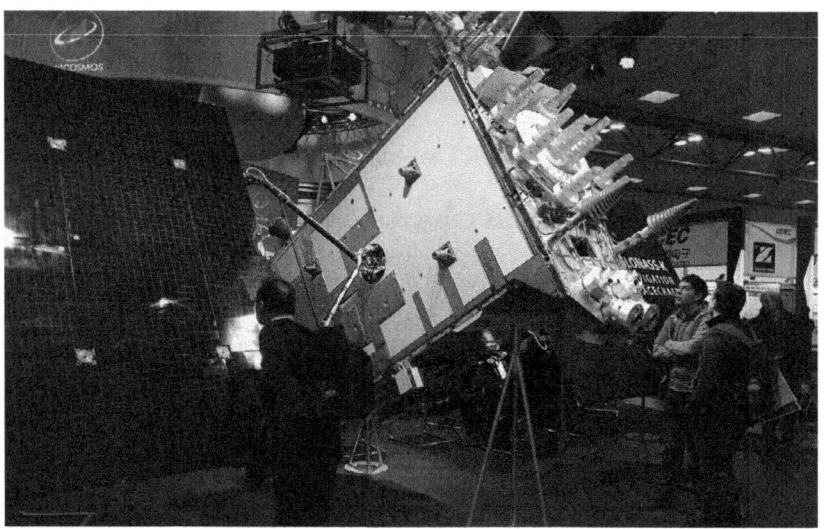
Model of GLONASS K1, a Russian navigation spacecraft. By Jürgen Treutler via Wikimedia Commons.

States' GPS. The success of civilian GPS had proven the market for navigation systems in automobiles, smartphones, and other devices. The first commercial automobile navigation device using GLONASS was released in 2007. However, it was larger, less effective, and more expensive than comparable GPS receivers. Early commercial initiatives also included the use of GLONASS in animal tracking collars and electronic bracelets for criminals. As the system improved, so did commercial and specialized devices. Russian law enforcement utilized GLONASS navigation for speedy location of crime scenes and locating police cars in need of backup. Statistics showed that the use of GLONASS among Russian police led to a decrease in the crime rate.

To encourage the adoption of GLONASS, the Russian government began taxing GPS-only devices coming into the country in 2010. The mandate excluded devices that supported both GPS and GLONASS technology. GLONASS became even more useful to the public after announcing it would pair up with GPS and other nationally recognized global navigation systems to provide a more powerful signal among navigation technology users.

Many major technology companies, including Sony and Garmin, began integrating GLONASS technology into their products. GLONASS quickly became a prominent part of the world's satellite navigation systems.

—*Cait Caffrey*

**Further Reading**

"About Satellite Navigation." *European Space Agency*. ESA, 14 Jan. 2013. Web. 6 May 2015.

Bergin, Chris, and William Graham. "Soyuz 2-1B Lofts GLONASS K-1 Satellite." *NASASpaceflight.com*. NASASpaceflight, 30 Nov. 2014. Web. 6 May 2015.

Clark, Stephen. "New Russian Navigation Satellite Launched into Orbit." *Space.com*. Purch, 28 Feb. 2011. Web. 6 May 2015.

"Global Navigation Satellite System (GLONASS) Overview." *PosiTim*. PosiTim, May 2010. Web. 6 May 2015.

"GLONASS Support in Our Latest Xperia™ Phones." *Developer World*. Sony Mobile Communications, 19 Jan. 2012. Web. 6 May 2015.

Moskvitch, Katia. "Glonass: Has Russia's Sat-Nav System Come of Age?" *BBC News*. BBC, 2 Apr. 2010. Web. 6 May 2015.

Razumovskaya, Olga. "GPS-Only Devices to Be Hit with Tax." *Moscow Times*. Moscow Times, 28 Oct. 2010. Web. 6 May 2015.

Russian Institute of Space Device Engineering. *Global Navigation Satellite System Interface Control Document*. Moscow: Russian Institute of Space Device Engineering, 2008. *Spacecorp.ru*. Web. 6 May 2015.

United Nations. Office for Outer Space Affairs. *Global Navigation Satellite Systems*. Vienna: United Nations, 2012. *UNOOSA*. Web. 6 May 2015.

Zak, Anatoly. "Uragan Satellites." *RussianSpaceWeb.com*. RussianSpaceWeb.com, 30 Nov. 2014. Web. 6 May 2015.

# GPS

## FIELDS OF STUDY
Astronomy; Space Technology

## SUMMARY
GPS, or Global Positioning System, is the satellite navigation system of the United States. GPS uses space-based satellites to transmit precise real-time positioning, navigation, and timing data to Earth-based receivers. Although GPS is a core tool of the military and intelligence sectors, the American government also makes the system freely available for civilian use.

## PRINCIPAL TERMS

- **satellite:** an object that orbits another object; most commonly, a machine that is launched into space to orbit Earth.
- **satellite navigation:** navigation, on Earth or in space, with the aid of satellites that pinpoint an object's position using radio signals.

### Satellite Navigation System
The Global Positioning System (GPS) is the satellite navigation system run by the United States. The system involves a series of space-based satellites that travel around Earth in fixed orbits. The satellites work together to collect data over a large swath of Earth. They then transmit that data via radio signals to Earth-based receivers. The satellite signals carry a time code and geographical data point that allow receivers to pinpoint their exact position, speed, and time anywhere on the planet.

GPS was originally conceived as a tool for the intelligence and military sectors. It was designed to help intelligence personnel gather more accurate information from different parts of the globe. It was designed to assist military personnel in conducting tactical operations with a greater degree of precision and accuracy. For example, the military could use GPS to fire weapons more accurately at otherwise inaccessible targets. GPS also helps deploy troops and supplies in remote regions and conduct search-and-rescue maneuvers under difficult conditions. In civilian life, GPS is used for navigation, archaeology, farming, and emergency response.

### How GPS Works
As many as thirty-two GPS satellites orbit in medium-Earth orbit (MEO) at altitudes of about 12,550 miles (about 20,000 kilometers). Initially, the system consisted of twenty-four satellites—four in each of six orbital planes around the globe. Additional satellites were added to improve the accuracy of the system, and to create redundancy, so that the system still functions accurately if some of the satellites fail. The

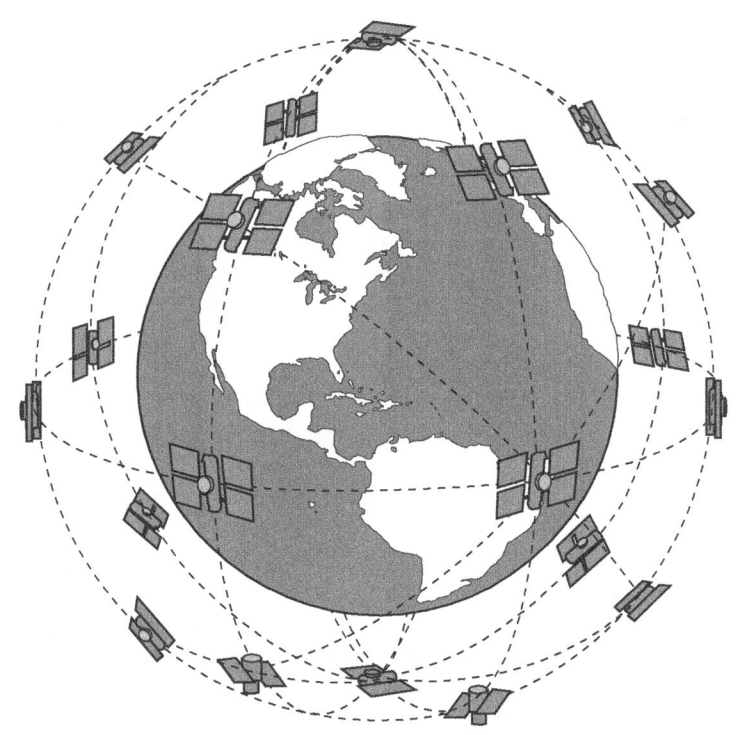

Many GPS satellites orbit Earth so that multiple satellites are visible from any point on the planet at any time, to reduce the chance of signal errors.

satellites cross the equator tilted at an angle of 55 degrees and circle Earth twice a day.

The satellites follow unique but overlapping orbits, which allows them to collect large amounts of data from wide areas. The satellites transmit their signals via radio waves back to Earth, where they are picked up by GPS-enabled receivers.

The satellites are programmed and synchronized to provide real-time positioning, navigation, and timing (PNT) information with tremendous accuracy—to within a few feet of receiver location. Positioning determines exact location and orientation anywhere in the world. Navigation calculates current and desired position (relative to another object or an absolute position). Navigation also allows for course, orientation, and speed corrections across the globe. Timing acquires and maintains accurate time, according to a set standard, anywhere in the world. Satellites use onboard atomic clocks to synchronize their timing.

GPS-enabled receivers use the satellite signals to calculate their distance, latitude, longitude, and speed based on the amount of time it takes the signals to travel from the satellites to the receiver. In most cases, the user of the receiver does not need to know anything about where the satellites are or how they or the receivers work. They merely need to turn on the device to benefit from the automated navigation system.

Because the satellites orbit above the clouds and dust in the atmosphere, their incoming signals are not blocked by objects in the atmosphere or on the planet's surface, such as mountains and tall buildings. However, receivers are passive instruments. They cannot transmit signals, so they require an unobstructed view of the sky to operate most effectively. This means they may not work well in areas where the sky is obstructed, and must be used outdoors. Additionally, GPS signals cannot pass through large bodies of water or into receivers that are underground.

## History of GPS

GPS was established in 1978 with the launch of the first satellite of the Navigation Signal Timing and Ranging Global Positioning System (NAVSTAR-GPS, commonly shortened to GPS). Over the next few decades, the system was expanded until it included as many as thirty-two satellites.

The core of GPS is operated and maintained by the US Department of Defense. Since the 1980s, the government has made GPS freely available for civilian use. The decision to make GPS freely available resulted from the realization that satellite navigation data had many practical applications. For example, the government realized that American ships and airplanes could use GPS to better determine their locations. It could also help them avoid accidentally straying into foreign territory, causing issues with hostile nations. Initially, a less accurate version of GPS was available for civilian use, to distinguish the civilian from the military system. This was called "selective availability." However, after civilian technology was developed that could work around selective availability, the government ended this feature starting in 2000. There remain two levels of GPS service, however: Standard Positioning Service (SPS) and Precise Positioning Service (PPS). SPS is accurate to within about 9 meters horizontally and 15 meters vertically, while PPS is accurate to within 2.7 meters horizontally and 4.9 meters vertically. PPS is available using an encrypted signal that is only accessible to the US military and the federal government, and certain other entities with government permission.

## GPS Applications

As GPS evolved and the technology became more sophisticated, thousands of different commercial devices and applications that use GPS have been created. GPS is used in applications from academic research, to agricultural pursuits, to simple navigation and everything in between.

One of the most common applications of GPS is navigation. GPS allows motorists, aircraft pilots, ship captains, and other transportation managers to navigate easily from one point to another. GPS is sophisticated enough to plot course corrections when they are needed. On the road, GPS allows freight operators to move freight across the nation using the shortest or least busy routes available. In the air, GPS is used to assist in the landing of aircraft and with air-collision avoidance systems. On the water, GPS is used for both recreational and commercial purposes. Recreational sailors and fishers can use GPS to manage marine navigation. Commercial ships can use GPS to manage their fleet as well as assist in steering and provide sophisticated hazard warning systems.

GPS technology also has many scientific applications. Archeologists can use GPS to determine exact location coordinates in the field. Based on on-site readings generated by GPS receivers, archeologists can determine where to dig. They can map their sites as single-point coordinates or as corridors with many points. Researchers can use GPS to measure their control points and quickly and accurately map their dig sites so that others know where the dig occurred.

Geological surveyors use GPS to measure the movement of Earth's tectonic plates. This can help geologists understand where earthquakes may occur and how much the plates have shifted after a quake. They can use GPS in two ways: by visiting an exact location periodically with a GPS receiver and taking measurements or by fixing a GPS receiver at a location and receiving a constant stream of measurements from that location. The first method is adequate for measuring the gradual seismic movement of tectonic plates. The second method is essential for gathering detailed information about sudden events, such as earthquakes.

Farmers use GPS to assist in precision farming techniques that protect the environment and maximize their profits. This approach, known as precision agriculture or satellite farming, involves the use of GPS-guided equipment to apply fertilizer, pesticides, and other substances only where and when they are needed. The GPS unit maps out the areas for application and also helps the farmers to understand how frequently the substances need to be applied. A GPS system can also be mounted on an agricultural drone—a small, robotic helicopter-style airborne unit—to guide aerial spraying operations.

Additionally, GPS can help farmers to determine where weed, insect, and/or disease infestations are occurring so these can be dealt with appropriately. GPS can be used to create field maps that exactly identify field borders, fence lines, canals, pipelines, and other locations such as wells, buildings, and landscape features.

Finally, while many applications of a core navigation system utilizing GPS mainly provide convenience or improved efficiency, others save lives. The ability to pinpoint addresses to a specific GPS coordinate enables first responders, such as fire, police, and ambulance personnel, to more exactly identify the location where help is needed. The use of onboard navigation systems that incorporate the GPS technology, coupled with these exact locations, means first responders reach sites more quickly and with fewer delays caused by misdirection.

—*Marie Keenan, MS*

**FURTHER READING**

"Archeology Program." *National Park Service*. NPS, n.d. Web. 10 July 2015.

"Beyond GPS: 5 Next Generation Technologies for Positioning, Navigation, & Timing (PNT)." *Defense-Aerospace.com*. DARPA, 24 July 2014. Web. 10 July 2015.

Cooksey, Diana. "Understanding the Global Positioning System (GPS)." *MSU Global Positioning System (GPS) Laboratory*. Montana State U, n.d. Web. 10 July 2015.

Dana, Peter H. "A Guide to the Global Positioning System." *Geographer's Craft Project*. U of Colorado at Boulder, 1999–2015. Web. 10 July 2015.

"GPS Info." *Naval Oceanography Portal*. US Navy, n.d. Web. 10 July 2015.

"The Global Positioning System." *GPS.gov*. Natl. Coordination Office for Space-Based Positioning, Navigation, and Timing, 11 Feb. 2014. Web. 10 July 2015.

Nave, C. R. "Global Positioning Satellites." *HyperPhysics*. Georgia State U, 2012. Web. 10 July 2015.

Stenmark, John. "Precise to a Fault: How GPS Revolutionized Seismic Research." *Earth: The Science behind the Headlines*. Amer. Geosciences Inst., 30 Apr. 2014. Web. 10 July 2015.

# HERTZSPRUNG GAP

### FIELDS OF STUDY
Astronomy; Astrophysics; Stellar Astronomy

### SUMMARY
Astronomers use the Hertzsprung-Russell diagram to chart the evolution of stars. The Hertzsprung gap is a section of this diagram where few stars are found. The Hertzsprung gap is important because it defines the stage in a star's evolution in which it transitions from burning fuel in its core to burning fuel in the shell around its core.

### PRINCIPAL TERMS

- **Hertzsprung-Russell diagram:** a tool used by scientists to classify and chart stars according to their temperature and luminosity.
- **spectral classification:** the categorization of stars according to the electromagnetic frequencies they emit.
- **star cluster:** a group of stars that share a common origin and are held together by gravity for a period of time.

### THE HERTZSPRUNG-RUSSELL DIAGRAM

The Hertzsprung-Russell diagram (H-R diagram), devised in the early twentieth century by Danish astronomer Ejnar Hertzsprung (1873–1967) and American astronomer Henry Norris Russell (1877–1957), is a chart that plots the brightness of stars against their temperature. It provides a way of categorizing stars based on where they are in their life span. Scientists can use the H-R diagram to study the age and evolutionary stage of different stars.

Stars plotted in the H-R diagram fall into several main areas. One such area is the main sequence, which follows a gentle diagonal curve from left to right. It begins with the hot, bright stars in the upper-left corner and runs down to the lower-right corner of the square, where the cooler, dimmer stars are found. Most stars, including the sun, are found in the main sequence. The red giants and supergiants are found in the upper-right corner of the H-R diagram. These stars are brighter but have lower surface temperatures than other stars. White dwarves are found in the lower-left corner. These stars are hot but small and not as bright as other stars.

There is a "gap" in the H-R diagram, between the main sequence and the red giants and supergiants, where very few of the observed stars in the universe can be found. However, this gap is not truly empty. Stars in the main sequence burn hydrogen in their core, resulting in their hot, bright status. Red giants and supergiants have exhausted the hydrogen in their core and have begun to burn the shell surrounding the core, causing them to be brighter but lower in temperature. When main-sequence stars are making the transition from burning the core to burning the shell, however, they are both dimmer and cooler, making them difficult to see. These are the stars that fall into the Hertzsprung gap. Because Hertzsprung was the first scientist to notice the lack of apparent stars in this area of the diagram, the gap was named for him.

### DIAGRAMMING THE STARS

Scientists compare the characteristics of similar objects in order to either determine if the objects are related or identify tendencies in their actions and reactions. In the early 1910s, Hertzsprung and Russell each independently devised a method for charting the characteristics of stars to make such comparisons easier. Hertzsprung created a chart comparing the luminosity (total energy output of an astronomical object) of stars to their temperature (as represented by color) to see how the two characteristics were related, while Russell plotted the stars' absolute magnitude (intrinsic brightness) against their spectral type. The spectral type of a star is determined by spectral

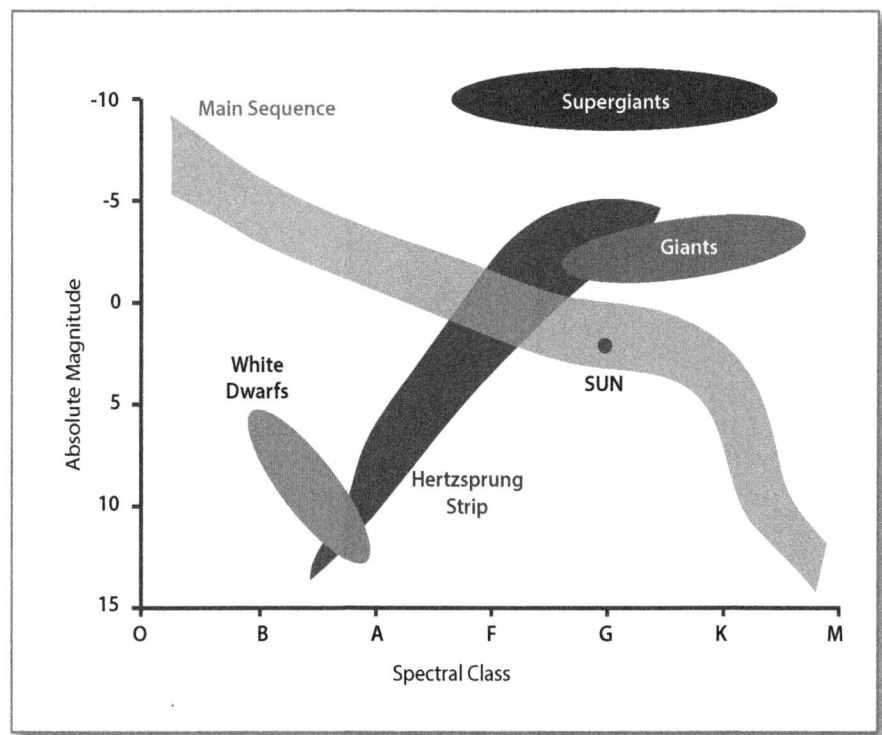

The Hertzsprung gap, also known as the instability strip, is a spectral class and absolute magnitude range on the Hertzsprung-Russell diagram that is minimally populated with stars. The natural evolution of stars causes them to pass quickly through this state. When a star during its evolution crosses the Hertzsprung gap, it means that it has finished core hydrogen burning but has not yet started hydro-gen shell burning. The RR Lyrae variables are an example of one of the few groups of stars found within the Hertzsprung gap.

classification, which categorizes stars based on the spectrum of electromagnetic radiation they emit.

The two methods produced similar charts. This is because magnitude is a function of luminosity, and a star's emission spectrum is determined mainly by its temperature. When scientists plot stars and star clusters in the H-R diagram, they use relative temperature, brightness, and color to determine both the age of the star or cluster and where it is in its evolution.

Stars may appear in one of two types of star clusters: open clusters or globular clusters. Open clusters are groups of stars that are separate enough for an observer with a telescope to see them as individuals. A few open clusters, including the Pleiades and the Hyades, are close enough to Earth to be seen without a telescope. These open clusters are also called galactic clusters because they are found on the arms of spiral galaxies. While gravity holds the cluster together, it is weak. Members of the cluster can be ejected from the cluster by the gravitational force of other space objects.

Globular clusters are groups of stars found mainly in the halo surrounding the galaxy. They are much older than other stars. The gravitational forces that hold globular clusters together are strong, so these clusters do not usually lose their stars to other forces.

## What Scientists Learn from Plotting Stars

Scientists know that all stars follow a similar pattern throughout their lifetime. The stages of this evolution are determined by the star's composition and how it burns matter to create light and energy. If they know the temperature and luminosity of a star, scientists can determine both its age and its stage in life.

The H-R diagram helps with this determination. As stars age, they move across the diagram over thousands or tens of thousands of years. Knowing if a star is in the main sequence, among the red giants or the white dwarfs, or in the Hertzsprung gap tells scientists more about a star than they could learn from simple observation.

—*Janine Ungvarsky*

## Further Reading

"The Hertzsprung-Russell Diagram." *Australia Telescope National Facility.* CSIRO, n.d. Web. 30 Mar. 2015.

"Hertzsprung-Russell Diagram." *Cosmos: The SAO Encyclopedia of Astronomy.* Swinburne U of Technology, n.d. Web. 15 Mar. 2015.

National Optical Astronomy Observatory. "Supergiant Census Supports Stellar Evolution Theory."*Astronomy Magazine.* Kalmbach, 23 Apr. 2012. Web. 30 Mar. 2015.

"Spectra and What They Can Tell Us." *Imagine the Universe!* NASA, Aug. 2013. Web. 30 Mar. 2015.

"Spectral Classification." *Cosmos: The SAO Encyclopedia of Astronomy.* Swinburne U of Technology, n.d. Web. 15 Mar. 2015.

"Star Clusters." *Australia Telescope National Facility.* CSIRO, n.d. Web. 30 Mar. 2015.

# HIGH-MASS STARS

### FIELDS OF STUDY

Astronomy; Cosmology; Stellar Astronomy

### SUMMARY

High-mass stars have hydrogen fusion cores and masses much greater than the sun. They burn brightly but have relatively short lives that end in powerful explosions called supernovas. Supernova explosions cast into the galaxy the elements created within the core of the stars. Scientists have determined that everything on Earth, including humans, contains elements that came from long-ago supernovas produced by high-mass stars.

### PRINCIPAL TERMS

- **black hole:** a region of space with a gravitational pull so strong that nothing, not even light, can escape it.
- **carbon-nitrogen-oxygen (CNO) cycle:** a six-step process that converts hydrogen into helium within the cores of high-mass stars.
- **neutron star:** a dense, rapidly spinning star made of the material that remains after a star becomes a supernova. It is formed from the neutrons created by the reaction of the star's remaining protons and electrons.
- **supernova:** a massive explosion that results when a dying star exhausts its fuel. The star's core collapses and explodes, releasing large amounts of energy and matter into the universe.

### HIGH-MASS VERSUS LOW-MASS STARS

Scientists use a number of techniques to classify stars in order to study them. One technique is to determine whether stars are high or low mass, which can be discovered by examining the amount of matter they contain.

High-mass stars contain significantly more matter than low-mass stars. Their mass may be up to eight times greater than that of the sun. Both high- and low-mass stars have cores in which nuclear fusion is constantly taking place, converting hydrogen to helium to create energy. However, the conversion process happens differently in high-mass and low-mass stars, although both processes take thousands of years.

High-mass stars convert hydrogen to helium mainly via the carbon-nitrogen-oxygen (CNO) cycle, so called because carbon, nitrogen, and oxygen isotopes act as catalysts, starting with carbon-12. They move quickly through the six-step process, giving them a comparatively shorter life span. Low-mass

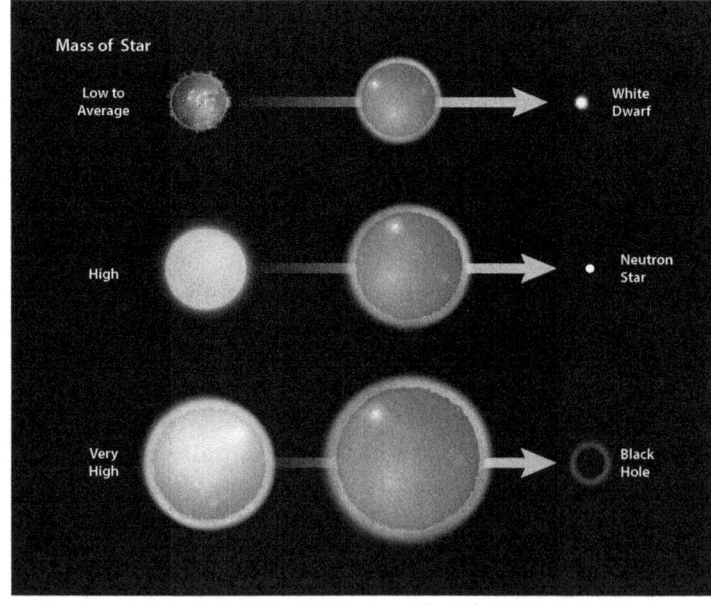

The fate of a star depends on its mass. Low- to average-mass stars evolve into white dwarfs. High-mass stars evolve into neutron stars. Very-highmass stars evolve into black holes.

stars use the proton-proton chain reaction and take a longer time to expend their energy. Both high- and low-mass stars eventually exhaust their energies and undergo change. The hydrogen in the core is used up, leaving a helium core surrounded by a hydrogen shell. When the helium core reaches a high enough temperature, the star begins fusing helium into carbon instead. This produces a carbon core surrounded by helium and hydrogen shells. What happens to the star next depends on whether it is high or low mass.

**How High-Mass Stars Become Supernovas**
The extra matter in high-mass stars means that they are more affected by gravity than low-mass stars. The carbon core contracts and heats up. The immense heat allows the carbon to fuse into oxygen, which then produces neon, then magnesium, then silicon, and finally iron. Iron does not generate enough energy for fusion, so the fusion process ends.

Once nuclear fusion has stopped, the star can no longer counter the force of gravity, and the core begins to collapse. As it does, the electrons in the iron atoms are forced ever closer to their nuclei, until they combine with protons to form neutrons. In an instant, the star shrinks from an object larger than the sun to one just a few kilometers in diameter. It then rebounds, causing a massive explosion known as a supernova that blasts the outer layers of the star into the universe.

**A Neutron Star Is Born**
Although the outer layers of a high-mass star are blown away in a supernova, the core remains. It is incredibly dense due to the combination of protons and electrons to form neutrons. This dense core forms a neutron star. The neutron core does not support fusion, so no energy is created. Without an outward flow of energy to counteract the pull of gravity, the star continues to shrink until it is only about 20 kilometers (12.4 miles) in diameter. Yet despite its smaller size, a neutron star contains more matter than the sun. Less than a teaspoon of this matter would weigh more than one hundred million tons on Earth. Thus, neutron stars have great gravitational pull despite their relatively small size.

When a high-mass star at least ten times the size of the sun becomes a supernova, a large neutron star forms with no energy output to counteract its immense gravity. This creates a black hole, a region of space with such incredible gravitational pull that it draws in all energy and matter that comes near. High-mass stars and their resulting supernovas are key ways that matter is distributed throughout the universe.

—*Janine Ungvarsky*

**Further Reading**
"Black Holes." *NASA Science.* NASA, 21 Jan. 2015. Web. 13 Mar. 2015.

"CNO Cycle." *Cosmos: The SAO Encyclopedia of Astronomy.* Swinburne U of Technology, n.d. Web. 18 Mar. 2015.

"The Death of Stars II: High Mass Stars." *Australia Telescope National Facility.* CSIRO, n.d. Web. 18 Mar. 2015.

"High Mass Star." *Space Book.* Las Cumbres Observatory Global Telescope Network, 2012. Web. 18 Mar. 2015.

"Neutron Stars." *Science and Space.* Natl. Geographic Soc., n.d. Web. 18 Mar. 2015.

"What Is a Supernova?" *NASA Education.* NASA, 4 Sept. 2013. Web. 18 Mar. 2015.

Young, Monica. "Cooking Up High-Mass Stars." *Sky & Telescope.* F+W Media, 4 June 2014. Web. 18 Mar. 2015.

# HUBBLE'S GALAXY CLASSIFICATION

### FIELDS OF STUDY

Extragalactic Astronomy; Observational Astronomy

### SUMMARY

Hubble's galaxy classification is a system devised by astronomer Edwin Hubble for categorizing different types of galaxies. The major types included are elliptical, spiral, barred spiral, lenticular, and irregular. Although modern astronomers consider some parts of Hubble's theories to be flawed, the galaxy classification system provided an important step toward a greater understanding of the universe.

### PRINCIPAL TERMS

- **bulge:** the central round part of a galaxy that contains old stars, dust, and gas.
- **disk:** the flat part of a spiral galaxy that forms the arms and surrounds the central bulge. Its stars are generally younger.
- **elliptical galaxy:** a massive group of stars and other space objects that is shaped like an ellipse, or a flattened circle.
- **halo:** the outer portion of a galaxy that contains old stars, star clusters, and dark matter.
- **irregular galaxy:** a massive group of stars and other space objects that has an unusual shape and does not share characteristics with other kinds of galaxies.
- **lenticular galaxy:** a massive group of stars and other space objects that shares traits of both elliptical and spiral galaxies.
- **spiral galaxy:** a massive group of stars and other space objects that takes on a pinwheel-like spiral form comprising a bright central bulge with swirling arms emerging from it.

### STUDYING THE GALAXIES

Hubble's galaxy classification was devised by American astronomer Edwin Hubble (1889–1953) around 1926. The major types classified by Hubble included elliptical galaxies, spiral galaxies, and barred spiral galaxies. He also noted additional types such as lenticular and irregular galaxies. This system marked a revolutionary advance in the study of the galaxies.

Long before Hubble, astronomers had little knowledge of galaxies, which are mostly extremely distant and difficult to see. Even when simple telescopes became available in the seventeenth and eighteenth centuries, they revealed little more than fuzzy patches of light in space. Early astronomers, puzzled, referred to these mysterious cloud-like bodies as "nebulae." Only much later, in the twentieth century, did new technology and scientific understanding lead to a greater ability to study these space features. At that time, astronomers realized that some of them were actually galaxies.

Most of these galaxies appeared to be held together by gravity, with a denser area in the middle and an outer region surrounding it called a halo. However, there was great diversity among the galaxies in size, shape, complexity, and color. Astronomers believed these changes were due to many factors, including the contents of the galaxies and their ages. They realized that galaxies were a vastly important feature of space with much to teach humans about the formation of the universe.

In the first decades of the twentieth century, the study of galaxies increased. Astronomers quickly identified two main varieties: the elliptical and the spiral galaxies. During Hubble's research on galaxies, he identified additional varieties, such as the barred spiral galaxy, the arms of which form a relatively straight line or "bar" where they emerge from the center. Hubble classified galaxies whose traits overlap those of both elliptical and spiral galaxies as lenticular. Finally, unusually shaped galaxies that did not share characteristics with other kinds of galaxies he termed "irregular." Hubble noted different ranges of shapes within each category, such as spiral galaxies with loose or tight spirals.

Hubble plotted these categories onto a diagram that was shaped like a horizontal letter Y. (Fellow astronomers called this diagram the "tuning fork" because of its two-pronged shape.) On the single flat plain of the diagram, Hubble placed the different varieties of elliptical galaxies. On the upper divergent line, he placed spiral galaxies; on the lower divergent line, barred spiral galaxies. Hubble placed lenticular galaxies at the point where the lines diverge. He generally did not include irregular galaxies on the diagram.

## Determining the Class of a Galaxy

Hubble's galaxy classification greatly simplified the process of studying and categorizing galaxies. Now astronomers could classify galaxies using simple, standardized criteria. Those basic criteria included galactic regularity or symmetry, central shape types, and the presence and arrangement of spiral arms on their disks. For each new galaxy, scientists have only to gather information on those features in order to determine whether the galaxy is elliptical, spiral, barred spiral, lenticular, or irregular.

If the galaxy being examined appears to have the shape of a flattened circle and has no evidence of spirals, it will most certainly be an elliptical galaxy. Hubble took that designation further by rating elliptical galaxies on their level of ellipticity ($E$), which is determined using the equation

$$E = 10 \times (1 - b/a)$$

where $a$ is the size of the larger axis of the ellipse and $b$ is the size of the smaller axis. Hubble's chart rates the least elliptical galaxies, which appear almost round, at E0, or zero ellipticity. Levels of ellipticity increase from there to E7, which is highly elliptical and resembles a football shape when observed from Earth.

If the galaxy exhibits any spiral arms, it will likely be classified as a spiral galaxy. Again, Hubble divided this group into classes based on their characteristics. Spiral galaxies of the Sa class have arms that tightly hug a large central bulge. Sb galaxies have arms slightly looser and longer around a medium-sized bulge. Sc galaxies have the loosest arms and the smallest central bulge.

If a spiral galaxy features a line or bar through the bulge from which the spirals emerge, it is classified as a barred spiral. Hubble denoted barred spiral galaxies as SB and noted their particular size and shape in a manner similar to regular spiral galaxies. SBa galaxies have tight spirals and large bulges. SBb galaxies have looser arms and medium bulges. SBc galaxies have the loosest arms and smallest bulges.

Galaxies that are shaped like spirals but have no visible spiral arms are classified as S0, or lenticulars. Lastly, if the galaxy does not fit into any of the above models, it is classified as an irregular galaxy. Irregular galaxies come in a countless variety of shapes and sizes and often appear asymmetric and disordered. Although irregular galaxies are not always included in Hubble's diagram, Hubble classified them as well, with the abbreviation Irr. Varieties that demonstrate some basic structure are designated Irr I, while Irr II varieties appear to be completely random in structure.

## Examples of Classified Galaxies

Using Hubble's criteria, astronomers have classified thousands of galaxies based on their sizes, shapes, and compositions. For example, Earth's

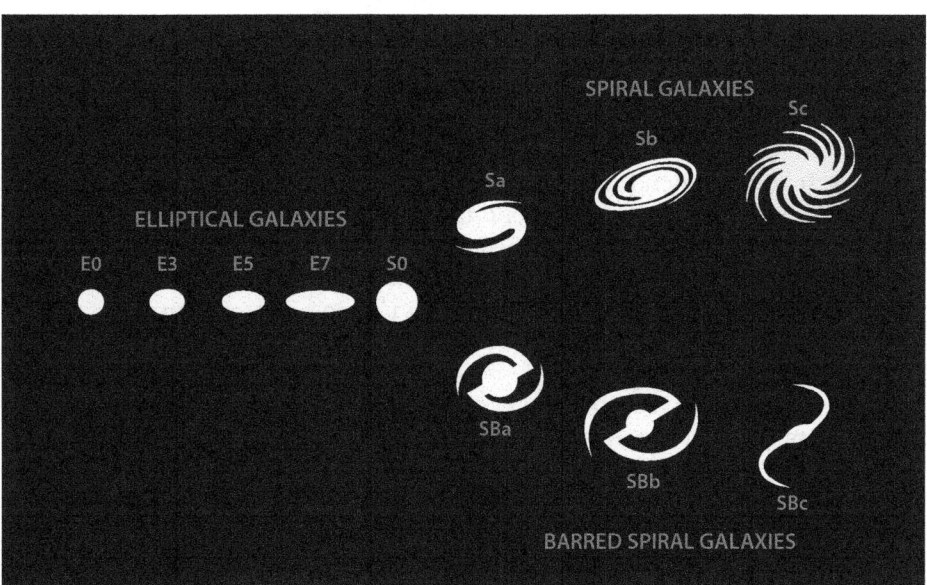

Edwin Hubble's classification scheme for galaxies includes elliptical, spiral, and barred spiral galaxies. Classes of elliptical galaxies range from round-looking galaxies (E0) to extremely elliptical galaxies (E7). Spiral galaxies may have few arms (Sa), or they may have many arms (Sc). Barred spiral galaxies are similar to spiral galaxies, except they all have a distinct bar passing through the galaxy's nucleus and the spiral arms extend from the ends of the bar.

galaxy, known as the Milky Way, is a barred spiral. Astronomers studying the arrangement of the galaxy noted that it has a center bulge leading to two bars that in turn lead to spiral arms. Other galaxies identified as barred spirals include New General Catalog 5101 (NGC 5101), a type SBa galaxy; NGC 1365, type SBb; and NGC 7479, type SBc. These galaxies share similar features but vary somewhat in structure, hence the different letter classifications.

Many other galaxies are regular spirals that do not have bars. The galaxy Messier 104 (M104), for instance, has a bright, large central bulge and large spirals surrounded by space dust. This shape has earned the galaxy the nickname the Sombrero galaxy. Spiral galaxies are also subcategorized by their details. The galaxy M81 is usually classified as Sa because its spirals are tightly wound around the central bulge. M51 has looser spirals, so it is classified as Sb. Finally, NGC 2997 has very loose spirals and a small bulge and is therefore classified as Sc.

If a regularly shaped galaxy is found to lack spiral arms, it is generally classified as an elliptical galaxy. Ellipticals are subcategorized by their shape. For example, the galaxy M59 is highly elliptical, resembling a football, and is therefore designated E5. Meanwhile, M89 is more circular than elliptical and therefore is rated E0, for zero ellipticity.

When astronomers discovered M82, they could clearly see an arrangement of stars and other space objects bound by gravity. However, the galaxy resembles a strange splash of light, lacking a clear shape or defining features. Therefore, astronomers classified it as an irregular galaxy. Due to its totally random-looking shape, M82 was subclassified as Irr II. An unusual-looking galaxy that has some semblance of order, such as the Large Magellanic Cloud, is categorized as Irr I.

## The Importance of Hubble's System

When Hubble first created his classification system, he suggested that the different forms of a galaxy might correspond to its age. Modern astronomers now believe that this is not necessarily correct. Galaxies form, age, and develop in different ways depending on their initial collapse, interactions with one another, and star formation. Astronomers have also determined that Hubble's system of classifying galaxies was somewhat subjective. For that reason, some astronomers have applied different classifications to the same galaxies due to differences of personal opinion.

Despite its flaws, Hubble's galaxy classification established a strong launching point from which researchers could embark on more in-depth studies of the galaxies. Over time, scientists continue to discover more about galactic types, formation, and evolution, thanks largely to the early innovations of Edwin Hubble.

—*Mark Dziak*

### Further Reading

Buta, Ronald J., Harold G. Corwin Jr., and Stephen C. Odewahn. *The de Vaucouleurs Atlas of Galaxies.* New York: Cambridge UP, 2007. Print.

"The Case Files: Dr. Edwin Hubble." *The Franklin Institute.* Franklin Inst., 2015. Web. 26 May 2015.

"Hubble Classification." *Cosmos: The SAO Encyclopedia of Astronomy.* Swinburne U of Technology, n.d. Web. 26 May 2015.

"Hubble's Classification Scheme." *Center for Educational Resources (CERES) Project.* NASA/Montana State U–Bozeman, n.d. Web. 26 May 2015.

"The Hubble Tuning Fork: Classification of Galaxies." *European Space Agency.* ESA, 15 Aug. 2013. Web. 26 May 2015.

O'Meara, Stephen James. *Deep-Space Companions: The Messier Objects.* New York: Cambridge UP, 2013. Print.

Palma, Christopher. "Hubble's Tuning Fork and Galaxy Classification." *Department of Astronomy and Astrophysics.* Pennsylvania State U, 2014. Web. 26 May 2015

Schombert, James. "Hubble Sequence." *Astronomy 123.* U of Oregon, n.d. Web. 26 May 2015.

# HYDROGEN CLOUD

## FIELDS OF STUDY

Astronomy; Observational Astronomy

## SUMMARY

A hydrogen cloud is an envelope of hydrogen that exists around the exterior of active comets. These clouds can be tens of millions of kilometers wide. They are not visible to the naked eye but can be detected in ultraviolet light. Scientists have learned more about hydrogen clouds through the exploration of active comets.

## PRINCIPAL TERMS

- **coma:** the atmosphere that forms around a comet when it passes near the sun.
- **comet:** a small solar system body mostly made up of mostly carbon and ice.
- **nucleus:** the body of a comet, made up of ice, dust, and carbon.

### Hydrogen Cloud Formation

A comet is made up of a nucleus, a coma, and a tail. The nucleus is made mostly of ice and carbon. In deep space, far away from the sun, the comet is a frozen mass and does not have a coma or a tail. When the comet travels closer to the sun, however, the surface ice begins to sublimate, meaning that it turns from a solid directly into a gas without first becoming a liquid. When sublimation begins, it creates an atmosphere around the nucleus. This atmosphere is the coma. The tail of the comet forms when solar winds and radiation push particles from the coma toward the back of the comet.

The coma of the comet is made up of a gas and dust. Gas that forms during sublimation jets and fans away from the nucleus, and it forces dust out from the nucleus. Often sublimation and radiation from the sun also cause a hydrogen cloud to form outside the coma of the comet. These clouds can be massive, measuring up to tens of millions of miles across. Some hydrogen clouds even dwarf the sun. The Hale-Bopp comet had a hydrogen cloud that was 100 million kilometers (roughly 62 million miles) in diameter, while the sun is less than 1.4 million kilometers (roughly 870,000 miles) in diameter. Hydrogen clouds are not visible to the naked eye.

### Attributes of Hydrogen Clouds

Hydrogen clouds extend far outside the coma because they are made up of hydrogen atoms. Because hydrogen atoms are much lighter than other atoms, they move out into space faster than the gas, dust, and particles that make up the coma. The hydrogen cloud that envelops a comet cannot be seen from Earth. Instead, pictures of this phenomenon are taken with spacecraft. Because the hydrogen cloud cannot be seen with naked eye, scientists take images of the clouds using ultraviolet radiation.

Hydrogen clouds vary in shape. Some are symmetrical and some are asymmetrical. This is likely due to varied pressure from solar winds and radiation, which also move gas and dust to create a comet's tail.

For many years, scientists suspected that water vapor would be present in comets' gas comas due to the sublimation of ice. In 1985, this idea was verified through observation. Scientists believe that the sun's ultraviolet radiation breaks down the water vapor molecules in the coma to separate the hydrogen atoms from the oxygen. The hydrogen atoms then form most of the surrounding hydrogen cloud. The breakdown of other molecules likely adds more hydrogen to the cloud.

### Study of Comets and Hydrogen Clouds

Comets have been observed by humans for thousands of years, but for many years humans did not know what they were. Many believed that comets were bad omens or had supernatural powers. However, in the sixteenth and seventeenth centuries, scientists such as Tycho Brahe (1546–1601) and Isaac Newton (1642–1727) made observations that helped people better understand comets.

In 1705, astronomer Edmond Halley (1656–1742) noticed that comets that had been observed in 1531, 1607, and 1682 had basically the same orbits. Halley determined that the three comets were in fact three appearances of the same comet, following a regular path and schedule. This comet was later named

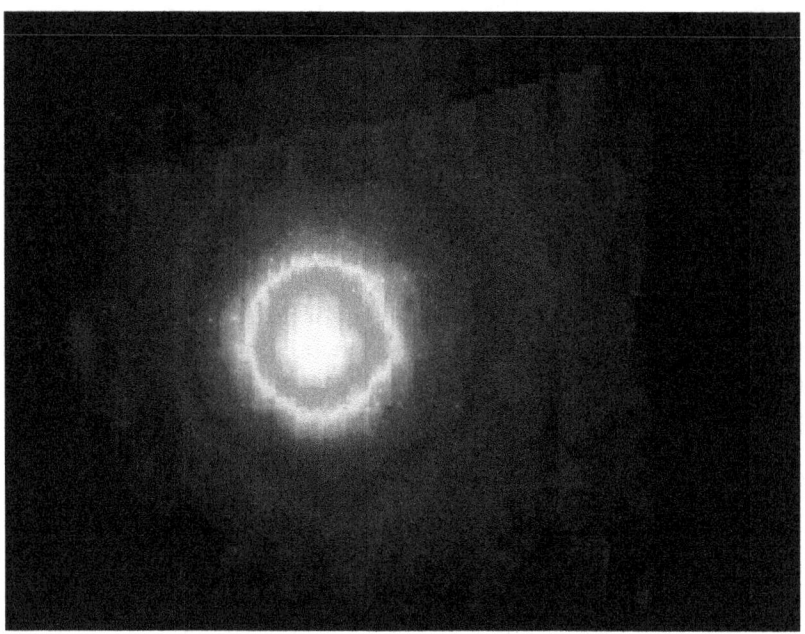

The hydrogen cloud of the comet Hale-Bopp, detected by UV light, extends much farther than the comet's tail (shown using visible light). NASA.

Halley's comet. Over the years, scientists identified more comets and developed theories about them.

Technological advances in the twentieth century allowed scientists to observe and study comets' hydrogen clouds. The clouds were first discovered by the Orbiting Astronomical Observatory 2 (OAO-2), launched in 1968. In February 1986, the *Pioneer Venus* orbital spacecraft was able to view Halley's comet with ultraviolet sensors. The images taken of the comet's hydrogen cloud gave scientists important information about the activity and makeup of the comet. They also revealed the huge extent of hydrogen clouds for the first time. In 2014, the Rosetta space probe conducted further comet research, including landing a spacecraft on a comet's surface for the first time.

## Importance of Comets and Hydrogen Clouds

Understanding hydrogen clouds is important to better understanding comets. Based on their observations, scientists believe that much of the hydrogen in a hydrogen cloud comes from the breakdown of water molecules. Scientists use this fact to help them determine the overall composition of comets.

Studying comets in general is also important because comets carry water, hydrogen, and carbon, all of which are vital to human life. Some scientists believe that some of the water on Earth came from comets. Furthermore, comets have the ability to greatly impact life on Earth, and understanding more about them will help scientists better understand their movements and their potential impacts on Earth.

—*Elizabeth Mohn*

### Further Reading

"About Comets." *Explore!* Lunar and Planetary Inst., 31 Oct. 2012. Web. 10 Mar. 2015.

"Anatomy of a Comet." *Smithsonian National Air and Space Museum.* Smithsonian Inst., 2002. Web. 10 Mar. 2015.

de Pater, Imke, and Jack J. Lissauer. *Planetary Sciences.* Updated 2nd ed. Cambridge: Cambridge UP, 2015. Print.

Dye, Lee. "An Aging Pioneer Has Halley Comet Show All to Itself." *Los Angeles Times.* Los Angeles Times, 27 Feb. 1986. Web. 10 Mar. 2015.

Lang, Kenneth R. *The Cambridge Guide to the Solar System.* 2nd ed. New York: Cambridge UP, 2011. Print.

Lang, Kenneth R. "Hydrogen Cloud of a Comet." *NASA's Cosmos.* Tufts U, 2010. Web. 10 Mar. 2015.

North, Gerald. *Astronomy in Depth.* Rev. and expanded ed. London: Springer, 2003. Print.

Shimizu, Mikio. "The Hydrogen Clouds of Comets." *Comets in the Post-Halley Era.* Ed. Ray L. Newburn Jr., Marcia Neugebauer, and Jurgen Rahe. Vol. 2. Norwell: Kluwer, 1991. 897–905. Print.

# INTERNATIONAL SPACE STATION (ISS)

## FIELDS OF STUDY

Aerospace Engineering; Astronautics; Space Technology

## SUMMARY

The International Space Station (ISS) is an artificial, habitable satellite that orbits Earth at an altitude of about 400 kilometers (250 miles). It has been designed, assembled, operated, and maintained by an international coalition that includes the space agencies of the United States, Russia, Japan, Canada, and Europe. At any time, up to six astronauts live and work on the space station, maintaining its operations and conducting important research studies that increase scientific knowledge of the solar system. ISS is hailed as a model for international scientific, technological, and diplomatic cooperation.

## PRINCIPAL TERMS

- **Canadian Space Agency:** Canada's government agency established in 1989 to manage Canadian space activities and research.
- **European Space Agency:** an intergovernmental organization made up of twenty-two European countries in charge of Europe's space program and research since 1975.
- **Japan Aerospace Exploration Agency:** Japan's government agency formed in 2003 to head the country's civilian space activities and research.
- **National Aeronautics and Space Administration:** the government agency established in 1958 to run and oversee the United States' space program and research.
- **Russian Federal Space Agency:** the government agency created in 1992 to officially coordinate Russian space activities and research.
- **Soyuz:** the Russian spacecraft used to carry astronauts to and from the International Space Station.

## An International Undertaking

The International Space Station (ISS) is the most complex space project ever undertaken. It is an enormous artificial, habitable satellite. A coalition of nations that includes the United States, Russia, Europe, Japan, and Canada designed, developed, built, operate, and maintain the station. ISS was established for the purpose of learning more about space in order to know more about Earth. The coalition's vision refers to the station as "a human outpost in space, bringing nations together for the benefit of life on Earth and beyond."

ISS is approximately the size of an American football field—about 109 meters (360 feet) long and 48 meters (160 feet) wide. It is so large that no Earth rocket had the power to launch it into space in one piece. Therefore, ISS was built in parts and assembled in space over a long period of time. The first piece of the station was launched from Earth in 1998. The last piece was added more than forty missions later, in 2011.

ISS is located in low Earth orbit at an altitude of about 400 kilometers (250 miles). It moves at a speed of about 28,000 kilometers (17,500 miles) per hour and orbits Earth every ninety minutes. Three to six astronauts live on the space station at any one time. These astronauts come from many different countries and stay on the station for missions that last an average of six months.

Designed as living and working quarters for astronauts, ISS has been occupied since 2000. It is constructed in a series of clustered modules. These are pressurized chambers that humans can live in. The modular structure has the added benefit of being customizable. Existing modules can be removed and new modules added as technologies improve and missions change.

The modules are grouped together in the center of the station and attached to a main truss (load-bearing support). The truss also supports the giant solar panels, heat radiators, and other exterior equipment that spread outside the station. This equipment provides the heat and power the station needs to support life.

The habitable modules also include a laboratory with a range of scientific equipment. The astronauts spend about thirty-five hours per week conducting experimental research in a range of disciplines, including physics, astronomy, biology, meteorology, and chemistry, among others. They also spend time operating the station's controls, cleaning and maintaining the equipment, and repairing anything that is not working optimally. Sometimes the astronauts take spacewalks outside the ISS to work on the outside of the station or make repairs.

ISS is a unique project not only because of the complex technology it needs to work, but also because it has been a greatly successful international effort. The main space agencies involved with ISS are the Canadian Space Agency (CSA), the European Space Agency (ESA), the Japan Aerospace Exploration Agency (JAXA), the Russian Federal Space Agency (Roscosmos), and the United States' National Aeronautics and Space Administration (NASA).

### Political and Technical Complexities

Representatives from the United States, Russia, Europe, Japan, and Canada set the legal framework for the ISS project in January 1998 by signing the International Space Station Intergovernmental Agreement (IGA). Through the IGA and subsequent agreements, officials created a cooperative partnership with shared responsibilities. These include the design, development, funding, operation, and use of the station. The agreement also established ISS as a peaceful venture in accordance with international laws and customs.

The ISS designers intended the station to serve as a laboratory and observation center in low Earth orbit. It was also appointed as a staging base for possible future missions to space bodies such as the moon, Mars, and asteroids. Later, the coalition agreed that ISS had other purposes and goals involving commerce, diplomacy, and education.

ISS operates as a joint station. Russian cosmonauts operate one side of the craft, and American astronauts operate the other. The American side includes modules and equipment from Canadian, Japanese, and European partners. The groups share responsibility for navigation, operations, and station command. They also share living space, meeting for meals, exercise times, and otherwise sharing resources. One of the greatest triumphs of ISS is the cooperation it fosters among the nations. The various groups work together to plan, coordinate, and monitor the different activities and missions conducted aboard.

ISS is connected to two main Earth-based mission control centers. These are operated by NASA and Roscosmos. NASA's center is located in Houston, Texas, while Roscosmos's center is located in Moscow, Russia.

### Unexpected Partnership

From the 1950s through the 1990s, the United States and the Soviet Union were engaged in the Cold War, with each side considering the other its major competitive threat. Part of the Cold War involved the space race. This was a competition between the two countries during which both sought to establish themselves as superior in space technologies.

During this time, both countries developed and launched a series of rockets and probes into space. This rivalry led to the development of both artificial satellites and manned spacecraft. The Soviet Union built the first simple space station by linking two of its craft in space. The United States developed the first reusable spacecraft through its shuttle program. These advances in technology helped scientists to learn more about the nature of Earth and space.

The space race was an important element of the larger competition between the two countries, as advances in space research could be put to military use. For example, space probes and satellites that supplied observational data could be used for spying purposes or to give one country an advantage in armed conflict.

The fall of the Soviet Union in 1991 and subsequent changes in political alliances and boundaries led to talks of an international partnership in space. These talks later came to fruition as the ISS.

### Life Onboard

Astronauts get to and from ISS via Soyuz TMA vehicles. This small Russian spacecraft launches and lands in the Republic of Kazakhstan. Historically it could take astronauts almost two days to reach the ISS, but recent trips have made the journey in about six hours. The Soyuz vehicle connects to ISS via specialized docks that allow the astronauts to enter and leave the space station. The return Soyuz trip is much faster and takes up to three astronauts about three and a half hours.

The inhabitants of ISS follow a designated schedule of tasks. Astronauts have compared the experience to life onboard a carefully controlled ship. Almost every day in orbit is planned by mission control. This keeps the astronauts occupied and ensures that the station continues to run in an optimal manner.

Astronauts spend many hours per day working in the laboratory or on research projects. They also complete many housekeeping and maintenance duties to make sure that the station continues to run as designed. Some days the astronauts spend hours removing and loading rubbish and moving fresh supplies into the station. On others the astronauts may spend the better part of the day preparing for brief space walks on the exterior of the station. These walks are conducted when observations or repairs need to be made. To stay safe during these walks, the astronauts wear special space suits that allow them to survive the hostile conditions of space.

Astronauts eat three meals and exercise at least two hours per day. This amount of exercise is needed to combat the effect of living in microgravity. Because of this microgravity, exercise is performed in unique ways. American astronauts ride an exercise bike that has no handles or seat. They can place a laptop anywhere in space around them to watch entertainment or speak to mission control while exercising.

Astronauts sleep in small, private units with sleeping bags attached to the walls. The attached sleeping bags anchor the astronauts to the wall and keep them from floating around during sleep. The bags are designed to help the astronauts feel contained so they can relax.

The International Space Station orbiting Earth. NASA.

**ISS in the Future**

ISS is an important part of NASA's space program, which is focused on designing and building capabilities to send humans farther into the solar system than ever before. ISS has made the goal of travel to Mars a real possibility because the station's microgravity environment lets scientists conduct research that would not be possible on Earth.

Through its work on ISS, NASA and other space agencies have developed a much deeper understanding of what it takes for humans to survive in space for long periods of time. Technologies developed for ISS have helped scientists to craft efficient water systems, advanced life support capabilities, and human-robotic interfaces that minimize the loss of human life. ISS has helped scientists build and test technologies such as automatic flight refueling and cargo supply systems.

Finally, ISS has been crucial in the creation of a sense of cooperation and goodwill among the participating space agencies and their nations. As scientists have teamed up and shared their research and resources, ISS has become a blueprint for international cooperation among the United States and its many allies.

—*Marie Keenan, MS*

**FURTHER READING**

"About the International Space Station." *European Space Agency*. ESA, 19 Nov. 2013. Web. 13 May 2015.

"About the Space Station: Facts and Figures." *National Aeronautics and Space Administration*. NASA, n.d. Web. 13 May 2015.

Fishman, Charles. "5,200 Days in Space." *Atlantic*. Atlantic Monthly, Jan.–Feb. 2015. Web. 13 May 2015.

"Higher Altitude Improves Station's Fuel Economy." *National Aeronautics and Space Administration*. NASA, 14 Feb. 2011. Web. 13 May 2015.

Sample, Ian. "Life aboard the International Space Station." *Guardian*. Guardian News and Media, 24 Oct. 2010. Web. 13 May 2015.

"Space Race." *Royal Air Force Museum*. Trustees of the Royal Air Force Museum, n.d. Web. 13 May 2015.

Tate, Karl. "The International Space Station: Inside and Out (Infographic)." *Space.com*. Purch, 20 Nov. 2012. Web. 13 May 2015.

"What Is the International Space Station?" *National Aeronautics and Space Administration*. NASA, 30 Nov. 2011. Web. 13 May 2015.

# INTERSTELLAR CHEMISTRY

**FIELDS OF STUDY**

Astronomy; Astrophysics; Astrochemistry

**SUMMARY**

Interstellar chemistry is the study of the quantity and chemical reactivity of the atoms and molecules that appear throughout interstellar space. Interstellar chemistry is a branch of astrochemistry, the study of chemical behavior in the universe. Scientists use spectroscopy and radioastronomy to understand the components of interstellar space. Interstellar chemistry helps astronomers understand how the stars and solar system formed. Studies in interstellar chemistry also assist scientists looking for life-sustaining planets in deep space.

**PRINCIPAL TERMS**

- **electromagnetic spectrum:** the standard range of all possible frequencies of electromagnetic radiation that classifies the colors of light (red, orange, yellow, green, blue, and violet) by their wavelength and energy. Beginning with red, each color emits more energy than the previous, while the wavelength shortens as it moves from red to violet.
- **high-resolution spectroscopy:** advanced spectroscopic technology capable of analyzing spectra in minute detail, giving greater information about the structure of the material that emitted it.
- **molecular radioastronomy:** the use of radio telescopes and remote sensing to detect signals from molecules in space.

**THE SPACE BETWEEN THE STARS**

The term "interstellar" refers to the area of space that lies between stars within a galaxy. Interstellar space is composed of gas, dust, dark clouds, and other matter referred to as the interstellar medium (ISM). Scientists discern the various levels of matter within the ISM based on whether the matter is ionic, atomic, or molecular. The ISM is mostly made of hydrogen and helium atoms but also small amounts of oxygen, carbon, and nitrogen. Ionized gas is also abundant throughout interstellar space. Analysts discovered that the thermal properties of the matter within the ISM create pressure throughout space. Cosmic microwave background radiation, magnetic fields, and cosmic rays also produce pressure in the ISM.

Prior to the invention of radio telescopes, which are large dish-shaped telescopes used to catch signals from space, scientists studied the ISM with basic visual telescopes. Data about the ISM's chemical behavior was attained by observing how the light from faraway stars affected interstellar matter. In the 1950s, scientists began using radio telescopes to analyze the ISM. Radio telescopes enabled them to study and distinguish the radio waves emanating from different ISM components. Radio waves could detect what phase of matter scientists encountered. For example, hydrogen atoms absorb and release small amounts of radio energy at a specific wavelength. By comparing nearby wavelengths, scientists can pinpoint the absorption or radiation levels of nearby hydrogen clouds. Satellite technology later enabled scientists to study the ISM with infrared satellite telescopes.

Scientists first discovered molecules in the ISM in the late 1960s, giving rise to the field of molecular radioastronomy. A group of astronomers determined carbon monoxide existed in the Orion Nebula in 1970. Before this, scientists had thought the ISM was atomic and too harsh for complex molecular compounds. Scientists located many more molecules over the next decades. As of 2015, more than 180 interstellar molecules had been discovered in the ISM.

#### USING LIGHT TO UNDERSTAND THE ISM

Scientists also turned to spectroscopy to better understand interstellar chemistry. Spectroscopy, or the study of the physical properties of light, was first examined in depth by English physicist Isaac Newton (1642–1727) as he studied optics during the late seventeenth century. Newton observed that white light dispersed into a rainbow of colors when shone through a prism. Newton noted the sources of white light originated from the sun, stars, and fire. Future experiments showed that the colors changed when interacting with various chemical elements. Light color was also determined by the wavelength of the light wave as well as how much energy the light emitted. By observing and measuring the spectra of various sources, such as the sun and flames, scientists assembled a standard solar spectrum that measured the wavelength, frequency, and energy of light waves. This standard range is called the electromagnetic spectrum.

The electromagnetic (EM) spectrum characterizes each color of light by how long or short its wavelength is and whether it emits high or low energy. Shorter wavelengths have higher energies, while longer wavelengths have lower energies. Radio waves, infrared, and microwave signals have long wavelengths. In the visible spectrum, red, orange, and yellow light has low energy and long wavelengths, while green, blue, and purple light have high energy and short wavelengths. Short wavelengths of high energy also characterize ultraviolet (UV) light, x-rays, and gamma rays. Most ISM emits light waves with short wavelengths and high energy. By studying the many EM wavelengths emitted or absorbed by atoms or molecules, astronomers are able to better understand interstellar chemistry. Spectroscopy can be applied to distinguish between specific forms of matter based on their spectral patterns. High-resolution spectroscopy uses high-resolution instruments to separate crowded spectra for a more accurate analysis of these light absorptions and emissions.

#### THE BEGINNINGS OF THE STARS

Scientists studying interstellar chemistry are often most concerned with understanding the processes that lead to ionization and energy balance in the ISM. This involves attempting to measure the abundance and density of elements to understand the physical limits of the ISM. Through continued observation, scientists have been able to solve detailed statistical balance equations relating to specific mediums.

Understanding the chemistry of interstellar space also gives scientists a glimpse into the star- and planet-making process. The collapse of dust clouds forms stars and solar systems. Earth's solar system formed from the collapse of a molecular dust cloud. Interstellar chemistry helps scientists understand the creation of the solar system and its planets, which has led to more accurate theories about how life developed on Earth.

#### CHEMICAL REACTIONS IN SPACE

Calculating ISM processes begins with analyzing dust grains that form from giant star explosions. The gases of the ISM eventually become ionized by cosmic rays or by UV radiation, inciting surface chemical reactions that create simple molecules. Other heat-inducing events such as shockwaves catalyze reactions between neutral atoms. Exposure to high-energy UV radiation forms complex molecules, such as dihydrogen ($H_2$). $H_2$ accounts for a large portion of gas-phase chemistry in the ISM. Common chemical reactions in interstellar chemistry include the combination of dihydrogen with carbon, nitrogen, or oxygen atoms. In the absence of hydrogen atoms, these atoms react to form compounds such as dioxygen ($O_2$), nitric oxide (NO), or carbon monoxide (CO). Scientists analyze radio and light waves to determine what wavelengths can be detected within these processes. The wavelengths and energy absorption or emission determine the quantity and relative mass of the atoms and molecules comprising the gas or dust. These same processes apply to the rest of interstellar

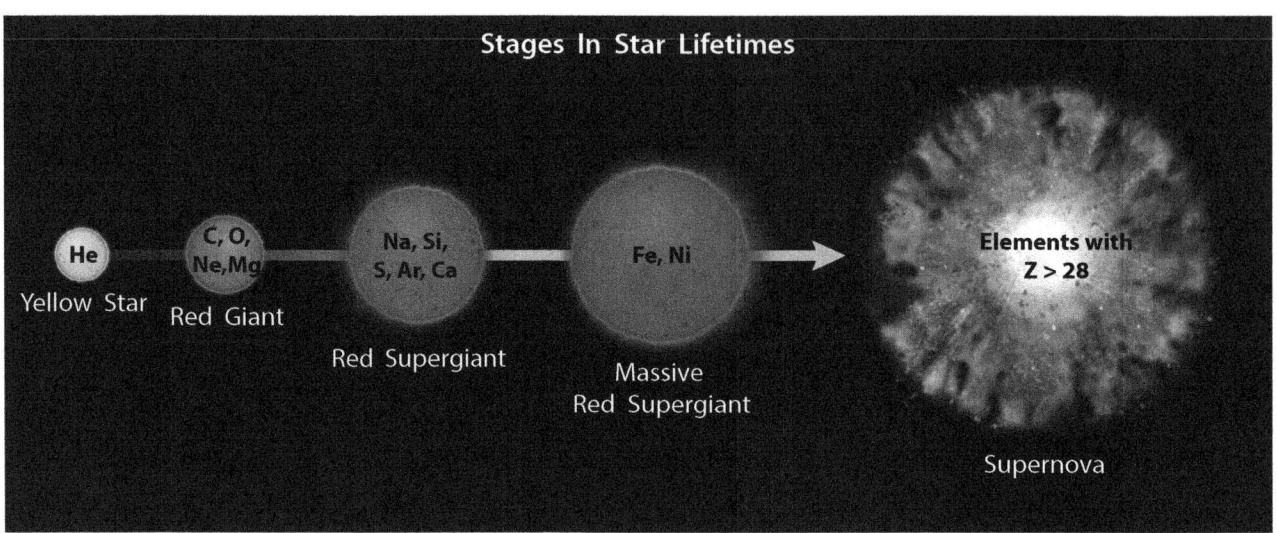

Chemical reactions within stars differ at each stage of their life cycle. A yellow star forms helium. Red giants form carbon, oxygen, neon, and magnesium. Red supergiants form sodium, silicon, sulfur, arsenic, and calcium. At later stages, massive red supergiants form iron and nickel. Supernovas produce heavier elements.

space as scientists probe space matter such as dark clouds, nebulae, and supernovas.

**Key Applications of Interstellar Chemistry**

Studies in interstellar chemistry have led to a better understanding of the composition of the solar system and the Milky Way galaxy. By comparing the wavelengths of varying components within the Milky Way, scientists can determine the most and least abundant matter in the galaxy. Understanding the chemical reactions that take place in interstellar space allows physicists to fathom the creation of stars and planets and, by extension, entire solar systems.

Interstellar chemistry plays an important role in deep-space exploration. Scientists use spectroscopy to detect complex interstellar molecules like those in Earth's solar system. Radio telescopes and spectrometers analyze chemical data from the farthest accessible depths of the universe. This information helps astrophysicists determine what components are needed to sustain life and what parts of space may be capable of hosting living organisms.

—*Cait Caffrey*

**Further Reading**

"Deep Space, Branching Molecules, and Life's Origins?" *Life, Unbounded*. Scientific Amer., 30 Sept. 2014. Web. 26 May 2015.

Hollas, J. Michael. "Low and High Resolution Spectroscopy." *High Resolution Spectroscopy*. Oxford: Butterworth-Heinemann, 2013. 85–86. Print.

"Interstellar Chemistry." *University of California Berkeley Astronomy Department*. U of California, n.d. Web. 26 May 2015.

"Molecules in Space." *University of Cologne Institute of Physics*. U of Cologne, 2015. Web. 26 May 2015.

"Molecules in the Interstellar Medium." *Department of Astronomy and Astrophysics*. Pennsylvania State U, n.d. PDF file.

Tielens, A. G. G. M. "The Galactic Ecosystem." *The Physics and Chemistry of the Interstellar Medium*. Cambridge: Cambridge UP, 2005. 1–25. Print.

"What Is Spectroscopy?" *Department of Astronomy and Steward Observatory*. U of Arizona, n.d. Web. 26 May 2015.

Ziurys, L. M., and A. J. Apponi. "Interstellar Molecules as Probes of Prebiotic Chemistry: A Laboratory in Radio Astronomy." *Arizona Radio Observatory*. U of Arizona, 2006. PDF file.

# INVERSE COMPTON SCATTERING

## FIELDS OF STUDY

Astronomy; Astrophysics; Cosmology

## SUMMARY

In the 1920s, Washington University professor Arthur H. Compton was studying x-rays when he discovered that the wavelength of x-rays and gamma rays changed as they interacted with matter. He determined that when a photon strikes a moving electron, the photon gains momentum while the electron loses some. Inverse Compton scattering, as this phenomenon came to be known, revealed that light is a stream of particles. Compton shared in the 1927 Nobel Prize for Physics for this important discovery.

## PRINCIPAL TERMS

- **accretion disk:** a relatively flat, rotating collection of debris and gas that forms around large celestial objects with great gravitational force, such as black holes and young stars.
- **astrophysics:** the branch of astronomy that deals with the physics of the universe, covering such areas as physical properties, composition, processes, and other phenomena.
- **black hole:** a region of space with such strong gravitational pull that not even light can escape.
- **photon:** the smallest unit of light and other electromagnetic radiation, a particle with no mass and no electrical charge.
- **Sunyaev-Zel'dovich effect:** an absence of microwaves in the cosmic background radiation, caused by x-rays interacting with hot gas.
- **thermal spectrum:** the continuous spectrum of electromagnetic radiation that an object emits in the form of heat, produced by the movement of the particles that make up the object.
- **x-ray astronomy:** the branch of astronomy that is concerned with detecting x-rays in space and studying the phenomena that produce them.

## DISCOVERY OF INVERSE COMPTON SCATTERING

In 1923, American physicist Arthur H. Compton (1892–1962) observed that when a photon strikes a stationary electron, some of the photon's energy is transferred to the electron, causing it to move. This phenomenon became known as the Compton effect. Compton also determined that the same phenomenon happens in reverse: when a photon strikes a moving electron, the photon gains some momentum, and the electron loses some. This phenomenon, known as inverse Compton scattering, creates high-energy photons by draining energy from electrons as they pass through areas of dense radiation.

These discoveries proved that electromagnetic (EM) energy is not just a wave but also a particle, and that these particles can be affected by encounters with matter. The application of these concepts led to the development of x-ray astronomy, in which scientists use x-rays to find celestial objects that would otherwise be difficult to detect in space.

## ENERGY CHANGES IN PHOTONS

Compton used a series of formulas to calculate the changes that result from collisions between electrons and photons. These formulas apply the photoelectric effect of light particles, the special theory of relativity, and the law of cosines. The result allows scientists to compute the change in wavelength caused by such a collision. The Compton scattering equation is as follows.

$$\lambda' - \lambda = h m_e c (1 - \cos\theta)$$

In the inverse Compton scattering process, photons gain energy. Low-energy photons are bombarded with high-energy electrons, resulting in low-energy electrons and high-energy photons.

In this equation, $\lambda$ is the initial wavelength of the EM radiation; $\lambda'$ is the final wavelength after collision; $h$ is the Planck constant, which has a value of about $6.626 \times 10^{-34}$ joule-seconds (J·s); and $m_e$ is the mass of one electron, about $9.109 \times 10^{-31}$ kilograms (kg). $C$ is the speed of light, or 299,792,458 meters per second (m/s), and $\theta$ is the angle at which the photons scatter.

One way to think of Compton scattering is to imagine a game of pool. The player strikes the cue ball so that it hits another, stationary ball on the pool table. When the two balls collide, the cue ball loses some momentum, while the stationary ball gains momentum and begins to move. This is similar to what occurs when photons and electrons collide.

## Black Holes

Inverse Compton scattering can be used to locate and study black holes. The powerful gravitational pull of a black hole draws gas and other matter into orbit around it, forming an accretion disk. As the matter in the disk spirals toward the center of the black hole, its speed and friction increase, producing a thermal spectrum of low-energy x-rays.

In the inner accretion disk, the particles move at near light speed, forming a corona of high-energy electrons. When the photons of the thermal spectrum collide with these electrons, the low-energy x-rays are scattered and become high-energy x-rays. These "hard" x-rays can be used to find objects in space that cannot be seen by conventional means.

## The Sunyaev-Zel'dovich Effect

The Sunyaev-Zel'dovich effect (SZE) is a distortion of the cosmic background radiation near regions of hot gas or plasma, such as galaxy clusters. It is named for Russian physicists Rashid Alievich Sunyaev (b. 1943) and Yakov Borisovich Zel'dovich (1914–87), who first proposed its existence.

Simply put, the SZE is an absence of cosmic background microwaves. It is caused by inverse Compton scattering and can be detected as a gap in the radiation spectrum. This effect is important in astrophysics because it can help locate galaxy clusters, which are otherwise difficult to detect. Astrophysicists can also use the SZE to study how these clusters have been affected over time by otherwise unobservable phenomena, such as dark matter and dark energy.

—*Janine Ungvarsky*

## Further Reading

"Arthur H. Compton and Compton Scattering." *Department of Energy Research and Development Accomplishments.* Dept. of Energy, 28 Jan. 2015. Web. 6 Apr. 2015.

"Arthur H. Compton—Biographical." *Nobelprize.org.* Nobel Media, 2014. Web. 6 Apr. 2015.

"Black Holes." *NASA Science.* NASA, 30 Mar. 2015. Web. 6 Apr. 2015.

Longair, Malcolm S. *High Energy Astrophysics.* 3rd ed. New York: Cambridge UP, 2011. Print.

"NASA-Led Study Explains Decades of Black Hole Observations." *NASA News and Features.* NASA, 14 June 2013. Web. 6 Apr. 2015.

Nave, Carl R. "Compton Scattering." *HyperPhysics.* Georgia State U, n.d. Web. 6 Apr. 2015.

"The Sunyaev-Zeldovich Effect." *NASA Science.* NASA, 6 Apr. 2011. Web. 6 Apr. 2015.

"A Sunyaev Zel'dovich Effect Primer." *SZA at Chicago.* U of Chicago, n.d. Web. 6 Apr. 2015.

# K

## (216) KLEOPATRA

### FIELDS OF STUDY

Astronomy; Astrophysics; Asteroid Impact Avoidance

### SUMMARY

Kleopatra is an asteroid that was discovered in 1880. It is a large asteroid that stretches 217 kilometers (135 miles) end to end and 65 kilometers (40 miles) across. In 2008, astronomers discovered two moons in orbit around Kleopatra, Alexhelios and Cleoselene. Astronomers once believed that all large asteroids were solid. However, to date, all asteroids that have been discovered to have moons are in fact porous rubble piles held together by gravity. Astronomers theorize that Kleopatra is the remains of two rocky, metallic asteroids that smashed together following the formation of the solar system about 4.6 billion years ago.

### PRINCIPAL TERMS

- **asteroid:** a rocky body that orbits the sun.
- **rubble-pile asteroid:** a nonsolid asteroid that is made up of rocks of various sizes held together by gravity.

### A Dumbbell-Shaped Asteroid

Kleopatra is an asteroid that is shaped like a dumbbell. It is long and thin, with the ends wider than the middle. An asteroid is a rocky body that orbits the sun. Kleopatra is one of billions of asteroids found between the orbits of Mars and Jupiter, an area referred to as the asteroid belt. Kleopatra's orbital period is 4.67 years. Kleopatra is orbited by two moons, Alexhelios and Cleoselene. Alexhelios is the outer moon, and Cleoselene is the inner moon.

The asteroid was named after the Egyptian queen Cleopatra VII (70–30 BCE), and the two moons were named after her twin children: Alexander Helios and Cleopatra Selene II. Each of Kleopatra's moons measures about 8 kilometers (5 miles) in diameter.

### A Rubble-Pile Asteroid

In 1994, astronomers discovered the first moon orbiting an asteroid. In 1998, astronomers using the Canada-France-Hawaii Telescope were the first to discover a moon orbiting an asteroid using a ground-based telescope with adaptive optics. Adaptive optics improves the resolution of telescopes on the ground so that astronomers can view celestial bodies more clearly. The discovery of moons in orbit around asteroids was significant. To date, all asteroids found with moons are not solid but are rubble-pile asteroids.

Kleopatra was first discovered by Austrian astronomer Johann Palisa in 1880. Astronomers previously thought that Kleopatra had a tubular shape similar to a cigar. It was not until 2000 that astronomers determined the asteroid's shape is more like a thigh bone or a dumbbell. In 2008, a team of astronomers focused on Kleopatra. Franck Marchis, a research astronomer at the University of California, Berkeley, and his collaborators sought to confirm Kleopatra's thigh-bone shape. They used an adaptive-optics system on the Keck II, an enormous telescope on the top of Mauna Kea in Hawaii. They confirmed the asteroid's unusual shape—and also discovered that the asteroid is orbited by two small moons and that the asteroid is not solid.

The discovery that Kleopatra is a rubble pile surprised Marchis and the team, because most asteroids of this type are much smaller. With a length of 217 kilometers (135 miles), Kleopatra is one of the largest rubble-pile asteroids; it is second only to Sylvia, which is 280 kilometers (174 miles) long. By charting the orbits of the moons and calculating the total mass of the system, astronomers were able to determine Kleopatra's density and estimated that about 30 to 50 percent of the asteroid is empty space.

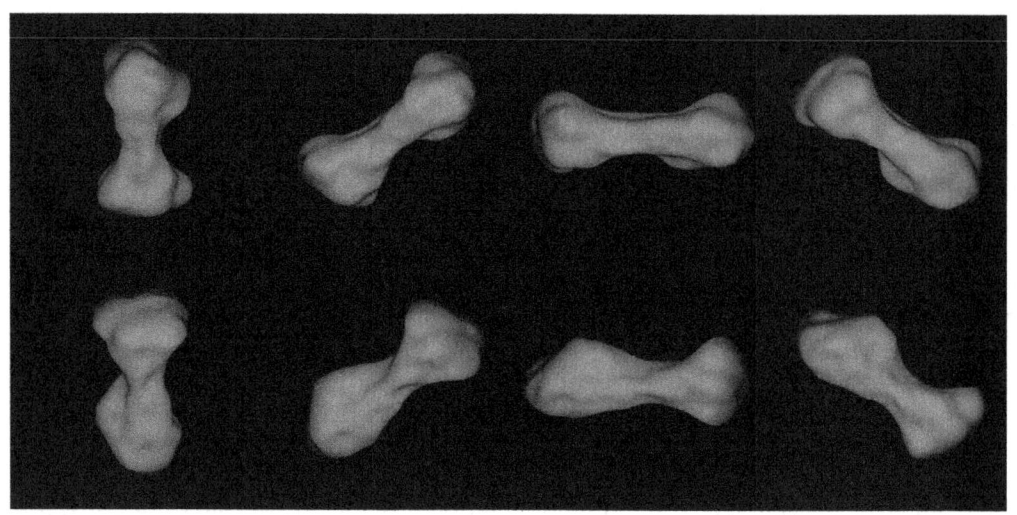

216 Kleopatra, asteroid, from eight views. JPL/NASA, public domain via Wikimedia Commons.

## FORMATION THEORIES AND RESEARCH EFFORTS

Scientists believe Kleopatra is actually the remains of two rocky, metallic asteroids that smashed together sometime after the formation of the solar system about 4.6 billion years ago. They theorize that the two moons are pieces of rubble that broke off from the asteroid, possibly when another collision caused Kleopatra to spin. Multiple collisions may also be responsible for Kleopatra's odd shape. While Alexhelios probably broke off one hundred million years ago, Cleoselene is thought to have formed more recently, possibly only ten million years ago.

Marchis and the team continue to study asteroids in the hopes of discovering their moons. They believe that many other large asteroids may also be rubble piles. They would like to figure out what percentage of all asteroids fit this description. They also hope that by studying rubble-pile asteroids, they will learn more about how planets form.

Astronomers also study asteroids to protect Earth from potential collision. Asteroids occasionally crash into planets, sometimes causing devastation. While it is unlikely that Kleopatra will ever be a threat, other rubble-pile asteroids might be. Near-Earth asteroid (29075) 1950 DA is on the National Aeronautics and Space Administration's (NASA) list of potential impact threats to Earth. This asteroid has about a one in four thousand chance of crashing into Earth in the year 2880. If it does, 1950 DA, which is approximately 1.2 kilometers (0.75 miles) in diameter, has the potential to create an explosion about 3.75 million times stronger than the nuclear bomb dropped on the city of Hiroshima, Japan, during World War II. In 2014, astronomers discovered that 1950 DA, like Kleopatra, is a rubble-pile asteroid. Understanding more about this type of asteroid can help scientists determine the best way to change its course to prevent collision with Earth.

—*Adrienne A. Kennedy*

## FURTHER READING

Baldwin, Emily. "Asteroid Kleopatra Spawned Twins." *Astronomy Now*. Pole Star, 23 Feb. 2011. Web. 16 Feb. 2015.

Choi, Charles Q. "Asteroids: Fun Facts and Information about Asteroids." *Space.com*. Purch, 21 Nov. 2014. Web. 16 Feb. 2015.

Choi, Charles Q. "Potentially Dangerous Asteroid Is Actually a Pile of Rubble." Space. com. Purch, 13 Aug. 2014. Web. 3 Mar. 2015.

"Kleopatra Gave Birth to Twins . . . Moons." *Cosmic Log*. NBC News, 23 Feb. 2011. Web. 16 Feb. 2015.

Redd, Nola Taylor. "Moons around Asteroid Reveal a Giant Rubble Pile." *Space.com*. Purch, 1 June 2011. Web. 16 Feb. 2015.

Sanders, Robert. "How Cleopatra Got Its Moons." *UC Berkeley News Center*. UC Regents, 22 Feb. 2011. Web. 16 Feb. 2015.

Schmadel, Lutz D. *Dictionary of Minor Planet Names*. 6th ed. Heidelberg: Springer, 2012. Print.

"Two Companion Moons Found Near Dog-Bone Asteroid." *SETI Institute*. SETI Inst., n.d. Web. 3 Mar. 2015.

# (216) KLEOPATRA I ALEXHELIOS AND II CLEOSELENE

### FIELDS OF STUDY

Astronomy; Astrophysics

### SUMMARY

Kleopatra is an asteroid that was discovered in 1880. It is a large asteroid that stretches 217 kilometers (135 miles) end to end. In 2008 astronomers discovered two moons orbiting Kleopatra, Cleoselene and Alexhelios. Astronomers once believed that all large asteroids were solid. However, to date, all asteroids that have been discovered to have moons are rubble piles held together by gravity.

### PRINCIPAL TERMS

- **asteroid:** a rocky body that orbits the sun.
- **Kleopatra:** an asteroid that is a rubble pile with two moons, Cleoselene and Alexhelios.

### AN ASTEROID WITH TWO MOONS

The asteroid Kleopatra is a rubble pile shaped like a dumbbell. An asteroid is a small rocky body that orbits the sun. Billions of asteroids, including Kleopatra, are found between the orbits of Mars and Jupiter, an area referred to as the asteroid belt. Kleopatra is orbited by two moons, Cleoselene and Alexhelios. Cleoselene is the inner moon and Alexhelios is the outer moon. The asteroid is named after the Egyptian queen Cleopatra VII (70–30 BCE), and the two moons are named after her twin children: Cleopatra Selene II and Alexander Helios.

### A RUBBLE PILE

Kleopatra is 217 kilometers (135 miles) long, which makes it a fairly large asteroid. While most large asteroids are solid chunks of rock or metal, Kleopatra is actually a pile of rubble held together by gravity. Astronomers estimate that 30 to 50 percent of the asteroid is empty space. Of the rubble-pile asteroids, Kleopatra is one of the largest, second only to Sylvia, which stretches 280 kilometers (174 miles) across.

When astronomers discovered Kleopatra in 1880, they thought the asteroid had a tubular shape similar to a cigar. However, astronomers in 2000 determined that the asteroid's shape was more like a thigh bone, with two wider ends and a thinner middle. In 2008, using the Keck II, a large telescope in Hawaii, scientists verified the thigh-bone shape of the asteroid and discovered its two moons, Cleoselene and Alexhelios. Each moon measures about 8 kilometers (5 miles) across.

### FORMATION THEORIES

Scientists believe that Kleopatra is actually the remains of two rocky, metallic asteroids that smashed together following the formation of the solar system about 4.5 billion years ago. They theorize that the asteroid's two moons are pieces of rubble that broke

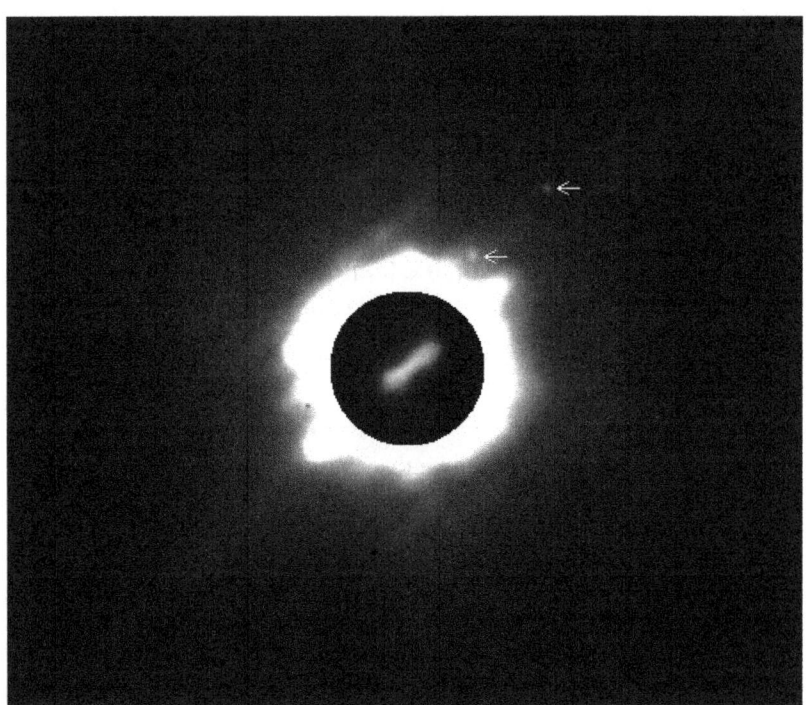

Moons orbitting asteroid Kleopatra. Alexhelios orbits farther from Kleopatra; Cleoselene orbits closer to Kleopatra. F. Marchis, et al., SETI Institute, UC Berkeley, public domain via Wikimedia Commons.

off, possibly when another collision caused Kleopatra to spin rapidly. Alexhelios may have broken off from the asteroid one hundred million years ago, and Cleoselene is thought to have formed more recently, possibly only ten million years ago.

—Adrienne A. Kennedy

**FURTHER READING**

Baldwin, Emily. "Asteroid Kleopatra Spawned Twins." *Astronomy Now*. Pole Star, 23 Feb. 2011. Web. 16 Feb. 2015.

Choi, Charles Q. "Asteroids: Fun Facts and Information about Asteroids." *Space.com*. Purch, 21 Nov. 2014. Web. 16 Feb. 2015.

"Kleopatra Gave Birth to Twins . . . Moons." *Cosmic Log*. NBC News, 23 Feb. 2011. Web. 16 Feb. 2015.

Redd, Nola Taylor. "Moons around Asteroid Reveal a Giant Rubble Pile." *Space.com*. Purch, 1 June 2011. Web. 16 Feb. 2015.

Sanders, Robert. "How Cleopatra Got Its Moons." *UC Berkeley News Center*. UC Regents, 22 Feb. 2011. Web. 16 Feb. 2015.

Schmadel, Lutz D. *Dictionary of Minor Planet Names*. 6th ed. Heidelberg: Springer, 2012. Print.

# KORONIS ASTEROIDS: 243 IDA

### FIELDS OF STUDY

Astronomy; Observational Astronomy; Sub-planetary Astronomy

### SUMMARY

243 Ida is a Koronis asteroid that orbits the sun between Mars and Jupiter. This asteroid and others in its family are irregularly shaped. The National Aeronautics and Space Administration (NASA) *Galileo* spacecraft observed 243 Ida in 1993. The spacecraft took images that showed 243 Ida has a natural satellite, named Dactyl. Studied frequently, 243 Ida has given scientists more information about asteroids. These objects can tell scientists about the formation of the early solar system and could have an impact on human life in the future.

### PRINCIPAL TERMS

- **aphelion:** the point in an object's orbital path that is farthest from the sun.
- **asteroid:** a small, rocky space object that orbits the sun or another celestial body.
- **Koronis asteroid:** a main-belt asteroid with an irregular shape that spins at about the same rate as the others in its family.
- **main-belt asteroid:** an asteroid that orbits the sun in the region between Mars and Jupiter.
- **perihelion:** the point in an object's orbital path that is closest to the sun.
- **s-type asteroid:** fairly bright asteroids that are usually made of stone (silicate), nickel, and iron.

### 243 IDA AND KORONIS ASTEROIDS

243 Ida is an asteroid, or a small, rocky celestial body, that orbits the sun between Mars and Jupiter. This asteroid is a Koronis asteroid. The members of the Koronis family of asteroids are main-belt asteroids located between the orbits of Jupiter and Mars. These asteroids were once thought to be a product of a collision that happened very long ago between two large objects in the early solar system. Further study has revealed that this is most likely a more diverse group of objects from areas all over the solar system.

Koronis asteroids are irregularly shaped. They also rotate at similar speeds. In the past, scientists believed that asteroids spun at random speeds and that their rotations were determined by the force of previous collisions. However, in the 1990s and early 2000s, scientists learned that many of the Koronis family asteroids actually rotate at almost the same rate. They later found that light from the sun influences the rotation of these asteroids.

Ida is one of the more researched Koronis asteroids. Scientists have determined that it is 58 kilometers (36 miles) long and 23 kilometers (14 miles) wide. Its rotation is four hours and thirty-nine minutes.

Ida is an s-type asteroid, meaning that it is made of stone (silicate) and metals such as nickel and iron. Ida's surface has many craters on it. This means that

243 Ida, asteroid, with its moon, Dactyl, orbiting. NASA/JPL.

it has been involved in many collisions and that the surface is geologically old.

## History of 243 Ida

In 1884 Austrian astronomer Johann Palisa (1848–1925) was searching for asteroids at the Vienna Observatory. On September 29, 1884, Palisa discovered 243 Ida. The asteroid was named by Moriz von Kuffner (1854–1939), an amateur astronomer from Vienna. The object was named after a nymph in Greek mythology who cared for the god Zeus when he was an infant.

Much new information about Ida was learned in the early 1990s when the spacecraft *Galileo* passed by it. *Galileo* was built to study Jupiter and its moons. The spacecraft observed Ida on its path to Jupiter. *Galileo* was closest to Ida on August 28, 1993. It traveled as close as 2,400 kilometers (1,500 miles) from the object. The spacecraft took images of the asteroid and transmitted them to Earth. In February 1994 Ann Harch—who was studying images from *Galileo*—noticed that Ida had a satellite. Scientists named the moon Dactyl, after a mythological creature that lived on Mount Ida.

Ida and Dactyl offered important information about asteroids. Scientists had long believed that some asteroids had their own moons, but they thought these moons were most likely uncommon. However, Ida was the second asteroid to be observed by a spacecraft and it had a moon. Scientists learned that asteroids with moons were probably more common than previously thought.

## Studying 243 Ida and Asteroids

Scientists study 243 Ida and other asteroids for a number of reasons. One main reason is to learn about the early solar system. Many asteroids are leftover material from when the solar system formed. Their makeup can help scientists better understand how the solar system came into being.

Another reason scientists study asteroids is that their orbits can change. Their periphelions can be closer or farther away from Earth. Because of that, scientists need to track the orbits of 243 Ida and other asteroids to monitor whether these objects will ever come close to Earth. Asteroid impacts have happened on Earth's surface in the past, and they will almost certainly occur again in the future. Understanding asteroids can help scientists predict if and when that might occur.

A third reason that scientists study Ida and other asteroids is that these objects could possibly supply humans with natural resources in the future. Many valuable resources could be mined from asteroids, so scientists are studying these objects to learn more about them. If humans use up resources on Earth, they may have to look to mining asteroids for valuable materials, such as metals.

—*Elizabeth Mohn*

## Further Reading

"Asteroids: Overview." *National Aeronautics and Space Administration*. NASA, 13 Jan. 2015. Web. 26 Mar. 2015.

Dymock, Roger. *Asteroids and Dwarf Planets and How to Observe Them*. New York: Springer, 2010. Print.

"Ida and Moon." *HyperPhysics*. Georgia State U, n.d. Web. 26 Mar. 2015.

"Images of Asteroids Ida & Dactyl." *National Aeronautics and Space Administration*. NASA, 22 Aug. 2005. Web. 26 Mar. 2015.

# KUIPER BELT OBJECTS (KBOS)

## FIELDS OF STUDY

Astrophysics; Observational Astronomy; Sub-planet Astronomy

## SUMMARY

Kuiper Belt objects (KBOs) are small icy space bodies that populate an area of the solar system known as the Kuiper Belt. This section of space is beyond the orbit of Neptune and contains more than one thousand known KBOs. They are thought to be composed of unspoiled material from the origin of the solar system. Thus, KBOs have important research potential.

## PRINCIPAL TERMS

- **dwarf planet:** a round celestial body that is much smaller than a planet and orbits the sun, but is unable to clear other objects from its orbit. Made of solid rock, ice, or both, dwarf planets are thought to be incapable of sustaining any known life.
- **Edgeworth-Kuiper Belt:** a ring of icy rock objects located beyond the orbit of Neptune; also known as the Kuiper Belt.
- **scattered disk objects:** small, icy planets with highly eccentric orbits at the farthest reaches of the solar system.

## ORIGINS OF KUIPER BELT OBJECTS (KBOS)

Also known as the Edgeworth-Kuiper Belt, the Kuiper Belt is a section of the solar system occupied by small ice and rock objects called Kuiper Belt objects (KBOs). They are also known as transneptunian objects (TNOs). These small, irregularly shaped space bodies are thought to contain untouched material from the formation of the solar system 4.6 billion years ago. The Kuiper Belt is named after the men who first theorized it existed, Kenneth Edgeworth (1880–1972) and Gerard Kuiper (1905–73). In 1992, its existence was confirmed, which led to Pluto being downgraded from a planet to a dwarf planet that is also considered a KBO.

## TYPES OF KBOS

KBOs are divided into three classes based on their orbit: classical, resonance, and scattered. Classical KBOs are subdivided into "cold" and "hot" classical KBOs. Cold classical KBOs have a semi-major axis, or mean distance from the sun, of 42 to 48 astronomical units (AU), or 390 million to 446 million miles. Their orbits are not eccentric—that is, they are fairly circular. Cold classical KBOs are also low inclination. In other words, they do not orbit at an angle. They are called "cold" not because of their temperature but because of their relatively calm movement. Their smaller size and reddish appearance lead scientists to think cold classical KBOs have a different origin than other types. Hot classical KBOs have less controlled orbits. Their more eccentric orbit and higher inclination path mean they are less likely to stay the same average distance from the sun. They are generally bigger and tend to be gray in color.

Resonance KBOs orbit the sun in a consistent ratio to Neptune. Most of these are in 3:2 resonance, orbiting the sun twice for every three orbits Neptune completes. Pluto is among the 3:2 resonant KBOs. Others travel farther from the sun and are in 2:1 resonance, making only one trip around the sun for every two that Neptune makes.

Scattered KBOs are also known as scattered disk objects (SDOs). They are space bodies made of ice and rock. They have very eccentric, or flattened, oval orbits that take them both closer to and farther from the sun than the other types of KBOs. Not all SDOs are KBOs, however. Some are farther from the sun or farther above or below the ecliptic (sun's path) than the Kuiper Belt proper.

Scientists have observed other objects with similar properties to the KBOs. However, their very eccentric orbits may keep them from being considered part of the Kuiper Belt.

## The Kuiper Belt and Pluto

When astronomers David Jewitt (b. 1958) and Jane Luu (b. 1963) confirmed the existence of the first KBO, 1992QB1, in 1992, it was the beginning of the end of planet status for Pluto. Scientists had long theorized that the belt of asteroids must be there. However, the great distance between Earth and the area that came to be known as the Kuiper Belt made it difficult to see any objects there with even the strongest telescope.

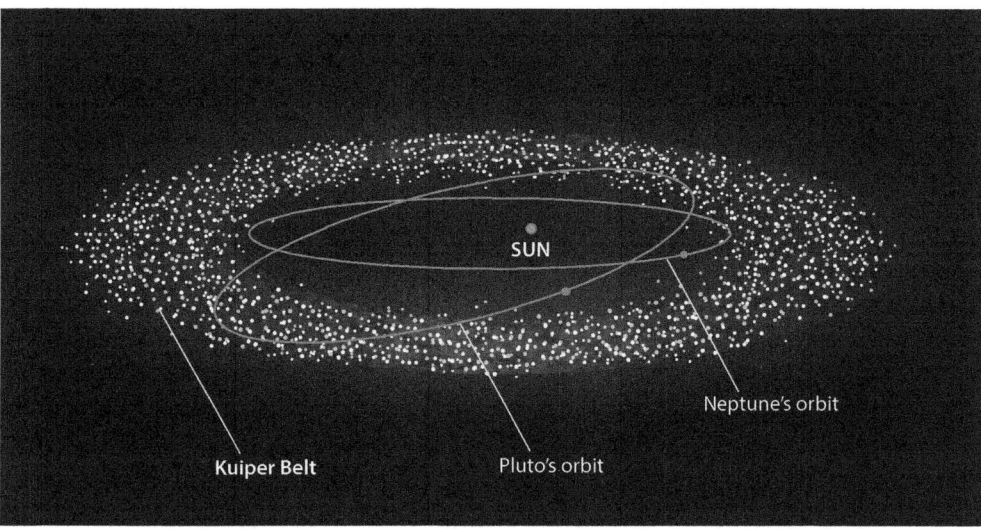

A model of the outer solar system. The Kuiper Belt is a region of the outer solar system beyond Neptune's orbit. Pluto's orbit extends into the Kuiper Belt.

Jewitt and Luu's success led to the discovery of other KBOs. This, in turn, led scientists to realize that Pluto had more in common with those objects than with the other eight primary planets. Pluto is larger than most KBOs and relatively reflective. This makes it easier to see from Earth than other KBOs, which is why it was spotted in 1930 and designated a planet. However, with the discovery of other KBOs of similar size and orbit, the International Astronomical Union demoted Pluto to a dwarf planet in 2006. Dwarf planets orbit the sun like the primary eight planets but cannot knock other objects out of their orbits. They are smaller than the primary planets and cannot maintain any known form of life. Other dwarf planets include Eris, which is about the same size as Pluto, and Ceres, the largest known asteroid in the asteroid belt.

## KBOs and the Origin of the Solar System

KBOs are valuable to scientists because they are thought to be made up of materials left over from the formation of the solar system about 4.6 billion years ago. They have likely gone unchanged for billions of years. Thus, their composition could tell scientists much about the early solar system.

The European Space Agency probe, Rosetta, and the National Aeronautic and Space Administration probe New Horizons were launched around 2004–5 to explore the Kuiper Belt for that reason. Both Rosetta and New Horizons approached KBOs, one a comet and the other Pluto, in late 2014.

—*Janine Ungvarsky*

## Further Reading

*Cosmos: The SAO Encyclopedia of Astronomy.* Swinburne U of Technology, n.d. Web. 9 Mar. 2015.

"Dwarf Planets." *National Geographic.* Natl. Geographic Soc., 1996–2015. Web. 9 Mar. 2015.

"Dwarf Planets: Overview." *Solar System Exploration.* NASA, 22 Jan. 2015. Web. 9 Mar. 2015.

"Kuiper Belt." *New Horizons: NASA's Mission to Pluto.* Johns Hopkins U, n.d. Web. 9 Mar. 2015.

"Kuiper Belt and Oort Cloud: Overview." *Solar System Exploration.* NASA, 19 Nov. 2014. Web. 9 Mar. 2015.

Lemonick, Michael D. "Kuiper Belt Missions Could Reveal the Solar System's Origins." *Scientific American.* Nature America, 14 Oct. 2014. Web. 9 Mar. 2015.

"Pluto and the Developing Landscape of Our Solar System." *International Astronomical Union.* IAU, n.d. Web. 9 Mar. 2015.

# LANGMUIR-HINSHELWOOD MODEL

## FIELDS OF STUDY

Astrochemistry; Astrophysics; Astrometry

## SUMMARY

The Langmuir-Hinshelwood model is one of two models commonly used to explain how two atoms form molecular hydrogen ($H_2$) in the interstellar medium (ISM), the material that fills the space between the stars. According to the Langmuir-Hinshelwood model, both atoms adsorb to, or adhere to the surface of, interstellar dust grains, where they combine to form a molecule. Scientists hope to learn more about the formation of life on Earth from studying dust grains in the ISM.

## PRINCIPAL TERMS

- **dust-grain model:** a model of molecular hydrogen formation in which the reaction takes place on the surfaces of interstellar dust grains.
- **Eley-Rideal mechanism:** a model of molecular hydrogen formation in which one hydrogen atom adsorbs to a grain of dust, then reacts directly with a second hydrogen atom as it passes.

## Modeling the Formation of $H_2$

The Langmuir-Hinshelwood model, also called the Langmuir-Hinshelwood mechanism, is used to explain how two atoms form molecular hydrogen ($H_2$). It is based on the work of American chemist and physicist Irving Langmuir (1881–1957) and English chemist Sir Cyril Norman Hinshelwood (1897–1967).

In the Langmuir-Hinshelwood model, two individual hydrogen atoms adsorb to, or stick to the surface of, a grain of interstellar dust. One or both of the two atoms then diffuse over the surface of the dust grain toward each other. When they meet, they undergo a chemical reaction in which they bond to form $H_2$. The newly formed molecule of hydrogen gas then desorbs, or is released, from the surface of the dust grain.

### Grains of Dust

The Langmuir-Hinshelwood model is a refinement of the dust-grain model of $H_2$ formation, which states that the formation of molecular hydrogen gas takes place on the surfaces of dust particles found in the interstellar medium (ISM). This is in contrast to a gas-phase reaction, which would use more energy.

The ISM is the matter that exists in space between the stars. About 99 percent of it is gas, mainly hydrogen. The remainder consists of solid particles, typically smaller than one millionth of a meter. These particles are called interstellar dust or dust grains. In addition to being the site of hydrogen gas formation, the ISM is also where new stars form.

The Langmuir-Hinshelwood model is not the only model of hydrogen gas formation. A second model, known as the Eley-Rideal mechanism, is also based on the dust-grain model. In this model, however, only one hydrogen atom adsorbs to the dust grain.

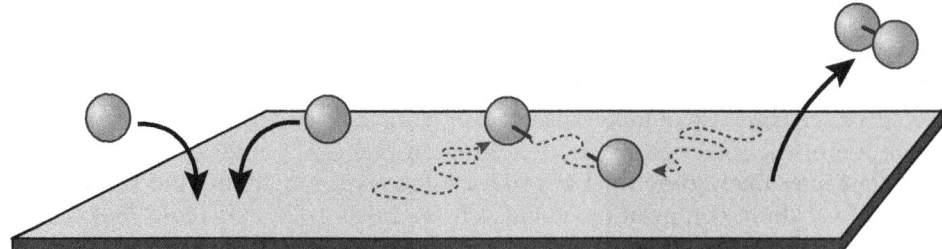

A proposed mechanism by which hydrogen molecules might form on dust grain surfaces. Two hydrogen atoms adsorb to the surface of the grain and diffuse across the surface. When they come into contact, they react to form a molecule of hydrogen, which can then desorb from the surface.

It then reacts directly with a second hydrogen atom as it passes by. The difference between the two is that in the Langmuir-Hinshelwood model, both atoms must adsorb to a dust grain before they react with one another; in the Eley-Rideal model, only one atom adsorbs.

### LEARNING MORE ABOUT THE ISM

It is difficult for astronomers to determine for certain how $H_2$ forms in the ISM in part because of its extreme temperatures. What they do know about dust grains and $H_2$ in the ISM stems largely from observations and laboratory experiments. Astronomers have determined that dust grains are composed mainly of silicates and carbon and are coated with a layer of ice. Scientists do not yet know definitively how hydrogen atoms on dust grains combine to form $H_2$. However, they hope to gain more information in the near future. The National Aeronautics and Space Administration's space probe Voyager 1 first reached the ISM in 2012, where it will continue to study dust grains until the end of its mission in 2025.

Information about the ISM may help scientists learn more about Earth. Some of the molecules necessary for life come from the ISM. By learning more about the ISM, scientists may discover how life began on Earth and whether it exists anywhere else.

—Adrienne A. Kennedy

### FURTHER READING

Cancellieri, Brian M., and Vladimir G. Mamedov, eds. *Interstellar Medium: New Research*. Hauppage: Nova Science, 2012. Print.

Dupraz, Sébastien, Trent Dupuy, and Sandrine Bottinelli. "Formation of Astrobiologically Important Molecules in Extraterrestial Environments." *Astrobiology Institute*. U of Hawaii NASA Astrobiology Inst., 2004. Web. 26 Mar. 2015.

"The Interstellar Medium." *University of New Hampshire Experimental Space Plasma Group*. U of New Hampshire, n.d. Web. 26 Mar. 2015.

"Irving Langmuir—Biographical." *Nobelprize.org*. Nobel Foundation, n.d. Web. 26 Mar. 2015.

Mann, Adam. "Voyager 1 Becomes First Man-Made Object to Reach Interstellar Space." *Wired*. Condé Nast, 12 Sept. 2013. Web. 26 Mar. 2015.

"Voyager: The Interstellar Mission." *NASA Jet Propulsion Laboratory*. California Inst. of Technology, n.d. Web. 26 Mar. 2015.

# LASER BEES

### FIELDS OF STUDY

Asteroid Impact Avoidance; Space Technology; Aerospace Engineering

### SUMMARY

A laser bee is a type of small spacecraft with a powerful built-in laser. Scientists are studying whether lasers from a number of these crafts can be aimed at an Earth-bound asteroid to cause it to change course. Laser bees could be an important way to protect Earth from a direct hit by an asteroid.

### PRINCIPAL TERMS

- **asteroid:** a small, rocky space object that orbits another larger celestial body such as the sun or a planet. Asteroids are categorized by their orbits.
- **laser:** stands for "light amplification by stimulated emission of radiation"; a device that creates a steady beam of visible radiation by stimulating electronic, ionic, or molecular particles to a higher energy level, causing them to emit some of that energy.
- **mirror bee:** a small spacecraft equipped with mirrors intended to reflect and focus the sun's light on a specific spot on an asteroid, which would superheat and explode. This would release a plume of gas that would act much like a propulsion system to redirect the asteroid. The mirror bee idea was the precursor to laser bees.

### DANGER OF ASTEROID IMPACTS

In mid-February 2013, people in the area of Chelyabinsk, Russia, were jolted when an asteroid exploded above them. The blast rattled buildings, broke windows, and injured more than a thousand people. This relatively small asteroid (17 meters, or about 56 feet in diameter) was the most powerful such collision with Earth's atmosphere since 1908. It drew much attention to the efforts to develop better ways to detect and redirect asteroids that could strike Earth.

Even before that incident, researchers were aware of the danger to Earth from a close approach by an asteroid. Many scientists believe that a large asteroid of about 9.7 kilometers (5.9 miles) in diameter struck Earth in the Yucatan Peninsula in southeastern Mexico 65 million years ago. Such an impact caused catastrophic changes that resulted in the extinction of the dinosaurs. For several decades, scientists have been considering ways to prevent such a mass extinction event from recurring. Some proposals called for the use of nuclear weapons or other explosives to destroy Earth-bound asteroids. However, others recognized that this could possibly create many smaller asteroids that could still cause massive destruction.

The idea of rerouting an oncoming asteroid to avoid Earth came to be seen as more feasible. One proposed method considered satellites that would use their own gravity to nudge the asteroid into a different course over time. Others put forth the use of explosive devices to knock the approaching asteroid off course. Another idea involved the use of a swarm of relatively tiny satellites equipped with mirrors to reflect sunlight onto the asteroid's surface. The vaporization caused by these mirror bees would, in theory, change the asteroid's orbit by creating propulsive jets of gas from the asteroid's surface.

As scientists explored the mirror bee concept, they quickly found that equipping each spacecraft with a laser, not a mirror, was more effective. Some scientists consider these laser bees to be the safest, most cost effective, and quickest way to prevent a direct hit on Earth by an asteroid.

### SMALL SATELLITES, BIG EFFECT

In 2008, researchers at the Universities of Strathclyde and Glasgow in Scotland began conducting work on the use of ablation to deflect an asteroid. Ablation is the process by which erosion removes material from an object. In theory, heating the surface of an asteroid would cause a gas plume to be ejected. This could act as a thruster, changing the course of the object in space.

The Scottish team, led by Massimiliano Vasile (b. 1970), investigated the use of swarms of small satellites to create ablation. Of nine different types of possible defense methods they evaluated, the "bee swarm" approach promised to be the fastest and most effective, short of using nuclear warheads. At first the project was based on using an effect like that of using a magnifying glass to start a fire. The initial experiments tested the use of satellites equipped with mirrors to focus sunlight on the surface of an asteroid. The focused energy would cause the asteroid to superheat and undergo sublimation (direct transition from solid to gas) in a selected area. This should release a jet of gas and material from the asteroid with enough force to propel it onto a course that will not cross paths with Earth.

These mirror bee satellites were promising, but the researchers soon discovered that satellites equipped with lasers could be even more effective and precise. They would also be simpler to set up, without requiring precise alignment

Artist's rendering of laser bee spacecraft "swarming" an asteroid.

with the sun. By surrounding a threatening asteroid with swarms of these small laser bees, scientists could direct repeated laser blasts to carefully manipulate the asteroid's course.

**PROCESS DETAILS**

The experiments by the Scottish researchers were carried out in a laboratory, not in space with real asteroids. However, they checked results against computer models and hold that the method should work in a real-life planetary defense situation. They propose that a small fleet of laser-carrying satellites would be deployed as soon as a threatening asteroid was in range. These laser bees would fly alongside the asteroid in a pattern designed to maximize the effectiveness of the laser blasts. The number of bees and their flight pattern would be set based on the size of the asteroid, its composition, and how much warning scientists have of its approach. Using a swarm of laser bees gives the scientists the most flexibility in attacking the asteroid and the greatest potential for generating a strong, overlapping beam of laser light to initiate redirection.

The laser bees would be directed to shoot the laser beam at a spot calculated to create the best opportunity to propel the asteroid in a different direction. The energy from the laser would heat the surface of the asteroid. This would convert the solid material into a gas that would act much like fuel does in a rocket, with the escaping gas pushing against the solid surface to propel it through space, but at lower speed. The low thrust would still be enough to move the asteroid off a collision course with Earth.

As of early 2015, the process had only been tested in laboratory conditions by directing lasers at rocks in a vacuum chamber. This provided valuable insight into the amount of gas released. Scientists working on the laser bee project also identified the possible issue of debris called "ejecta" being released along with the gas during laser ablation. Scientists are studying ejecta to determine its impact on the diversion outcome in a real-life scenario and to find ways to mitigate it.

One advantage of this approach is that the small craft can be more easily deployed in the event of an asteroid approach. The unmanned satellites do not require astronauts to undertake a risky mission. Ablation also does not create large amounts of debris that could still pose a threat to Earth, which exploding an oncoming asteroid might do. The laser bee process also uses the asteroid itself as a fuel source for redirection, simplifying the deployment of the asteroid diversion mission.

Possible drawbacks include the possibility that larger asteroids might be more difficult to deflect with laser bees and some uncertainty about possible contamination by ejecta.

The research at the Universities of Strathclyde and Glasgow was supported by the nonprofit Planetary Society, founded in 1980 by space experts including Carl Sagan (1934–1996). In 2013, with the support of the Planetary Society and the European Commission, the Scottish research team launched a four-year project called Stardust. Stardust is aimed specifically at training scientists to detect and monitor asteroids and space debris. The program also seeks to develop the best practices for deflecting and possibly exploiting asteroids.

**IMPORTANCE OF ASTEROID DETECTION**

On the day of the Chelyabinsk asteroid incident in 2013, scientists had their eyes on a much larger asteroid that was expected to pass near Earth but not pose any threat. The smaller space body that struck the atmosphere over Chelyabinsk was undetected until it hit. Fortunately, many of the smaller asteroids (which became known as "meteors" once they enter Earth's atmosphere) are destroyed by the atmosphere and never reach Earth's surface. But those that are large enough can cause a great deal of destruction.

By early 2015, scientists for the National Aeronautics and Space Administration (NASA) had detected more than twelve thousand near-Earth asteroids (NEAs). Many are considered potential hazards. These could pose a threat to Earth because they are fairly close to the planet or its orbit. But they have been studied enough that scientists can predict their paths and determine that they do not pose an imminent threat for at least one hundred years. Undetected asteroids pose the greatest risk.

NASA's and other agencies' asteroid watch programs constantly monitor space for any approaching threats. They also continue to seek ways to prevent an asteroid strike, such as the use of laser bees and other technology that can prevent a collision. In addition, researchers are planning missions to some NEAs to study them more directly. These missions could provide insight into the makeup of these asteroids and help refine the processes for deflecting or removing them from Earth's path, making systems such as the laser bees even more effective.

—*Janine Ungvarsky*

**FURTHER READING**

Griggs, Mary Beth. "Laser Bees Could Save Us from Asteroids." *Smart News*. Smithsonian Inst., 12 July 2013. Web. 4 June 2015.

"Laser Ablation Experiments." *University of Strathclyde Engineering*. U of Strathclyde, 2015. Web. 4 June 2015.

"Laser Bees." *Planetary Society*. Planetary Soc., 2015. Web. 4 June 2015.

"Laser Bees Concept." *University of Strathclyde Engineering*. U of Strathclyde, 2015. Web. 4 June 2015.

Malik, Tariq. "NASA Maps Dangerous Asteroids that May Threaten Earth." *Space.com*. Purch, 14 Aug. 2013. Web. 4 June 2015.

"Number of Known Accessible Near-Earth Asteroids Doubles since 2010." *Near Earth Object Program*. NASA, 6 Feb. 2015. Web. 4 June 2015.

Phillips, Tony. "What Exploded over Russia?" *NASA Science News*. NASA, 26 Feb. 2013. Web. 4 June 2015.

Seitz, David E. "NASA Announces Next Steps on Journey to Mars: Progress on Asteroid Initiative." *Asteroid Redirect Mission*. NASA, 25 Mar. 2015. Web. 4 June 2015.

Wall, Mike. "Deflecting Killer Asteroids away from Earth: How We Could Do It." *Space. com*. Purch, 7 Nov. 2011. Web. 4 June 2015.

# LAUNCH VEHICLE AERODYNAMICS

**FIELDS OF STUDY**

Aerospace Engineering; Space Technology; Astronautics

**SUMMARY**

Launch vehicle aerodynamics is used in the design of space launch vehicles used to send a spacecraft into orbit or beyond. The study of aerodynamics usually involves examining several forces such as drag, thrust, and lift. Launch vehicle aerodynamics includes testing on models of launch vehicles that will be used in space travel. The National Aeronautics and Space Administration (NASA) performs such testing, which often consists of different types of wind tunnel tests. Engineers use these tests to prepare proposed launch vehicles for space missions.

**PRINCIPAL TERMS**

- **aerodynamics:** the study of the way in which air or gases act on an object; often used when designing aircraft, automobiles, buildings, and bridges.
- **payload:** the carrying capacity of an aircraft or spacecraft. With spacecraft, the payload may include astronauts, equipment, a satellite, and/or a space probe.
- **wind tunnel tests:** aerodynamics tests that involve moving air past an object in a tunnel to study the resulting effects.

**A BRIEF HISTORY OF LAUNCH VEHICLES**

A space launch vehicle is a rocket that is used to carry a payload into outer space. Many different launch vehicles have been used in space exploration. The United States and Russia are two of the leading nations in the manufacture of launch vehicles. The United States began using the Delta family of launch vehicles for space missions in the 1960s. This family includes several different types of expendable, or single-use, launch vehicles, some of which were still in use as of 2015. From 1959 to 2005, the United States also used the Titan family of launch vehicles for space exploration missions.

Other notable launch vehicles that have been produced by the United States include the Atlas family, Saturn V, and the reusable Space Transportation System (STS), which is commonly known as the space shuttle. As trends in space missions shifted in the early twenty-first century, the United States developed several launch vehicles that are smaller in size than previous launch vehicles. Pegasus and Taurus are examples of these smaller launch vehicles designed to carry payloads of advanced satellites and other small spacecraft. Russia's successful launch vehicles include the widely-used Soyuz and the larger Proton (both originally developed by the Soviet Union). In 2014, the nation performed a test flight of a new Angara launch vehicle.

All launch vehicles, whether expendable or reusable, are meant to lift a spacecraft into low-Earth orbit (LEO) or beyond. To do so they must accelerate beyond Earth's atmosphere, working against both gravity and atmospheric friction. To achieve the required thrust, launch vehicles are powered by chemically propelled rockets. Many considerations, including rocket construction, fuel types, and launch site and conditions, are taken to minimize the energy needed to reach outer space. As launch vehicles must travel through the atmosphere as efficiently as possible, aerodynamics helps determine their optimal design.

**BASICS OF AERODYNAMICS**
Aerodynamics is used to study how objects such as aircraft and rockets are affected by air or other gases. Several forces come into play with aircraft aerodynamics, including that of launch vehicles. Drag is one of the main aerodynamic forces. This force acts against an aircraft's motion as it travels through the air. Therefore, drag is generally an unwanted force, as the aircraft must counteract or overcome the force. Different forces affect drag, including the air pressure on the front of the aircraft, the suction on the back, and the friction on the sides.

A value known as the drag coefficient (cd) can be used to identify the amount of drag on an aircraft. The drag coefficient is dependent on a number of factors, including the shape, surface roughness, and speed of the aircraft, the turbulence of the air, and the density of the air. The drag coefficient is typically calculated using either wind tunnel tests or computer models. Modern aircraft have a drag coefficient between 0.01 and 0.03.

An aircraft can counteract or overcome drag by producing thrust. Thrust is generated by an aircraft engine and is the force that actually moves the aircraft. If an aircraft is flying horizontally at a steady speed, the thrust is barely counteracting the drag. Thrust must be increased in order to accelerate.

Lift is yet another aerodynamic force. The wings of a standard airplane produce lift, which is the force that prevents the plane from falling from the sky. Lift involves the speed of the air that flows over the top of the wings compared to that along the bottom of the wings. Air flows slower along the bottom, which causes the air to push up on the wings more forcefully than the air that is pushing down on the wings. Traditional launch vehicles, however, do not use wings. Instead they rely on vertically-launched rockets to reach the upper atmosphere (where drag is lower) as fast as possible without using lift. The lack of wings allows rockets to be lighter and more aerodynamic. Still, future spaceplanes may use winged designs to allow operation in the atmosphere along with the ability to reach escape velocity.

**WIND TUNNEL TESTS**
Before an aircraft such as a launch vehicle is built, aerodynamics testing of the aircraft is performed. Typically, a model of the proposed aircraft is tested

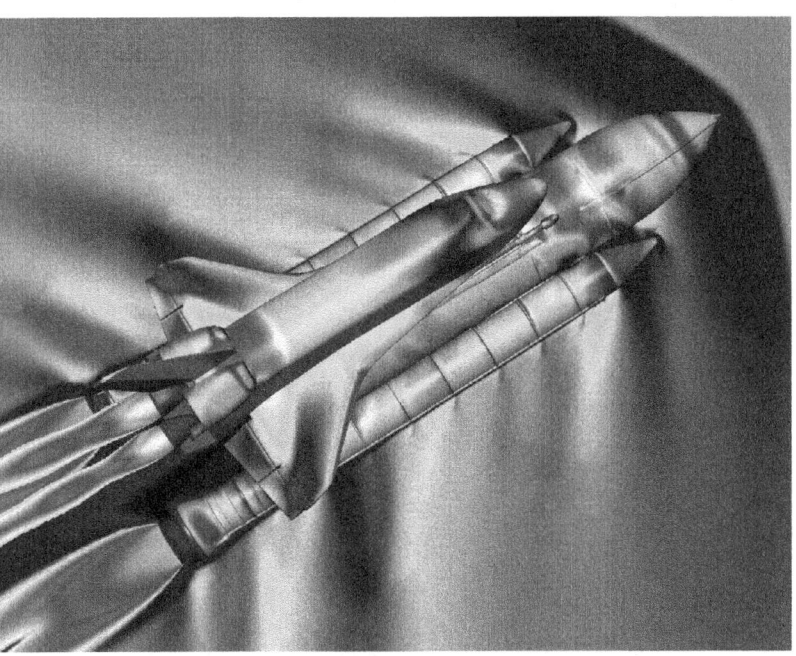

The US space shuttle launch vehicle with color indicating pressure points (red is high pressure) and the surrounding flowfield (white is high density air). This information is used to evaluate aerodynamic forces on the launch vehicle. Contributed by Ray J. Gomez, NASA Johnson Space Center, Houston, Texas.

using wind tunnel tests. These tests involve moving air past the model in a controlled environment such as a tunnel and determining the forces that are acting on the model. Several different types of wind tunnel tests can be used. One type directly measures aerodynamic forces and moments. A machine known as a force balance is used to mount the model in the tunnel. This force balance measures lift and drag.

In another type of wind tunnel test, pressure data is collected using pressure taps, which allows component performance to be calculated. Aircraft inlet performance and airfoil drag can also be determined in this type of wind tunnel test. A third type of test involves collecting diagnostic information through instrumentation including static pressure taps, total pressure rakes, and hot-wire velocity probes. Such instrumentation provides various kinds of flow information. Yet another type of wind tunnel test collects diagnostic information by making the flow of gases around the craft visible. The visualization methods used include laser sheet, free stream smoke, and surface oil flow.

Example of Launch Vehicle Aerodynamics Testing

NASA conducts extensive aerodynamics testing on their space launch vehicles. Typically, this testing allows engineers to predict vehicle control, trajectories, and payload performance. One example of such testing was during the design of the Space Launch System (SLS) in 2012. The SLS is the United States' successor to the space shuttle and first exploration-class rocket since Saturn V. It is expected to allow for human exploration outside of Earth's orbit.

The aerodynamics testing of SLS was performed at NASA's Langley Research Center in Hampton, Virginia. Scientists tested the aerodynamics of a model of the SLS in Langley's

Transonic Dynamics Tunnel (TDT), simulating the environment to which the full-size SLS rocket would be exposed in flight. As part of the wind tunnel testing, 360 pressure transducers were placed on the surface of the ten-foot-long model. The transducers allowed measurement of the rapidly changing forces of unsteady flow encountered in the atmosphere, providing the NASA team with a huge amount of data. The data was then used to prepare the SLS for future space missions. For example, data collected allowed the team to determine the launch vehicle's capability to handle the bending forces and accelerations it would be exposed to and make final design tweaks to improve it.

The intense energy required for liftoff and the nature of Earth's atmosphere mean that powerful forces act upon any craft launched into space. Aerospace engineers must ensure that a craft can structurally withstand these forces, with little margin for error. In this way understanding aerodynamics is crucial not only to a launch vehicle's ability to reach orbit efficiently, but to the safety of the vehicle's payload.

—*Michael Mazzei*

**FURTHER READING**

Blevins, John A, et al. "An Overview of the Characterization of the Space Launch System Aerodynamic Environments." *Aerospace Research Central.* American Inst. of Aeronautics and Astronautics, 13 Jan 2014. Web. 8 June 2015.

"Chapter 14. Launch Phase." *Basics of Space Flight.* Jet Propulsion Laboratory/NASA, 2013. Web. 8 June 2015.

Congiu, Sasha. "SLS Model 'Flies' through Langley Wind Tunnel Testing." *NASA.* NASA, 28 Nov. 2012. Web. 8 June 2015.

David, Leonard. "Russia Reignites Its Rocket Industry with New Angara Booster." *Space. com.* Purch, 19 Aug. 2014. Web. 8 June 2015.

Lucas, Jim. "What Is Aerodynamics?" *Live Science.* Purch, 20 Sept. 2014. Web. 8 June 2015.

"Wind Tunnel Testing." *NASA.* NASA, 5 May 2015. Web. 8 June 2015.

# LEO

## FIELDS OF STUDY

Stellar Astronomy; Observational Astronomy

## SUMMARY

The constellation Leo is a pattern of stars and part of the zodiac. Leo represents a lion and consists of a sickle, which resembles a backward question mark, and a triangle grouping of stars. Scientists study Leo and other constellations to learn more about the galaxy and the universe. Scientists have located a number of interesting objects while studying Leo, including a possibly terrestrial exoplanet that is a few times larger than Earth.

## PRINCIPAL TERMS

- **celestial equator:** the imaginary line above Earth's equator that halves the celestial sphere; it is equally distant from the celestial poles.
- **constellation:** a region of space defined by a pattern of stars that can be seen in the night sky from Earth.
- **declination:** the north-south position of a celestial body relative to the celestial equator expressed in degrees of arc.
- **International Astronomical Union:** an association of professional astronomers from all over the world who define astronomical constants while promoting research, education, and discussion on important astronomical topics.
- **right ascension:** the east-west position of a celestial body defined in relation to the celestial equator and expressed in hours and minutes, not degrees of arc.

## Leo and the Zodiac

Leo is a constellation that represents a lion from Greek mythology. This constellation is one of the easiest to recognize in the night sky, and it also one of the brightest. Constellations are regions of space defined by patterns of stars in the sky that people have identified over many centuries, typically naming them after figures represented in cultural stories and myths. The Greeks described more than half of the total constellations that have been identified. To make the designations of constellations more formal, Eugène Delporte assigned them official boundaries on behalf of the International Astronomical Union (IAU). The IAU recognizes eighty-eight constellations.

Leo is also part of the zodiac, which is a collection of thirteen constellations located within a specific part of the sky. Ancient cultures believed this group of constellations was important because the sun seemed to pass through them. This path, called the "ecliptic," is an illusion caused by Earth's orbit of the sun. However, the constellations, including Leo, are noted by their distance relative not to the ecliptic but to the celestial equator. Leo, for instance, has a declination of about +20 degrees from the celestial equator and a right ascension of eleven hours.

## Attributes of Leo

The constellation Leo resembles a lion viewed from the side. The stars that make up Leo form a shape that looks like a backward question mark, or sickle, connected to a triangle. In this star pattern, the backward question mark forms the lion's head, mane, and body. The triangle serves as the animal's haunches.

Stargazers in the Northern Hemisphere can see the Leo constellation best in March, April, and May. The constellation can be viewed only until July. After that point, people can see the constellation again in September or October. To the west of Leo lies the constellation Cancer, and the constellation Virgo lies to the east of it.

Leo is made up of a number of different stars and even galaxies. The brightest star in the constellation is alpha Leonis, which is also known as Regulus. This blue-white star represents the lion's heart. Also one of the brightest stars in the entire night sky, it is located about seventy-seven light-years from Earth. Regulus is located in the sickle of the constellation. Other stars in the sickle include Algieba (at the lion's mane), Adhafera, Ras Elased Borealis, and Ras Elased Australis.

The brightest star in the triangle of the constellation is beta Leonis, or Denebola. In Arabic, this name means "lion's tail." Denebola is located about thirty-six light-years away, which means that it is a relatively close star to Earth's solar system.

Some of the galaxies inside the Leo constellation include Messier 65, Messier 66, and New General Catalog 3628, which make up a formation called the Leo Triplet. The galaxy M66, which is a bright spiral galaxy, can be seen with binoculars or a telescope when the sky is clear and the viewer is in a dark place. Viewers can sometimes see M65 with binoculars as well when the conditions are right. The Leo constellation also includes an object called the Leo Ring. This is an enormous cloud made up of hydrogen and helium that orbits two large galaxies.

**HISTORY OF LEO**

The Leo constellation is one of the best-known constellations in the night sky. People from many different cultures—including the Persians, the Turks, and the Syrians—all recognized this constellation, but researchers believe the Mesopotamians were the first people to recognize the constellation. Since they did not have calendars, such ancient cultures used constellations to tell stories and to track seasons and different times of the year. Farmers often relied upon the apparent annual cycle of the stars to plant and harvest accordingly. Most of the constellations these cultures identified represented animals or people and were usually related to stories or myths from their culture.

Many people associate the constellation Leo with the Nemean lion from Greek mythology that Hercules killed during the twelve labors he was forced to undertake. According to that myth, Hercules tried to kill the lion with weapons, but the animal's skin could not be punctured. Hercules then strangled the lion and placed it in the heavens, which is how it became one of the constellations. Another story that might have been related to the lion is the tragedy of Pyramus and Thisbe. In this story, two lovers' fates are changed because they see a lion.

Since ancient times, records have shown that humans have witnessed a persistent, periodic phenomenon occurring within the Leo constellation. Approximately every thirty-three years, the comet Tempel-Tuttle enters the inner solar system, leaving a stream of debris behind. Some of these streams have crossed Earth's orbit, resulting in a vibrant and sometimes dense meteor shower. As this shower always takes place within the Leo constellation, it is known as the Leonid meteor shower, and scientists and stargazers continue to be captivated by and track this event.

**STUDYING LEO AND OTHER CONSTELLATIONS**

Scientists study the stars and other objects inside the constellations to learn more about the galaxy and the universe. Since there are so many stars in the sky, scientists can easily name and identify stars by associating them with the constellations they are found in.

Research of the Leo constellation has led to a number of important discoveries. In 2008, scientists announced that they found a possibly terrestrial exoplanet about five times the mass of Earth orbiting

Star chart of the Northern and Southern Hemispheres in November. Lines connect the stars of the constellation Leo.

one of Leo's stars located about thirty light-years from Earth. Though the exoplanet is larger than Earth, it was still the smallest exoplanet known at the time. Over one year later, scientists discovered that the binary star Algieba in the constellation has a huge exoplanet—an object that is eight times the size of Jupiter—orbiting it. Scientists were excited about both of these discoveries because they hope to find more small exoplanets that could be similar to Earth.

—*Elizabeth Mohn*

**FURTHER READING**

"The Constellations." *International Astronomical Union.* IAU, n.d. Web. 9 Apr. 2015.

"Constellations in the Zodiac." *National Aeronautics and Space Administration.* NASA, n.d. Web. 9 Apr. 2015.

Eratosthenes, and Hyginus. *Constellation Myths.* Trans. Robin Hard. Oxford: Oxford UP, 2015. Print.

Kambic, Bojan. *Viewing the Constellations with Binoculars: 250+ Wonderful Sky Objects to See and Explore.* New York: Springer, 2010. Print.

"Leo, the Lion." *StarDate.* U of Texas McDonald Observatory, n.d. Web. 9 Apr. 2015.

McClure, Bruce. "Leo? Here's Your Constellation." *EarthSky.* EarthSky Communications, 10 Apr. 2014. Web. 9 Apr. 2015.

Nave, R. "Ecliptic Plane." *HyperPhysics.* Dept. of Physics and Astronomy, George State U, 2012. Web. 10 Apr. 2015.

"New Rocky Planet Found in Constellation Leo." *ScienceDaily.* ScienceDaily, 10 Apr. 2008. Web. 9 Apr. 2015.

Schneider, Howard. *National Geographic Backyard Guide to the Night Sky.* Washington, DC: Natl. Geographic Soc., 2009. Print.

Zimmermann, Kim Ann. "Leo Constellation: Facts about the Lion." *Space.com.* Purch, 1 Aug. 2012. Web. 9 Apr. 2015.

# LIBRA

**FIELDS OF STUDY**

Stellar Astronomy; Observational Astronomy

**SUMMARY**

The constellation Libra is a grouping of stars and part of the zodiac. The constellation represents a balance scale, with two pans hanging from a center beam. Humans have used constellations such as Libra for millennia to track the seasons and tell the time of year. Scientists study constellations today mainly to name and track the stars that make up the patterns. By studying constellations and their stars, scientists can learn more about the galaxy and the universe.

**PRINCIPAL TERMS**

- **constellation:** a pattern of stars that can be seen in the night sky from Earth.
- **International Astronomical Union:** a worldwide professional association of astronomers who conduct research and seek to educate others about astronomy. It sets the rules for naming celestial bodies and features on them and defines scientific constants of importance to astronomy.

**LIBRA AND OTHER CONSTELLATIONS**

Constellations are patterns of stars that can be seen in the night sky from Earth. Humans have been identifying constellations for tens of thousands of years. People from different cultures grouped the stars in different constellations and gave them different names. The ancient Greeks described forty-eight constellations. European scientists in the sixteenth and seventeenth centuries saw other constellations. In 1930, Eugène Delporte listed these constellations for the International Astronomical Union (IAU). The IAU list of eighty-eight constellations is used by astronomers around the world. They use the constellations to identify and keep track of the stars.

Libra is one of the constellations in the zodiac. The zodiac is a group of twelve or thirteen constellations that appear in a specific section of the night sky. People from ancient civilizations thought these constellations were special because it appears as though the sun travels them. Although modern scientists know that Earth moves around the sun, giving the illusion that the sun is moving, people still enjoying looking for the constellations in the zodiac in the night sky.

**ATTRIBUTES OF LIBRA**

People in the Northern Hemisphere can view Libra best during June and July. It is located in the southwest quadrant in the Northern Hemisphere in the summer between Virgo in the west and Scorpius to the east. It is also close to the constellations Hydra, Ophiuchus, and Lupus.

The brightest stars in the Libra constellation are beta Librae (Zubeneschamali), alpha Librae (Zenelgenubi), and sigma Librae (Brachium). The drawn image of the constellation is a balance scale, with a central triangle or beam from which hang two pans. According to the IAU, alpha Librae, beta Librae, and gamma Librae (Zubenelakrab) form a triangle that represents the top of Libra's scale. Other astronomers draw the triangle through alpha Librae, beta Librae, and sigma Librae. The constellation also includes upsilon Librae, tau Librae, and a number of other stars. Libra is one of the dimmest constellations in the zodiac.

**HISTORY OF LIBRA**

People have been viewing the stars and identifying grouping of stars for tens of thousands of years. People first developed constellations for a number of reasons. Scientists believe that people identified some constellations according their religious beliefs. Other people developed constellations to help them with farming. These people noticed that the constellations were visible in the same spot in the sky at particular times of year. At this time, people did not have calendars, so farmers used the constellations to mark when to plant and harvest their crops.

The Sumerians called the constellation Zib-ba An-na, meaning "balance of heaven." Zib-ba An-na sounded like the Arabic and Akkadian words that mean both "weighing scale" and "scorpion."

The constellation was once called Chelae Scorpionis ("claws of a scorpion"). In fact, the names of Libra's stars Zubeneschamali and Zubenelgenubi translate as "northern claws" and "southern claws," respectively.

Researchers believe that Roman astronomers identified Libra in the first century BCE. In Latin, *libra* means "weighing scales." To the Romans, the constellation Libra depicted the scales held by the goddess of justice. It also stood for balance because at that time the sun shone in front of the constellation on the autumnal equinox, when there are equal amounts of day and night. Libra is the only constellation in the zodiac that represents an inanimate object instead of an animal or a person.

**STUDYING LIBRA AND OTHER CONSTELLATIONS**

When scientists study the various stars in the constellations, they can make interesting discoveries. In the early 2000s, scientists studied stars in the constellation Libra. Based on their research, scientists thought they made an important discovery. Between 2007 and

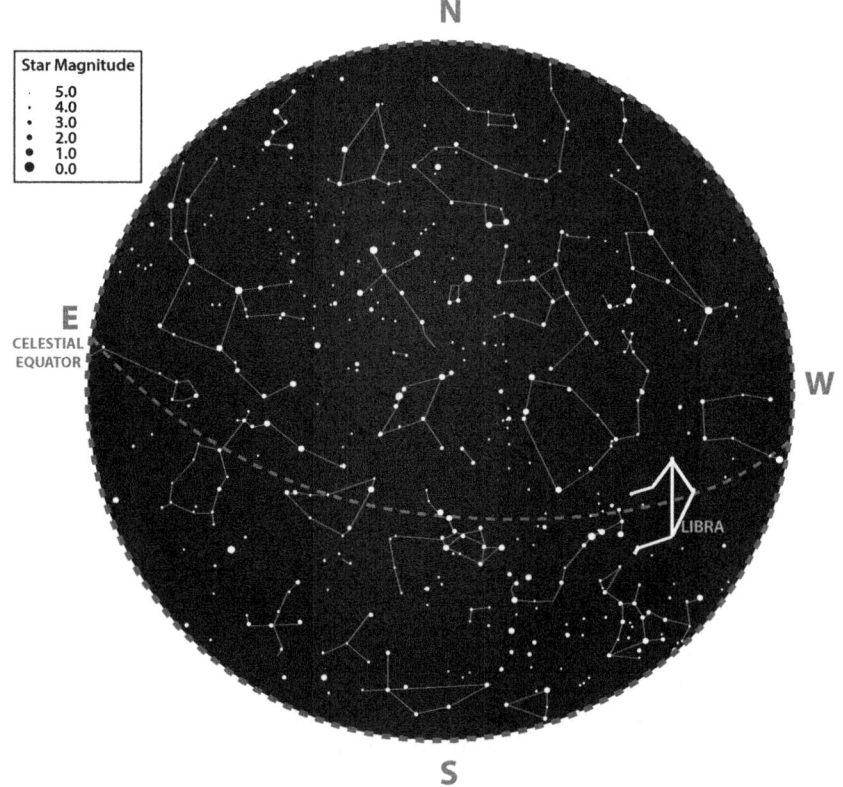

Star chart of the Northern and Southern Hemispheres in June. Lines connect the stars of the constellation Libra.

2010, scientists announced that they had found two exoplanets (planets outside Earth's solar system) that might be able to support life. They named the exoplanets Gliese 581d and Gliese 581g. Scientists thought Gliese 581g was a few times the size of Earth but at just the right distance from its star so that it could have liquid water, which is a key ingredient for life. Many scientists thought this planet may have life on it. However, in 2014, further research raised doubt that Gliese 581d and Gliese 581g might not actually exist. Despite some people being discouraged by the later findings, scientists continue to study this star system in the hope of learning more about the galaxy and the universe.

—Elizabeth Mohn

**FURTHER READING**

"The Constellations." *International Astronomical Union*. IAU, n.d. Web. 27 Mar. 2015.

"Constellations in the Zodiac." *NASA*. NASA, 27 Mar. 2015. Web. 27 Mar. 2015.

Dearden, Lizzie. "One of the Most Earth-Like Planets in Our Galaxy Gliese 581d 'Really Does Exist,' Astronomers Say." *Independent*. Independent.co.uk, 7 Mar. 2015. Web. 30 Mar. 2015.

Kambic, Bojan. *Viewing the Constellations with Binoculars: 250+ Wonderful Sky Objects to See and Explore*. New York: Springer, 2010. Print.

"Libra, the Scales." *StarDate.org*. U of Texas McDonald Observatory, 2015. Web. 24 Mar. 2015.

Schneider, Howard. *National Geographic Backyard Guide to the Night Sky*. Washington, DC: Natl. Geographic Soc., 2009. Print.

Schultz, Colin. "Gliese 581g, the First Exoplanet Found That May Have Been Able to Host Life, Doesn't Actually Exist." *Smithsonian Magazine*. Smithsonian.com., 7 July 2014. Web. 24 Mar. 2015.

Zimmermann, Kim Ann. "Libra Constellation: Facts about the Scales." *Space.com*. Purch, 17 June 2013. Web. 24 Mar. 2015.

# M

## MARS

### FIELDS OF STUDY

Planetary Science; Planetary Geology; Orbital Mechanics

### SUMMARY

Mars is the fourth terrestrial planet, and the last of the inner planets. It is slightly more than half of the size of Earth in diameter, but just one-tenth of the mass of Earth. It has very little atmosphere, but often has large dust storms. Surface temperatures range from -140°C to +20°C. Mars' orbit is somewhat eccentric, and the planet has a relatively low albedo. The largest known mountain and valley are found on Mars.

### PRINCIPLE TERMS

- **Amor asteroid group:** a group of near-Earth asteroids that come within 0.3 astronomical units (AU) of Earth but do not cross its orbit, although most do cross Mars's orbit.
- **asteroid belt:** an oval-shaped ring in which millions of asteroids orbit, located between the orbits of Mars and Jupiter, about 2 to 4 astronomical units (AU; 186 to 370 million miles) from the sun.
- **astronomical unit:** a unit of length equal to the average distance between the Earth and the sun, about 150 million kilometers (93 million miles); often used to measure distances within the solar system.
- **planetary rover:** mobile robotic science lab sent to explore and study the surface of other planets.
- **radiation hardening:** process of making electronic components resistant to radiation in space.
- **thermal considerations:** how exposure to heat or cold affects components in spacecraft or robotics.

### MARS AND ITS PLACE

The planet Mars is the fourth of the four terrestrial planets, along with Mercury, Venus and Earth. Like the other terrestrial planets Mars is a solid rocky planet. It is considerably smaller than Earth, having a diameter of 6,786 kilometers, as compared to the 12,741 kilometer of Earth.

The average density of Mars is 3.95 g.cm-3, as compared to Earth's average density of 5.52 g.cm-3. The primarily carbon dioxide atmosphere of Mars is very thin, with pressures of just 0.005 times the average atmospheric pressure at Earth's surface. The orbit of Mars is more eccentric, having an eccentricity of 0.0934, as compared to the nearly circular orbit of Venus, which has an eccentricity of 0.0068. The average distance of Mars from the Sun is 2.28 X 108 kilometers, or 1.5 astronomical units. The orbit of Mars marks the limit of the 'inner planets'. Between Mars and Jupiter, the next farthest planet in the Solar system, is the asteroid belt and Jupiter is the first of the 'outer planets'. Mars' orbit is somewhat compromised by the Amor asteroid group, whose eccentric asteroids cross the orbit of Mars with some regularity and which have doubtless been responsible for some of the craters that mark the surface of that planet. Mars is a rather dark planet, with an albedo of 0.15, similar to that of Mercury. Polar ice caps are also apparent, and have been found to consist of water ice covered by carbon dioxide ice during the Martian winters. The planet appears as a reddish 'star', due to the formation of iron oxide and other oxides on the surface. The lack of a protective atmosphere allows solar and cosmic radiation to be absorbed by any water molecules in the atmosphere. The water molecules are subsequently split apart into one oxygen atom and two hydrogen atoms. The oxygen atom typically reacts with any available iron or other element, and over the lifetime of the Martian surface this has resulted in the formation of the "red planet". Mars has two moons in orbit around it.

### QUALITIES

Since Earth is between the Sun and the orbit of Mercury, viewing the planet is not restricted to within specific angles relative to the Sun, and observation of the movement of Mars through the sky reveals that the relative motion of Mars and Earth causes it to appear to reverse direction in its orbit. The surface of Mars is marked with several craters, though no canals, and shows a great deal of evidence of erosion characteristic of running water. Dust storms are common, although the dust is extremely fine, unlike the sand grains that are typically carried by dust storms on Earth. Having a diameter only slightly more than half the diameter of Earth, it might be expected that the planet should have about one-eighth of the mass of Earth. It is found, however, that the density of Mars is much lower than that of Earth, and its mass is only about one-tenth that of Earth. Accordingly, the escape velocity of Mars is 5.02 kilometers per second, as compared to Earth's escape velocity of 11.12 kilometers per hour. Mars is much like Earth with regard to its axial tilt and the length of the day. The axial tilt of Mars is 25.02°, as compared to the value of 23.4°

The planet Mars. By NASA / USGS

for Earth's axial tilt. At 24 hours and 40 minutes, the length of the Martian day is very nearly the same as the length of the day on Earth. However, the larger orbit of Mars requires 1.88 years for one complete orbital transit. Temperatures on Mars range from lows of -140°C (-220°F) in the dark or at night, to a high of just 20°C (68°F) at midday.

There are no tectonic plates in the crust of Mars like those of Earth's crust. The smaller size and mass of the planet allowed it to cool in such a way that the crust became too thick for tectonic plates to form, and the internal pressure from shrinkage as the planet continued to cool was instead released by volcanic activity. There are a number of extinct volcanic mountains on the surface, each one situated over a stationary hot spot. In the absence of tectonic plates eruptions would have continued in the same places rather than the volcanic activity being spread along fault lines like it is on Earth.

The largest mountain known, an ancient volcano called Olympus Mons, is found on Mars. Its base has a diameter of more than 640 kilometers (400 miles), encompassing more than four times the extent of the island of Hawaii over the sea floor, and it stands more than twice the height of Mt. Everest.

A similarly gigantic feature called the Valles Marineris, a valley much like the Grand Canyon only large enough to extend completely across the continental United States, is found on Mars. The formation of Valles Marineris, however, is not understood. It is unlikely that it was formed by the action of running water like the Grand Canyon, and theories for its formation include both the glancing impact of some large body that gouged out the formation and the direct impact of a large body on the opposite side of the planet such that the pressure generated inside of Mars as a result was sufficient to cause the crust to break open.

## Mars in History and Mythology

The planet Mars has been known since ancient times. It has apparently always been associated with blood due to its red color, which is visible to the naked eye. The Greeks associated the planet with their god of war, presumably because of its blood-like color, and named it Ares after that same god. When Greek culture was superseded by the Roman empire, the Romans used the name Mars, which was the name of the Roman god of war. The two moons of Mars are named Phobos (Fear) and Deimos (Panic) after the two servants of Mars as the god of war.

Galileo and other astronomers have viewed the planet by telescope, and the relatively low power and resolution of earlier telescopes allowed for numerous erroneous theories about the planet to arise. In 1802, Gauss believed the planet to be not only habitable, but inhabited. In 1863, Secchi drew a colored map of Mars that showed several linear features that he called channels, or canali. In 1877, Schiaparelli added many more canali in his own drawings of the planet, but interpreted Secchi's canali to mean canals instead of channels. In 1894, Lowell showed some 500 canals' extending over the surface of Mars, and he proposed that they had been constructed purposely to carry water from the polar regions to more arid regions of the planet. Other astronomers, however, could not find any of these canals, but did see craters.

Close observation of Mars by satellite and landing probe began in 1965 with the Mariner program. Most recently, the surface of the planet is being explored "first hand by remote control" using planetary rovers that carry out a number of scientific tests and return photographic images and analytical data to Earth.

The lack of atmosphere means that radiation hardening is an essential part of the design of a Mars planetary rover in order to protect the devices from damage while on the surface of Mars. Similarly, the extreme temperature variations that the devices will experience determine the thermal considerations that must also be included in the design. Research and exploration of Mars is an ongoing endeavor, with the intention of having human explorers on the planet in the foreseeable future.

## Measuring Mars

The orbits of all planets are not circular, but elliptical. Accordingly, the regular motions of planets such as Mars in their orbits relate to the mathematics of ellipses. The 16th century astronomer Johannes Kepler recognized this and was able to combine observations of the motions of the planets with the theory that the planets followed elliptical orbits around the Sun. Kepler's Third Law relates the orbital period of a planet to its average distance from the Sun, as

$$C = (d_{ave})^3 / (P_{orb}^2)$$

where $d_{ave}$ is the average distance from the Sun and Porb is the planet's orbital period. $C$ is found to have a constant value of 1 $AU^3 year^{-2}$.

—*Richard M. Renneboog M.Sc.*

## Further Reading

Barlow, Nadine G. *Mars: An Introduction to Its Interior, Surface and Atmosphere*. New York, NY: Cambridge University Press, 2008. Print.

Carr, Michael. *The Surface of Mars*. New York, NY: Cambridge University Press, 2007. Print.

Chapman, Mary. *The Geology of Mars.: Evidence from Earth-Based Analogs*. New York, NY: Cambridge University Press, 2007. Print.

Miller, Ron. *Mars*. Minneapolis, MN: Twenty-First Century Books, 2006. Print.

Rapp, Donald. *Human Missions to Mars: Enabling Technologies for Exploring the Red Planet*. 2nd ed. Chichester, UK: Praxis/Springer, 2016. Print.

Zubrin, Robert. and Richard Wagner,. *The Case for Mars: The Plan to Settle the Red Planet and Why We Must*. New York, NY: The Free Press, 2011. Print.

Zubrin, Robert. *Mars Direct:. Space Exploration, the Red Planet, and the Human Future*. New York, NY: Penguin, 2013. Print.

# MASS ATTENUATION COEFFICIENTS

## FIELDS OF STUDY

Astrophysics; Theoretical Astrophysics

## SUMMARY

Mass attenuation coefficients are measurements of how waves of energy and sound are absorbed or scattered by various materials.

## PRINCIPAL TERMS

- **electromagnetic radiation:** energy produced by the interaction of electric and magnetic fields that travels in the form of electromagnetic waves.
- **wavelength:** the distance between a point on one wave and the same point on the next wave.

### ATTENUATION AND PENETRATION

Mass attenuation coefficients are measurements of how electromagnetic radiation is absorbed or scattered by various materials. "Attenuation" is the loss of energy from waves or particles due to interaction with another object. For example, x-rays attenuate when they come in contact with lead, and visible light attenuates when it comes in contact with darkened glass.

X-rays and gamma rays are two types of electromagnetic radiation. They are used in medical imaging and cosmology because they can penetrate better than other types of waves. The type of electromagnetic wave is determined by the wavelength. Gamma rays have the shortest wavelengths, followed by x-rays. Many waves, including visible light, can scatter when they hit an object. However, x-rays and gamma rays penetrate more deeply into the objects. Penetration is the opposite of attenuation.

Electromagnetic radiation has characteristics of both waves and particles, and it can be treated as either depending on the situation. Particles of electromagnetic radiation are called photons. The extent to which radiation will penetrate an object depends on the energy of its photons. It also depends on the attributes of the object it is penetrating, such as density, thickness, an atomic number. All of these factors affect a material's mass attenuation coefficient.

A related value is the linear attenuation coefficient, which also measures how much a particular type of energy can penetrate a certain material. However, it does not take into account the density of the material. The mass attenuation coefficient, which can be found by dividing the linear attenuation coefficient by the density of the material, can be used as a standard measurement.

### CONTRIBUTING FACTORS

In astrophysics, mass attenuation coefficients are most often used when studying x-rays and gamma rays. Various different interactions cause x-rays to attenuate, including Compton scattering, Rayleigh scattering, the photoelectric effect, pair production, and triplet production. Scientists calculate linear attenuation coefficients by taking all of these probable interactions into account.

Attenuation can change depending on the density of the material the x-ray is passing through. For example, an x-ray passing through water vapor will attenuate differently from one passing through ice. Even though the two materials are chemically the same, they have different densities. Because of this, scientists often choose not to work with the linear attention coefficient. Instead, they work with the mass attenuation coefficient, which is measured in square centimeters per gram.

### DETERMINING VALUES

Scientists often work with tables of pre-established mass attenuation coefficients. These standard values were developed based on both observations and theoretical knowledge about various substances. Commonly used mass attenuation coefficients include lead, water, iron, and air.

To determine a mass attenuation coefficient via direct observation, a scientist would follow these steps: (1) direct a beam of photons with, all of the same energy, at a specific material; (2) measure the intensity of the photons before they enter the material; (3) measure the intensity of the photons after they exit the material; (4) compare the measurements to determine how many photons were absorbed and scattered; (5) take the density and thickness of the material into account.

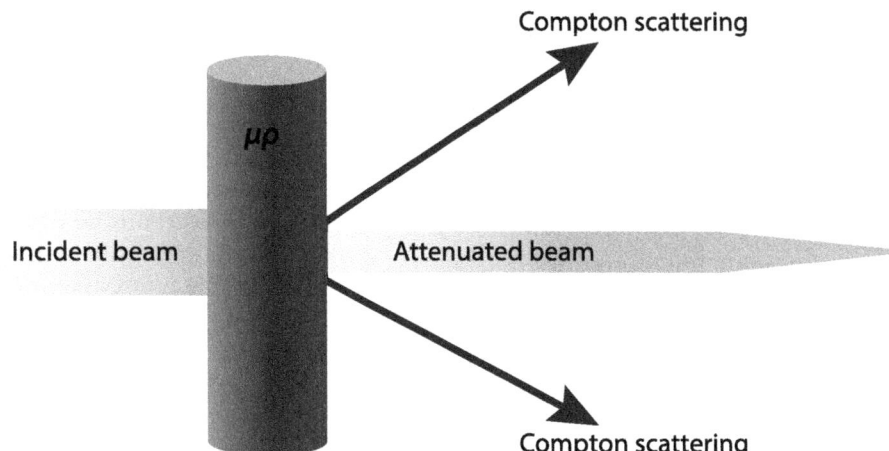

The intensity of a wave is altered by the attenuation, or absorption, that it undergoes as it passes through matter. The mass attenuation coefficient is equal to the attenuation coefficient ($\mu$) divided by the density of the matter ($\rho$), and differs for each element.

### Applications of Mass Attenuation Coefficients

The amount of energy that is absorbed by a particular material will determine where that energy goes and what it affects. Each layer of an object will attenuate the same percentage of photons. This means that the first layer will attenuate the most photons, while fewer and fewer photons will be attenuated with each successive layer that is penetrated. This phenomenon has a number of practical applications. For example, because x-rays can be harmful to humans, scientists must be careful when working with them. If they know how different materials attenuate x-rays, they can determine how much of which material is needed to shield people from harm.

Another application of mass attenuation coefficients is in astronomy. For example, gamma-ray astronomers observe gamma rays in the universe. Gamma rays travel at the speed of light and are unaffected by electric or magnetic fields, which makes them ideal tools for scientific observation. Their interactions with various materials in space help scientists gather data on celestial objects and the processes that may have shaped them. Studying how different materials attenuate gamma rays may also shed light on the origins of the radiation. Other forms of electromagnetic radiation, such as x-rays and bremsstrahlung, can be studied in this manner as well.

—*Elizabeth Mohn*

### Further Reading

Beutel, Jacob, Harold L. Kundel, and Richard L. Van Metter. *Handbook of Medical Imaging: Physics and Psychophysics.* Bellingham: SPIE, 2000. Print.

Holder, Jamie. "Gamma-Ray Astronomy." *Department of Physics and Astronomy.* University of Delaware, 2008. Web. 8 Apr. 2015.

Sprawls, Perry. "Radiation Penetration." *Sprawls Educational Foundation.* Sprawls Educ. Foundation, n.d. Web. 6 Apr. 2015.

"Transmitted Intensity and Linear Attenuation Coefficient." *NDT Education Resource Center.* Iowa State U, n.d. Web. 6 Apr. 2015.

"X-Ray Mass Attenuation Coefficients." *Physical Measurement Laboratory.* Natl. Inst. of Standards and Technology, n.d. Web. 6 April 2015.

"X-Ray Test Facility Calculations." *Extreme Universe Laboratory.* Moscow State U, n.d. Web. 6 Apr. 2015.

# MERCURY

### FIELDS OF STUDY

Planetary Science; Planetary Geology; Orbital Mechanics

### SUMMARY

The planet Mercury is the closest planet to the Sun. In this article, its various features are presented, along with its historical background. Mercury has the second most eccentric orbit. A study of the motions of the planet enables a better understanding of the geology of Earth.

### PRINCIPLE TERMS

- **aphelion:** the point in an object's orbital path that is farthest from the sun.
- **asteroid:** a rocky, irregularly shaped space body that is thought to be a remnant of the formation of the solar system.
- **crater:** a bowl-shaped depression on the surface of a planet, moon, or other astronomical body, usually caused by either volcanic activity or a high-speed impact with another, smaller body.
- **eccentricity:** the extent to which a celestial body's orbit deviates from a perfect circle.
- **magnetic field:** an area around an object that has a magnetic influence.
- **perihelion:** the point in an object's orbital path that is closest to the sun.

### MERCURY AND ITS PLACE

The planet Mercury is one of the four terrestrial planets, along with Venus, Earth and Mars. Like the other terrestrial planets Mercury is a solid, rocky planet. It is comparable in size to a number of other solar satellites, and is closest in size to Callisto, one of the moons of Jupiter. Mercury is 4787 kilometers in diameter compared to 4800 kilometers in diameter for Callisto. Other bodies of similar diameter include Ganymede, another moon of Jupiter having a diameter of 5262 kilometers; Titan, a moon of Saturn having a diameter of 5150 kilometers; Earth's Moon, which has a diameter of 3476 kilometers; and Pluto, whose diameter is 2260 kilometers. Mercury has essentially no atmosphere, although there are minute traces of helium and argon, and the surface bears numerous craters from impacts with asteroids. Mercury follows the closest orbit about the sun of any of the Solar system planets. The average diameter of Mercury's orbit is 57.9 million kilometers.

### QUALITIES

Because of this proximity to the Sun, Mercury is only visible in the western sky just after sunset and in the eastern sky just before sunrise. Mercury's orbit has the second highest eccentricity of all the planets from aphelion to perihelion. Only the orbit of Pluto is more eccentric, which is to say, more elliptical in shape.

The planet spins on its axis with a period of rotation of 58.65 Earth days, and travels through one complete orbit in 87.97 days. The planet has a large mass imbalance, perhaps due to a major impact in its past, that is affected by the gravitational forces of the Sun. As the planet spins in the Sun's gravitational field, the difference in gravitational attraction between the more and less massive parts of the planet induce a torque on the planet such that it now completes 1.5 rotations in each complete orbit. Earth, in comparison, completes 365.245 complete rotations in each complete orbit.

Mercury's mass is just 5.5 percent that of Earth's, but it has a very similar average density of 5.43 times the density of water, compared to Earth's average density of 5.52. This suggests that Mercury has a very large iron core that was probably formed through impacts with iron meteorites as the planet grew. This core is believed to be about three-quarters of the size of the planet, and contains enough iron to meet the demands of Earth industry for several thousand years. Despite this high quantity of iron in its compositon, Mercury nevertheless has only a very small magnetic field, just 1 percent that of Earth's. The magnetic poles of the planet are aligned very closely with its spin axis.

The lack of atmosphere on Mercury means that there is no protection of the planet from incoming meteorites, resulting in a surface scarred with abundant impact craters. In addition, the lack of atmosphere means there is no protective layer against the light coming from the Sun, producing extreme temperatures on Mercury. In the daylight, surface temperatures reach, and perhaps exceed, 450°C (840°F), but at night, temperatures drop to -150°C (-240°F) or lower.

Mercury's overall albedo (overall reflectivity) is just 0.11, compared to 0.75 for Earth. The albedo at Mercury's poles, however, is significantly higher, which suggests that the polar regions of the planet have a permanent ice presence, especially within craters that offer protection from the sunlight. Presumably, the temperatures inside of polar craters does not rise above -150°C (-240°F).

Even though the mass of Mercury is just 5.5 percent that of Earth's mass, the gravity of the planet is 38 percent that of Earth's gravity. Mercury's diameter is just 39 percent that of Earth's, and the escape velocity of Mercury at 4.25 kilometers per second is also 39 percent that of Earth's.

The largest known feature on the surface of Mercury is the Caloris Basin, a named derived the Latin word for heat. Caloris Basin is 1400 kilometers wide, and is believed to have been formed by gigantic lava flows from an impact that ruptured the surface of the planet, perhaps the same impact that caused the mass imbalance within the planet. There are also lines of towering cliffs up to 4 kilometers high stretching in long lines across the planet.

The crust of Mercury is very thick in comparison to the size of the planet, too thick, in fact, to consist of large tectonic plates. It is thought that the lines of cliffs were formed by the shrinking of the planet as it cooled. Material that was not taken into the core exceeded the surface area available, causing it to buckle and crack so that huge cliffs were raised up.

## Mercury in History and Mythology

Being the closest planet to the Sun, Mercury has the fastest orbit of all of the planets and almost all of the other objects that orbit that star. It has been known since ancient times. To early astronomers, the planet was thought to be two different stars because of the times at which it can be seen. To the ancient Greeks, the planet was given the name Hermes, the name of the swift messenger of the gods. When Greek art and philosophy was overtaken by the Roman Empire, the Romans gave the planet the name Mercury, its own name for the messenger. Galileo and other astronomers since then have been able to view the planet through telescopes, and direct observation of Mercury by satellite and landing probe began in March of 1974.

## Measuring Mercury

Because of the great distances involved in space, geometric calculations are very important. The relatively short distances to planets, as compared to the much greater distances between stars, can be calculated using simple Pythagorean theory and triangulation. Precise calculations of elliptical orbits require more advanced mathematical principles such as differential calculus. The relative motions of Earth and another planet require that they be considered in any calculation if a rendezvous with a satellite is to be successful. Triangulation is also used to determine the size of an object such as a planet or asteroid. If the dimensions of an object are well known, as well as its distance from another object

Internal structure of Mercury: 1. Crust: 100–300 km thick 2. Mantle: 600 km thick 3. Core: 1,800 km radius. By NASA-APL such as a planet, then the method can be used to determine the size of the planet and various features on it by proportion.

—Richard M. Renneboog M.Sc.

**FURTHER READING**

Balogh, André, Leonid Ksanfomality, and Rudolf von Steiger, eds. *Mercury*. New York, NY: Springer Science+Business Media BV, 2008. Print.

Elkins-Tanton, Linda T. *The Sun, Mercury, and Venus*. New York, NY: Chelsea House, 2006. Print.

Grego, Peter. *Venus and Mercury and How to Observe Them*. New York, NY: Springer Science+Business Media, 2008. Print.

Mahoney, T.J. *Mercury*. New York, NY: Springer Science+Business Media, 2014. Print.

Rothery, David A. *Planet Mercury: From Pale Pink Dot to Dynamic World*. New York, NY: Springer, 2015. Print

Strom, Robert G. and Ann L Sprague. *Exploring Mercury: The Iron Planet*. Chichester, UK: Praxis/Springer, 2003. Print.

# MESSIER CATALOG

**FIELDS OF STUDY**

Extragalactic Astronomy; Historical Astronomy

**SUMMARY**

French-born astronomer Charles Messier used telescopes to seek out comets in the Northern Hemisphere. He noticed other celestial objects in addition to the comets. These objects looked like blurs in the sky. To help astronomers identify these objects, Messier decided to categorize them. During his life, the astronomer classified more than one hundred objects in a catalog. The Messier Catalog became an important tool to astronomers and is still used in modern times.

**PRINCIPAL TERMS**

- **deep-sky object:** a space object that exists beyond the solar system and cannot be seen with the naked eye.
- **Messier object:** an astronomical deep-sky object cataloged by French astronomer Charles Messier.

## Who Was Charles Messier?

Charles Messier was born in Salm, now part of the Lorraine region of France, on June 26, 1730. He was the tenth of twelve children born to Nicolas Messier and Françoise Grandblaise. Nicolas died when Messier was eleven years old, changing the dynamic and financial status of the family. An accident forced him to leave school and be tutored at home by his oldest brother. His brother taught him administrative, organizational, and observational skills, the latter of which served Messier well later in life.

Messier became interested in astronomy at a young age. Witnessing a solar eclipse in 1748 heightened this interest. In 1751 he began work with Joseph-Nicolas Delisle (1688– 1768), the astronomer to the French Navy. Messier's job was to keep records of space observations, and Delisle taught him how to

Charles Messier, French astronomer, published a catalogue of astronomical objects. Painting by Ansiaume (1729–1786), public domain via Wikimedia Commons.

measure exact positions of observations. Messier's first documented observation was of the Mercury transit in 1753.

Messier became skilled at tracking astronomical objects such as comets, celestial bodies that form tails as they orbit the sun. While tracking comets in 1758, he noticed other astronomical objects that existed beyond the solar system. He later categorized these as deep-sky objects. Messier became a member of the Royal Society of London in 1764. He published the first edition of his catalog of deep-sky objects in 1771. Around that time, he was named astronomer to the French Navy. Messier continued to seek out comets and catalog deep-sky objects in the years that followed. At the time of his death on April 12, 1817, he had cataloged 103 Messier objects.

### CLASSIFYING DEEP-SKY OBJECTS

Messier discovered deep-sky objects because of a mistake. In the late seventeenth century, English astronomer Edmond Halley (1656–1742) had examined historical records of comet sightings and cataloged his findings in *A Synopsis of the Astronomy of Comets* (1705). He found similarities between the orbits and parameters of three comet sightings in 1531, 1607, and 1682 and concluded that they were most likely the same comet returning to Earth every seventy-five or seventy-six years. Halley predicted the return of the same comet in 1758, but he died before he could see his prediction come true. The comet was later named Halley's comet in his honor.

Messier began searching for the return of the comet around 1758, while he was working as an assistant to Delisle. However, Delisle made a mistake in his calculations, which led Messier to look in the wrong quadrant of the sky. While Messier was exploring the night sky in late August or early September 1758, he saw a fuzzy cloud between the stars of the Taurus constellation. He continued to study the blurry patch in the sky and determined that it did not move like the stars around it. This helped him rule it out as a comet. He determined that it was a nebula, or a cloud of gas and dust between the stars. Messier decided to catalog the astronomical object and named it Messier 1 (M1).

Messier continued to search for Halley's comet using Delisle's incorrect calculations. A German amateur astronomer, Johann Georg Palitzsch, spotted the comet on December 25, 1758. Messier used different calculations and eventually located Halley's comet on his own nearly a month later.

Messier continued to add his and other astronomers' findings to his journal, which became known as the Messier Catalog or Messier Album. The second entry, Messier 2 (M2), is a globular cluster that had been discovered by Italian astronomer Jean-Dominique Maraldi II (1709–88) in 1746. (A globular cluster is a near-spherical collection of old stars bound together by gravity.) Thereafter, Messier tasked himself with finding and recording deep-sky objects, or Messier objects as they came to be known. He said he undertook the project so that other astronomers would not mistake these objects for actual comets. Messier devoted himself to searching the night sky. By the time he died in 1817, he had cataloged 103 objects, 41 of which he had found himself. He also identified fifteen comets during his studies of the sky, of which twelve bear his name.

### EXAMPLES OF MESSIER OBJECTS

Most Messier objects appear faint and blurry. They typically cannot be seen without the aid of binoculars or a telescope. They are classified into several categories, including clusters, galaxies, nebulae, and supernova remnants.

The first recorded Messier object, M1, is a supernova remnant, left behind after the explosion of a star. It is more commonly known as the Crab Nebula. M3, located in the Canes Venatici constellation, is another globular cluster. M31, the Andromeda galaxy, is the closest galaxy to the Milky Way. M40 is a double star, or two stars that appear very close together when seen from Earth. M42, the Great Orion Nebula, is located in the Orion constellation. M45, the Pleiades, is an open cluster located in the Taurus constellation. An open cluster, also called a galactic cluster, is a group of same-aged stars within a cloud of dust and gas. M64, the Black Eye or Sleeping Beauty galaxy, is a spiral galaxy located in the Coma Berenices constellation. It has an area of dust that darkens its center.

### MODERN RELEVANCE OF THE MESSIER CATALOG

During the twentieth century, seven new Messier objects were added to the Messier Catalog, bringing the total to 110 objects. The final entry, M110, was recorded in 1967. It is a dwarf spherical galaxy found in the Andromeda constellation and a satellite galaxy of Andromeda galaxy.

Modern amateur astronomers still use the Messier Catalog to locate deep-sky objects in the Northern Hemisphere. Some even participate in the Messier Marathon, an event in which aspiring astronomers stay up to try to observe all 110 Messier objects in just one night. All that is needed is a pair of binoculars or a telescope and a clear night sky. The best time for viewing in the Northern Hemisphere is early March to early April.

Messier observed the sky from Paris, France, at a latitude of 49 degrees north. Consequently, people who do not have a northern vantage point cannot view all of the Messier objects. For example, the objects M81 and M82 are at a declination of +69 degrees, which means that viewers below a latitude of 21 degrees south cannot see them above the horizon.

—Angela Harmon

**FURTHER READING**

"Charles Messier (June 26, 1730–April 12, 1817)." *Students for the Exploration and Development of Space.* SEDS-USA, n.d. Web. 5 May 2015.

Dickinson, David. "Why This Weekend Is Perfect for a Messier Marathon." *Universe Today.* Fraser Cain, 5 Mar. 2013. Web. 5 May. 2015.

"Edmond Halley (1656–1742)." *BBC History.* BBC, 2014. Web. 7 May 2015.

Gilmour, Jess K. *The Practical Astronomer's Deep-Sky Companion.* London: Springer, 2003. Print.

Levy, David H., and Wendee Wallach-Levy. *Cosmic Discoveries: The Wonders of Astronomy.* New York: Prometheus, 2001. Print.

O'Meara, Stephen James. *Deep-Sky Companions: The Messier Objects.* 2nd ed. New York: Cambridge University Press, 2014. Print.

# MILKY WAY'S STRUCTURE

**FIELDS OF STUDY**

Astronomy; Observational Astronomy

**SUMMARY**

The Milky Way is the galaxy that contains the Earth's solar system. It is a barred spiral galaxy, which means it consists of a flat, bar-shaped central disk surrounded by spiraling arms. The Milky Way is home to hundreds of billions of stars. Its disk measures roughly one hundred thousand light-years across.

**PRINCIPAL TERMS**

- **bulge:** a large, relatively dense group of stars at the center of a spiral galaxy.
- **dark halo:** a hypothetical massive cloud of dark matter that surrounds a galaxy's disk.
- **galaxy:** a huge collection of stars, gas, dust, and dark matter that is held together by gravity.
- **Hubble classification:** a galactic classification system based on shape, invented by Edwin Hubble; also called the Hubble sequence.
- **spiral galaxy:** a galaxy consisting of a flat central disk surrounded by spiraling arms.
- **stellar halo:** a spherical mass of stars, star clusters, dust, and gas that surrounds certain types of galaxies.
- **thick disk:** the part of a galactic disk that contains older stars, typically more than ten billion years old.
- **thin disk:** the part of a galactic disk that contains younger stars, typically less than ten billion years old, as well as the dust and gas from which new stars are formed.

**THE MAKEUP OF THE MILKY WAY**

The Milky Way is the galaxy that includes Earth's solar system. The Milky Way is a barred spiral galaxy, according to the Hubble classification system. This system, also known as the Hubble sequence, is used to categorize galaxies according to their shape. The four main categories are spiral, elliptical, lenticular (lens-shaped), and irregular. Roughly 77 percent of all observed galaxies are spiral galaxies.

The Milky Way is an old galaxy, believed to have formed about 13.2 billion years ago. Earth's solar system was created only 4.6 billion years ago. The Milky Way most likely contains between two hundred billion and four hundred billion stars.

At the center of the Milky Way is its bulge. The bulge is so named because it is a large, somewhat

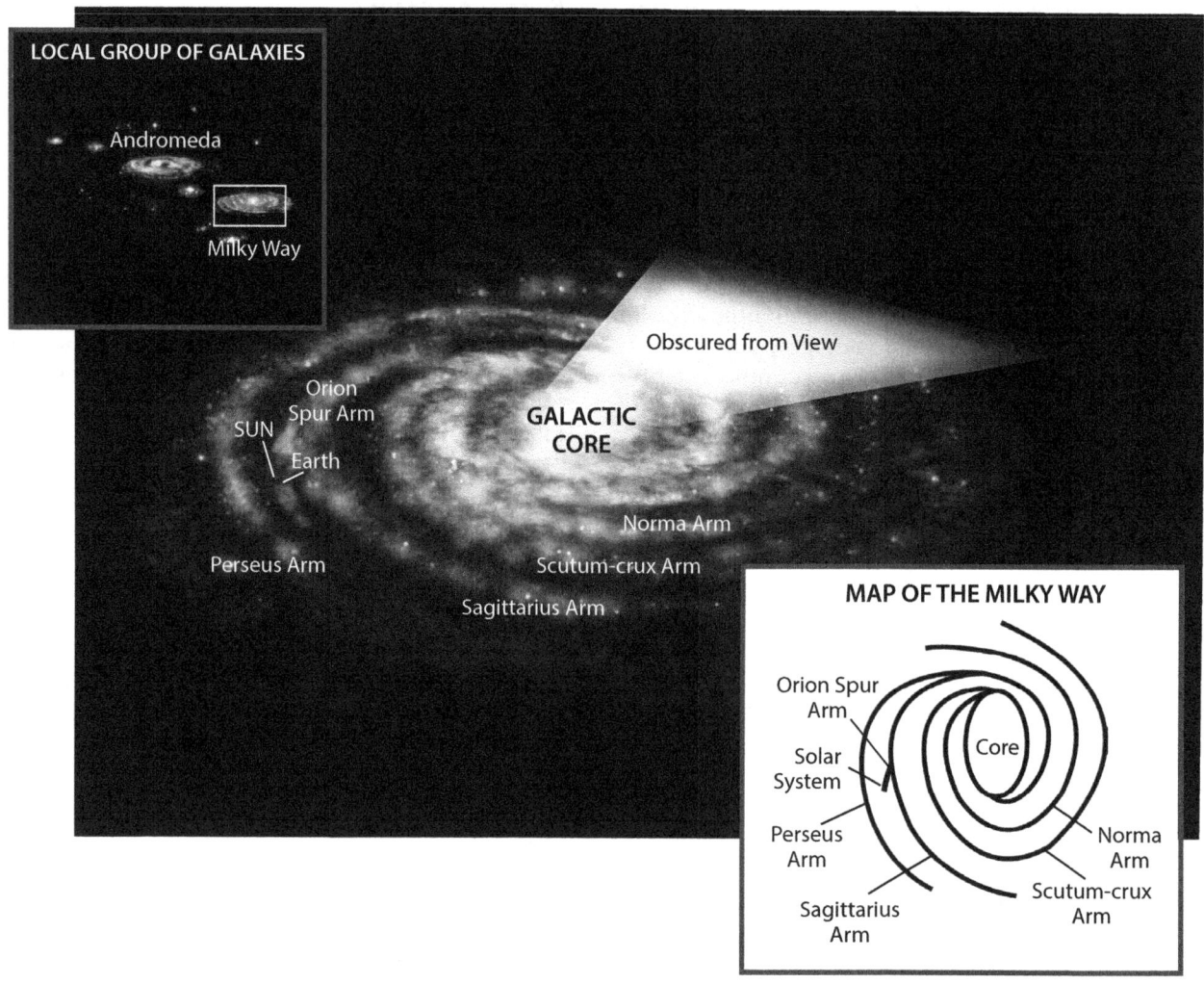

Artist's rendering of the Milky Way galaxy. Insets show its location in relation to other galaxies and a map of the arms of the galaxy, with our solar system labeled on the Orion Spur Arm.

spheroid mass of stars and gas that bulges above and below the plane of the galactic disk. The Milky Way's bulge is estimated to contain about ten billion stars. Most of these stars are older, generally seven billion years or more, although there are some younger stars as well. The bulge of the Milky Way is approximately ten thousand light-years across and is roughly peanut-shaped.

Observations have revealed that in the center of the Milky Way's bulge is an object of incredible mass, about 4.5 million times the mass of the sun. This object, named Sagittarius A*, is generally thought to be a supermassive black hole. Scientists believe that most, if not all, large galaxies have a supermassive black hole at their center, around which the entire galaxy rotates. NASA's Chandra X-Ray Observatory has recorded near-daily x-ray flares from Sagittarius A*. In January 2015, it detected an x-ray flare about four hundred times brighter than usual, the largest such flare to date.

The disk of the Milky Way is the part of the galaxy that holds most of the gas and dust. This gas and dust is also known as the interstellar medium, because it is matter that exists in the space between the stars. The Milky Way's disk is approximately one hundred thousand light-years across. The galaxy's spiral arms are attached to the disk. Scientists had long debated whether the Milky Way had two main spiral arms or four. A study published in late 2013 confirmed that there are in fact four main arms, plus a number of smaller arms called spurs. The arms of a spiral galaxy are what give it its pinwheel-like shape.

In many spiral galaxies, including the Milky Way, the disk is made up of two distinct regions: the thin disk and the thick disk. The Milky Way's thin disk contains about 80 percent of all normal (non-dark) matter in the galaxy. Due to the presence of the interstellar medium, it is a site of active star formation. The stars in the thin disk range in age from newborn to ten billion years old. The thick disk is made up of older stars and contains no interstellar medium. The majority of stars in the thick disk are older than ten billion years.

The stellar halo around the Milky Way and other spiral galaxies is a sphere made up of old stars, globular clusters (groups of stars), dust, and gas. This halo is believed to be surrounded by the dark halo, a huge, invisible area made up of dark matter. Dark matter is matter that has never been directly observed but is believed to exist due to its gravitational effects. It may account for as much as 84.5 percent of all matter in the universe. Scientists believe that the dark halo may be ten times as massive as the rest of the Milky Way.

#### Human Knowledge of the Milky Way

People have studied the sky and pondered the stars throughout all of human history. For much of this time, people did not understand the nature of the Milky Way or its relationship with Earth. During the twentieth century, however, humanity's knowledge of galaxies changed dramatically.

Previously, many scientists had believed that Earth's sun was near the center of the Milky Way. In the late 1910s, American astronomer Harlow Shapley (1885–1972) calculated that the sun was actually more than fifty thousand light-years from the galactic center. While this was an overestimate—the true distance is around twenty-eight thousand light-years—it was a fundamental shift in scientists' understanding of the galaxy. Then, in 1924, American astronomer Edwin Hubble (1889–1953) confirmed that what was then known as the Andromeda Nebula was in fact the Andromeda Galaxy, as it was too distant to be part of the Milky Way. These two discoveries transformed the way people saw the galaxy and the universe. By the early twenty-first century, scientists were able to calculate that at least 170 billion galaxies, and possibly more than twice that number, existed in the known universe.

#### Importance of Studying the Milky Way

Scientists study the Milky Way for a number of reasons. A better understanding of the galaxy could reveal more information about the origins of the universe. It could also shed new light on dark matter, dark energy, and other phenomena that scientists do not fully understand.

—*Elizabeth Mohn*

#### Further Reading

Croswell, Ken. "Star Struck." *National Geographic*. Natl. Geographic Soc., Dec. 2010. Web. 13 Mar. 2015.

Gregersen, Erik, ed. *The Milky Way and Beyond: Stars, Nebulae, and Other Galaxies*. New York: Britannica, 2010. Print.

"Hubble's Classification Scheme." *CERES Project*. Montana State U, n.d. Web. 13 Mar. 2015.

Lang, Kenneth R. *A Companion to Astronomy and Astrophysics: Chronology and Glossary with Data Tables*. New York: Springer, 2006. Print.

"NASA's Chandra Detects Record-Breaking Outburst from Milky Way's Black Hole." *NASA*. NASA, 5 Jan. 2015. Web. 13 Mar. 2015.

Neuman, Scott. "New Research Affirms That Milky Way Has Four Spiral Arms." *The Two-Way*. NPR, 17 Dec. 2013. Web. 16 Mar. 2015.

"Questions About: Galaxies." *HubbleSite*. Space Telescope Science Inst., n.d. Web. 13 Mar. 2015.

Redd, Nola Taylor. "3D Map of Milky Way Galaxy Reveals Peanut-Shaped Core." *Space. com*. Purch, 28 Sept. 2013. Web. 13 March 2015.

"Timeline." *Everyday Cosmology*. Observatories of the Carnegie Inst. for Science, n.d. Web. 13 Mar. 2015.

# *MIR*

## FIELDS OF STUDY

Space Technology; Aerospace Engineering; Astronautics

## SUMMARY

*Mir* was a modular space station launched in stages between 1986 and 1996. Begun by the Soviet Union, it continued operations under Russia and international cooperation. It operated for fifteen years in orbit about 402 kilometers (250 miles) above Earth. During that time, it provided the first opportunity for prolonged stays in space. It gave scientists a way to study the effects of microgravity on human, plant, and animal life.

## PRINCIPAL TERMS

- **extravehicular activity (EVA):** all phases of a mission that take place while the crew member is in a specially designed pressurized suit in an unpressurized environment. These are commonly called space walks when occurring outside of a spacecraft. EVA can be a planned part of a mission or an unplanned addition to deal with a problem or urgent need.
- **Soviet Union:** the Union of Soviet Socialist Republics (USSR), a communist state established in the aftermath of the Russian Revolution of 1917. Following World War II, the Soviet Union and the United States raced for supremacy in space as part of the Cold War. The Soviets were the first to put a satellite in orbit, the first to put a human in orbit around Earth, and the first to launch a space station.
- **space station:** an orbiting space vehicle intended for long-term support of a human crew. Space stations include facilities for living and research in addition to the systems necessary to run the station. Crew members can enter and leave the space station while it remains in orbit by means of secondary craft used for transport.

## HISTORY OF SPACE STATIONS

*Mir* was a groundbreaking space station, the first allowing extended space habitation. It was made possible by a relatively short but frantic period of technological development and space research. In the 1950s and 1960s, the world's two superpower countries—the United States and the Soviet Union—were locked in the Cold War. Tensions were high between the two countries, and each was competing to gain advantage in any way possible. One arena for this Cold War competition was outer space, opened to exploration for the first time by rocket technology.

The Soviets took the early lead when they launched the first satellite into Earth's orbit with *Sputnik 1* in 1957. They then followed that up by putting the first person into Earth orbit in April 1961. A short time later, US president John F. Kennedy issued a challenge to the fledgling National Aeronautics and Space Administration (NASA) when he promised that the United States would put a man on the moon by the end of the decade.

While the Americans focused on the *Apollo* missions that would ultimately put not one but several men on the moon, the Soviets struggled with their attempts at lunar missions. Ultimately the Soviet Union instead focused much of its attention and resources on manned space stations. It launched the *Salyut 1*, the first space station in history, on April 19, 1971. The earliest designs for space stations could not be refueled or resupplied, so they were not meant for long-term use. Several other *Salyut* vessels failed. However, the Soviets did later launch three more *Salyut* stations that housed a total of five crews before turning their attention to long-term space station development.

As the Cold War died down, interest began to grow in combining knowledge and resources between the two superpowers. The joint *Soyuz-Apollo* mission in 1975 saw spacecraft from the two countries dock to each other and their crews meet. This was a precursor to the joint missions that would occur once the long-term *Mir* space station was launched.

### *MIR*'S GROUNDBREAKING MISSION

The *Mir* space station was a modular space vehicle that made its way to space in pieces over a ten-year span. Its name derives from a Russian word that translates as "world" or "peace" but also refers to a system of government where the peasants owned and worked their own land. The core module was launched on February 20, 1986. It included all primary operating systems as well as crew quarters, areas for food service and hygiene, and storage. The first crew members arrived at Mir in mid-March 1986.

More modules began arriving just over a year later. These were attached to parts of the core module and to each other. The various sections and occasional rearrangements gave the full structure a somewhat haphazard look. Four modules came together at one point on their shorter ends, which some described as looking like four buses had collided in a four-way intersection.

Each module had a specific purpose. *Kvant I*, launched in 1987, included laboratories for astronomy and biotechnology. It also included a docking station used by some vehicles visiting the space station. *Kvant II*, launched in 1989, included an airlock and preparation area for extravehicular activity (EVA), including space walks. *Kristall*, launched in 1990, included solar panels and the docking station for the space shuttle. *Spektr*, launched in 1995, included remote-sensing equipment to study x-rays and gamma rays on Earth. This unit was damaged when an unmanned supply craft struck it in 1997. *Priroda*, launched in 1996, included remote-sensing equipment to study ecological issues on Earth.

The *Mir* mission was planned for five years, but the orbiter remained in service for fifteen, breaking all prior records for a vehicle's useful time in space. Russia continued its operations after the fall of the Soviet Union in 1991. During its fifteen years in space, *Mir* hosted a total of 125 crewmembers from the Soviet Union, Russia, the United States, and other countries. A total of ninety-five vehicles docked with the space station. Among them were nine space shuttles, twenty-two other manned vehicles, and sixty-four unmanned cargo vehicles.

Several cosmonauts who were stationed on *Mir* set records for time in space. A few set progressively

*Mir Space Station* orbiting Earth. NASA, via Wikimedia Commons.

longer records of 326 days or more, but Valeri Polyakov (b. 1942) set the record with 438 days on *Mir* between 1994 and 1995. The record for overall time spent in space was set in 1999 by Sergei Avdeyev (b. 1956), who spent a combined 747.5 days in space over three missions.

*Mir* stayed in space three times longer than expected, but the space station also experienced some significant misfortunes, including damage incurred when an unmanned supply ship crashed into some of *Mir's* solar panels in 1997. A fire that same year caused some frightening moments for the crew and led to improved space station safety regulations. The fire erupted during a routine oxygen replenishment maneuver. In order to boost the available oxygen aboard *Mir*, its crewmembers used perchlorate canisters to generate oxygen by a chemical reaction. But in February 1997, this routine procedure resulted in a fire that burned wildly while the crew struggled to put on gas masks and get fire extinguishers off the wall. They watched as metal burned and melted like wax from the intense heat of the blaze. Ultimately they were able to put it out without any injuries to the crew.

With some troubling issues developing, *Mir* began to require more and more costly upkeep. It needed regular boosting to maintain its low-Earth orbit. Russian space scientists began to focus their time and resources on a new project, the International Space Station (ISS). In 2000 *Mir's* final crew began preparing the space station for deorbiting, and on March 23, 2001, *Mir* was allowed to fall into Earth's atmosphere and burn up. Although many expressed concern that wreckage could crash into cities, the reentry was directed in an area over the South Pacific where it would cause no damage.

**LESSONS FROM *MIR***
*Mir* set a number of records during its time in space and was the base of operations for a number of important experiments. The most important might be the study of the effect of prolonged exposure to microgravity on human beings. Many of those who visited the space station spent far more time in space than had ever done before. By studying the long-term health of those who spent time on *Mir* and returned to Earth, scientists can better understand what must be done to prepare for future long-range planetary exploration missions or space habitation.

Experiments were also done with plant life on *Mir*. The first crops to be grown from seed to plant and back to seed were planted aboard the space station. Research into growing food crops in the microgravity of space will be important if humans are to attempt planetary exploration or establish permanent living spaces on other planets.

*Mir* also gathered vast amounts of information about stars and planets. Equipment in its astronomy laboratory was used to study quasars, neutron stars, and galaxies. Studies were also done of ecological conditions of Earth, including the condition of the ozone layer of Earth's atmosphere and the temperature of the oceans. This will help in understanding climate conditions on Earth and identifying potential problems and solutions.

Even the problems aboard *Mir* provided learning experiences. Repairs to the exterior necessitated EVA, helping to refine such techniques for future space missions. The near-disastrous fire resulted in several improvements to the fire prevention and response methods used on later spacecraft. Changes were made to the way perchlorate canisters are manufactured and insulated. Stricter procedures about keeping areas near the canisters clear of combustible debris were set. Additionally, fire suppression technology and techniques were improved. Lastly, the focus on emergency preparation and drills to minimize panic that could cost lives was increased.

Many of these lessons had a direct impact on the next generation of space exploration, including the International Space Station. In continuous operation since November 2000, the International Space Station picks up where *Mir* left off, building on the multinational cooperation and the lessons that are part of *Mir's* legacy. The work done on *Mir* will provide the building blocks for any missions to farther reaches of the solar system and possibly beyond.

—*Janine Ungvarsky*

**FURTHER READING**
Aleksandrov, Aleksandr Pavlovich, and Richard Fullerton. "Extravehicular Activity (EVA)." *Phase 1 Program Joint Report.* Ed. George C. Nield and Pavel Mikhailovich Vorobiev. NASA, Jan. 1999. *NASA.* Web. 4 June 2015.
"Extravehicular Activity (EVA)." *Man-Systems Integration Standards.* NASA, 7 May 2008. Web. 4 June 2015.
"History." *History of Shuttle-Mir.* NASA, 20 Aug. 2013. Web. 4 June 2015.
Howell, Elizabeth. "Fire! How the Mir Incident Changed Space Station Safety." *Universe Today.* Universe Today, 25 Feb. 2013. Web. 4 June 2015.
Howell, Elizabeth. "Mir Space Station: Testing Long-Term Stays in Space." *Space.com.* Purch, 5 Feb. 2013. Web. 4 June 2015.
"Mir FAQs: Facts and History." *European Space Agency.* ESA, 21 Feb. 2001. Web. 4 June 2015.
"Mir Space Station." *NASA History Program Office.* NASA, n.d. Web. 4 June 2015.
"Russian Space History." *Space Station.* PBS, Houston Public Television, 1999. Web. 4 June 2015.

# MOON IMPACTS (GENERAL)

## FIELDS OF STUDY

Astronomy; Observational Astronomy; Sub-planet Astronomy

## SUMMARY

Moon impacts are the result of meteors striking the moon's surface. Impacts generally create flashes of light, some of which are visible from Earth, and leave behind craters. Scientists study moon impacts because they provide important information about the age of the moon and the types and sizes of meteors near Earth.

## PRINCIPAL TERMS

- **crater:** a bowl-shaped depression on the surface of a planet, moon, or other astronomical body, usually caused by either volcanic activity or a high-speed impact with another, smaller body.
- **trajectory:** the curved path followed by an object as it moves through space.

### CRATERS

Since the beginning of the solar system, about 4.6 billion years ago, every planet and orbiting body in the solar system, including Earth, has been struck countless times by meteors and the debris of comets. Wind, precipitation, surface water, and plant growth have obscured the effects of these impacts on Earth. However, the moon has retained the evidence of billions of years of space collisions in the form of more than thirty thousand craters.

A crater is formed by the high-speed impact of a smaller space object into a larger one. Although meteors are rarely spherical, craters are nearly always roughly circular. This is because shock waves cause the displaced matter to fly out equally in all directions from the point of impact. The force of the impact is so great that the impactor is often shattered and sometimes even vaporized. The size of the crater is determined by the impactor's speed and trajectory before the collision.

Craters have several parts. The floor, or bottom, is generally below the surface level. A central peak can form in the center of a large crater when a large amount of displaced matter causes the crater's edges to collapse back in on itself. The walls, or the sides of the crater bowl, can be steep and sometimes irregular as gravity pulls on the loose material. The rim, or top edge of the crater, is usually pushed higher than the surrounding surface by the impact. Ejecta is the rocky debris thrown clear of the crater. Rays are brighter streaks of ejecta that can sometimes be tossed far from the impact.

### MOON IMPACT CRATER TYPES

Scientists classify impact craters into several groups. Categories include simple craters, complex craters, impact basins, multi-ring basins, irregular craters, and degraded craters. In most cases, the size and shape of the crater is affected by the size, mass, and speed of the impactor. An object that is larger or moving faster creates a bigger crater. Sometimes craters are twenty times the width of the impactor.

Simple craters are small with relatively smooth walls. They resemble round bowls. Complex craters are bigger and have peaks, terraces or multiple layers to the floor, and several rings. The large, dark areas of the moon that are visible from Earth are impact basins that are at least 298 kilometers (about 185 miles) across and at least 11 kilometers (7 miles) deep. Multi-ring basins have a number of mountains circling a large impact basin. Irregular craters result when several meteors hit at the same time, creating multiple impact craters. Meteors with a low trajectory sometimes cause irregular craters. Degraded craters are those that have been affected by weathering, additional impacts, or the sliding of material due to gravity.

### HISTORY OF LUNAR CRATERS AND EARTH CRATERS

The moon has more than thirty thousand impact craters, while Earth has fewer than two hundred. More than 99 percent of the lunar surface is three billion years old or older. About 80 percent of Earth's surface is less than two hundred million years old. The moon's surface is much older than Earth's because the moon has no water, atmosphere, or tectonic activity, such as earthquakes and volcanoes. These

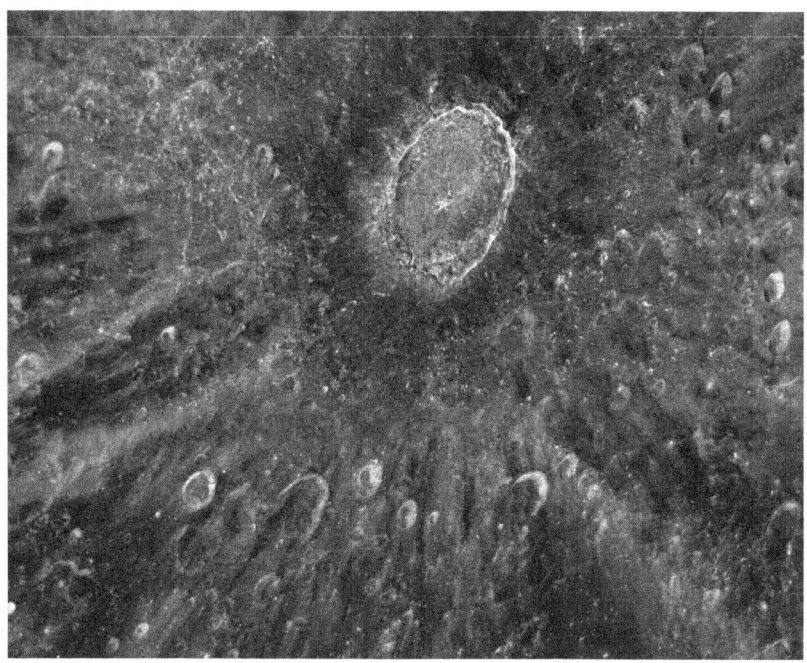

The crater Tycho is the result of one of the moon's most violent impacts. The pattern of this relatively new crater gives scientists insight into the physics of this and other impact events. NASA/ESA/D. Ehrenreich.

moon. In April 2014, NASA allowed its *Lunar Atmosphere and Dust Environment Explorer* (*LADEE*) satellite to crash on the moon's surface after it had completed its mission. *LADEE* crashed at a relatively low speed of 1,699 meters per second (3,800 miles per hour). It created a crater about ten feet wide. In October 2014 another NASA craft, the *Lunar Reconnaissance Orbiter* (*LRO*), was able to identify the crater among all the existing craters on the moon. Studying the results of this planned crash helps scientists better understand the craters that occur naturally on the moon and elsewhere.

The flashes from many moon impacts are visible from Earth with a telescope. In February 2014, a flash caused by a meteor the size of a small car was even visible without any special equipment.

—*Janine Ungvarsky*

forces have eroded Earth's surface. Thus, many of Earth's craters have worn away over time.

Unlike Earth, the moon's surface has not been worn away by erosion. This is why its surface shows many more impact craters than Earth's surface. Larger lunar craters were left by larger impactors early in the moon's history. As time went on, the impactors became smaller, and the craters they left became smaller as well.

**STUDYING MOON IMPACTS**
The larger bodies in the solar system have settled into consistent orbits, so the larger collisions that pocked the moon and planets have become less frequent. Meteors approach Earth on a daily basis, but they are generally small and burn up in the atmosphere before reaching the surface. Scientists who want to study meteors turn to the moon. The moon lacks an atmosphere to deflect or destroy meteors before impact.

The National Aeronautics and Space Administration (NASA) has several efforts in place to monitor meteors, which are known as near-Earth objects (NEOs) when they are close to Earth. NASA also tracks and observes meteors that impact the

**FURTHER READING**
Chappell, Bill. "Meteorite Impact on Moon Sets Record as Brightest Ever Seen." *The Two-Way*. Natl. Public Radio, 24 Feb. 2014. Web. 9 Mar. 2015.

"Crater." *Cosmos: The SAO Encyclopedia of Astronomy*. Swinburne University of Technology, n.d. Web. 9 Mar. 2015.

"Exploring Impact Craters with Google Earth." *Las Cumbres Observatory Global Telescope Network*. LCOGT, 2012. Web. 9 Mar. 2015.

"Impact Cratering." *Explore!* Lunar and Planetary Inst., 3 Dec. 2013. Web. 9 Mar. 2015.

"Lunar Impacts." *Marshall Space Flight Center*. NASA, 13 Nov. 2014. Web. 9 Mar. 2015.

Neal-Jones, Nancy. "NASA's LRO Spacecraft Captures Images of LADEE's Impact Crater." *Lunar Reconnaissance Orbiter*. NASA, 28 Oct. 2014. Web. 9 Mar. 2015.

"Structure of the Moon." *Natural History Museum*. Trustees of the Natural Hist. Museum, London, n.d. Web. 9 Mar. 2015.

# MORGAN-KEENAN CLASSIFICATION SYSTEM (MK OR MKK)

## FIELDS OF STUDY

Stellar Astronomy; Observational Astronomy; Astrochemistry

## SUMMARY

The Morgan-Keenan classification system (abbreviated MK or MKK) is a system used to classify stars by their surface temperatures. The MKK system grew out of systems used by other researchers, including Annie Jump Cannon and her team at the Harvard College Observatory. The star classes, from hottest to coldest, are O, B, A, F, G, K, and M.

## PRINCIPAL TERMS

- **dwarf star:** a low- to medium-mass star that becomes a white dwarf when it burns out.
- **giant:** a star larger than a dwarf star.
- **luminosity:** the essential brightness of a celestial object; the amount of energy it emits.
- **luminosity class:** a secondary classification scheme that, in addition to temperature, indicates a star's brightness.
- **spectral patterns:** lines in the electromagnetic spectra of stars that represent the presence or absence of wavelengths.
- **supergiant:** the largest type of star in the universe, and usually the hottest.
- **white dwarf:** the remnant left after a small- or medium-sized star burns out.

## Classifying Stars

Early astronomers relied on observation of stars visible to the human eye. Greek astronomer Hipparchus (ca. 190–ca. 120 BCE) classified stars using perceived luminosity, based on how soon the stars became visible during twilight. The first stars to be discerned were the brightest and were placed in the first class. This classification was flawed for many reasons. Human observation is not absolute and relies on the vision of the observer. Atmospheric conditions vary due to weather, temperature, and pollution levels. The atmosphere filters out some ultraviolet wavelengths and distorts light and spectral patterns. In time, better methods of observing celestial bodies were developed, including the use of telescopes. Increasingly sophisticated telescopes allow modern researchers to observe objects far beyond Earth's solar system and gather a great deal of data about them.

During the late nineteenth century, Edward Charles Pickering (1846–1919), director of the Harvard College Observatory, launched a program to compile data on the stars and create a classification system. Pickering's largely female staff pored over data in the Henry Draper Catalogue of spectra. One staff member, Williamina Fleming (1857–1911), studied the spectra of more than ten thousand stars and placed them in twenty-two classes. Antonia Maury (1866–1952) then developed an even more detailed system.

Annie Jump Cannon (1863–1941), studying the bright stars of the Southern Hemisphere, created a simplified system using components of earlier versions. She divided stars into seven spectral classes: O, B, A, F, G, K, and M. The International Union for Cooperation in Solar Research, a forerunner of the International Astronomical Union (IAU), adopted this system in 1910. For their 1943 book *An Atlas of Stellar Spectra*, William W. Morgan (1906–94), Philip Keenan (1908–2000), and Edith Kellman (1911–2007) laid out a more refined classification based on the work of Cannon and others. The Morgan-Keenan classification includes spectral classes as well as luminosity classes, ranging from supergiants to white dwarfs. According to the MKK system, as explained in *An Atlas of Stellar Spectra*, a star's approximate spectral type should be determined first; next, its luminosity class should be identified; and finally, it should be compared with stars of similar luminosity to determine its accurate spectral type.

## Classified from Hot to Cold

Modern astronomers categorize stars based on the spectral patterns of the light they emit. The gases at the surface of a star are usually cooler and thinner than those in lower layers. These surface gases allow most of the light from the star's interior to pass through them. Researchers are then able to detect, measure, and analyze the spectrum of this light.

The spectral pattern of each element shows distinctive emission and absorption lines, which represent the characteristic wavelengths of light that a particular element is able to emit and absorb. The interior of a star is so hot that it emits light at almost all wavelengths. When this light passes through the star's surface gases, the elements in those gases absorb it at their characteristic wavelengths, producing a near-continuous spectral pattern interrupted by black absorption lines. These absorption lines enable scientists to identify the elements present in the surface gases. In addition, the peak wavelength of the continuous spectrum and the intensity of the absorption lines both depend on the star's effective (surface) temperature. Thus, a star's spectral pattern is essentially determined by its temperature.

Though two stars may have the same effective temperature, one might be brighter than the other. This difference in luminosity required greater refinement of the early star classification system, and a luminosity class designation was created in addition to the spectral class. Luminosity classes range from extremely luminous supergiants to white dwarfs.

One key aspect of the MKK classification system is its use of the Kelvin scale as a measure of temperature. William Thomson, the first Baron Kelvin (1824–1907), was a British physicist who developed the absolute temperature scale, better known as the Kelvin scale. He studied molecules and realized they stopped moving at absolute zero (–273.15 degrees Celsius or –460 degrees Fahrenheit). In the Kelvin scale, 0 kelvin is equal to absolute zero.

In the MKK classification system, a star's letter designation indicates the spectral characteristics of its light. The scheme is arranged from hottest to coolest. O-class blue stars are the hottest stars, with temperatures ranging from 28,000 to 50,000 kelvins, and emit mostly ultraviolet radiation. The hottest stars tend live the shortest, from one to ten million years. B-class blue-white stars are 10,000 to 28,000 kelvins. Rigel, a star in the Orion constellation, is a B-class blue supergiant. Most of the hotter stars are giants, because a star's temperature is directly related to its mass. Most stars visible to the naked eye are A-class white stars (7,500 to 10,000 kelvins). Sirius, in the constellation Canis Major, is an A-class star. F-class stars burn white-yellow at 6,000 to 7,500 kelvins, while G-class yellow stars are 4,900 to 6,000 kelvins. All F- and G-class stars are dwarf stars, or stars of average size. K-class orange stars, such as alpha Centauri B, range from 3,500 to 4,900 kelvins. M-class dark red stars are the coolest stars, burning at 2,000 to 3,500 kelvins. They include Antares, in the constellation Scorpio, and Betelgeuse, in the constellation Orion. Some researchers have added a new classification, L, for very low-mass stars.

The MKK classification also assigns stars a number from zero to nine to represent their temperature, with lower numbers indicating hotter stars. These numbers represent differences of 10 percent within the spectral letter. A roman numeral between I (supergiants) and V (main-sequence stars) further defines a star's size and luminosity. (In rare cases, 0 is used to indicate an extreme supergiant, sometimes

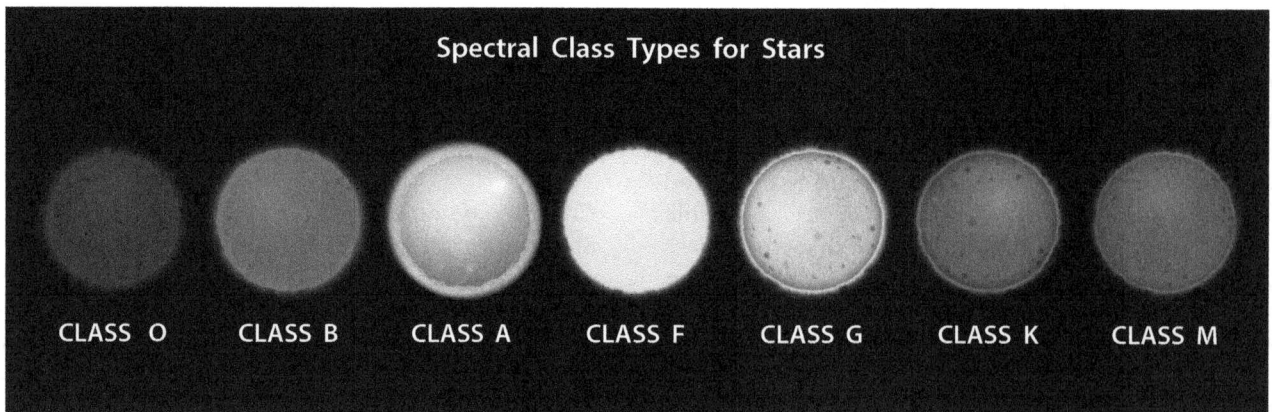

The MKK system was developed by William Wilson Morgan, Philip Childs Keenan, and Edith Kellman in 1943. Classes O, B, A, F, G, K, and M correspond to specific composition spectra, surface temperatures, and colors. O is dark blue, B is light blue, A is bluish-white, F is white, G is yellow, K is orange, and M is red.

called a hypergiant.) The sun's designation, for example, is G2V.

## LOOKING AT THE SUN

The sun, a yellow dwarf star, is among the larger stars in the G class. It has a surface temperature of about 5,800 kelvins (5,600 degrees Celsius or 10,000 degrees Fahrenheit). At this temperature, the light that reaches Earth's surface through its atmosphere is mostly yellow light. From space, however, it would appear white. The sun's core temperature, where the nuclear reactions are taking place, may be 15.6 million kelvins (about 15 million degrees Celsius or 27 million degrees Fahrenheit). Other yellow dwarf stars include alpha Centauri, 51 Pegasi, and tau Ceti.

A star's surface brightness often provides clues as to its age. The sun is a population I star. This generation of stars is the youngest and includes stars that are a few billion years old. Population I stars contain a large amount of "metal" (elements with a higher atomic number than helium), about 2 to 3 percent. The largest percentages are found in the youngest stars. Population II stars are older. They contain very little metal, only about 0.1 percent. These stars are between two billion and fourteen billion years old. The oldest stars, the first to exist in the universe, are believed to have been population III stars. Some astronomers posit that because solar winds carried off elements and these stars exploded as supernovas, later generations, including the sun, contain some of their metals.

Within the different classes, stars vary in age. Although the sun is 4.6 billion years old, it is considered a fairly young star. It has burned about half of its hydrogen and should continue to burn for another five billion years. In time, however, it will no longer be a G2V star. When the sun finishes burning its hydrogen, it will swell into a red giant, far larger than its current size, and engulf some of the nearer planets. Then it will expel its outer layers, and the remaining core will become a white dwarf.

## NEW CLASSES

With advances in technology, astronomers have discovered stars and other celestial objects that defy classification. To define these discoveries, some researchers have added new classes to the original MKK system.

The L, T, and Y classes include extremely cold stars. For example, brown dwarfs start off the same as other stars but never achieve hydrogen fusion. They are cooler than M-class stars and fall into a middle ground between being a star and being a giant gas planet such as Jupiter.

Some L-class brown dwarfs burn hydrogen. Methane-rich T-class brown dwarfs have surface temperatures between about 700 and 1,300 kelvins (430–1,030 degrees Celsius, or 800–1,880 degrees Fahrenheit). Y-class brown dwarfs are even cooler. Researchers believe these stars might be no warmer than a human body.

Thus, the MKK classification, although largely unchanged since the 1940s, continues to evolve in response to expanding knowledge of the stars and their properties. It remains a vital means of conveying information about the relative sizes, temperatures, compositions, and brightness of the stars.

—*Josephine Campbell*

## FURTHER READING

Choi, Charles Q. "Earth's Sun: Facts about the Sun's Age, Size and History." *Space.com*. Purch, 20 Nov. 2014. Web. 10 June 2015.

"History of Photometry in Astronomical Observations." *McCormick Museum*. University of Virginia, 2009. Web. 10 June 2015.

Howell, Elizabeth. "Brightest Stars: Luminosity & Magnitude." *Space.com*. Purch, 19 June 2013. Web. 10 June 2015.

"Introductory Astronomy: Stellar Populations." *Department of Astronomy*. University of Maryland, 10 Feb. 2009. Web. 10 June 2015.

MacRobert, Alan. "The Spectral Types of Stars." *Sky & Telescope*. F+W Media, 1 Aug. 2006. Web. 10 June 2015.

Maisel, Merry, and Laura Smart. "Annie Jump Cannon." *Science Women*. San Diego Supercomputer Center, 1997. Web. 10 June 2015.

Morgan, W. W., Philip C. Keenan, and Edith Kellman. *An Atlas of Stellar Spectra*. Chicago: University of Chicago Press, 1943. Digital file.

"Star Classification." *Astrophysical.org*. Instituto Scientia, 2004–14. Web. 10 June 2015.

"Spectral Classes." *Australia Telescope National Facility.* CSIRO, n.d. Web. 10 June 2015.

Tate, Karl. "Brown Dwarfs: Strange Failed Stars of the Universe Explained (Infographic)." *Space.com.* Purch, 29 Jan. 2014. Web. 10 June 2015.

# N

## *NEAR SHOEMAKER*

### FIELDS OF STUDY

Sub-planet Astronomy; Space Technology

### SUMMARY

In 1996 the *NEAR Shoemaker* spacecraft was sent to analyze the asteroid 433 Eros. Although an error caused the spacecraft to fly by, a year later *NEAR Shoemaker* successfully entered an orbit around Eros and sent a great deal of information back to Earth. At the end of the mission, the spacecraft made a soft landing on the asteroid's surface and sent images taken during its descent. Thanks to the NEAR mission, scientists know that asteroids are solid, not loose rubble, and gained a better understanding of their makeup.

### PRINCIPAL TERMS

- **C-type asteroid:** one of three broad classes of asteroids. C-type asteroids are mainly composed of black carbon.
- **Eros:** a very large S-type asteroid, the first NEA discovered, and the first asteroid to be landed on by a spacecraft.
- **NASA:** abbreviation for the National Aeronautics and Space Administration; the US government agency responsible for developing aviation and space technologies and conducting both manned and unmanned space missions to research Earth, its solar system and galaxy, and the universe.
- **near-Earth asteroid (NEA):** a small, irregularly shaped celestial body with an orbit that brings it close to Earth. NEAs travel within 0.3 astronomical units (AU) of Earth's orbit and within 1.3 AU of the sun. An astronomical unit is equal to about 149.6 million kilometers (about 93 million miles).
- **S-type asteroid:** one of three broad classes of asteroids. S-type asteroids are mainly composed of silicate (stone) and metals such as iron and nickel.

### THE *NEAR SHOEMAKER* MISSION

The Near-Earth Asteroid Rendezvous Shoemaker (*NEAR Shoemaker*) mission began on February 17, 1996, with the launch of a spacecraft of the same name. This was the first launch of NASA's Discovery program. It was also the first mission designed to orbit an asteroid, with the goal of increasing understanding of asteroids' makeup. The spacecraft entered orbit around the asteroid 433 Eros on February 14, 2000, where it remained for a year. Then, on February 12, 2001, the mission team, led by Robert Farquhar (b. 1932), landed *NEAR* on the surface of the asteroid—a first in space exploration.

Over the course of the mission, *NEAR* provided 160,000 images and a great deal of new information about asteroids. In fact, it transmitted ten times as much data as expected. Scientists the world over benefitted from the new evidence. The findings were published in prominent journals such as *Nature and Science*. The accomplishments earned the team the Trophy for Current Achievement from the Smithsonian National Air and Space Museum in 2001. The mission also generated new questions, which NASA hoped would be answered in future Discovery missions.

### FIRST MISSION TO AN ASTEROID

Asteroids were thought to be leftovers from the origins of the universe. Researchers believed that the makeup of space objects could tell them more about how it began. A small *Delta II* rocket launched *NEAR* into space toward its destination, an asteroid named 433 Eros some 2.38 astronomical units (AU) or 355 million kilometers (220.5 million miles), from Earth. The mission's objective was for the spacecraft to orbit Eros for a year and measure its shape, density, and physical and geological properties. It was also to record the asteroid's gravity and magnetic fields, map its topography, and analyze minerals and other elements on the surface.

For much of the long voyage, the spacecraft's instruments were in a state of minimum activity, as it had few tasks to perform. However, on its way to 433 Eros, *NEAR Shoemaker* flew close to another asteroid, 253 Mathilde, and captured images of 60 percent of its surface. From these images, scientists concluded that Mathilde was a heavily cratered, completely black C-type asteroid.

As *NEAR Shoemaker* finally approached Eros in December 1998, the mission operators were unable to get it close enough to allow it to be caught by Eros's gravity. A software problem caused them to abort the attempt to place it in orbit. For more than twenty-four hours, they lost contact with *NEAR*. The spacecraft was scheduled to enter orbit around Eros on January 10, 1999. After it flew past the asteroid, it had to be reprogrammed and turned around to make another pass at Eros. It was more than a year before the spacecraft finally made it into orbit around the asteroid.

An artist's rendition of the *NEAR Shoemaker* spacecraft. NASA, public domain Wikimedia Commons.

Once in place, *NEAR* did the job for which it was sent. Between February 14, 2000, and early February 2001, the spacecraft orbited the asteroid. It collected a variety of data, including thousands of images. In the meantime, the mission team received permission to attempt to land the spacecraft on Eros. On February 12, 2001, in a controlled descent, *NEAR* became the first spacecraft ever to touch down on an asteroid. Its cameras captured high-resolution images from as close as 120 meters (394 feet) during the descent. Its instruments sent data back to Earth until February 28, when the mission was completed. All told, *NEAR* had traveled 3.7 billion kilometers (2.3 billion miles).

After *NEAR Shoemaker* had stood on Eros for nearly two years, the mission team made an effort to communicate with the spacecraft one more time. Its solar panels had been exposed to sun for several months, and in spite of the frigid temperatures on the asteroid, the operations team hoped *NEAR Shoemaker* would respond. They gathered on December 10, 2002, when the asteroid would be passing within about 0.9 AU of Earth. Eros had been more than twice as far away when *NEAR Shoemaker* first landed on it.

The scientists listened for a signal but heard nothing. They sent commands to one of the computers on the spacecraft telling it to transmit data to show that it was still functioning. When there was no reply, they tried a second computer. Still no return signal came. Finally, after twelve hours, the team called off the experiment. The *NEAR Shoemaker* mission was over.

**THE *NEAR SHOEMAKER* SPACECRAFT**

The *NEAR Shoemaker* spacecraft was designed and built by the Johns Hopkins University Applied Physics Laboratory, which also managed the mission. The octagonal spacecraft was made of aluminum panels and, when finished, weighed about 800 kilograms (1,775 pounds). It was fitted with four solar panels, each measuring about 1.8 meters by 1.2 meters (6 feet by 4 feet), and an antenna that was attached to the top. The solar panels could generate up to 1,800 watts of power, depending upon the distance from the sun. A rechargeable battery stored this power. The spacecraft was named in honor of Eugene Shoemaker (1928–97), a well-known astronomer and geologist. Shoemaker was a leader in astrogeology at the United States Geological Survey and NASA. His background in geology led to a strong interest in impact craters and the asteroids believed to have caused them.

*NEAR Shoemaker* carried with it many sophisticated instruments, such as a multispectral camera, which records various wavelengths of light, such as ultraviolet or infrared. The images helped identify elements and compounds by the way they reflected or absorbed light. The spacecraft also was fitted with an x-ray/gamma-ray spectrometer, a laser range finder, and other measuring and recording devices.

As it orbited Eros, the spacecraft transmitted several films, including one showing a full rotation of the asteroid. The researchers could see each end of the battered asteroid, its boulder patches, and a crater over five kilometers (three miles) wide. NASA's decision to allow *NEAR* to attempt a landing on the asteroid delivered even closer pictures and more evidence of its makeup.

### 433 Eros

433 Eros is a near-Earth asteroid (NEA) measuring about 33 kilometers (20.5 miles) across its widest part. It is considered near Earth because it is within 0.3 AU of Earth's orbit. Its shape has been compared to a potato, a shoe, and a peanut. Eros is an S-type asteroid, meaning it is made of stone and metals. The number appearing before its name indicates its place among known asteroids in the order in which their orbits of were calculated.

Before *NEAR Shoemaker* allowed a close-up view of Eros, scientists believed asteroids were made of iron or space rubble. The data collected by the spacecraft instead suggested that the asteroid was rock similar to Earth's crust. It also supported the idea that asteroids could be made up of materials from the beginning of the universe. Eros is believed to have broken away from a larger space body. Its surface is pocked with more than 100,000 craters. Eros also has more than a million boulders the size of houses, patches containing smaller boulders and rocks, and dry ponds covered by powdery dust. These also suggest a history of collisions. Understanding the nature of asteroids such as Eros may one day help scientists if an asteroid ever needs to be deflected away from Earth.

—*Nancy Comstock*

### Further Reading

"Asteroids: Overview." *Solar System Exploration*. NASA, 13 Jan. 2015. Web. 6 Mar. 2015.

"Frequently Asked Questions." *Near Earth Asteroid Rendezvous*. Johns Hopkins U Applied Physics Laboratory, 2000. Web. 5 May 2015.

"It's Not Over Yet." *Science News*. NASA, 14 Feb. 2001. Web. 5 May 2015.

Marsden, Brian. "Eugene Shoemaker (1928–1997)." *Jet Propulsion Laboratory*. NASA, 18 July 1997. Web. 5 May 2015.

"NEAR Shoemaker." *National Space Science Data Center*. NASA, 26 Aug. 2014. Web. 5 May 2015.

"NEAR Shoemaker's Silent Treatment." *Johns Hopkins University Applied Physics Laboratory*. Johns Hopkins U, 12 Dec. 2002. Web. 5 May 2015.

"Missions to Asteroids." *Solar System Exploration*. NASA, 9 May 2014. Web. 5 May 2015.

"Smithsonian Selects NEAR Mission for 2001 Aerospace Trophy." *Applied Physics Laboratory*. Johns Hopkins U, 14 Nov. 2001. Web. 5 May 2015.

# NEOWISE

### FIELDS OF STUDY

Astronomy; Asteroid Impact Avoidance; Space Technology

### SUMMARY

The National Aeronautics and Space Administration's Wide-field Infrared Survey Explorer (WISE) mission used infrared telescopes to find asteroids and other space objects in the solar system and the Milky Way galaxy that were invisible to light-dependent instruments. A follow-on project, NEOWISE focuses on gathering additional data and preparing for potentially threatening near-Earth objects, such as undiscovered asteroids. The data revealed that there are fewer such objects than previously believed. Future projects, such as a proposed mission to Mars, would also benefit from the data gathered by NEOWISE.

### PRINCIPAL TERMS

- **asteroid:** a relatively small, irregularly shaped, rocky space object that orbits the sun.

- **infrared:** electromagnetic wavelengths that are slightly longer than red visible light waves but shorter than microwaves.
- **near-Earth object:** a celestial object that orbits the sun within 1.3 astronomical units (AU) and that potentially poses a collision hazard to Earth.
- **Wide-field Infrared Survey Explorer telescope:** an instrument used to collect images and data from throughout the Milky Way.

## THE NEOWISE MISSIONS

Following the successful Wide-field Infrared Survey Explorer (WISE) mission, which ran from December 2009 until October 2010, scientists at the National Aeronautics and Space Administration (NASA) decided to continue to use the spacecraft to explore the galaxy. By that time, the cryogen, or frozen hydrogen coolant, that had kept some of the instruments working had run out. NASA personnel used the two remaining operational cameras to complete the scan of the main asteroid belt and continue to seek asteroids and comets. The spacecraft was put into a state of hibernation after four months.

In September of 2013, NASA reactivated the WISE spacecraft and renamed it Near- Earth Object Wide-field Infrared Survey Explorer (NEOWISE). This marked the beginning of a three-year planned study on near-Earth objects (NEOs). The focus on NEOs resulted in new data that included millions of measurements and images collected through 2014. Data from December of 2013 to December of 2014 was released online for public use in March of 2015.

An extension of WISE, NEOWISE was designed to find asteroids, determine their size, and place them into family groups. The project provided data on numerous celestial objects in the asteroid belt between Jupiter and Mars. It revealed 34,000 previously unknown asteroids and twenty-one comets, in addition to some hundreds of NEOs.

Because some asteroids were considered potentially threatening to Earth, NASA continued to closely study their sizes, orbits, and probable future paths. NASA requested and Congress approved increased spending for the Near-Earth Object (NEO) Observations Program in 2012 and again in 2014. The collection of data on 158,000 additional objects was followed up with close studies of the discoveries. Scientists are using the data from NEOWISE to compute objects' physical properties for inclusion in NASA's Planetary Data System.

## MISSION BACKGROUND

While there were many early attempts to detect infrared emissions in the galaxy, most provided insufficient information due to the limitations of early equipment. More success was achieved with the launch of the *Infrared Astronomical Satellite* (*IRAS*), a project to map the sky that began on January 25, 1983. It was a mission shared by the United States, the United Kingdom, and the Netherlands. *IRAS* carried 62-pixel cameras, which was state-of-the-art technology at the time. Even so, the satellite found some 500,000 sources of infrared light, including six comets and more than two thousand asteroids. It also revealed 12,000 variable stars and the core of the Milky Way galaxy. Variable stars change in brightness.

As successful as *IRAS* was, improved technology later inspired another look at the galaxy. The WISE mission, which NASA launched in in 2009, was driven by an increased concern for near-Earth asteroids and the possibility of collisions. Its telescope and four cameras provided overlapping images that were hundreds of times more sensitive than those taken by *IRAS*. After scanning the sky twice, WISE had transmitted 2.7 million images. But the mission was designed to run for only a limited time, because the cryogen used to maintain the sensitivity of some of the instruments would run out.

When the WISE mission ended, NASA decided to extend the hunt for near-Earth objects using the two remaining operational cameras. Named NEOWISE, the extension of the mission also was expected to complete a full scan of the solar system's main asteroid belt.

## DATA FINDINGS

Although examining and interpreting the images and data sent back from WISE and NEOWISE is an ongoing project, some of the early findings surprised NASA scientists. First, infrared images showed that there are fewer midsize NEOs than previously believed. The number of asteroids larger than one hundred meters wide was reduced from 35,000 to just under 20,000.

NEOs, such as near-Earth asteroids (NEAs), are defined as objects that orbit within 1.3 astronomical units (AU) of the sun and whose orbits swing toward

An artist's rendition of NEOWISE, Near-Earth Objects Wide-field Infrared Survey Explorer, orbiting Earth. By NASA/JPL-Caltech, public domain via Wikimedia Commons.

Earth. One astronomical unit is equal to about 149.6 million kilometers (about 93 million miles). While the new numbers are still estimates, they are based on a more accurate representation of the asteroid population. Because the instruments WISE used did not rely on visible light, they were able to detect many more objects and judge their sizes more accurately. For example, to a person using a telescope that relied on light, a small, very reflective object would look the same size as a larger, darker mass. The infrared survey supplied more information on the numbers, sizes, and reflective properties of objects in space.

However, the number of large asteroids, one kilometer (0.62 miles) wide or larger, did not change significantly. The previous estimate had been 1,000 and the new finding was 981. Most of those had already been found, including all those 10 kilometers (6.2 miles) or larger, which would be a serious threat to life on Earth should they collide with the planet.

NEOWISE data also helped researchers study asteroid families. When a large space object is smashed in a collision with another body, the resulting pieces move together through space. The original mass is called the "parent body," and all the fragments belong to a single family. The NEOWISE data helped scientists group about one-third of the examined asteroid population into seventy-six families. Of those, twenty-eight were newly discovered families. The study also showed that some belonged to a different family than originally believed.

### NEOWISE and Future Missions

An additional purpose for NEOWISE was to provide information to use for future NASA missions. For example, NEOWISE collected data on NEOs that could be targets for the possible Asteroid Redirect Mission, which would involve the capture of a boulder from an asteroid. The boulder would then be placed in orbit around the moon for later study. That project was part of a much longer-term proposal for a journey to Mars. The boulder capture would test the robotic spacecraft and advanced solar electric propulsion (SEP) needed for the Mars mission.

NASA also planned to test techniques for intercepting asteroids on an Earth-bound path. Humankind has long been aware of the possibility of a collision between Earth and a large asteroid. It is believed that such a collision was responsible for the extinction of dinosaurs from the planet. That makes the information on the size and location of space objects provided by NEOWISE vital in helping researchers plan for future threats and possible interceptions.

—*Nancy Comstock*

### Further Reading

"Discoveries by IRAS." *Infrared Processing and Analysis Center*. California Inst. of Technology, 2000. Web. 6 May 2015.

Major, Jason. "Recommissioned NEOWISE Discovers Near-Earth Asteroid." *Discovery News*. Discovery Communications, 7 Jan. 2014. Web. 8 May 2015.

"Mission Overview." *NEOWISE*. NASA, 30 Apr. 2015. Web. 6 May 2015.

"NASA Asteroid Hunter Spacecraft Data Available to Public." *Jet Propulsion Laboratory*. California Inst. of Technology, 26 Mar. 2015. Web. 6 May 2015.

"NASA Space Telescope Finds Fewer Asteroids near Earth." *Jet Propulsion Laboratory.* California Inst. of Technology, 29 Sept. 2011. Web. 6 May 2015.

"NEOWISE: The Wide-Field Infrared Survey Explorer (WISE)." *Near Earth Object Program.* NASA, 7 May 2015. Web. 6 May 2015.

"New Asteroid Families Discovered." *Science News.* NASA, 29 May 2013. Web. 6 May 2015.

"Top Ten WISE Factoids." *Jet Propulsion Laboratory.* California Inst. of Technology, n.d. Web. 6 May 2015.

# NEPTUNE

### FIELDS OF STUDY

Planetary Science; Planetary Geology; Orbital Mechanics

### SUMMARY

The planet Neptune is discussed as the eighth planet of the Solar system and the last of four gas giant planets. Its structure is essentially identical to that of Uranus, but it has a 'normal' rotation. The higher density and smaller size of Neptune indicate that it has a more massive core of rock and water-ice than Uranus. Neptune also radiates more energy than it receives, indicating an unknown internal energy source.

### PRINCIPLE TERMS

- **astrophysics:** the branch of astronomy that deals with the physics of the universe, covering such areas as physical properties, composition, processes, and other phenomena.
- **convection:** the transfer of heat via the movement of molecules in a fluid.
- **Edgeworth-Kuiper Belt:** a ring of icy rock objects located beyond the orbit of Neptune; also known as the Kuiper Belt.
- **gravitational force:** the attractive force or pull between objects due to their mass.
- **Kuiper Belt objects (classical and resonant):** small, icy space bodies in the Kuiper Belt, also known as trans-Neptunian objects; most KBOs are classical, with orbits of low eccentricity and inclination; resonant KBOs orbit the sun in a consistent ratio to Neptune's orbit.
- **shepherd moons:** two or more relatively large bodies orbiting within the rings of a planet; the synchronicity of their respective orbits controls, or shepherds, the position of adjacent small particles, often producing bands of particles that appear as 'braids'.

### Neptune and Its Place

Neptune is the fourth, and last, of the outer, or gas giant, planets. It is remarkably similar to Uranus in almost every way. Neptune has a diameter of 49,500 kilometers, as compared to the 50,800 kilometer diameter of Uranus. The average distance of Neptune from the Sun is 4.497 X 109 kilometers, or 30.06 astronomical units (AU). Neptune's mass of 1.03 X

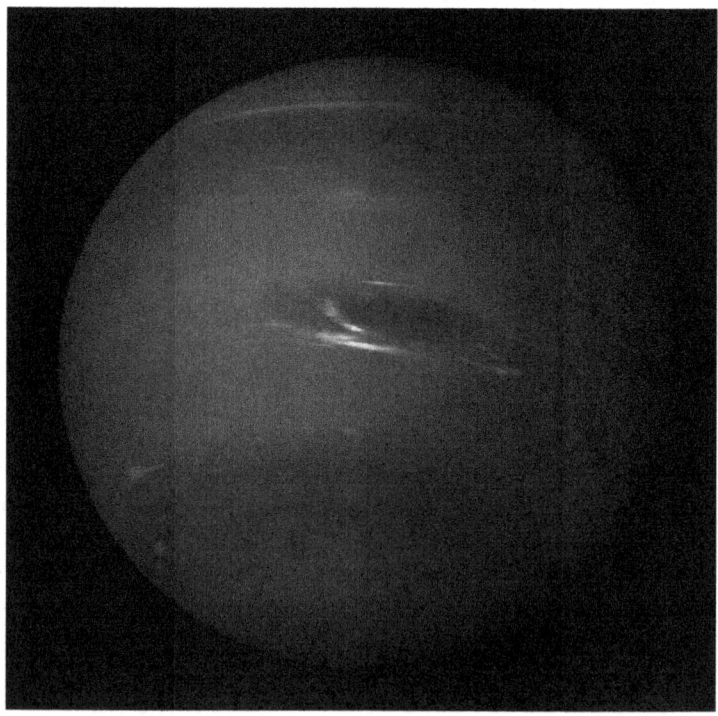

The picture shows the Great Dark Spot and its companion bright smudge; on the west limb the fast moving bright feature called Scooter and the little dark spot are visible. By NASA - JPL

1026 kilograms is some 17.2 times the mass of Earth, and is about 20 percent greater than that of Uranus. The overall density of Neptune is 1.76 g.cm-3, and so is also greater than that of Uranus (1.2 g.cm-3).

The period of rotation of the planet is fast, which seems to be a common characteristic of gas giant planets. Neptune rotates once on its axis in just 16.05 hours, but requires 164.79 years to complete one orbit around the Sun.

The orbit has an eccentricity of 0.0086, and in that respect is very similar to the nearly circular orbit of Venus, which has an eccentricity of just 0.0068. The distance of Neptune from the Sun is such that it is little affected by energy received from the Sun, but the planet has an albedo of 0.29, and so is more reflective of light than Uranus (0.34). Neptune has at least eight orbiting moons, and a set of faint planetary rings. Beyond Neptune is the Edgeworth-Kuiper belt, where the planet Pluto is found among the myriads of rock and ice objects called Kuiper Belt objects.

## Qualities

Like Uranus, Neptune is a deep blue color due to the methane in its upper atmosphere. Otherwise, very few features can be discerned on the uniformly blue surface of the planet. There are parallel bands of atmospheric motion that have been detected, as well as a large dark blue spot. The dark spot has the same relation to Neptune's size as the Great Red Spot has to Jupiter's size, and is believed to be a storm system produced by convection. Winds associated with this storm system are the highest winds known, reaching speeds as high as 2,000 kilometers per hour.

The parallel bands in the atmosphere mark the most extreme differential rotation of any atmosphere in the Solar system. The period of rotation of the upper atmosphere is 18 hours at the equator, but decreases to just 12 hours near the poles. Wind speeds in the bands reach 1120 kilometers per hour, a leisurely pace compared to the winds measured within the dark spot.

One feature of interest that makes Neptune distinctly different from Uranus in appearance is the wispy white clouds of what are believed to be methane crystals that have been detected. The presence of their shadow on the surface below shows that these clouds are suspended above the main atmosphere layer of Neptune. Neptune's rotational axis is tilted at a more normal angle of 28.8 degrees, unlike the 98-degree tilt of Uranus's rotational axis. Such an angular tilt would be expected to produce seasonal temperature variations in the planet. However, the temperature of the planet is found to be fairly constant at -216°C (-360°F) throughout.

The internal structure of Neptune is believed to be very similar to the internal structure of Uranus, consisting of a solid, rocky core surrounded by water or water ice, and an outer atmosphere of hydrogen and helium with a minor component of other materials. The smaller size and higher density of Neptune as compared to Uranus is an indication that the rocky core and the water layer occupy a larger percentage of the volume of Neptune than the same features occupy within Uranus. The magnetic field of Neptune was measured by the Voyager 2 mission, and found to have a rotational period of 16 hours and 3 minutes. This is believed to be the basic period of rotation of the planet itself. Like Uranus, Neptune has also found to emit more energy than it receives, but there is no verifiable theory for the source of the additional energy.

Two moons are visible around Neptune from Earth, and the fly-by of the Voyager 2 mission revealed the presence of at least six additional smaller moons. Also revealed was the presence of a series of dark, lumpy rings around the planet, composed of rocks and dust. The structure and behavior of the rings is a challenge to the study of astrophysics, especially with respect to the presence of shepherd moons within the rings, although the smaller size and lower complexity of the ring system around Uranus facilitate an understanding of how shepherd moons function. In essence, the gravitational force of shepherd moons as they progress through their orbit influence the positions of neighboring particles in the ring system. The perturbation produced by the passage of the shepherd moon causes the particles to respond to the gravitational field of the moon, but the movement of the moon in its orbit takes it past the particle before there can be any contact, and over time a stable configuration of the particles results. In some cases, the configuration resembles lines of particles in a braid, while in others the result is a clear path in the orbit of the shepherd moon. Two shepherd moons are required to produce the braid structure in a planetary ring.

### Neptune in History and Mythology

Neptune was known as the god of the sea in Roman mythology, the equivalent to Poseidon in the Greek pantheon. The planet Neptune, though, would not have been known to the ancient astronomers and the name is a recent convention. That a planet existed beyond Saturn and Uranus was suspected by Bouvard in 1820, based on minor variations in the positions of those planets compared to the positions that were calculated. Its probable position was calculated by Adams and LeVerrier in 1846, and was found by Galle very near that position on September 23, 1846, just three weeks after LeVerrier's calculations had been published in the scientific community. The planet came under direct observation by satellite in the *Voyager 2* fly-by mission, which has been the source of most of the information known about Neptune. Currently, space-borne telescopes such as the Hubble telescope provide any new information about Neptune.

### Measuring Neptune

The diameter of a planet is often estimated by use of the "small angle formula". The method is reasonably accurate for angles less than 5°. Measurements made by telescope can provide an accurate value for the angular width of a planetary disk such as that of Neptune, especially given the sampling size of digital electronic devices and the fine motor control that digital control provides. The small angle formula is

$$W = d\theta / 57.3$$

where $W$ is the width of the object being measured, $d$ is the distance to the object, and $\theta$ is the angular width of the object.

—*Richard M. Renneboog M.Sc.*

### Further Reading

Blunck, Jürgen *Solar System Moons. Discovery and Mythology* New York, NY: Springer, 2010. Print.

Elkins-Tanton, Linda T. *Uranus, Neptune, Pluto and the Outer Solar System* New York, NY: Chelsea House, 2006

Faure, Gunter and Mensing, Teresa M. *Introduction to Planetary Science. The Geological Perspective* New York, NY: Springer, 2007. Print.

Miller, Ron *Uranus and Neptune* Minneapolis, MN: Twenty-First Century Books, 2003. Print.

Schmude Jr., Richard *Uranus, Neptune, Pluto and How to Observe Them* New York, NY: Springer Science+Business Media, 2008. Print

# NEUTRINO ASTROPHYSICS

### FIELDS OF STUDY

Astrophysics

### SUMMARY

Neutrino astrophysics is the study of the neutrino particle. Neutrinos are fundamental particles that have no electrical charge and rarely interact with other particles. Neutrino astrophysics began with the discovery of the neutrino in the early twentieth century. This branch of science has uncovered information about many phenomena in the universe and even the origins of the universe.

### PRINCIPAL TERMS

- **big bang theory:** the generally accepted theory explaining the origins of the universe.
- **neutrino:** a fundamental particle that has no charge and very little mass.
- **solar neutrino:** a neutrino created by nuclear reactions in the sun.
- **stellar neutrino:** a neutrino created by a star other than the sun.
- **supernova:** the immensely forceful explosion produced when a massive star reaches the end of its life cycle and collapses under its own mass.

### Neutrinos and Neutrino Astrophysics

Astrophysics is a scientific discipline that uses the laws of physics and chemistry to explain how objects in the universe work. It is related to cosmology, which is the study of large bodies in the universe and the universe itself, and astronomy, which is the study of objects in the universe. Neutrino astrophysics focuses specifically on the neutrino, a particle that may help scientists better understand the origins of the universe.

Neutrinos are fundamental particles, which means that they cannot be broken into smaller particles. They are tiny particles that have almost no mass. They have no charge, unlike electrons and protons. Neutrinos are the second most abundant particles in the universe, after photons. They travel close to the speed of light. Scientists once believed that neutrinos, like photons, did not have any mass. However, tests completed in the 1990s and 2000s indicated that neutrons do have minute amounts of mass. Three different types of neutrinos exist: electron neutrinos, muon neutrinos, and tau neutrinos.

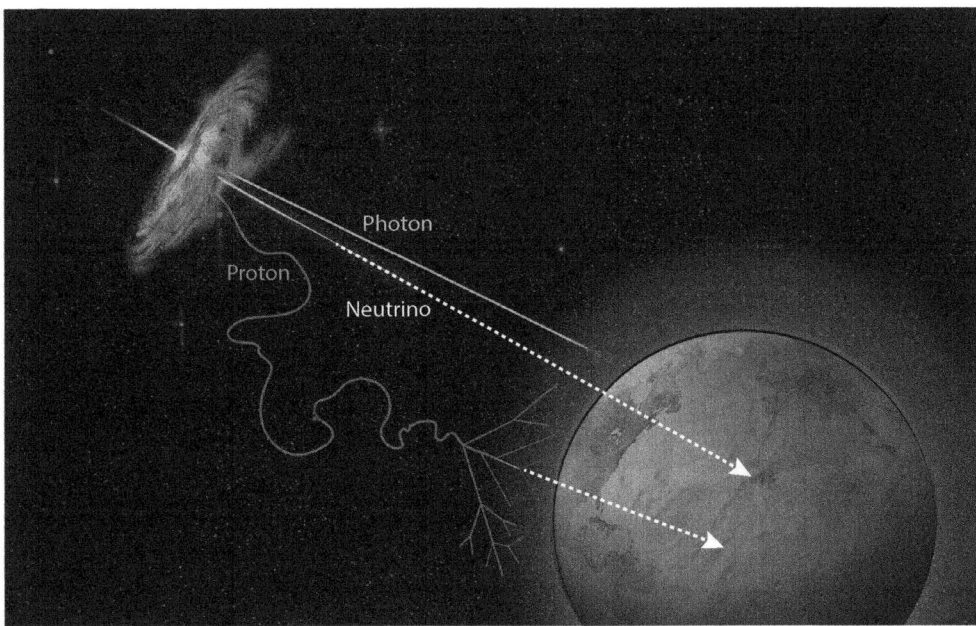

Neutrinos take a straight path to Earth and are able to pass through Earth's atmosphere. However, protons have mass and charge; thus, they are pulled through space by the gravity and electromagnetic forces of nearby objects, causing their path to vary. Photons take a straight path, but Earth's atmosphere is opaque to both protons and photons.

According to the standard model of the big bang theory, neutrinos formed within the first second after the big bang, even before the formation of atoms. Neutrinos can be found everywhere in the universe. They move through matter very easily, even through materials that stop other particles, such as lead. In fact, neutrinos that hit Earth's surface can pass all the way through the planet and come out the other side.

Because they do not interact with other particles, neutrinos have no charge, and they also have very little mass. This makes them extremely difficult to detect. Scientists have built huge neutrino detectors in order to identify and track neutrinos.

Neutrinos can be produced in several different ways. Stellar neutrinos are produced by nuclear fusion reactions within the cores of stars. When the star in question is the sun, the particles are called solar neutrinos instead. Neutrinos can also be produced by supernovas, specifically by the compression of protons and electrons within the dying star to form a neutron core. Studying stellar neutrinos can teach scientists about supernovas, black holes, and other phenomenon. Stellar neutrinos are more difficult to study than solar neutrinos.

### HISTORY OF NEUTRINO ASTROPHYSICS

In 1930, theoretical physicist Wolfgang Pauli (1900–1958) predicted the existence of neutrinos. While studying radioactive beta decay, he realized that the products of the decay did not have enough mass and energy to satisfy the law of conservation of mass and energy. Pauli predicted that another, as yet undetected particle must also be emitted during beta decay. However, he had no proof, and he doubted whether such a particle, even if it did exist, could be detected. Physicist Enrico Fermi (1901–54) later named this particle "neutrino," or "little neutral one" in Italian. In 1941, nuclear physicist Wang Ganchang suggested that neutrinos could be detected using a process called "beta capture." This proposal was proved correct by physicist James Sircom Allen (1911–82) the following year.

Physicists continued to search for ways to track and count neutrinos emitted from the sun. To this end, astrophysicists Raymond Davis Jr. (1914–2006) and John N. Bahcall (1934–2005) devised the Homestake experiment, which began operating at the Homestake Mine in South Dakota in 1970. They installed a tank of chlorine-rich fluid deep in the mine. They believed that the effect of the neutrinos on the chlorine atoms would allow them to count the number of neutrinos being released by the sun. In fact, the experiment only detected about one-third of the neutrinos that the scientists had expected to find. This difference between the observed and actual numbers of neutrinos became known as the "solar neutrino problem."

In the early 2000s, scientists concluded that a phenomenon called the Mikheyev-Smirnov-Wolfenstein (MSW) effect was to blame for the solar neutrino problem. This effect caused neutrinos to change from one type of neutrino to another. The Homestake experiment had only looked for electron neutrinos, the type of neutrino produced by the sun. Meanwhile, two-thirds of the solar neutrinos had converted to muon and tau neutrinos before passing through the earth and thus were not captured.

### STUDYING NEUTRINO ASTROPHYSICS

Neutrino astrophysics is an important branch of astrophysics that scientists hope will reveal more about the universe and its origins. Although there is still much to learn about these particles and the roles the play, some important discoveries have already been made. Some of the topics to which neutrino astrophysics has contributed include fundamental particles, the standard model of the universe and its origins, black holes and other sources of high-energy particles, the causes and effects of supernovas, and the nature of dark matter and dark energy.

—*Elizabeth Mohn*

### FURTHER READING

Bahcall, John N. "Solving the Mystery of the Missing Neutrinos." *Nobelprize.org*. Nobel Media, 28 Apr. 2004. Web. 15 Apr. 2015.

Cartlidge, Edwin. "Neutrinos Point to Rare Stellar Fusion." *Physicsworld.com*. IOP, 9 Feb. 2012. Web. 6 Apr. 2015.

Giunti, Carlo, and Chung W. Kim. *Fundamentals of Neutrino Physics and Astrophysics*. New York: Oxford University Press, 2007. Print.

Murayama, Hitoshi. "The Origin of Neutrino Mass." *Physics World* May 2002: 35–39. Print.

Setton, Dolly. "Neutrinos: Ghosts of the Universe." *Discover*. Kalmbach, 31 July 2014. Web. 15 Apr. 2015.

Soper, Davison E. "Homestake Gold Mine Experiment." *The Electronic Universe*. University of Oregon, 22 Oct. 2007. Web. 6 April 2015.

# NEUTRON STAR

### FIELDS OF STUDY

Astronomy; Cosmology; Stellar Astronomy

### SUMMARY

A neutron star is a dense, rapidly spinning star made of the matter left behind after a large star has experienced a supernova, or massive explosion. Gravity compresses the remaining iron core, causing the protons and electrons to combine to form neutrons. The resulting object is a neutron star. Scientists study neutron stars to learn about the radio waves they release and how the behavior of stars impacts the surrounding galaxy.

### PRINCIPAL TERMS

- **density:** mass per unit volume.
- **pulsar:** a neutron star formed from the remains of a supernova explosion that gives off radio waves in intermittent bursts or pulses.
- **supernova:** a massive explosion that results when a dying star has exhausted its thermonuclear fuel. The star's core collapses and explodes in the biggest type of blast known in space.

### AN EXPLOSIVE BIRTH

Over millions or billions of years, stars use up the energy-producing material in their cores and begin to die. Some die in a spectacular explosion known as a supernova. Supernovas occur when the star's

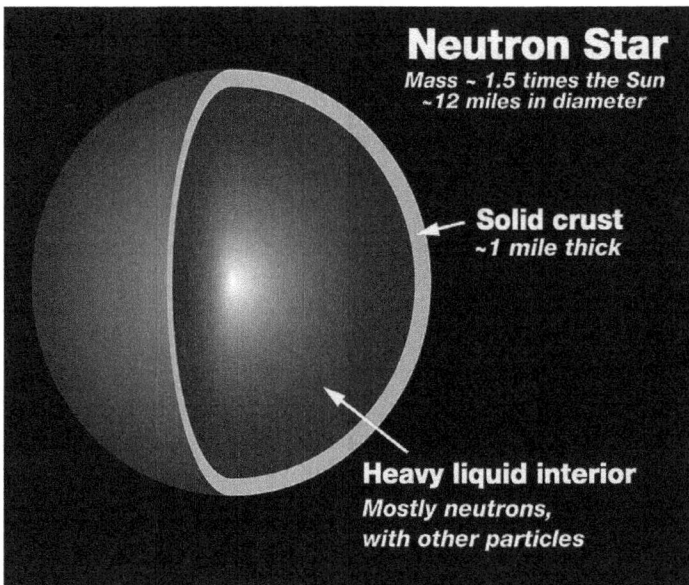

Model of a neutron star. Its mass is about 1.4 times the sun, it has a solid crust about 1 mile thick, its diameter is about 20 kilometers, and it has a heavy liquid interior of mostly neutrons, with other particles. NASA, public domain via Wikimedia Commons.

gravitational force overwhelms the energy-depleted core, causing it to collapse. Much of the star's matter explodes into the universe at speeds approaching millions of kilometers or miles per hour. The force of this blast not only sends bits of matter from the star into space, it also creates entirely new kinds of matter.

Not all the matter that makes up a star will be blown away in the supernova. The collapse of a massive star is the result of the carbon in its core fusing to form iron. Unlike the elements that previously made up the core, the iron does not fuse to produce a new element, and thus it does not generate enough energy to counteract the star's gravitational force. As a result, the protons and electrons in the core are crushed together, becoming neutrons. This neutron core, combined with trace amounts of other matter, becomes a neutron star.

Though they had been theorized to exist years before, the first neutron star was not discovered until 1967, when Jocelyn Bell, a graduate student at Cambridge University, detected the first series of radio-wave pulses using a radio telescope. Further study revealed that these signals were coming from neutron stars.

## CHARACTERISTICS OF NEUTRON STARS

Because the new neutron star's core does not support fusion, no energy is created. Without an outward flow of energy to counteract the pull of gravity, the star shrinks, compressing the matter in its core, until it is typically only about 20 kilometers (12.5 miles) in diameter. The high-mass star that was once several times the size of the sun shrinks until it is about the size of a midsize city on Earth.

Due to its smaller size, the star has an extremely high density. It actually contains more matter than the sun. This incredible mass means that neutron stars have great gravitational force despite their relatively small size. The new neutron star rotates at a much faster rate than most space objects, anywhere between several times per minute and hundreds of times per second. In contrast, Earth's sun rotates approximately once per month.

This increased rate of rotation is due to a phenomenon known as the conservation of angular momentum. Because the mass of the star remains relatively constant as its radius decreases, the speed of its rotation must increase to compensate. This is the same phenomenon that allows figure skaters to spin faster on the ice by drawing their arms closer to the body. If it were not for the intensity of its gravity, the neutron star's rapid rotation would strip off its outer layers.

## PULSARS

Neutron stars release radio waves. Some stars emit the waves in a steady stream. Others release them in regular intermittent bursts. A star that emits waves in intermittent bursts is known as a pulsar.

Because they have exhausted their energy source, neutron stars are generally very dim. This faintness, added to their relatively small size, makes them difficult to observe in space. However, the bursts of radio waves given off by pulsars make them uniquely visible. These pulses are affected by the dense, spinning star's magnetic forces and appear in beams, much like a beam from a lighthouse. This allows scientists to determine the distance of pulsars and nearby objects from Earth. The pulses of light are so regular that they are timed more precisely than the most accurate atomic clocks. As a result, scientists can use pulsars to help refine the timing of scientific clocks.

### Reasons to Study Neutron Stars

In addition to improving Earth's atomic clocks, scientists study pulsars and other neutron stars to enhance their understanding of other celestial objects. The supernovas that created neutron stars also played a role in the formation of the planets and other bodies in our universe, as well as black holes and similarly difficult-to-detect objects in space. Scientists have discovered elements cast off by supernovas all over the universe, including in humans and other living beings on Earth.

By studying neutron stars in general and pulsars in particular, scientists hope to learn about the origins of the universe. They also hope to learn the location of potentially dangerous objects, such as massive black holes.

*—Janine Ungvarsky*

### Further Reading

"An Introduction to Pulsars." *Australia Telescope National Facility*. CSIRO, n.d. Web. 31 Mar. 2015.

Lang, Kenneth R. *The Life and Death of Stars*. New York: Cambridge University Press, 2013. Print.

"Neutron Stars." *National Geographic*. Natl. Geographic Soc., n.d. Web. 31 Mar. 2015.

"Neutron Stars and Pulsars." *Kavli Institute for Particle Astrophysics and Cosmology*. KIPAC, n.d. Web. 31 Mar. 2015.

"Pulsar." *Cosmos: The SAO Encyclopedia of Astronomy*. Swinburne University of Technology, n.d. Web. 31 Mar. 2015.

"What Is a Supernova?" *NASA Education*. NASA, 4 Sept. 2013. Web. 31 Mar. 2015.

# NUCLEOSYNTHESIS

### FIELDS OF STUDY

Cosmology; Theoretical Astronomy

### SUMMARY

Nucleosynthesis is the process by which new atomic nuclei, and thus new elements, are created. The two main types of nucleosynthesis are big bang nucleosynthesis (BBN) and stellar nucleosynthesis. BBN occurred first, during the formation of the universe, and created the lightest elements. These elements later underwent stellar nucleosynthesis to create heavier elements. Nucleosynthesis accounts for all of the elements in the universe and is the reason that life on Earth was able to develop.

### PRINCIPAL TERMS

- **big bang nucleosynthesis (BBN):** the coming together of free neutrons and protons to create hydrogen isotopes, and the subsequent fusion of hydrogen nuclei to form other light elements, within the first three minutes of the big bang.
- **big bang theory:** the theory that the universe began in an extremely hot, dense state and then began a process of expansion that continues today.
- **fusion:** the joining together of two or more atomic nuclei to form a new, heavier nucleus of a different element.
- **stellar nucleosynthesis:** the fusion of light elements into heavier elements, such as carbon, oxygen, and iron, within the interiors of stars.

### Basics of Nucleosynthesis

Nucleosynthesis is a process in which protons and neutrons fuse together to form new atomic nuclei. These nuclei then capture free electrons to form neutral atoms of new elements. The two principal types of nucleosynthesis are big bang nucleosynthesis (BBN) and stellar nucleosynthesis. BBN refers to the initial fusion of protons and neutrons to create nuclei of light elements. Stellar nucleosynthesis refers to the fusion of lighter nuclei within stars to create nuclei of heavier elements.

Big bang nucleosynthesis is so called because it took place during the first few minutes of the big bang. In contrast, stellar nucleosynthesis is an ongoing process that takes place within every star and continues for as long as that star is burning fuel. The elements created during BBN were mainly hydrogen and helium, plus trace amounts of lithium, beryllium, and possibly boron. Stellar nucleosynthesis is responsible for the creation of all elements between carbon (atomic number 6) and iron (atomic number 26) in the periodic table. Naturally occurring elements heavier than iron are formed during supernova explosions via a process called neutron capture. This phenomenon is sometimes called supernova nucleosynthesis.

For nucleosynthesis to take place, nuclei must collide with one another at very high speeds. Nucleosynthesis also requires extremely high temperatures—millions or even billions of degrees, depending on the element. For example, the fusion of hydrogen into helium requires a temperature of at least 3 million kelvins (5.4 million degrees Fahrenheit). But that temperature is low compared to carbon fusion, which requires a temperature of at least 600 million kelvins (1.1 billion degrees Fahrenheit). The more protons an element has in its nucleus, the higher the minimum temperature required for fusion.

**THE BIG BANG THEORY**

Although scientists may never know for certain how the universe was created, the big bang theory is widely accepted and well supported by the available evidence. According to this theory, the universe began as an extremely small, hot mass of near-infinite density. Then, approximately 13.8 billion years ago, that mass rapidly expanded, becoming cooler and less dense as it grew. One of the biggest misconceptions about the big bang is that it was an explosion in space. Rather, the term usually refers to the initial moment of expansion. It did not take place "in space" because before the big bang, there was no such thing as "space." Everything that would become the universe was contained within that infinitesimal mass.

The modern understanding of the big bang theory was developed and refined based on two main scientific concepts: the general theory of relativity and the cosmological principle. The general theory of relativity was published in 1915 by German-born physicist Albert Einstein (1879–1955). It builds on the law of universal gravitation, which Isaac Newton (1643–1727) developed in the late seventeenth century. While Newton's theory identifies mass as the source of gravity, Einstein proposed that gravity is generated by a combination of energy and momentum, while mass is just a different way of measuring energy. Put simply, wherever matter exists in the universe, its combined energy and momentum causes space-time to curve around it. The effects of gravity, such as planetary orbits, are the result of objects traveling along these curves.

In 1917, Einstein and several others started trying to use this idea to explain the entire universe. They found that the equations for general relativity did not permit the universe to exist in a static, unchanging state, as was previously thought to be the case. Instead, the universe had to be either expanding or shrinking. Einstein thought this was a flaw in his equations and tried to correct it by adding a constant value, known as the cosmological constant. However, in 1929, American astronomer Edwin Hubble (1889–1953) used his observations of the light from distant galaxies to prove that the universe is in fact expanding. Einstein accepted Hubble's proof and renounced the cosmological constant.

The fact that the universe is expanding implies that it had a definite beginning. This was the first step toward a theory of the origin of the universe. However, this implication raised another problem: it seemed to contradict a long-standing principle of cosmology. The cosmological principle states that the matter in the universe is uniformly distributed in all directions when viewed on a large enough scale. Originally, this was thought to be true with respect to time as well; in other words, scientists believed that the universe as a whole had not changed significantly over time. If this were not the case, then the relative uniformity of the universe would have to be accounted for in some way.

The problem was that the universe is too large for more distant regions to have a causal connection to one another. If the universe had changed over time, one would expect different parts to have evolved in different ways, because it would take too long for information to travel from one region to another. The fact that this is not the case implies that at some point in the past, all parts of the universe must have been close enough together to all be affected at once. Resolving the conflict with the cosmological principle provided yet more evidence in support of the big bang theory.

The cosmological principle is still being tested today, particularly with regard to the distribution of galaxies. Together, the theory of general relativity and the cosmological principle allow for precise predictions and observations regarding the universe. For example, they allow scientists to conclude that the universe must have one of three shapes: it may be flat and infinite; it may be negatively curved and infinite, resembling a saddle; or it may be positively curved and finite, resembling a ball.

## Big Bang Nucleosynthesis

Nucleosynthesis first occurred shortly after the big bang, when the expansion of the universe allowed it to cool enough to produce light elements. Russian physicist George Gamow (1904–68) and his doctoral student Ralph Alpher (1921–2007) first proposed the idea in Alpher's doctoral dissertation in 1948. They claimed that the process was responsible for creating all of the elements in the universe. It was later determined that big bang nucleosynthesis accounts only for the light elements, and heavy elements were produced through a different process.

Big bang nucleosynthesis occurred within just a few minutes of the big bang. Scientists believe that immediately after the big bang, the universe was so hot and dense that no individual particles could exist. All matter was condensed into a form of matter called quark-gluon plasma. This plasma, which can exist only at extremely hot temperatures—roughly $1.8 \times 10^{12}$ kelvins, or one hundred thousand times hotter than the core of the sun—consists of elementary particles called quarks and gluons. Less than one second after the big bang, the quark-gluon plasma lost enough energy that the quarks and gluons could join together to form protons and neutrons. Because the nucleus of a hydrogen atom is just a single proton, hydrogen was the first element created in the universe.

As quarks continued to combine, the ratio of neutrons to protons in the universe reached approximately 1:6, or one neutron for every six protons. However, free neutrons decay, which brought the ratio closer to 1:7. About one hundred seconds after the big bang, the temperature of the universe had cooled enough for deuterons to form. A deuteron is the nucleus of a heavy isotope of hydrogen called deuterium (hydrogen-2). It consists of one proton and one neutron.

The deuterons began to react to form other nuclei. The most common product of these reactions was the helium-4 nucleus (two protons and two neutrons). Other products included "light" helium (helium-3, two protons and one neutron) and tritium (hydrogen-3, one proton and two neutrons). Additionally, a small number of helium nuclei joined together to create lithium-7 (three protons and four neutrons). The entire process of big bang nucleosynthesis lasted less than twenty minutes. At the end, the matter in the universe consisted of approximately 75 percent hydrogen, slightly less than 25 percent helium-4, about 0.01 percent deuterium and helium-3, and infinitesimal amounts of lithium-7 and other isotopes.

## Stellar Nucleosynthesis

Stellar nucleosynthesis is the other main type of nucleosynthesis. It is the process by which heavy elements such as carbon, oxygen, and iron are created. Stellar nucleosynthesis takes place in stars, and unlike big bang nucleosynthesis, it continues to occur. English astronomer and physicist Arthur Eddington (1882–1944), who was an adamant supporter of Einstein, first proposed the idea in 1920. However, English astronomer Fred Hoyle (1915–2001) is credited with developing the concept into a theory

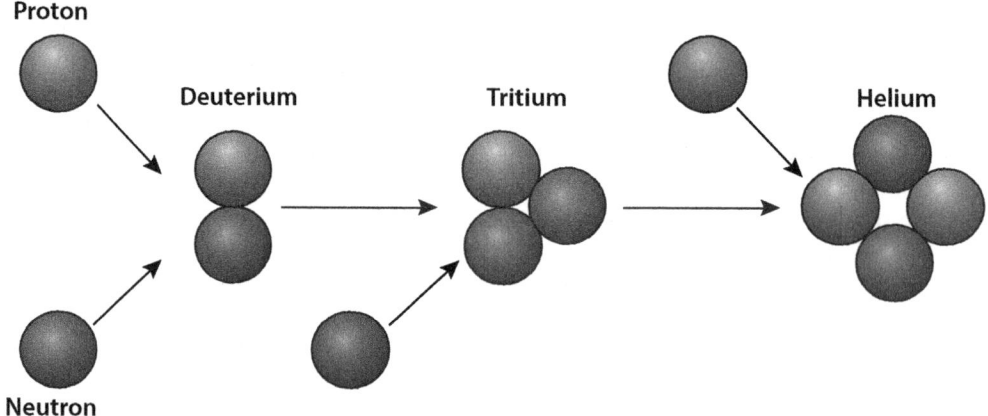

Protons and neutrons fuse together to form atomic nuclei. A hydrogen ion (proton) fuses with a neutron to form the nucleus of a deuterium (hydrogen-2) atom. If the deuterium nucleus fuses with another neutron, it becomes tritium (hydrogen-3). When this tritium nucleus fuses with another proton, it becomes a helium nucleus.

following World War II. Notably, Hoyle did not believe in the big bang theory, because he thought that the idea of the universe having a beginning was too irrational. Instead, he helped formulate an alternative theory known as the "steady state theory." This theory proposed that matter was being constantly created in the space between galaxies, resulting in the universe's ongoing expansion. While the steady state theory was ultimately rejected as new evidence supported the big bang theory instead, Hoyle's theory of stellar nucleosynthesis is well supported and remains widely accepted.

All stars are fueled by nuclear fusion. When a star first forms, it consists primarily of hydrogen. For the majority of its life cycle—a period known as the "main sequence"— the star will fuse the hydrogen in its core into helium. Once the majority of its hydrogen has been burned, the star will evolve away from the main sequence. What happens next depends on the mass of the star. If it is greater than 0.5 solar masses (one-half the mass of the sun), the core will become hot enough for helium to fuse into carbon. If the star is more than about 8 solar masses, it will then go on to fuse carbon into a succession of heavier elements until it reaches iron. Iron is the heaviest element that can be produced through stellar nucleosynthesis, because to fuse iron would consume more energy than it would release. Elements heavier than iron can be produced only in supernovas.

## Significance of Nucleosynthesis

Much like the big bang theory, nucleosynthesis is an important part of cosmology. It is the process that created all of the elements in the universe, thus making human life—and, in fact, all forms of life—possible. Stellar nucleosynthesis created the elements that are essential for life to exist, particularly carbon, nitrogen, oxygen, and iron. When the earliest stars ejected some of their mass, or when they reached the ends of their lives and went supernova, these elements were dispersed into space for the first time. They became part of the interstellar medium, out of which planets would later form. As a result, when Earth formed about 4.6 billion years ago, it already contained the elements necessary for life to eventually evolve. In a sense, then, it can be said that human beings are made of stardust.

Scientists at the Relativistic Heavy Ion Collider (RHIC) in New York and the Large Hadron Collider (LHC) in Switzerland, the two most powerful particle accelerators in the world, continue to investigate nucleosynthesis and the conditions of the big bang. In 2005, researchers at the RHIC reported that for the first time ever, they had created a plasma similar to the quark-gluon plasma that was present at the beginning of the universe. It was such a remarkable achievement that the results were only verified in 2010. The following year, LHC researchers announced the production of a plasma even hotter than the one produced by the RHIC. Scientists hope that studying these plasmas will help them better understand the origin of the universe.

—*Michael Mazzei and Randa Tantawi, PhD*

**Further Reading**

Chaisson, Eric J. "Origin of Heavy Elements." *Cosmic Evolution*. Harvard-Smithsonian Center for Astrophysics, 2013. Web. 2 June 2015.

"Cosmology: The Study of the Universe." *Wilkinson Microwave Anisotropy Probe*. NASA, 21 Dec. 2012. Web. 2 June 2015.

Moskowitz, Clara. "Hottest Particle Soup May Reveal Secrets of Primordial Universe.'" *LiveScience*. Purch, 13 Aug. 2012. Web. 2 June 2015.

"Nucleosynthesis." *Cosmos: The SAO Encyclopedia of Astronomy*. Swinburne University of Technology, n.d. Web. 2 June 2015.

Oakes, Kelly. "On the Origin of Chemical Elements." *Basic Space*. Scientific Amer., 2 Aug. 2011. Web. 16 July 2015.

Prantzos, Nikos, and Sylvia Ekström. "Stellar Nucleosynthesis." *Encyclopedia of Astrobiology*. Ed. Muriel Gargaud et al. Vol. 3. Heidelberg: Springer, 2011. 1584–92. Print.

Than, Ker. "Densest Matter Created in Big-Bang Machine." *National Geographic*. Natl. Geographic Soc., 26 May 2011. Web. 16 July 2015.

"Tests of Big Bang: The Light Elements." *Wilkinson Microwave Anisotropy Probe*. NASA, 16 Apr. 2010. Web. 2 June 2015.

Weiss, Achim. "Big Bang Nucleosynthesis: Cooking Up the First Light Elements." *Einstein Online*. Max Planck Inst. for Gravitational Physics, 2006. Web. 2 June 2015.

Wright, Edward L. "Big Bang Nucleosynthesis." *Ned Wright's Cosmology Tutorial*. UCLA Astronomy & Astrophysics, 26 Sept. 2012. Web. 2 June 2015.

# NUCLEUS (COMET ANATOMY)

## FIELDS OF STUDY

Astronomy; Astrophysics; Cosmology

## SUMMARY

The nucleus is the frozen body of a comet. When a comet's orbit takes it close to the sun, the icy nucleus begins to melt and forms a cloud. Comets reside in either the Kuiper Belt or in the Oort Cloud. Comets are believed to have existed since the formation of the solar system.

## PRINCIPAL TERMS

- **comet:** an icy body made primarily of frozen water and gases, as well as dust and rocks, that orbits the sun.
- **dust coma:** the cloud of dust particles that come off a comet's nucleus when its gas molecules sublime.
- **dust tail:** a visible stream of dust that flows off a comet's nucleus and forms an arc behind it as the comet approaches the sun.
- **ion tail:** a stream of ions that flows out behind a comet when it approaches the sun and its gas coma interacts with the solar wind.
- **gas coma:** the cloud of molecules that are sublimed from the comet's nucleus when it approaches the sun and begins to heat.

## THE BODY OF A COMET

The nucleus is the solid body of a comet. It is made primarily of ice along with dust and rocks. Comets orbit the sun and have existed since the formation of the solar system. Short-period comets come from the Kuiper Belt beyond the planet Neptune. Long-period comets come from the Oort Cloud far outside of Pluto's orbit, about 5,000 to 100,000 astronomical units (AU) from the sun. One AU is about 150 million kilometers (93 million miles).

## A COMET'S TRANSFORMATION

A comet has an elliptical, or oval-shaped, orbit. When a comet is far from the sun, the frozen nucleus is the only part that exists and the comet usually extends

Comet Hale-Bopp; the nucleus is not visible through the surrounding bright white coma. By E. Kolmhofer and H. Raab, Johannes-Kepler-Observatory, Linz, Austria, via Wikimedia Commons.

only about 16 kilometers (10 miles). When the comet's orbit takes it closer to the sun, it begins to melt and the ice, rocks, and dust sublime. When a solid sublimes, it becomes a gas without first turning into liquid. When the solids in a comet sublime, they form a coma, or cloud, around the comet. The coma grows and can become very large: 96,000 kilometers (60,000 miles) or more across.

As the comet gets closer to the sun, the gas coma and the dust coma are pushed behind the nucleus. The moving dust coma forms a dust tail that reflects sunlight and can been seen in the night sky. A comet's ion tail is sometimes hundreds of millions of kilometers long.

A comet with a smaller orbit is visible from Earth more often. Halley's Comet travels around the sun every seventy-six years and will again be visible from Earth in 2061. Comet Hale-Bopp, on the other hand, has a very large orbit of over 2,500 years. It was last visible in 1995.

### Studying the Cometary Nucleus

Scientists are trying to learn as much as possible about comets and their nuclei. Over twenty space missions have been sent to explore comets and learn about the formation and evolution of the solar system. When Halley's Comet orbited close to Earth, five unmanned spacecraft flew near it to gather data. Scientists learned that its nucleus is about half dust and half ice, 80 percent of which is water and 15 percent is carbon monoxide. Scientists believe that many other comets are similar to Halley's.

—*Adrienne A. Kennedy*

### Further Reading

"Anatomy of a Comet." *Rosetta.* NASA Jet Propulsion Laboratory, n.d. Web. 30 Mar. 2015.

"Comets Overview." *Solar System Exploration.* NASA, 20 Oct. 2014. Web. 30 Mar. 2015.

"Kuiper Belt & Oort Cloud." *Solar System Exploration.* NASA, 19 Nov. 2014. Web. 30 Mar. 2015.

Meierhenrich, Uwe. *Comets and Their Origin: The Tools to Decipher a Comet.* Weinheim: Wiley, 2015. Print.

"Mission Science Goals." *Rosetta.* NASA Jet Propulsion Laboratory, n.d. Web. 30 Mar. 2015.

Petersen, Carolyn Collins. *Astronomy 101.* Avon: F+W Media, 2013. Print.

# O

## OB STARS

### FIELDS OF STUDY

Astronomy; Astrophysics; Historical Astronomy

### SUMMARY

Astronomers classify stars using a letter system based on spectral type. OB stars, which fit the O and B spectral classifications, are the hottest stars on the spectral scale. The ultraviolet radiation emitted by OB stars helps form interstellar gases, which in turn create conditions that lead to the formation of new stars. Groups of OB stars are often found in clusters, or star associations. Scientists study OB stars because their presence shows areas of new star birth and also provides a record of past star creation.

### PRINCIPAL TERMS

- **HII region:** an interstellar region formed when gas clouds consisting primarily of atomic hydrogen are ionized by the ultraviolet radiation given off by OB stars.
- **OB star association:** a loose collection of O and B stars. One of the best-known OB star associations is Orion B1 in the Milky Way galaxy.
- **spectral classification:** a method of classifying stars based on various elements of their emission spectra, as determined by such characteristics as surface temperature, luminosity, and size.
- **Strömgren sphere:** a spherical region of ionized atomic hydrogen (HII) surrounding a young OB star in an HII region.
- **ultraviolet radiation:** short-wavelength electromagnetic radiation just beyond the visible light spectrum.

### Star Classifications

In the early twentieth century, astronomers at the Harvard Observatory developed a method of classifying stars based on their surface temperatures. This method was called spectral classification because they used the stars' emission spectra to determine their temperature. The Harvard spectral classification divided stars into seven classes: O, B, A, F, G, K, and M. O stars have the highest surface temperature, and M stars have the lowest. Later, each class was divided into numbered subclasses. "OB stars" refers collectively to the short-lived O and B stars. These two types are typically found together.

In 1943, American astronomers William Morgan (1906–94), Philip Keenan (1908–2000), and Edith Kellman (1911–2007) published the Yerkes spectral classification. It was based on the Harvard method, but it measured luminosity as well as temperature. The name was a reference to Yerkes Observatory, where the three were working at the time. Morgan and Keenan later refined the system, which came to be known as the Morgan-Keenan (MK) classification. Astronomers still use the MK system today.

### Characteristics of OB Stars

O and B stars are the two hottest classes of stars. B stars have a surface temperature between 10,000 and 25,000 kelvins (K), and O stars have a surface temperature above 25,000 K. They are on the blue end of the visible light spectrum, close to ultraviolet light, which is invisible to humans. As a result, O and B stars produce a great deal of ultraviolet (UV) radiation.

O stars are massive and emit such strong radiation that they can ionize any hydrogen within a thousand light-years. B stars are also high mass and emit large quantities of UV radiation, but they contain more helium than O stars. O and B stars are often found clustered together in what astronomers call OB star associations. These associations are believed to form in giant molecular clouds. One of the most studied OB associations is Orion OB1, which includes the young stars of the Orion Nebula and other stars of the Orion constellation.

O and B stars are very bright and strong, but they burn through their energy quickly. They live for a relatively short time, perhaps ten million years. In contrast, Earth's sun has been burning for about five billion years.

## HII Regions

In the mid-1900s, Swedish astronomer Bengt Georg Daniel Strömgren (1908–87) discovered a relationship between the density of the interstellar gas surrounding a star, the star's temperature, and the range of the area where hydrogen was ionized by the star's UV radiation. The spheres of ionized atomic hydrogen gas surrounding young OB stars became known as Strömgren spheres. These spheres are a type of HII region, which refers to any interstellar gas cloud ionized by UV radiation. The "HII" designation refers to the fact that the ionized gas is primarily atomic hydrogen; this is to distinguish it from either neutral, un-ionized atomic hydrogen (HI) or molecular hydrogen (H2).

HII regions are commonly found in conjunction with OB star associations, because the massive OB stars release large amounts of the UV radiation that creates these regions. The stellar winds generated by OB stars create holes in the molecular cloud that surrounds the region, and ionized gases rush into these holes. The resulting shock waves compact the gas and lead to the formation of a new cluster of stars. This star-formation process repeats when some of the new stars become massive enough to create HII regions of their own. While only a small percentage of ionized gas is actually compressed into stars, most of the stars in the Milky Way are thought to have formed via OB star associations and HII regions.

—*Janine Ungvarsky*

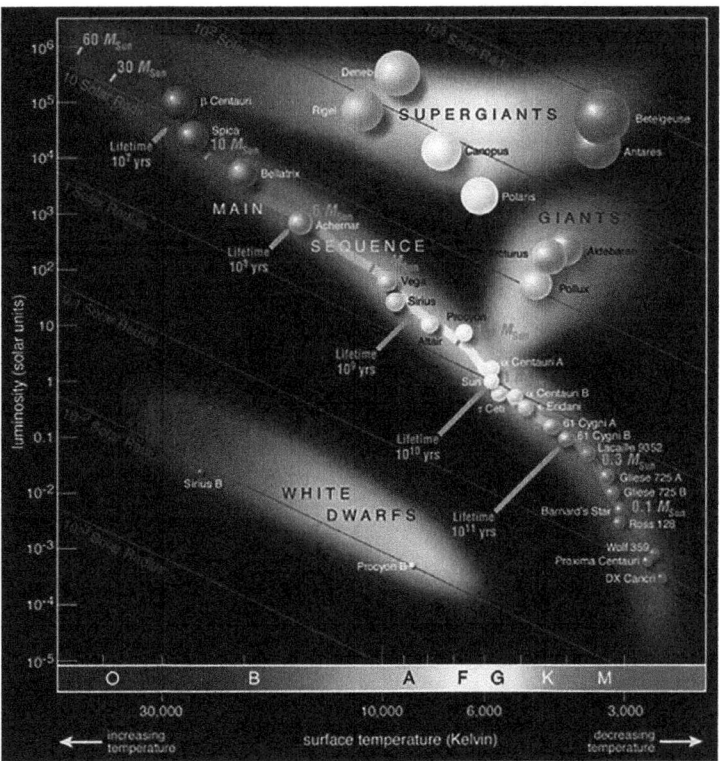

The Hertzsprung-Russell diagram indicates where stars fall on the Morgan-Keenan classification system and the absolute magnitude scale. O- and B-type stars are the hottest of main sequence stars and have a higher luminosity. By ESO (eso.org), via Wikimedia Commons.

## Further Reading

Allen, Jesse S. "The Classification of Stellar Spectra." *LHEA Team and Group Web Sites*. NASA, 1998. Professor Paul Crowther, University of Sheffield. Web. 8 Apr. 2015.

Hrovat, Mary. "1905, Hertzsprung Notes Relationship between Star Color and Luminosity." *Great Events from History: The 20th Century, 1901–1940*. Ed. Robert F. Gorman. Vol. 1. Pasadena: Salem, 2007. *eBook Academic Collection*. Web. 8 Apr. 2015.

"HII Region." *Cosmos: The SAO Encyclopedia of Astronomy*. Swinburne University of Technology, n.d. Web. 25 Mar. 2015.

"Hubble Peeks inside a Stellar Cloud." *Image of the Day Gallery*. NASA, 24 Apr. 2012. Web. 25 Mar. 2015.

Kaler, James B. "The Natures of the Stars." *Stars*. University of Illinois, 19 June 2013. Web. 25 Mar. 2015.

Ratcliffe, Martin, and Alister Ling. "The Deep Sky: The Region Surrounding Orion's Belt Is Studded with a Broad Array of Celestial Wonders." *Astronomy* Mar. 2004: 63. Print.

"Strömgren, Bengt George Daniel." *A Dictionary of Astronomy*. Ed. Ian Ridpath. 2nd ed. rev. New York: Oxford University Press, 2012. 453. Print.

"Spectral Classification." *Cosmos: The SAO Encyclopedia of Astronomy*. Swinburne University of Technology, n.d. Web. 25 Mar. 2015.

"Ultraviolet Waves." *Mission:Science*. NASA, 13 Aug. 2014. Web. 25 Mar. 2015.

# ORBIT PLOTTING

## FIELDS OF STUDY

Orbital Mechanics; Astrometry; Astronomy

## SUMMARY

Every object in space, whether natural or artificial, follows an orbit, or path around another object or point in space. The study of the movement of objects in space falls under the branches of celestial and orbital mechanics. An object's orbit may be plotted using an equation. Studying orbits has led scientists to better understand the movement of Earth and other bodies.

## PRINCIPAL TERMS

- **anomaly:** the angular distance between the position of a celestial body and its point in orbit nearest to the body it revolves around.
- **eccentricity:** in astronomy, a number that indicates whether an orbit is more circular or more elliptical; a circular orbit has an eccentricity of 0.
- **ellipse:** an oval or elongated circular shape.
- **gravitational force:** the attractive force or pull between objects due to their mass.
- **orbital mechanics:** the study of the movements of artificial satellites and spacecraft; also called flight mechanics or astrodynamics.
- **semimajor axis:** the distance from the center of an ellipse and the farthest point of its perimeter.

## What Is an Orbit?

Every object in space follows an orbit, which is a regular and repeating path around another object. Objects in orbit are called satellites. Natural satellites include objects such as moons and planets, while artificial satellites include spacecraft such as space vehicles.

The study of the motions of celestial objects is called celestial mechanics, and it uses applications from physics to study objects in space. Orbital mechanics, or flight mechanics, is a branch of celestial mechanics that deals with the movement of artificial satellites. This field studies how objects such as spacecraft move while under the influence of gravitational force (attraction between two masses), atmospheric drag (resistance), and thrust (push forward).

Orbiting objects typically move in an oval pattern known as an ellipse. Orbit plotting is the process of calculating the elliptical path that a particular orbiting object will follow.

## Orbital Motion

The origins of orbital mechanics date back to German astronomer Johannes Kepler (1571–1630). Kepler, using observations made by Danish astronomer Tycho Brahe (1546–1601), developed his three laws of orbital motion. Kepler's first law says that one object (such as a planet) moves around another object (such as the sun) along an ellipse. The second law states that an imaginary line connecting the two objects sweeps out equal areas of the ellipse in equal amounts of time. This means, for example, that Earth slows its orbit as it moves farther from the sun and speeds up again as it moves closer. Kepler's third law states that the square of an object's orbital period ($T$) is proportional to the cube of the orbit's semimajor axis ($a$). This relationship is written as

$$T^2 \propto a^3$$

Not long after, English physicist Isaac Newton (1642–1727) proposed three laws of motion and the law of universal gravitation. His first law of motion states that if no force acts on an object, then the velocity of the object will remain constant. Thus, stable objects remain in place, and objects in motion continue to move in a straight line. According to Newton's second law, applying force will change the velocity of an object based on the extent and direction of the force applied. The third law says that a pair of equal but opposite forces, such as push and pull, act on interacting objects. The law of universal gravitation states that two objects attract each other with a gravitational force ($F_g$) that is directly related to the product of their masses ($m_1$ and $m_2$) divided by the distance between them ($r$) squared:

$$F_g = \frac{Gm_1 m_2}{r^2}$$

The constant *G* is the gravitational constant and is equal to $6.67384 \times 10^{-11}$ m³/kg·s².

When the gravitational pull on one object (such as a satellite) by another (such as Earth) balances precisely with that object's inertia, the object can enter orbit. However, if the object's momentum is greater, it will escape and continue moving in a straight line. If the gravitational force is greater, the objects will collide.

**PLOTTING AN ORBIT**
The body that an object orbits is called the primary. The primary is located at one of two points within the ellipse, known as foci. The foci are located on the major axis at equal distances from the center point. The major axis is the longest diameter of the ellipse, running lengthwise between its two farthest points. The minor axis is perpendicular to the major axis and is the shortest diameter of the ellipse. The sum of the distances from any point along the perimeter to each focus is constant and equal to the major axis.

The semimajor axis is one-half of the major axis, measured from the center point to the perimeter. It represents the orbiting object's average distance from the primary. The eccentricity, which is always between 0 and 1, is the distance between the foci divided by the length of the major axis. The anomaly is the angular distance between the position of the object and its nearest point to the primary.

To plot an orbit, the major and minor axis lengths of the ellipse as well as the foci must be known. The following equation can be used to find the foci:

$$F = \sqrt{x^2 - y^2}$$

Here, *F* is the distance between one focus and the center, *x* is the semimajor axis (major radius), and *y* is the semiminor axis (minor radius).

The equation for an ellipse is given as

$$\frac{x^2}{a^2} + \frac{y^2}{b^2} = 1$$

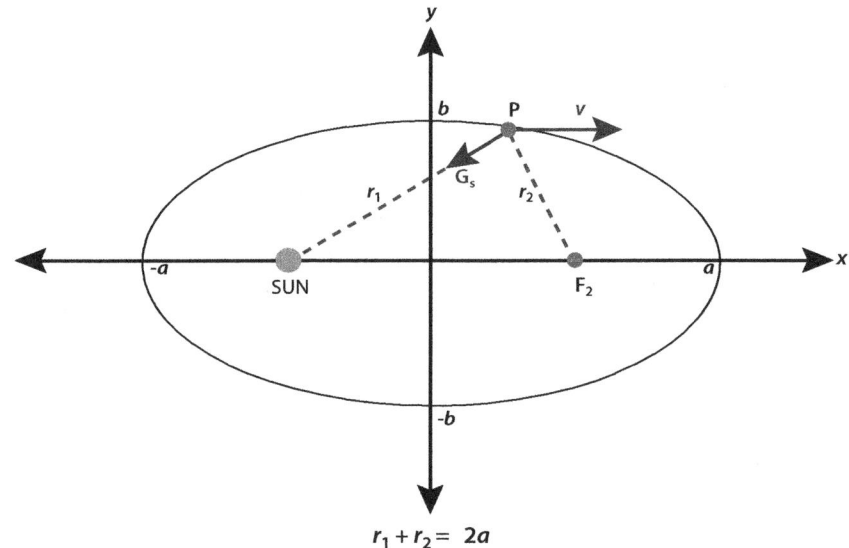

Illustration of Kepler's first law. A point (P) moves in an ellipse so that the sum of its distance from two foci (the sun and F2) inside the ellipse is constant. The planets move in elliptical orbits with the sun at one of the foci.

where *a* is the semimajor axis and *b* is the semiminor axis, for points along the *x* and *y* axes of the Cartesian coordinate system. Thus, by assigning coordinates to anomalies from different instants in time, various aspects of orbital motion, such as the position or velocity of an object traveling along an ellipse, can be plotted on a Cartesian coordinate graph.

**PERFORMING ORBITAL PLOTTING**
Here is a practical exercise in orbital plotting.

The semimajor axis of an ellipse is 14.2 units in length, and the semiminor axis is 8.7. Using the equation for finding the foci of an ellipse, determine the distance from the center to each focus.

The equation for finding the foci of an ellipse is

$$F = \sqrt{x^2 - y^2}$$

Substitute the lengths of the semimajor axis and the semiminor axis for x and y, then calculate.

$$F = \sqrt{14.2^2 - 8.7^2}$$
$$F = \sqrt{201.64 - 75.69}$$
$$F = \sqrt{125.95}$$
$$F = 11.2$$

Each focus of the ellipse is 11.2 units from the center.

**SIGNIFICANCE**

Scientists study orbital and celestial mechanics to better understand the movement of planets and other bodies in space. This knowledge has enabled scientists to successfully launch and maintain the orbits of all manner of artificial satellites, which provide invaluable information about Earth, its neighbors, and the greater solar system.

*—Angela Harmon*

**FURTHER READING**

Braeunig, Robert A. "Orbital Mechanics." *Rocket & Space Technology.* Author, 2013. Web. 4 June 2015.

"Foci (Focus Points) of an Ellipse." *Math Open Reference.* Math Open Reference, 2009. Web. 4 June 2015.

"How Do Objects Travel in Space?" *Qualitative Reasoning Group.* Northwestern U, n.d. Web. 4 June 2015.

"Mathematics of Satellite Motion." *The Physics Classroom.* Physics Classroom, n.d. Web. 4 June 2015.

Nave, Carl R. "Kepler's Laws." *HyperPhysics.* Georgia State U, 2012. Web. 18 June 2015.

"PHYS 3.2: Gravitation and Orbits." *PPLATO: Promoting Physics Learning and Teaching Opportunities.* University of Reading, 1996. Web. 4 June 2015.

# ORION

**FIELDS OF STUDY**

Stellar Astronomy; Observational Astronomy

**SUMMARY**

The constellation Orion is easily found in the night sky. It is named for a figure in Greek mythology and is said to be in the shape of a hunter. Its most recognizable feature is a line of three bright stars that make up the hunter's belt.

**PRINCIPAL TERMS**

- **celestial equator:** the imaginary line above Earth's equator that halves the celestial sphere.
- **constellation:** a region of space defined by a pattern of stars that can be seen in the night sky from Earth.
- **declination:** a space object's angular distance north or south of the celestial equator, expressed in degrees of arc.
- **International Astronomical Union:** an association of professional astronomers from all over the world who define astronomical constants while promoting research, education, and discussion on important astronomical topics.
- **right ascension:** a space object's angular distance from the vernal equinox, measured eastward along the celestial equator and expressed in hours, minutes, and seconds.

**FINDING THE HUNTER**

A constellation is an area of space defined by group of stars that people have interpreted as a pattern. Humans have identified and studied constellations for tens of thousands of years. Modern astronomers of the International Astronomical Union recognize eighty-eight official constellations, of which Orion is one.

Though Orion can be seen from every part of the world due to its position on the celestial equator, it is only visible during the winter months in the Northern Hemisphere (summer in the Southern Hemisphere). The later months of the year are often referred to as the hunting season, which makes Orion, named after a mythological hunter, particularly appropriate for this time of year. Ancient peoples recognized that some stars and constellations were only visible at certain times of year and used that information to keep track of the seasons. Throughout history, many people also found spiritual meaning in the patterns of the stars. Today, astronomers study constellations such as Orion to name and track stars and deep-space objects and to learn more about their origins.

Orion can be clearly seen in the night sky from October to March in either hemisphere. It is most easily observed between latitudes 85 degrees north and 75 degrees south. It has a declination of +4.58 degrees and a right ascension of 5.59 hours. Orion is recognizable by the arrangement and intensity of its stars.

Many budding astronomers have found Orion easy to locate because of a distinctive row of bright stars known as Orion's Belt, which is the constellation's most prominent feature. The three stars of Orion's Belt are epsilon Orionis (also known as Alnilam), delta Orionis (Mintaka), and zeta Orionis (Alnitak). All three belong to a spiral arm of the Milky Way galaxy. Orion's shoulders are also stars: alpha Orionis (Betelgeuse) is the right shoulder, and gamma Orionis (Bellatrix) the left. Lambda Orionis (Meissa) forms Orion's head; iota Orionis (Na'ir al Saif), the tip of his sword; kappa Orionis (Saiph), the right knee or foot; and beta Orionis (Rigel), the left knee. The Greek letters in star names usually rank a constellation's stars in order of brightness. However, the brightest star in Orion is not Betelgeuse but beta Orionis (Rigel).

One element of the constellation appears to be a star but is really made of dust and ionized gases, including helium and hydrogen. This is the Orion Nebula, the midpoint of the hunter's sword.

Most of the major stars of Orion are young blue giants. Though they appear to be grouped together, most are only in the same field of vision when viewed from Earth. Bellatrix is roughly 252 light-years away, while Alnilam is much farther away, at 1,340 light-years. The Orion Nebula is very far away, at about 1,600 light-years. Betelgeuse is different from the other stars of Orion. It is a red giant three hundred times broader than the sun and about twenty times as massive. Its distance from Earth is just 522 light-years.

**THE HUNTER'S HISTORY**

Myths from many world cultures surround this constellation and its stars. Stories about Orion are found in the most ancient Greek works, indicating that it was among the earliest constellations recognized by early humans. Even before the Greeks, the Sumerians saw the figure as another hero, the demigod Gilgamesh. In these tales, Gilgamesh fights and slays the Bull of Heaven, a neighboring constellation that the Greeks reimagined as Taurus the bull.

According to several well-known Greek myths about the constellation, Orion's parents were the sea god Poseidon and the Cretan princess Euryale. In one such tale, Orion claimed that he was the world's mightiest hunter and would kill all the beasts. This angered Gaia, the earth mother, who sent a scorpion after him. The scorpion killed Orion, after which Zeus, king of the gods, lifted the hunter into the sky. Though many myths about Orion's death have been told, most agree that he was killed by a scorpion. This constellation, Scorpius, is located far from Orion. According to some lore, as the scorpion rises in the eastern sky, Orion escapes by disappearing over the western horizon.

Many stories say that Orion is doomed to chase his prey across the sky but never catch it. Two other constellations, Canis Major and Canis Minor, are closely associated with Orion in Greek mythology. These

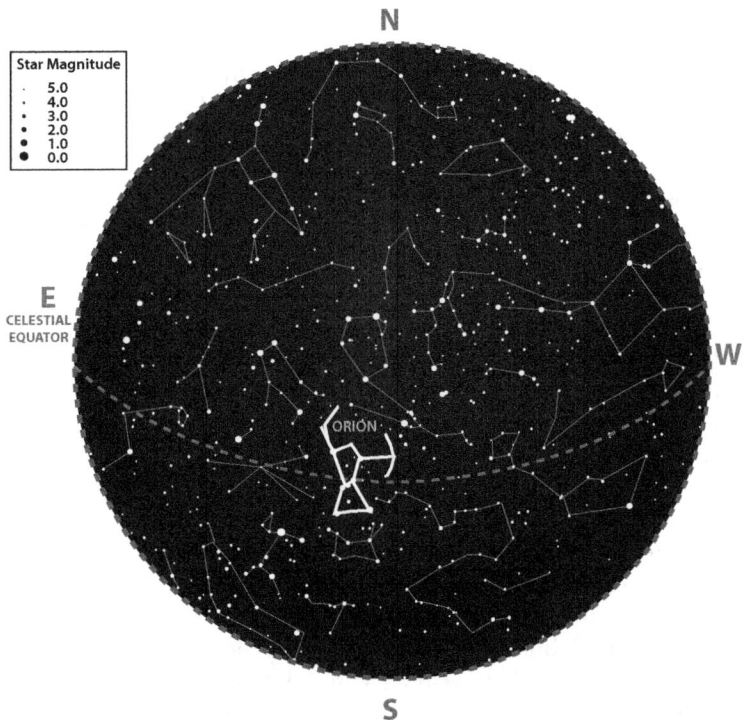

Star chart of the Northern and Southern Hemispheres in November. Lines connect the stars of the constellation Orion.

nearby groupings of stars are known as the hunter's dogs. They follow their master in pursuit of Lepus, the hare.

Another myth about Orion claims he pursued the Pleiades, the seven daughters of the Titan Atlas. Zeus rescued the maiden nymphs by placing them in the sky as the Pleiades star cluster, in the constellation Taurus, which Orion follows nightly in pursuit. Yet another says that Orion was killed by his lover, the goddess Artemis, when her brother, Apollo, tricked her into shooting him during an archery demonstration. The bereaved goddess put her beloved among the stars.

Arab astronomers saw Orion's stars as part of a larger female figure that included the Gemini constellation as well. The name of this figure, al-Jauza, loosely translates as "central one," possibly because of its location on the celestial equator. Betelgeuse is one of her hands.

Chinese astronomers viewed Orion much as the Greeks did, as a warrior or hunter figure. The Chinese name for the constellation is Shen, which means "three stars." They saw this grouping as part of a vast hunting scene.

**ORION NEBULA**

Though Orion's Belt is easily locatable by even the most casual observers, it is another area of the constellation that draws the most intense scrutiny. Scientists study the Orion Nebula, formally called Messier 42 (M42), because they are interested in learning about the birth of stars and the origins of other celestial bodies and even the universe. The nebula is described as a birthplace of stars, and scientists have observed the creation of stars there.

The Orion Nebula is one of the most photographed areas of space. Scientists have found many nebulae, which are clouds of interstellar dust and gas, in the Milky Way galaxy. However, the Orion Nebula is one of only a handful visible from Earth with the naked eye on dark, moonless nights. Backyard astronomers can get a good look at this nebula with binoculars or a basic telescope, especially if they are far from well-lit areas such as cities.

The Orion Nebula is about 1,600 light-years from Earth and about 15 light-years across. Thousands of stars are being born there. Scientists have observed the development of an open star cluster, which some call the Orion Nebula Star Cluster. Gravity loosely ties these young stars together.

Early astrophysicists developed a theory of star formation that was fairly orderly. It theorized that stars formed one by one in isolation when the delicate balance of gravity and pressure in a gas cloud was disturbed. Late in the twentieth century, infrared images revealed a less tidy birth story. After the Spitzer Space Telescope was launched in 2003, the resulting images revealed that stars are created in chaos and through violent reactions. Most stars develop in groups, as is happening in the Orion Nebula.

Another discovery in the Orion Nebula is a series of gas clumps that scientists have dubbed Orion Bullets. These gas clumps were first detected in 1983. Researchers say the clumps of gas contain iron atoms and are about ten times the size of Pluto's orbit around the sun. More detailed images taken by the Gemini South Observatory in Chile suggest that the bullets were ejected from the nebula's star-forming areas at hundreds of kilometers per second.

The Orion Nebula is significant for scientists and amateur astronomers alike. Many people are able to study the four brightest stars in the nebula, which they call the Trapezium, with basic telescopes. These infant stars are about a million years old.

*—Josephine Campbell*

**FURTHER READING**

Byrd, Deborah. "Orion the Hunter Easy to Spot in January Night Sky." *EarthSky*. Earthsky Communications, 9 Jan. 2015. Web. 1 May 2015.

Frank, Adam. "The Violent, Mysterious Dynamics of Star Formation." *Discover*. Kalmbach, 26 Mar. 2009. Web. 4 May 2015.

McClure, Bruce. "Orion Nebula Is a Place Where New Stars Are Born." *EarthSky*. Earthsky Communications, 30 Jan. 2015. Web. 4 May 2015.

"The Orion Bullets." *Astronomy Picture of the Day*. NASA, 10 Jan. 2013. Web. 6 May 2015.

"Orion, the Hunter." *StarDate*. University of Texas McDonald Observatory, n.d. Web. 6 May 2015.

Ridpath, Ian. "Orion the Hunter." *Star Tales*. Author, n.d. Web. 1 May 2015.

# P

## 4450 PAN

### FIELDS OF STUDY
Astronomy; Observational Astronomy

### SUMMARY

The asteroid 4450 Pan is a small, rocky minor planet that orbits the sun. Its orbit crosses the paths of Venus, Earth, and Mars. It is a near-Earth object, which means it orbits relatively close to Earth. The asteroid was discovered in 1987 by astronomers Carolyn and Eugene Shoemaker. It was named for the Greek god Pan. Because 4450 Pan is a near-Earth object, scientists study it closely to observe whether it could have any impact on Earth.

### PRINCIPAL TERMS

- **asteroid:** a small, rocky minor planet that orbits the sun.
- **near-Earth asteroid:** an asteroid whose orbit brings it within 1.3 astronomical units (AU) of the sun.

### ATTRIBUTES OF 4450 PAN

Pan is an asteroid, which means it is a rocky space object that orbits the sun. Asteroids are too small to be called planets. Instead, they are called minor planets. The term "minor planet" refers to any astronomical object other than a planet or a comet that orbits the sun. The solar system has tens of thousands of asteroids. Pan is an Apollo asteroid, which is a type of asteroid that crosses Earth's orbital path on a certain axis. It also a near-Earth object (NEO)—specifically, a near-Earth asteroid (NEA)—which means that gravity has pulled it into an orbit close to Earth. Scientists predict that 4450 Pan will be roughly 0.42 astronomical units (AU) from Earth in 2020.

Pan has a rotation period of 56.5 hours. It has an elliptical orbit around the sun. Its orbit also crosses the paths of Venus and Mars.

### DISCOVERY AND NAMING

Pan is not bright enough to be seen with the naked eye. It was discovered on September 25, 1987, by Carolyn Shoemaker and her husband, Eugene Shoemaker, at the Palomar Observatory in California. During their careers, the Shoemakers discovered a number of other asteroids and other bodies, including the comet Shoemaker-Levy 9 (D/1993 F2).

Asteroids can be named after many different things. Pan was named after the mythical Greek god of nature, who was known as a hunter and a fisher. In art, Pan is depicted as half-man, half-goat, with horns on his head.

### STUDYING ASTEROIDS

Scientists identify and study asteroids such as 4450 Pan for a number of reasons. Asteroids were formed when the solar system was first forming. Because of this, scientists are interested in studying asteroids to learn more about the early solar system.

As a near-Earth asteroid, 4450 Pan is of particular interest to scientists. Asteroids that are closer to Earth have a higher likelihood of impacting Earth. Therefore, scientists study their movements closely.

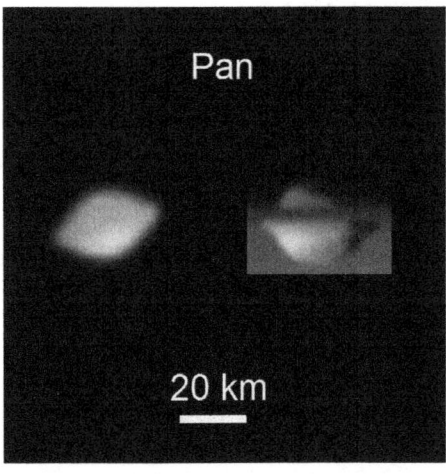

4450 Pan, asteroid. NASA/JPL/Space Science Institute.

233

Furthermore, humans might someday mine asteroids for minerals. If this happens, the asteroids nearest Earth will be easiest to access.

—Elizabeth Mohn

**FURTHER READING**

Chapman, Mary G. "Gene Shoemaker: Founder of Astrogeology." *USGS Astrogeology Science Center.* Dept. of Interior, n.d. Web. 25 Mar. 2015.

"4450 Pan (1987 SY)." *JPL Small-Body Database Browser.* California Inst. of Technology, 29 Aug. 2003. Web. 25 Mar. 2015.

"4450 Pan (1987 SY)." *Small-Body Radar Database.* University of California, Los Angeles, 9 Dec. 2014. Web. 25 Mar. 2015.

"Frequently Asked Questions." *Near-Earth Object Program.* NASA, 16 Mar. 2015. Web. 16 Mar. 2015.

"Object Classifications." *PDS: Small Bodies Node.* University of Maryland, 26 Aug. 2011. Web. 25 Mar. 2015.

Schmadel, Lutz D. *Dictionary of Minor Planet Names.* 6th ed. Heidelberg: Springer, 2012. Print.

# PARTICLE ACCELERATION

**FIELDS OF STUDY**

Astrophysics; Cosmology; Stellar Astronomy

**SUMMARY**

Particle acceleration is any process that increases the energy and speed of particles, or small units of matter, such as protons and electrons. This can be achieved by means of a device or an object, natural or artificial, that is able to accelerate the particles to nearly the speed of light. Particle acceleration has important scientific, medical, industrial, and security applications.

**PRINCIPAL TERMS**

- **cosmic rays:** extremely high-energy charged particles, primarily atomic nuclei, that travel through space at near light speed.
- **gamma rays:** a form of electromagnetic radiation with high energy and a very short wavelength.
- **particle accelerator:** a machine designed to increase the speed and energy of subatomic particles to extremely high levels.

**INCREASING SPEED AND ENERGY**

In the early twentieth century, scientists believed that the radiation on Earth came from rocks on the surface. To prove this theory, scientists measured radiation from the top of the Eiffel Tower and other heights, trying to prove that radiation decreased with altitude. They learned that while radiation levels did decrease, the rate was far less than anticipated. Seeking answers, Austrian physicist Victor Hess (1883–1964) took to a hot-air balloon in August 1912 to measure the radiation in the atmosphere. At first, he found no clear differences. However, when the balloon reached a height of about 5,300 meters (17,388 feet), Hess discovered that the radiation levels were three times as high as at sea level. He concluded that radiation must be reaching Earth from outside its atmosphere. This particular radiation became known as cosmic rays.

Cosmic rays were originally thought to be a form of electromagnetic radiation, hence the name "rays." However, in the late 1920s, they were found to consist of highly energetic charged particles, awakening researchers to a wide range of previously unknown particles smaller than an atom. These included antiparticles, or particles with the same properties as ordinary particles but the opposite electrical and magnetic charges, such as antiprotons and positrons (antielectrons). Scientists began trying to determine what forces in the universe could create such high-speed, high-energy particles.

While this research was going on, other scientists began looking for ways to replicate these speed and energy levels in particles on Earth. The first manufactured particle accelerators were designed in the late 1920s and built in the early 1930s. By the 1950s, researchers were regularly using such devices to learn more about fundamental particles and matter. However, this achievement did not stop them from trying to track down the natural particle accelerators in space.

## Natural Particle Accelerators

The key to finding the source of cosmic rays was another form of radiation, discovered more than a decade before Hess's balloon ride revealed the presence of radiation from beyond Earth's atmosphere. In 1900, Paul Villard (1860–1934), a French chemist and physicist, was conducting research into x-rays and cathode rays. He identified a new type of ray with high energy and a short wavelength. These rays stood out to Villard because they were able to "see" deeper than an x-ray and because they were not affected by either electric or magnetic fields. These newly found rays were called gamma rays.

In the decades that followed, researchers found that gamma rays could point the way to natural accelerators. Because gamma rays are unaffected by magnetic fields, it is easier to trace them back to their source. Study of gamma rays has led scientists to explore the powerful shock waves from supernovas and supermassive black holes as possible natural accelerators. In 2005 and 2006, scientists working with *Suzaku*, a joint Japanese Aerospace Exploration Agency (JAXA) and United States National Aeronautics and Space Administration (NASA) x-ray observatory, were studying a white dwarf star known as AE Aquarii, which might be a natural particle accelerator. The star also shows some qualities of a pulsar, releasing light in bursts. This meant that pulsars might also be natural sources of cosmic rays. Researchers continue to search the universe for objects with the power to imbue particles with great energy and speed.

## Manufactured Particle Accelerators

Beginning in the 1930s, researchers have had access to increasingly advanced manufactured particle accelerators. These accelerators produce streams of charged particles—protons, electrons, and sometimes even atoms—that have high energy and travel at near the speed of light.

There are two main types of particle accelerators, linear and circular. In a linear accelerator, the particles travel in a straight line; in a circular accelerator, they travel in a loop or ring. The type of accelerator used depends on the experiment or procedure being conducted. Both linear and circular accelerators work in similar ways. The particles are placed in a metal beam pipe that is vacuum sealed to keep out dust and air, which could interfere with the experiment. Electromagnets are used to push and steer the particles as they move through the beam pipe. Electric fields are placed along the pipe and can have their current switched from positive to negative. This creates radio waves that help speed up clusters of particles as needed for the experiment. Most accelerators include targets, such as a piece of metal or other particles, where a beam of particles can be directed to cause a collision. They also have detectors that record the results. These collisions are orchestrated in order to study what happens when different types of particles encounter various targets. Because of this, some particle accelerators are also called colliders.

The largest and perhaps best-known collider is the Large Hadron Collider (LHC). After beginning operation in 2008, the LHC became one of several accelerators that are part of the European Organization for Nuclear Research (CERN). The LHC's accelerator ring is twenty-seven kilometers (seventeen miles) long and is located in an underground tunnel near Geneva, Switzerland. Researchers announced in 2012 that data from the LHC showed strong evidence of the elusive Higgs boson, a type of elementary particle that was believed but not proved to exist. The next year, the LHC was shut down for two years to undergo repairs. A few of its 1,232 main superconducting dipole magnets were replaced, and more than ten thousand electrical connections to the magnets were reworked to increase safety and performance. All the work done prepared the LHC to create more collisions with greater power.

The LHC is only one of a dozen or so accelerators in the CERN facility. In 2010, the US Department of Energy estimated that more than thirty thousand particle accelerators were in use around the world. This estimate includes smaller accelerators that are used for medical or industrial purposes rather than for research.

Large accelerators are often located underground, both for practical reasons—they take up a lot of space—and for safety due to the power of the particle beams they create. These large facilities are mainly focused on research, seeking to learn why the universe is the way it is. Scientists theorize that the universe began when a singularity—an infinitesimal, infinitely dense collection of matter—expanded outward with explosive force. This "explosion" is known as the big bang. By studying how particles react when they collide with other objects or each other, scientists seek to understand how the big bang happened, what it means for the universe, and what other unknowns await discovery.

### EVERYDAY APPLICATIONS OF PARTICLE ACCELERATION

Not all particle accelerators are the size of small towns or used to study the collisions of the tiniest parts of the universe. Smaller accelerators are used around the world in a variety of different ways. These include sealing consumer packaging, inspecting cargo, and helping diagnose and treat illnesses.

Particle acceleration plays two main roles in medicine: diagnostic and therapeutic. Many of the radioisotopes used in diagnostic testing are created by particle acceleration, and accelerators can also be used to generate x-rays or gamma rays for diagnostic imaging. Other accelerated particles are be used to treat cancer and other illnesses, to develop new drugs to treat serious illnesses, and to help in the study of DNA.

Thousands of products used every day by people worldwide are created or enhanced through particle acceleration. Particle accelerators are used to harden the materials used in manufacturing, improve medical procedures and food service by destroying pathogens, and seal the packaging of consumer goods, among other tasks. The semiconductors necessary for many of the electronic items that have become part of daily life are made with the help of particle acceleration. Accelerated particles may also play a role in a cleaner environment; studies have shown them to be an effective way to clean up everything from nuclear waste to polluted water and air.

Additionally, the national security of a number of countries, including the United States, has benefited from particle accelerators. Inspecting cargo, determining the characteristics of various materials, and inspecting nuclear fuels are all done with the help of accelerated particles. These particles can also play a role in national defense through the use of laser technology.

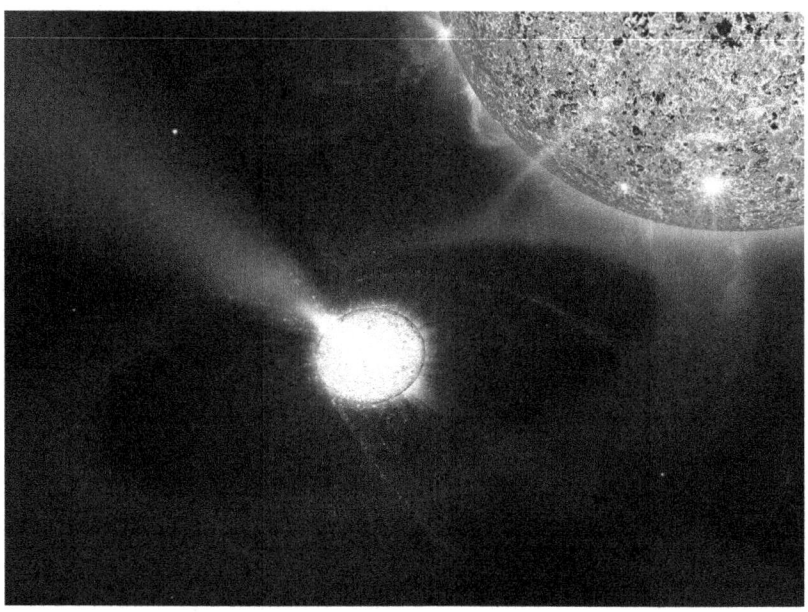

Artist's rendering of a white dwarf giving off pulsar-like pulsations, powered by its rotation and particle acceleration. Casey Reed.

Scientists and researchers continue to study and use particle acceleration in hopes of understanding how the universe came to be the way it is and what could happen to it in the future. In the meantime, particle accelerators also help improve everyday safety and quality of life.

*—Janine Ungvarsky*

### FURTHER READING

Achenbach, Joel. "The God Particle." *National Geographic.* Natl. Geographic Soc., Mar. 2008. Web. 12 June 2015.

"The Benefits of Particle Accelerators for Society." *Accelerators for America's Future.* US Dept. of Energy, 30 June 2010. Web. 12 June 2015.

"Cosmic Rays: Particles from Outer Space." *CERN.* CERN, n.d. Web. 12 June 2015.

Dotson, Ben. "How Particle Accelerators Work." *Energy.gov.* US Dept. of Energy, 18 June 2014. Web. 12 June 2015.

Greene, Brian. "How the Higgs Boson Was Found." *Smithsonian.* Smithsonian Inst., July 2013. Web. 12 June 2015.

"The Large Hadron Collider." *CERN.* CERN, n.d. Web. 12 June 2015.

Naeye, Robert, and Rob Gutro. "White Dwarf Pulses like a Pulsar." *Goddard Space Flight Center.* NASA, 2 Jan. 2008. Web. 12 June 2015.

Noyes, Dan. "Cosmic Rays Discovered 100 Years Ago." *CERN.* CERN, 7 Aug. 2012. Web. 12 June 2015.

# PHORCYS: (65489) CETO I PHORCYS

### FIELDS OF STUDY

Sub-planet Astronomy; Astrophysics; Cosmology

### SUMMARY

(65489) Ceto I Phorcys is the companion object of the small solar system body 65489 Ceto. Because the two objects orbit a common center of mass and are close in size, they form a binary system. Their orbit around the sun goes well beyond Neptune, classifying the pair as a binary trans-Neptunian object. It is also a binary centaur.

### PRINCIPAL TERMS

- **binary centaur:** two small solar system bodies that orbit around a common center of mass while simultaneously following a larger, nonresonant, unstable orbit around the sun that crosses the path of one or more outer planets.
- **binary trans-Neptunian object:** two minor planets that orbit around a common center of mass while simultaneously following a larger orbit around the sun at a greater average distance than Neptune.
- **Ceto:** an asteroid in the outer solar system that forms a binary system with Phorcys.
- **tidal force:** the effect of variation in the gravitational pull exerted by celestial bodies on other space objects nearby, causing distortions in shape.

### Discovery of Phorcys

(65489) Ceto I Phorcys is the companion object of the asteroid Ceto in the outer solar system. Although Phorcys is officially Ceto's moon, the two are of a similar size and orbit a common center of gravity, making them a binary system. Ceto and Phorcys are considered a binary centaur because they follow an unstable orbit that crosses the orbital paths of both Uranus and Neptune. Because part of their orbit takes them beyond Neptune, they are also considered a binary trans-Neptunian object (TNO).

Ceto was discovered in 2003, but its binary nature was not recognized at the time. Phorcys's proximity to Ceto made it difficult to see, even with the Hubble Space Telescope. However, a research team noticed that Ceto followed an orbit that could only be explained by the presence of another object in close proximity exerting a tidal force on it. Image analysis led to the discovery of Phorcys in 2006.

Phorcys's discovery was important because it helped researchers calculate the likely size and physical composition of both objects. Data has shown Phorcys to be a relatively small space object, about 132 kilometers (82 miles) in diameter—close in size to Ceto, which has an estimated diameter of 174 kilometers (108 miles). Phorcys is composed of rock and ice, with a low internal temperature and an extremely rough surface. It is relatively dense, and its mutual orbit with Ceto is nearly circular.

### Pair Naming

Ceto was named for the Greek sea goddess who represented the unknown dangers of the sea. Following the conventions of the International Astronomical Union (IAU), Phorcys was named for a closely related

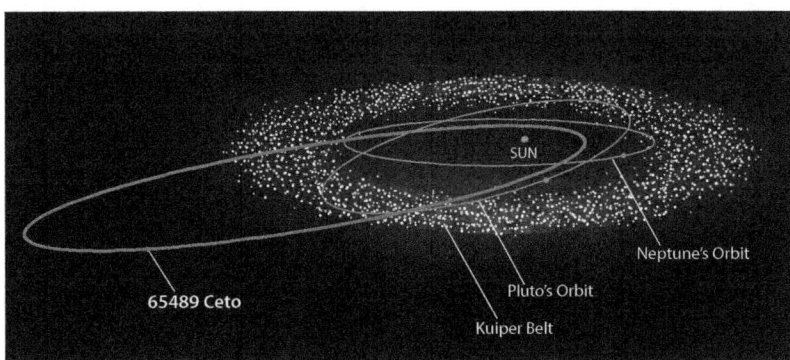

Orbit of 65489 Ceto, asteroid; Phorcys is a companion object of Ceto.

figure from Greek mythology. Phorcys was Ceto's husband and an ancient sea god.

—*Marie Keenan, MS*

**FURTHER READING**

"Catalog of Hubble Space Telescope Small Bodies Observations." *PDS: Small Bodies Node.* University of Maryland, 10 Oct. 2013. Web. 25 Mar. 2015.

Grundy, W. M., et al. "The Orbit, Mass, Size, Albedo, and Density of (65489) Ceto/Phorcys: A Tidally-Evolved Binary Centaur." *Icarus* 191.1 (2007): 286–97. Print.

Johnston, William Robert, comp. "(65489) Ceto and Phorcys." *Johnston's Archive.* Johnston, 20 Sept. 2014. Web. 25 Mar. 2015.

"Naming Astronomical Objects." *International Astronomical Union.* IAU, n.d. Web. 25 Mar. 2015.

"65489 Ceto (2003 FX128)." *JPL Small-Body Database Browser.* California Inst. of Technology, 9 Dec. 2014. Web. 25 Mar. 2015.

"Small Bodies: Profile." *PDS: The Planetary Data System.* California Inst. of Technology, 10 May 2005. Web. 25 Mar. 2015.

# PHOTOMETRY

**FIELDS OF STUDY**

Astronomy; Stellar Astronomy; Space Technology

**SUMMARY**

Photometry is the science of measuring the brightness of celestial objects, including stars. It is important as a means of knowing more about very distant objects. The data can help scientists understand what bodies are made of, their temperatures, and other information. The information gathered may be very general or, with more measurements, highly detailed.

**PRINCIPAL TERMS**

- **electromagnetic radiation:** the flow of waves generated by periodic electric and magnetic field variations; examples include gamma rays, radio waves, visible light, and x-rays
- **flux:** the amount of electromagnetic radiation from an object passing through a detector
- **luminosity:** a celestial object's total energy output per second
- **optics:** science of the properties and behaviors of light
- **visible light spectrum:** portion of the electromagnetic spectrum visible to the human eye

**PHOTOMETRY**

Photometry is the science of measuring the brightness of celestial objects. It is used to learn about distant objects in space that cannot be observed easily otherwise. The measurements taken may be very basic, and provide very general information, such as a star's temperature. With more focused study and technology, scientists are often able to learn more specific information, such as an object's elemental composition.

Ancient astronomers first studied stars that were visible to the human eye. They measured the brightness of these stars using magnitude. Greek astronomer Hipparchus (ca. 190–120 BCE) divided the stars he catalogued into six classes. The magnitude reflected how soon the stars became visible during twilight. The first stars to be discerned were the brightest, and were placed in the first class. However, this classification relied on an individual's eyesight and observation. It did not provide a measurement that could be compared, and visibility was affected by changes in the atmosphere. *The Almagest* (ca. 170 CE) by Ptolemy of Alexandria was thought to be based on the star catalogue of Hipparchus. *The Almagest* served as the basis of astronomy until the sixteenth or seventeenth century. Starting in the early seventeenth century astronomers tried to measure the brightness of stars by comparing their sizes as seen through a telescope, but this again was imperfect. For example, atmospheric conditions affected the appearance of stars. The atmosphere blocks out many wavelengths, is not transparent under the clearest conditions, absorbs and scatters some light, and acts as a prism. Even the telescope used by an astronomer might perform differently under varying conditions.

The invention of astronomical photography in the nineteenth century allowed observers to maintain records of what they were seeing for future reference. By comparing multiple images of the same object over time, researchers could learn which were most accurate, and use the information to compare with other objects for more precise measurements. As early photography became more sophisticated, these images provided researchers with a wealth of information.

In 1856 English astronomer Norman Robert Pogson (1829–91) calculated magnitude in terms of the density, or flux, of the radiation it emits. Flux is the number of photons passing through a unit of area within a unit of time. Pogson found that the magnitude scale corresponds to the logarithmic response of human retina receptors. He also found that the difference between five levels of magnitude is about 100 in flux.

In 1910 Edward Pickering suggested a method for heating or lighting photographic plates. He did so to measure the amount of heat or light they let through. Pickering matched these measurements with stellar brightness. He was thus able to calculate the brightness of celestial objects. Harlan Stetson and Jan Schilt each furthered Pickering's work by creating plate photometers. Their photometers could measure the intensity of light that passed through a photographic plate.

Beginning in the late twentieth century, telescopes were launched into space to beam back even more information. Unhindered by Earth's atmosphere, these telescopes and their successors can capture data using x-rays and gamma rays. The new technology opened up whole new areas of study for scientists. Through optics, or the study of how electromagnetic radiation waves act and interact, researchers could view things previously unseen. These space-based telescopes include the Hubble Space Telescope, launched in 1990, which measures visible, UV, and near-infrared wavelengths; the Chandra X-Ray Observatory, launched in 1999 to measure x-rays; the Spitzer Space Telescope, launched in 2003 to measure infrared wavelengths; the Herschel Space Observatory, which has been measuring far-infrared radiation since 2009; and the Planck Observatory, launched in 2009 to measure microwave wavelengths.

## STUDYING LIGHT

Electromagnetic radiation includes visible light as well as many other kinds of radiation on what is called the electromagnetic spectrum. Electromagnetic radiation includes gamma rays, x-rays, ultraviolet (UV) radiation, visible light, infrared radiation, and radio waves.

Radiation travels in waves. Wavelength is measured from the crest, or highest point, of one wave to the crest of the next. Usually, smaller wavelengths mean the radiation has greater energy. For example, gamma rays have very small wavelengths, and very high energy, while radio waves may be very large—sometimes even longer than a football field—and have much lower energy. The wavelengths of the visible light spectrum, which includes the radiation detectable by the human eye, fall between these extremes. Red has the longest wavelength in the visible light spectrum, while violet has the shortest wavelength. The energy of visible light is also somewhere in the middle range. X-rays and UV wavelengths are shorter than those of visible light, while infrared and microwave wavelengths fall between visible light and radio waves.

When scientists know the flux of a celestial object, and the approximate distance to the object, they can calculate the total energy output, or luminosity, of the object. Researchers can also glean the size, temperature, and other properties of the object.

Though stars emit a great deal of radiation, scientists can also study objects that are far colder or hotter. Planets, molecular clouds, and even interstellar dust emit radiation in the longer wavelengths. Ionized gas clouds emit radiation in the ultraviolet range of the spectrum.

Because hotter objects radiate energy with shorter wavelengths, this appears to the human eye as color within the visible spectrum. Scientists can use this information to measure the temperatures of stars. For example, among stars we can see, red stars are the coolest, while blue stars are the hottest. Other stars are so cold they produce almost no visible light, so they can only be observed through infrared telescopes. At the opposite end of the spectrum, astronomers can find the hottest stars by searching for ultraviolet light. UV telescopes can filter out other wavelengths and focus on the nurseries where stars are being born.

## Classifying the Stars

The temperature of a star is related to its apparent brightness. Astronomers measure the luminosity, or the total amount of energy a star emits from it surface, because energy in this case refers to light. Astronomers refer to stars in terms of both apparent magnitude— how bright it appears from Earth—and absolute magnitude, which refers to how bright it appears at a standard distance of 10 parsecs (32.6 light years). Stars that are far away and very bright might appear fainter than closer, less-bright stars. Classifying stars by absolute magnitude eliminates this difference. The modern five-magnitude scale classifies stars by a brightness ratio of 100, using Vega as a reference point. Vega, which is 25 light years from Earth, has an absolute magnitude of 0.6. A star with an absolute magnitude of 1 is 100 times as luminous as a star with an absolute magnitude of 6. However, classification is not so simple, because the astronomers must also know which wavelength is being measured for the classification. Some stars might be highly luminous when measuring x-rays, but emit much less infrared radiation. Astronomers also rely more on newer measurements. As technology has improved, so has the accuracy of classifying celestial bodies.

The Sun is just 93 million miles from Earth and is the closest star. Its absolute magnitude is 4.2. One of our nearest neighbors at just 4.3 light years from Earth, Alpha Centauri (Rigil Kentaurus) in the Centaurus constellation has an absolute magnitude of 4.4, and is about 1.3 times as luminous as our Sun. Rigel, in Orion, is 1,400 light years away, and has an absolute magnitude of −8.1. Betelgeuse, also 1,400 light years from Earth, has an absolute magnitude of −7.2.

Viewed from Earth, many stars appear to be different colors. This is due to the amount of radiation in the visible light spectrum that the stars emit. The surface temperature of the Sun is about 5,800 kelvins (9,932 degrees Fahrenheit). At this temperature, it produces mostly yellow light. Scientists know this without sending a probe to take the star's temperature because they can measure the light waves. They also know that the star Rigel, which looks blue-white, is about 11,000 kelvins (19,000 degrees Fahrenheit), while Betelgeuse, at a mere 3,600 kelvins (about 6,000 degrees Fahrenheit), appears blue.

## Seeking Answers among the Stars

Scientists even use measurements of wavelengths to see beyond and behind celestial bodies. For example, they can study something cold, such as a gas, which is obscured by interstellar clouds. These long wavelengths cut right through dense matter that would prevent it from being seen by any other means. Microwave telescopes can detect wavelengths left over from the Big Bang, the event that gave birth to the universe and all celestial objects.

The BRITE (Bright Target Explorer) nanosatellite photometry project began in 2005 with the funding of an Austrian BRITE nanosatellite by the University of Vienna. On February 25, 2013, the first two of a planned six nanosatellites were launched. The sixth BRITE nanosatellite was launched on August 19, 2014. The satellite constellation is intended to study the brightest of stars. Austria, Canada, and Poland are participating in the joint effort.

Exoplanets, planets orbiting stars in far-off galaxies, are being located and studied using photometry. For

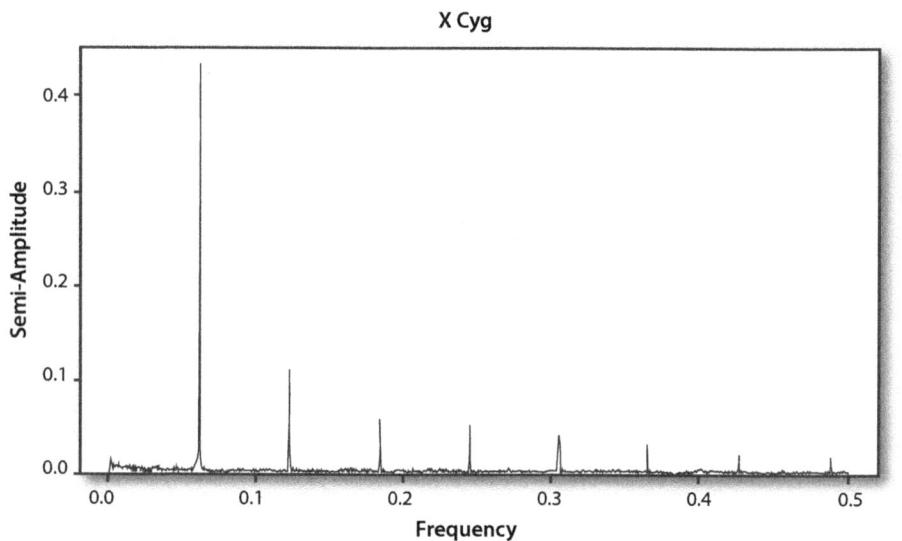

Spectrum graph of a Cepheid-type variable star, showing peaks at specific frequencies corresponding to the harmonics of the star's oscillations.

example, scientists can detect the briefest dimming of a star when a planet crosses it in orbit because they know the luminosity of the star and can accurately measure any changes. A passage between Earth and a far-off star is called a transit. When such dimming is detected regularly, scientists know a planet is likely in transit around the star. The degree of the dimming helps them to know the general size of the planet. When they detect distant planets, scientists can focus their efforts on them to learn about their temperatures, atmospheres, and other information. They do this by waiting for the planet to disappear behind the star. Then they compare the light spectrum of the star when the planet is in transit with the spectrum when the planet is hidden behind it to get the planet's spectrum. This research allows astronomers to find and explore new places from Earth. For example, in 1998 astronomers found a large planet orbiting Gliese 876, a red dwarf not far from Earth. In 2001 a second large planet was found. In 2005 astronomers discovered a much smaller third planet orbiting Gliese 876. They deemed it the most Earth-like exoplanet to be found so far.

Scientists are looking for exoplanets on which humans might survive. They are also looking for planets that support forms of life. The first planet to be discovered orbiting a star like the Sun was found in 1995. As of January 2015 NASA's Kepler Telescope has found more than one thousand exoplanets since its launch in 2009.

Other research in the field of photometry is seeking to understand the origin of the universe and the celestial bodies. For example, quasars emit tremendous amounts of energy while at the same time sucking matter into the black holes at their centers. Scientists hope that data gathered about quasars and other objects will help explain the physical processes of the Big Bang.

—*Josephine Campbell*

**FURTHER READING**

"Chapter 4: Introduction to Stellar Photometry." *Keele Astrophysics Group*. Keele U, n.d. Web. 28 May 2015.

Crockett, Christopher. "What Is the Electromagnetic Spectrum?" *EarthSky*. Earthsky Communications, 19 May 2014. Web. 22 May 2015.

"Electromagnetic Radiation & Electromagnetic Spectrum." *Chandra X-Ray Observatory*. Harvard-Smithsonian Center for Astrophysics, 2 Dec. 2008. Web. 22 May 2015.

"History of Photometry in Astronomical Observations." *McCormick Museum*. University of Virginia, 2009. Web. 22 May 2015.

Howell, Elizabeth. "Brightest Stars: Luminosity & Magnitude." *Space.com*. Purch. 19 June 2013. Web. 22 May 2015.

NASA. "NASA's NExSS Coalition to Lead Search for Life on Distant Worlds." *Scientific Computing*. Advantage Business Media, 28 Apr. 2015. Web. 29 May 2015.

Netting, Ruth. "Wavelengths of Visible Light." *Mission: Science*. NASA's Science Mission Directorate, 13 Aug. 2014. Web. 22 May 2015.

"Photometry." *Encyclopaedia Britannica*. Encyclopaedia Britannica, 2015. Web. 22 May 2015.

Romanishin, W. "An Introduction to Astronomical Photometry Using CCDs." *Astrophysics Research at OU*. University of Oklahoma, 22 Oct. 2006. Web. 28 May 2015.

Stolte, Daniel, and Christian Veillet. "Supermassive Black Hole Discovered with Mass of 12 Billion Suns." *Scientific Computing*. Advantage Business Media. 27 Feb. 2015. Web. 29 May 2015.

Thompson, Andrea. "Major Space Telescopes." *Space.com*. Purch. 18 May 2009. Web. 28 May 2015.

"Transit Photometry: A Method for Finding Earths." *Planetary Society*. Planetary Society, n.d. Web. 29 May 2015.

# PISCES

## FIELDS OF STUDY

Astronomy; Observational Astronomy

## SUMMARY

Pisces is a constellation, or pattern of stars, in the Northern Hemisphere. The large yet dim constellation is also part of the astronomical zodiac, along with twelve other constellations. As part of the zodiac, Pisces plays a particularly important role in the March equinox. The constellation resembles two fish that are joined together by a cord. Different myths, including those within Greek and Roman mythology, surround the constellation. The stars that make up Pisces include eta Piscium, gamma Piscium, alpha Piscium, beta Piscium, and Van Maanen's Star. Pisces also contains the spiral galaxy known as Messier 74.

## PRINCIPAL TERMS

- **constellation:** a pattern of stars, developed by humans, that can be seen in the night sky from Earth.
- **declination:** a space object's angular distance north or south of the celestial equator.
- **International Astronomical Union:** an association of professional astronomers from all over the world who define astronomical constants while promoting research, education, and discussion on important astronomical topics.
- **right ascension:** a space object's longitudinal arc along the celestial equator, measured eastward from the vernal equinox.

## PART OF THE ZODIAC

Pisces is one of the eighty-eight constellations that are officially recognized by the International Astronomical Union (IAU). It is also one of thirteen constellations of the astronomical zodiac. These are the constellations through which the sun appears to travel over the course of a year. The sun's apparent path through the zodiac is called the ecliptic.

The sun passes in front of Pisces from about March 12 to April 19 each year. Pisces is considered the first constellation of the zodiac because this period includes the spring equinox, also called the vernal equinox, which typically marks the beginning of the tropical year. The twelve other constellations that make up the astronomical zodiac are Aquarius, Aries, Cancer, Capricornus, Gemini, Leo, Libra, Ophiuchus, Sagittarius, Scorpius, Taurus, and Virgo.

Pisces is Latin for "fish," plural. The Pisces constellation is located in the first quadrant of the Northern Hemisphere, northeast of the constellation Aquarius and northwest of the constellation Cetus. It is also bordered by the constellations Andromeda, Aries, Pegasus, and Triangulum.

Pisces is the fourteenth largest of the eighty-eight constellations, occupying a V-shaped area of 889 square degrees. It has a right ascension of 0.85 hours and a declination of +11.08 degrees. The constellation resembles two fish that are swimming in opposite directions, joined together by a cord.

Pisces is visible between the latitudes of 90 degrees north (the North Pole) and 65 degrees south. It is best seen between November 6 and November 9 around 9:00 p.m., although it is difficult to locate with the naked eye because of its dim stars and its large area. Typically, a dark country sky is necessary to see the constellation. To find Pisces, it helps to first find the marker called the Great Square of Pegasus, which is part of the constellation Pegasus. The Circlet of Pisces, also called the head of the Western Fish, is located to the south of the Square of Pegasus. The Eastern Fish, which appears as though it is jumping upward, is located to the east of the Square of Pegasus. From the Southern Hemisphere or the northern tropics, however, the Eastern Fish looks like it is diving downward.

The principal stars that make up Pisces are eta Piscium, gamma Piscium, alpha Piscium, beta Piscium, and Van Maanen's Star. Also called Alpherg or Kullat Nunu, eta Piscium, is a bright giant star and is the brightest star in the constellation. It is 316 times brighter than the sun and lies 294 light-years from Earth. Gamma Piscium is a yellow giant that is approximately 130 light-years away. Alpha Piscium, also known as Alrisha or Alrescha (the cord), is a pair of white dwarf stars 139 light-years from Earth. It is located where the two fish of Pisces are tied together at their tails. Beta Piscium, also called Fum al Samakah (mouth of the fish), is located about 492

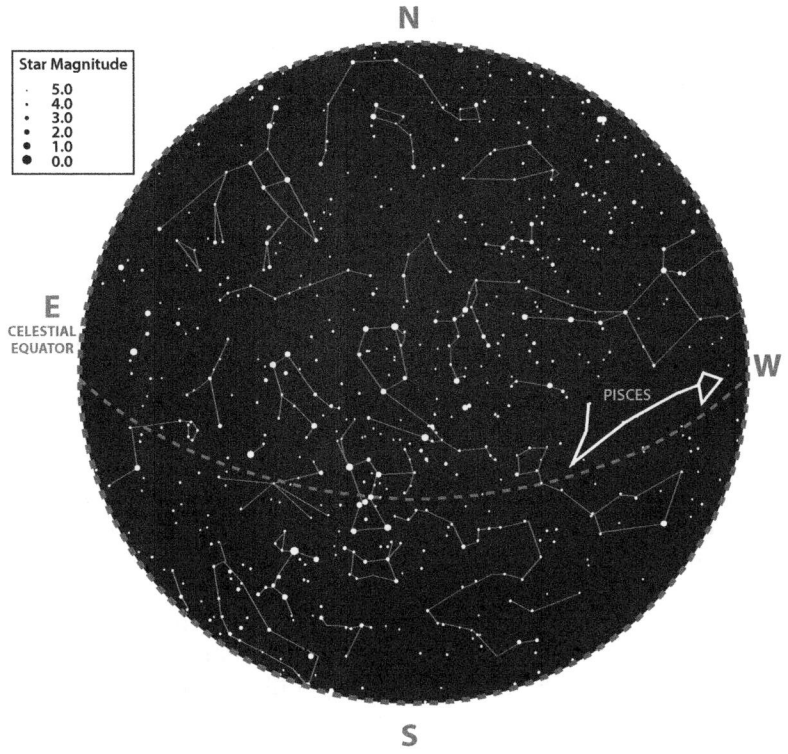

Star chart of the Northern and Southern Hemispheres in November. Lines connect the stars of the constellation Pisces.

light-years away. Van Maanen's Star was discovered by and named after the Dutch astronomer Adrian van Maanen in 1917. Only 14 light-years away, Van Maanen's Star is the closest single white dwarf to the sun.

### History and Myths

Pisces was first described by the Babylonians, who also saw the constellation as two fish connected by a cord. The Greek astronomer Ptolemy (ca. 100–ca. 170) included Pisces in his list of forty-eight constellations.

There are several different myths about Pisces. It is believed that Greek and Roman myths about the goddess Aphrodite (Venus to the Romans) were inspired by the myth of Atargartis, the Syrian goddess of love and fertility who appeared as half woman, half fish. In one Greek myth, two Syrian river fish, Ikhthyes or Ichthyes, are associated with Aphrodite and her son Eros. While trying to escape the monster Typhon (Typhoeus), Aphrodite and Eros change into fish and enter the Euphrates River. They tie themselves together with a cord so they will not lose each other.

A different Greek myth states that the two fish led Aphrodite and Eros to safety by carrying them on their backs. Similar Roman myths associate the two fish with Venus and Cupid, the Roman versions of Aphrodite and Eros.

### Significance

Ancient people believed that zodiac constellations such as Pisces were especially significant. Pisces was not always considered the first constellation of the zodiac; the sun used to seem to pass through the constellation Aries on the vernal equinox instead. However, due to precession—the gradual change in the orientation of Earth's axis—the sun began appearing in Pisces on the equinox in 68 BCE and will continue to do so until 2597 CE. This period is commonly referred to as the Age of Pisces. In 2597, the sun will seem to pass through the constellation Aquarius on the vernal equinox, and the Age of Aquarius will begin.

Modern astronomers are particularly interested in another distinguishing feature of Pisces: Messier 74 (M74), which is a grand-design spiral galaxy, or a spiral galaxy with a clear, organized spiral structure. M74 was discovered by Pierre Méchain (1744–1804) in 1780 and was later cataloged by Charles Messier (1730–1817). It is located between alpha Arietis and eta Piscium and is about equal in size to the Milky Way galaxy. The supernovas SN 2002ap and SN 2003gd have been observed in M74.

—*Michael Mazzei*

### Further Reading

McClure, Bruce. "Pisces? Here's Your Constellation." *EarthSky*. Earthsky Communications, 5 Nov. 2014. Web. 4 May 2015.

"Messier 74." *SEDS Messier Database*. SEDS-USA, 14 Aug. 2013. Web. 8 May 2015.

Plotner, Tammy. "Pisces." *Universe Today*. Fraser Cain, 30 Dec. 2008. Web. 4 May 2015.

Ridpath, Ian. "Pisces: The Fishes." *Ian Ridpath's Star Tales*. Author, n.d. Web. 4 May 2015.

Ridpath, Ian, and Wil Tirion. *Stars and Planets: The Most Complete Guide to the Stars, Planets, Galaxies, and the Solar System.* 4th ed. Princeton: Princeton University Press, 2007. Print.

Zimmermann, Kim Ann. "Pisces Constellation: Facts about the Fishes." *Space.com.* Purch, 5 June 2013. Web. 4 May 2015.

# PLANET SPIN-ORBIT COUPLING

## FIELDS OF STUDY

Astrophysics; Theoretical Astronomy; Observational Astronomy

## SUMMARY

The spin-orbit coupling of planets in relation to the sun, their moons, and each other refers to the combination of each body's spin, radius of orbit, and speed. In most cases, this results in the same face of the objects always facing each other. This affects the gravitational forces of the orbiting bodies and the tidal forces they exert on each other.

## PRINCIPAL TERMS

- **angular momentum:** the product of the mass of a rotating object, its velocity, and the radius of its axis of rotation, which in the case of an orbiting body is the orbital radius.
- **spin-orbit coupling:** the rotation rate of a celestial object in relation to the rate at which it orbits another body; also known as orbital resonance.
- **sunspots:** cooler, darker areas of the sun that appear in pairs or groups and result from the sun's magnetic field pushing through its surface.
- **tidal force:** the effect of the gravitational force of one celestial body on another.
- **vector:** a quantity having both magnitude and direction.

## ORIGINS OF THE CONCEPT

The orbital relationship of celestial bodies to each other has long been of interest to scholars and scientists, beginning with the work of English physicist Isaac Newton (1642–1727) on gravity. French mathematician and physicist Pierre Simon de Laplace (1749–1827) developed some of the earliest theories about the effects of gravity on the interactions of celestial bodies. He is credited with much of the early work on the interactions between the orbits of Jupiter and Saturn. He also defined early theories of lunar orbit and tides.

In 1786, Laplace showed that the curves (or eccentricities) and angles of the planets' orbits remain constant and that any deviations are minor and self-correcting. He determined that the eccentricity vector is constant regardless of where in the orbit it is calculated. Together with Italian French mathematician Joseph Louis Lagrange (1736–1813), Laplace laid

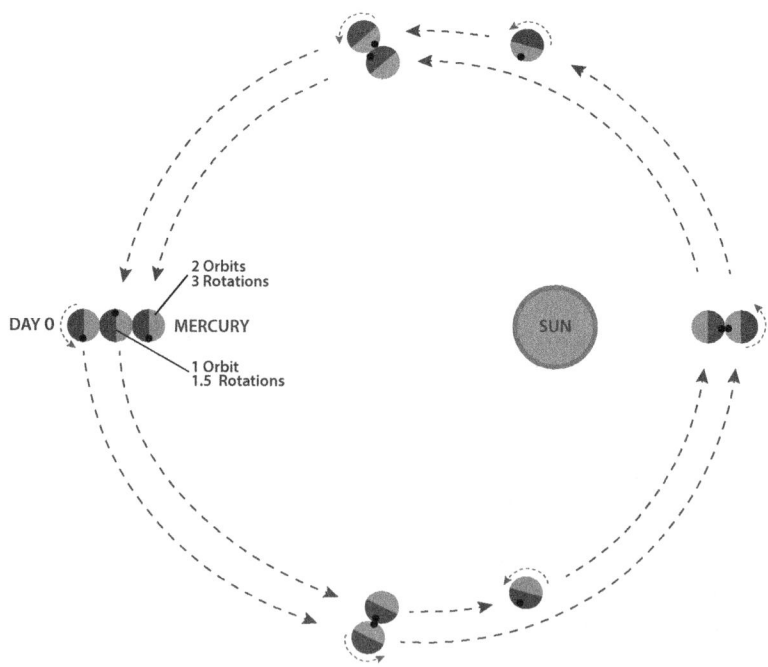

The synchronization of planetary rotations (spin) and revolutions (orbit) create unique patterns in each planet's motion. Mercury rotates three times in the time it takes to orbit the sun two times. Because of this, from any particular point on the planet's surface (indicated here by a black dot at sunrise) it appears that the sun sets once every two Mercury years.

the foundation for the study of positional astronomy, including the phenomenon of spin-orbit coupling.

The way planets and moons orbit each other and the sun is governed by specific laws of physics. The property of angular momentum states that any orbiting body must maintain a set ratio with the body it orbits. This ratio is determined by multiplying the object's mass by its velocity by the radius of its orbit. Since each planet or moon also rotates on an axis, spin momentum must also be factored in to this ratio.

The laws of physics require the angular momentum to remain constant. This principle is known as the conservation of angular momentum. In a circular orbit, all the values would remain constant. However, in an elliptical orbit, the radius of orbit varies, but the mass does not. To maintain the angular momentum, the object's velocity must also change. Thus, an orbiting body travels faster when it is closer to the body it is orbiting and slower when it is farther away.

These fixed relationships mean that the sun, the planets, and their moons will maintain stable orbits for a very long time. This ratio also locks the bodies into a synchronicity of orbit. This can cause one body to act on the other or both to act on each other.

The gravitational effects of one orbiting body on another can create changes in its orbiting partner. One well-known example of this is the way the moon's gravitational effect on Earth causes Earth's tides. The tidal forces exerted by the moon cause the surface water to bulge away from Earth. These forces are also at work when a smaller star in a binary system pulls matter away from the larger one.

### PLING

While the tidal effect of the moon on Earth is perhaps the best-known example of the interaction between orbiting bodies, there are others. For instance, the sun also affects tides on Earth. The sun's gravity adds to that of the moon when they align and lessens the moon's effect a bit when they are at right angles. The first situation causes stronger tides on Earth, and the second causes weaker ones.

The planets' gravitational forces can also affect the sun. Once every eleven years, the midway point in a full solar cycle, there is a fiery burst of solar activity, including flares, sunspots, and coronal mass ejections (large plasma eruptions that affect the solar wind). In 2008, Australian researchers proposed a theory that linked this activity to the effects of Jupiter and Saturn on the sun. The tidal forces of these giant planets on the sun are too weak to cause so much activity. However, the researchers theorized that the gravitational forces of the two planets either slow down or speed up the sun's orbital motion. When this change occurs, the sun's rotation rate also changes. This demonstrates that Jupiter, Saturn, and the sun are connected in a spin-orbit coupling resonance. The ratio of this spin-orbit coupling is believed to be 9:8, with the nine alignments of the two planets being equal to eight solar cycles.

Other planets have different coupling ratios. For instance, Earth and its moon are on a 1:1 cycle. The moon rotates on its axis at precisely the same rate that it rotates the earth, once every twenty-seven days, seven hours, and forty-three minutes. This is why observers on Earth always see the same side of the moon. Similarly, when Venus reaches its inferior conjunction (passes between Earth and the sun), the same side always faces Earth, because Venus's rotation period is nearly the same as its orbital period.

### FUTURE STUDY

The slight but constant pull of objects in spin-orbit coupling relationships can affect those objects. The speed of the moon's orbit has been changed after years in orbit around Earth, for instance. Researchers have also determined that Earth's rotation period slows by sixteen seconds every million years and that the moon moves 120 centimeters farther away from Earth every year. Eventually, in the distant future, Earth's day (rotation period) will be equal to the moon's orbital period, currently about twenty-seven days. Then, just as the moon always keeps the same side toward Earth, so too will Earth keep the same side toward the moon.

Continued study into these phenomena will help scientists understand the changes that the celestial bodies in the solar system are undergoing and what how they might affect human life in the future.

—*Janine Ungvarsky*

### FURTHER READING

Bond, Peter. *Exploring the Solar System*. Oxford: Wiley, 2012. Print.

Head, Marilyn. "Planetary Lineup Excites the Sun." *ABC Science*. ABC, 2 July 2008. Web. 24 Apr. 2015.

Lissauer, Jack J., and Imke de Pater. *Fundamental Planetary Science: Physics, Chemistry and Habitability*. New York: Cambridge University Press, 2013. Print.

Mangum, Jeff. "The Earth-Moon Spin-Orbit Resonance." *Ask an Astronomer*. Natl. Radio Astronomy Observatory, 24 Oct. 2013. Web. 24 Apr. 2015.

"Pierre Simon de Laplace (1749–1827)." *High Altitude Observatory*. UCAR, n.d. Web. 24 Apr. 2015.

"Tides and Tidal Forces." *Royal Museums Greenwich*. Natl. Maritime Museum, n.d. Web. 24 Apr. 2015.

# PLASMA PHYSICS

### FIELDS OF STUDY

Astrophysics; Astrochemistry

### SUMMARY

Plasma physics is the branch of physics that studies the behavior of plasmas. Plasma is one of the four main states of matter, along with solid, liquid, and gas. Plasma is ionized gas that conducts electricity. A gas becomes ionized when its interparticle bonds are broken, freeing negatively charged electrons and creating positively charged ions. Plasma is the most abundant form of matter in the universe. Plasma physics emerged from many fields of study, including nuclear fusion and radio technology.

### PRINCIPAL TERMS

- **plasma:** a fluid-like state of matter consisting of highly ionized gas that contains positively charged ions and negatively charged electrons. Plasmas can propagate waves and are highly electrically conductive.

### THE FOURTH STATE OF MATTER

Physics recognizes four fundamental states of matter: solid, liquid, gas, and plasma. Each state of matter has a distinct form and specific physical properties. Whether a substance is solid, liquid, or gas is determined by the strength of the forces holding its constituent particles together. These interparticle (interatomic or intermolecular) forces are strongest in the solid state, keeping the particles packed closely together and holding them rigidly in place. They are weakest in the gas state, allowing the particles to move freely about. The greater the kinetic energy of the particles, the weaker the forces holding them together.

Plasma is different from the other three states because it is not characterized by the strength of its interparticle forces. Plasma is formed by ionizing a gas—that is, by causing some or all of the gas particles to lose electrons and become ions. In fact, a plasma can be described most simply as an ionized gas. Like gas, plasma does not have a fixed shape or volume. Unlike gas, it conducts and responds to electromagnetic forces.

Ionization can be achieved either with high temperatures or by the application of electricity. Heating gas to tens of thousands of kelvins causes the gas particles to collide forcefully. If the gas is molecular, this causes the molecules to dissociate into their constituent atoms. As the temperature continues to rise, the atoms collide with increasing force. Eventually the collisions become forceful enough to knock electrons out of their atomic orbits. The resulting collection of positively charged ions and negatively charged unbound electrons forms a plasma. These individual electrically charged particles make plasma very electrically conductive. However, the plasma as a whole is electrically neutral, as the positive and negative charges balance each other.

The level of ionization within a plasma is called its "electron density" or "plasma density," measured in terms of electrons per unit volume. The electron density of a plasma depends on the ionization process used. Heating a gas produces only a small amount of ionization. The processes of photoionization and electric discharge are far more efficient.

Photoionization occurs when an atom interacts with one or more photons, or particles of light, whose energy is equal to or greater than the atom's ionization energy. The ionization energy of an atom is simply the minimum energy needed to remove one of its electrons. Photons from the sun ionize Earth's upper atmosphere, creating what is called the ionosphere.

Ionization by electric discharge involves exposing a partially ionized gas to an electric field. This causes the already-freed electrons to collide with the still-neutral atoms with enough energy to ionize them. Newly freed electrons continue to collide with other atoms, perpetuating the ionization process to increase the plasma density. When the electric field is removed, the electrons and ions start to recombine, and the plasma reverts to its former state.

## UNIQUE PROPERTIES OF PLASMA

Plasma is different from the other main states of matter because it is not held together by the usual interparticle forces. Rather, the ions and electrons of a plasma are held together by the coulomb force. This force, also known as the electrostatic force, is the force of attraction or repulsion between two electrical charges.

Because many if not all of the particles in a plasma are electrically charged, plasmas have properties not present in solids, liquids, or gases. One defining characteristic of plasmas is that each particle can affect numerous other nearby particles, not just the one that is closest. This means that a plasma, unlike a gas, will display collective behavior. For example, plasmas can generate electromagnetic fields. They are good electrical and thermal conductors, capable of withstanding high currents of electricity and extreme temperatures. Plasmas can also sustain many kinds of wave phenomena, such as electromagnetic waves and electrostatic waves.

## A MULTITIERED HISTORY

Many scientific fields contributed to the modern understanding of plasma physics. Nineteenth- and twentieth-century scientists interested in the phenomena of electric discharge, magnetohydrodynamics, and kinetic theory helped lay its foundations. Developments in radio technology, astrophysics, nuclear fusion, electric and magnetic fields, and laser technology also helped scientists figure out the behavior of plasma.

The term "plasma" was first used in a medical context in the eighteenth century. Blood plasma is the fluid that carries red and white blood cells through the body. American physicist and chemist Irving Langmuir (1881–1957) first described ionized gas as "plasma" in 1927, likening the electrons and ions carried in the gas to the cells suspended in blood plasma. Langmuir's studies of electric discharges in ionized gas led him to discover the plasma sheath, a boundary layer that forms between a plasma and a solid surface. He also observed the electron variations that create waves in plasma, now called "plasma oscillations" or "Langmuir waves." Langmuir's research formed the academic basis for most modern plasma-processing techniques as well as for plasma research in general.

The studies of Swedish physicist Hannes Alfvén (1908–95) in magnetic fields and wave phenomena contributed greatly to the understanding of plasmas. In 1942, Alfvén introduced the concept of magnetohydrodynamic waves, or electromagnetic waves that propagate through electrically conducting fluids. This concept led to the new field of magnetohydrodynamics, which studies how conducting fluids interact with magnetic fields.

In the 1950s, the development of the hydrogen bomb introduced the idea of using controlled thermonuclear fusion as a potential power source. The extremely high temperatures required for thermonuclear fusion cause the gas undergoing the reaction to become plasma. As a result, researchers began to investigate the possibility of containing the fusion reaction using magnetic fields to control the plasma. These studies rapidly expanded scientific knowledge

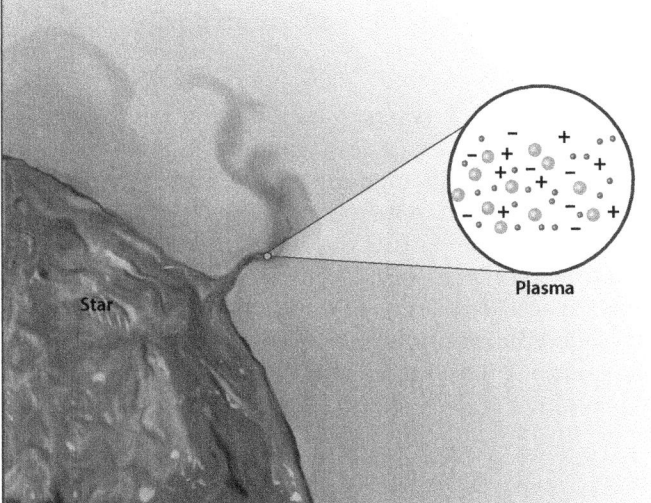

Plasma is composed of free electrons and positively charged ions. The extremely high temperatures in stars cause increased ionization of particles.

of thermonuclear plasma and the nature of plasma in general.

The principles of both electric discharge and magnetohydrodynamics led to the establishment of the kinetic theory of the plasma state. According to this theory, plasma resembles a gas in the way its particles undergo random motion. However, while gas particles interact solely through collisions, plasma particles can interact through electromagnetic forces as well. In many cases, this electromagnetic interaction predominates.

Early satellite technology led to a broader understanding of space plasma and the discovery of radiation belts circling Earth. In 1958, the first US satellite, Explorer 1, was launched into space under the direction of American astrophysicist James Van Allen (1914–2006). *Explorer 1* confirmed the existence of a layer of highly energetic charged particles—that is, a plasma—trapped within Earth's inner magnetosphere. This layer was named the Van Allen radiation belt. Earth has two persistent Van Allen belts in its magnetosphere, an outer belt and an inner one. Other such belts may form temporarily before disappearing.

In the 1960s, the invention of high-powered lasers ushered in the field of laser plasma physics, or the formation of plasma by focusing a laser on a solid or gaseous target. These laser plasmas have certain unique properties, such as densities closer to those of solids than of gases. Over the next few decades, continued studies in plasma physics led to greater comprehension of astronomical and terrestrial particle behavior.

## THE IMPACT OF PLASMA PHYSICS

Plasma physics applies to countless scientific fields of study. A few examples include astrophysics, electromagnetism, and fusion power. Plasma is found everywhere in the universe. It is the most plentiful state of matter in existence, found in stars and much of interstellar space. Earth is surrounded by terrestrial plasma called the ionosphere. Lightning and auroras, or streams of colored light in the sky, are also terrestrial plasmas.

Artificial plasmas can be created in laboratories for commercial purposes. They are used in neon signs, fluorescent lamps, and high-definition televisions. The ability of plasmas to withstand high temperatures makes them useful tools in industrial metalworking and cutting. Industrial plasma is used to manufacture a variety of products, including batteries, glass, plastics, food packaging, and textiles. Plasma technology is also used to clean materials such as glass, metals, plastics, and polymers at the molecular level prior to the production process. Advanced studies in plasma physics led to the development of plasma torches, which ignite and hold the flame of supersonic aircraft (aircraft that exceed the speed of sound). Plasma physics is a highly applicable area of science with an ever-increasing number of potential uses.

—*Cait Caffrey*

## FURTHER READING

"Basics." *Perspectives on Plasmas.* Plasmas Intl., n.d. Web. 19 May 2015.

Bittencourt, J. A. *Fundamentals of Plasma Physics.* 3rd ed. New York: Springer, 2004. Print.

Fitzpatrick, Richard. "Brief History of Plasma Physics." *Home Page for Richard Fitzpatrick.* University of Texas at Austin, 31 Mar. 2011. Web. 19 May 2015.

Glanz, James. *The Pervasive Plasma State.* College Park: Amer. Physical Soc., 1996. *Division of Plasma Physics.* Web. 26 May 2015.

Hutchinson, Ian H. "Introduction." *Introduction to Plasma Physics.* Massachusetts Inst. of Technology, 2001. Web. 19 May 2015.

"Ionization and Plasmas." *Astronomy 162: Stars, Galaxies, and Cosmology.* University of Tennessee, Knoxville, n.d. Web. 19 May 2015.

Sturrock, Peter Andrew. Plasma Physics: An Introduction to the Theory of Astrophysical, Geophysical and Laboratory Plasmas. New York: Cambridge University Press, 1994. Print.

# PLUTO

## FIELDS OF STUDY

Planetary Science; Planetary Geology; Orbital Mechanics

## SUMMARY

The planet Pluto is described as the ninth planet of the Solar system, using results obtained from the *New Horizons* mission in 2015. Pluto has an atmosphere of nitrogen, carbon monoxide, and methane, with sufficient internal heat to drive seismic activity. Surface features of the planet are between 4 billion and 10 million years old. There are at least five moons orbiting Pluto.

## PRINCIPLE TERMS

- **binary trans-Neptunian object:** two minor planets that orbit around a common center of mass while also following a larger orbit around the sun at a greater average distance than Neptune.
- **dwarf planet:** a round celestial body that is much smaller than a planet and orbits the sun, but is unable to clear other objects from its orbit. Made of solid rock, ice, or both, dwarf planets are thought to be incapable of sustaining any known life.
- **eccentric orbit:** an orbit that is not a perfect circle, causing the distance between the orbiting object and the body it orbits to vary.
- **Edgeworth-Kuiper Belt:** a ring of icy rock objects located beyond the orbit of Neptune; also known as the Kuiper Belt.
- **International Astronomical Union:** an association of professional astronomers from all over the world who define astronomical constants while promoting research, education, and discussion on important astronomical topics.
- **Kuiper Belt objects (classical and resonant):** small, icy space bodies in the Kuiper Belt, also known as trans-Neptunian objects; most KBOs are classical, with orbits of low eccentricity and inclination; resonant KBOs orbit the sun in a consistent ratio to Neptune's orbit.

## PLUTO AND ITS PLACE

Pluto is an anomaly as a planet of the solar system. It more closely resembles one of the terrestrial planets, and it is certainly not one of the gas giant planets. It is known to have a mass of $1.3 \times 10^{22}$ kilograms, and orbits the Sun at an average distance of $5.9 \times 10^9$ kilometers, or 39.44 astronomical units (AU).

Pluto rotates on its axis once in 6.39 Earth days, and completes one orbit around the Sun in 248.5 years. The orbit has a very large eccentricity of 0.250, indicating that it is a highly elongated ellipse, and is inclined to the plane of the ecliptic by 17.2 degrees. This eccentric orbit has fueled debate over whether Pluto actually is a planet or just perhaps the largest of the Kuiper Belt objects to be found in the Edgeworth-Kuiper Belt. According to the International Astronomical Union, the body that decides the classification of astronomical objects, Pluto has not been declared anything but a planet, despite claims by other groups that it should be classed as a dwarf planet at most.

The surface temperature of Pluto is just $-230°C$ ($-380°F$). The overall density of Pluto of $2.03$ g.cm-3 is considerably greater than any of the gas giant planets, though lower than that of the terrestrial planets. The planetary albedo of 0.50 is about halfway between those of Mars (0.154) and Earth (0.75).

Only one moon in orbit around Pluto is visible from Earth. The *New Horizons* mission in 2015 added substantially to the information about Pluto, including the identification of at least four small moons in addition to the one that was known.

## QUALITIES

It is somewhat surprising that much of the traditional information about Pluto was born out by the *New Horizons* mission, though the planet provided numerous details that were quite unexpected.

The mass and size of Pluto were not known until the discovery of Pluto's moon, called Charon, in 1978. Their relative masses could then be calculated and related to the masses of the other planets through their relative gravitational effects. The pair were subsequently classed as a binary trans-Neptunian object.

That Pluto has a methane-containing atmosphere was known from spectroscopic analysis of light from a distant star passing through the thin atmosphere when the two bodies were in occultation. *New Horizons,* however, found that the atmosphere of Pluto consists of carbon monoxide and nitrogen, in addition to the methane.

The blue haze of the atmosphere was found to be due to Rayleigh scattering of light by layers of soot-like hydrocarbon particles rather than methane. These particles aggregate slowly to form a sooty equivalent of snowflakes that fall and accumulate on the surface, giving it a dark red hue.

Perhaps the greatest surprise found in Pluto by *New Horizons* is that it is an active planet that apparently has sufficient internal heat to produce "cryovolcanism" and other features that seem to indicate a considerable body of water or water ice beneath the solid outer crust of methane, nitrogen, carbon monoxide and other ices.

The age of some of the major surface features of Pluto is estimated to range from 4 billion years to as little as a mere 10 million years. The surface features suggest that Pluto may still have sufficient internal heat to maintain an ocean of ammonia-rich water deep within the planet. Large mountains and cliffs were also observed on Pluto's surface, with indications of seasonal erosion effects.

*New Horizons* also identified four previously unknown moons in orbit around Pluto, in addition to Charon. This, of course, removes the classification of binary trans-Neptunian object from the Pluto-Charon pair, and turns any argument about the status of Pluto as a planet firmly back into the direction of "planet."

Charon itself is very similar to Pluto in regard to its internal structure, and is now thought to have, or to have had, an underground ocean similar to that of Pluto. Currently, the best theory for the manner of Charon's formation is that it, and the other four companion moons that were discovered, resulted from an impact that broke an original planet into at least as many pieces. Charon is also spherical in shape, like a planet, which would indicate that the impact would have been in the distant past, permitting Charon and Pluto to reorganize into their current shapes under the influence of their relatively low respective gravities. During that period, one of the current moons, which has been named Kerberos, seems to have been formed by joining together of two of the pieces produced by the impact, since it now has a 'dumbbell' shape

**PLUTO IN HISTORY AND MYTHOLOGY**

Pluto, in Roman mythology, is the god of the Underworld, and was known to the Greeks as Hades. He was one of the trio with Jupiter and Neptune who were the sons of the Titan Cronus (known to the Romans as Saturn). The planet known today as Pluto could never be known to ancient astronomers, however, since it is not visible from Earth.

The discovery of Pluto is one of the classic stories of scientific investigation. Following the discovery of Neptune, calculations could not fully account precisely for the variations in the orbit of Uranus, causing Percival Lowell (1855-1916) and his colleagues to suspect that another planet existed beyond the orbit of Neptune. Nor was the orbit of Neptune as regular as would be expected. Lowell searched from 1905 through 1916, but was unable to locate the "missing" planet. It was not until 1930 that Tombaugh, continuing

## DISCOVERY OF THE PLANET PLUTO

January 23, 1930      January 29, 1930

Original plates from Clyde Tombaugh's discovery of Pluto. Lowell Observatory Archives.

Lowell's search, located Pluto close to where Lowell had predicted it should be.

The planet was located as a pinpoint of light that shifted position in photographic images made over a period of eight nights. The movement of the pinpoint was observed using a device called a "blink comparator," which was the only tool available for such studies in the time before computers and graphics manipulation programs. A blink comparator allows the user to flip quickly between two images, and any differences between the two images are seen as movement as a result.

**MEASURING PLUTO**

Pluto's great distance from Earth and its corresponding inaccessibility reduce the study of the planet seemingly to a level of mere curiosity, and in many ways the practicality of such study is missing. However, the development of methods and technology that permit such study are of great importance closer to home in that they permit much more effective and robust methods to be applied in more accessible locations such as Mars and the Moon, and even on Earth.

Examination of the region of space in which Pluto is found also provides a better understanding of the type of objects that are to be found there, of which could one day find their way into the inner part of the Solar system, where they could conceivably do some serious damage on Earth. Disturbed from their stable orbit in the Edgeworth-Kuiper belt, an object could be drawn by the force of gravity between that object and Earth.

—*Richard Renneboog, MSc*

**FURTHER READING**

Blunck, Jürgen. *Solar System Moons: Discovery and Mythology.* New York, NY: Springer, 2010. Print.

Elkins-Tanton, Linda T. *Uranus, Neptune, Pluto and the Outer Solar System.* New York, NY: Chelsea House, 2006

Faure, Gunter, and Teresa M. Mensing. *Introduction to Planetary Science: The Geological Perspective.* New York, NY: Springer, 2007. Print.

McFadden, Lucy-Ann, Paul R. Weissman, and Torrence V. Johnson, eds. *Encyclopedia of the Solar System.* 2nd ed. San Diego, CA: Academic Press, 2007. Print.

Schmude Jr., Richard. *Uranus, Neptune, Pluto and How to Observe Them.* New York, NY: Springer Science+Business Media, 2008. Print

# Q

# QUARK STAR

## FIELDS OF STUDY

Stellar Astronomy; Astrophysics

## SUMMARY

Quark stars are stars that are too large to be classified as neutron stars but not large enough to collapse into black holes. Although quark stars have not yet actually been observed, scientists have enough evidence to believe that they exist. Quark stars are important because they may provide information about quark particles, which have yet to be replicated in laboratories.

## PRINCIPAL TERMS

- **neutron star:** a dense, quickly spinning star made of the matter left behind after a star has experienced a supernova, or intense, bright explosion.
- **quark:** an elementary particle that is part of every hadron (proton or neutron). It is found in pairs or triplets.
- **quark-nova:** an explosion believed to occur after a high-mass star experiences a supernova, resulting from a second collapse of its core.

## Death of a Star

Over millions or billions of years, high-mass stars use up the energy-producing material in their core and die in a spectacular explosion known as a supernova. Supernovas result when the star's gravitational force overwhelms its energy-depleted core, causing the core to collapse until the matter inside becomes so dense that the pressure forces it apart again. Much of the star's matter then explodes out into the universe at speeds approaching ten thousand kilometers per hour.

Not all of the star's matter is blown away in the supernova, however. A supernova occurs when the star's carbon core has fused into iron, at which point fusion would consume more energy than it would produce. Instead, the protons and electrons in the core are compressed so much that they merge and become neutrons. The resulting object is called a neutron star. Neutron stars are generally about 10 to 20 kilometers (6.25 to 12.5 miles) in diameter.

If a high-mass star is more than three times the mass of Earth's sun, it will leave behind matter from its core. This core remnant cannot produce any energy. Thus, it will experience extreme pressure from the star's gravitational forces and eventually collapses into a black hole.

Scientists long believed that a supernova produced by a high-mass star could result in only one of two options, a neutron star or a black hole. But modern astrophysicists theorize that a third option may exist: quark stars. A quark star would be composed of small particles known as quarks, or elementary particles that make up certain types of subatomic particles. Two quarks (or, more accurately, one quark and one antiquark) form a type of particle called a meson. Three quarks form a type of particle called a baryon, a category that includes neutrons. The quarks that form mesons and baryons are held together by the strong interaction, one of four fundamental forces in the universe. Forcing these quarks apart would produce resistance similar to that experienced when stretching a piece of elastic. In a quark star, the quarks would exert enough outward pressure to prevent the star's total collapse.

## The Third Option

When the material left over after a supernova is too massive to form a neutron star, it becomes difficult for the star to resist collapsing from its own gravity. At this point, the quarks come into play. Each neutron contains two down quarks and one up quark. (There are six types of quarks; the other four are strange quarks, charm quarks, bottom quarks, and top quarks.) As the gravitational pressure builds, some down quarks become strange quarks. The star is

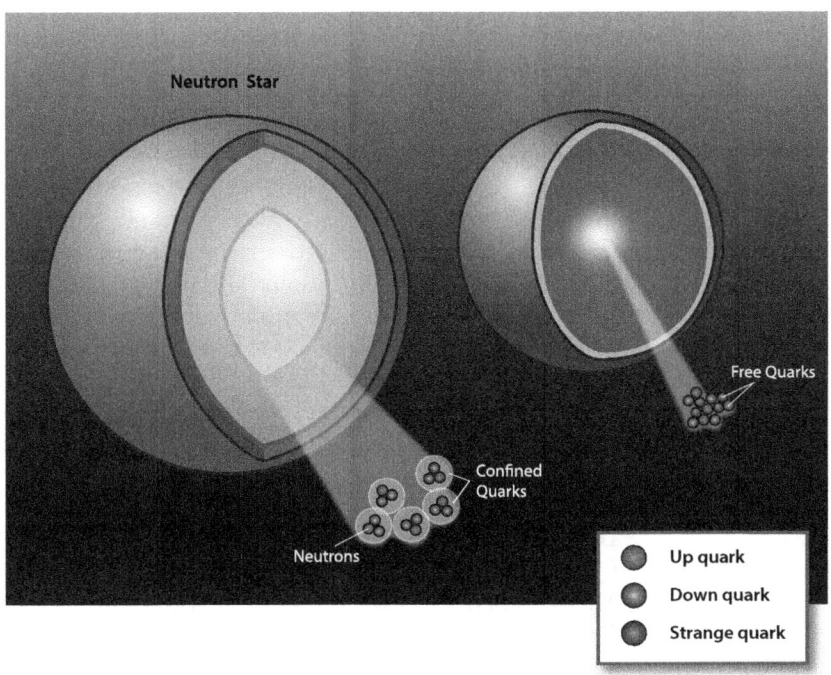

Neutron stars have up quarks and down quarks confined as neutrons. Strange quark stars have up quarks, down quarks, and strange quarks freely arranged.

then considered a quark star or, if it contains strange quarks, a strange star.

Some scientists have theorized that quark stars may be formed by quark-novas. They propose that just days after a massive star explodes in a supernova, it experiences a second explosion—the quark-nova—and becomes a quark star. It is possible that the gases and debris from the supernova may obscure this second explosion. Scientists have not observed quark-novas and have no direct proof that they occur. However, there is some evidence that the phenomenon may exist. In particular, scientists have noticed that some stars thought to result from supernovas contain less iron and more titanium than expected. If these stars underwent quark-novas, this might explain the discrepancy.

### The Importance of Quark Stars

Astronomers have not identified any quark stars, but they continue to look for them. Some scientists theorize that if they find large neutron stars with a mass 2.5 times greater than that of Earth's sun, it might actually be a quark star. This discovery would be particularly valuable to astrophysicists. Because strange quarks cannot be replicated in a laboratory, a quark star would be an important source of information for particle physicists as well.

—*Janine Ungvarsky*

### Further Reading

Bernoskie, Brandi, Heather Deiss, and Denise Miller. "What Is a Supernova?" *NASA Education*. NASA, 4 Sept. 2013. Web. 6 Apr. 2015.

Hall, Shannon. "Quark Nova Spotted in Cas A?" *Sky & Telescope*. F+W Media, 29 May 2014. Web. 6 Apr. 2015.

"High Mass Star." *Space Book*. Las Cumbres Observatory Global Telescope Network, 2012. Web. 6 Apr. 2015.

Nave, Carl R. "Quarks." *HyperPhysics*. Georgia State U, n.d. Web. 7 Apr. 2015.

"Neutron Stars: Incomprehensible Density." *Science and Space*. Natl. Geographic Soc., n.d. Web. 6 Apr. 2015.

O'Neill, Ian. "Why Are Quark Stars So Strange?" *Discovery News*. Discovery Communications, 19 Jan. 2010. Web. 6 Apr. 2015.

"Quark." *Cosmos: The SAO Encyclopedia of Astronomy*. Swinburne University of Technology, n.d. Web. 6 Apr. 2015.

# QUASAR (QUASI-STELLAR RADIO SOURCE)

## FIELDS OF STUDY

Extragalactic Astronomy; Cosmology; Theoretical Astronomy

## SUMMARY

Quasars are powerful astronomical objects characterized by huge bursts of energy from small points in space. They are among the most distant objects observed in the universe. They are also the most luminous known objects in the universe. It is believed that quasars are generated by the accretion disks of supermassive black holes in forming galaxies.

## PRINCIPAL TERMS

- **active galactic nuclei (AGN):** a central region of some galaxies that emits more electromagnetic radiation than normal, thought to be caused by supermassive black holes.
- **black hole:** an area of space-time with infinite density, from which light cannot escape.
- **gamma rays:** a high frequency form of electromagnetic radiation composed of energetic protons.
- **radio emission:** electromagnetic radiation with a longer wavelength than infrared light.
- **redshift:** the phenomenon in which the wavelength of electromagnetic radiation increases, moving towards the red end of the spectrum. Calculations of redshift are used to determine the distances of distant astronomical objects.

## ACTIVE GALACTIC NUCLEI

Quasars, or quasi-stellar radio sources, are the brightest known objects in the universe. They can be ten to ten thousand times brighter than the entire Milky Way galaxy. They are also among the furthest objects from Earth ever observed. Quasars are a type of active galactic nuclei (AGN), or a central area of a galaxy that is much brighter than normal. This extreme brightness is caused by the heating of materials around the edge of a supermassive black hole at the center of a galaxy. While scientists believe all galaxies have central black holes, not all produce AGNs. Only when the black hole is actively "feeding" on matter can an AGN be formed.

There are several types of AGNs in addition to quasars. Most notable are blazars, radio galaxies, and Seyfert galaxies. However, most scientists believe that these classifications are in fact all variations of the same phenomenon. It is only the angle at which these distant objects are viewed from Earth that makes them appear different. Quasars and other AGNs give off huge amounts of electromagnetic radiation, which may include visible light, radio waves, gamma rays, and other emissions.

## DISCOVERY OF QUASARS

Quasars were discovered in 1963 by the astronomer Maarten Schmidt (b. 1929) using the telescopes of Mt. Palomar Observatory. Schmidt was examining sources of radio emissions, focusing on an unusual object named 3C 273. The object did not follow the patterns of known phenomena. It looked like a star through normal telescopes. However, when viewed with a radio telescope, the object was seen to emit incredibly powerful radio signals, unlike normal stars. It also produced other strange spectral emissions.

Schmidt then realized that the mysterious emissions were simply those of hydrogen gas whose wavelengths had been shifted. This phenomenon, known as redshift, happens to all types of radiation over extremely long distances. Objects moving farther away shift toward the red end of the spectrum (longer wavelength), while objects moving closer shift toward the blue end (shorter wavelength). Schmidt used this observation to calculate that 3C 273 would be billions of light-years from Earth. To be visible at that distance, 3C 273 must be brighter than a million galaxies.

For this reason, Schmidt concluded that 3C 273 could not be a star. However, the object and others like it failed to match the profile of any known celestial object. For this reason, they became known as "quasi-stellar radio sources." This was later shortened to the word "quasars." Over time, many more quasars were discovered. Although it was found that not all quasars produced radio emissions, the name stuck. Yet it would not be until the 1980s that scientists reached a consensus on what quasars actually were.

## Black Holes and Quasars

Scientists eventually determined that quasars are linked to another type of object: black holes. Black holes are infinitely dense areas of space-time formed when incredibly massive stars burn all their fuel. The nuclear reaction sustaining the star ceases, and the gravity from the star's dense core quickly pulls the entire star into the center. This causes the core to become even denser, increasing its gravitational pull and causing nearby objects to be pulled into it. As yet more objects are pulled into the core, the process increases in speed and power. Eventually, the area becomes so dense that not even light can escape its gravitational pull. At this point, the absorption and compression process speeds up so much that the area becomes infinitely dense, which creates a small area with an infinitely strong gravitational pull. This area is known as a "singularity."

Black holes are sometimes surrounded by glowing clouds of gas and compressed matter called "accretion disks." This matter orbits the black hole as it spirals toward the singularity at the center. The intense, inescapable gravity near this point compresses the spiraling matter, which causes huge amounts of friction. The friction generates so much heat that the entire accretion disk begins to glow, sometimes glowing more brightly than a star. Some of these accretion disks are even visible from Earth.

Most astronomers believe that a supermassive black hole is located at the center of almost every galaxy. It is believed that the supermassive black holes found in the center of young galaxies are the sources of quasars. Scientists theorize that matter is condensed much more in young galaxies than in their older counterparts. For this reason, far more matter is pulled into the supermassive black hole at the center of the galaxy, creating an AGN. Most of this material is compressed inside the black hole's accretion disk.

Because of the friction associated with huge amounts of matter being compressed into a relatively small area, the accretion disk begins to emit large amounts of energy. Up to 10 to 20 percent of the captured matter's mass is converted directly into energy. For comparison, nuclear fusion reactors, such as Earth's sun, convert less than 1 percent of their

Intense amounts of energy are emitted from certain galaxy centers called quasars. The energy produced from these galactic nuclei surrounding supermassive black holes is more than 100 times that of a normal galaxy.

mass to energy. This extremely efficient energy conversion explains how such comparatively small objects can produce more energy than entire galaxies. When that energy is emitted in a luminous jet that is oriented generally toward Earth, its point of origin is classified as a quasar. If the jet aims directly at Earth, it is seen as a blazar. Seyfert galaxies are thought to be oriented sideways relative to Earth, so that no jets are observed.

## Studying Quasars

Because light has a finite speed, the quasars that are detected on Earth probably existed billions of years ago. For this reason they are valuable to scientists studying the origins of the universe and the life cycle of galaxies. It is theorized that older galaxies, such as the Milky Way, may have had quasars at their centers at one point. As these galaxies grew, matter would have become less concentrated, eventually cutting off the fuel supply for the AGN. Astronomers believe that because of how spread out matter is in the current universe, it would usually be impossible for new quasars to form in established galaxies.

However, galaxy collisions might be one exception to this rule. When two galaxies collide, sometimes the supermassive black holes at their centers merge together. This might pull large chunks of the galaxies into the black hole, creating an accretion disk large enough to cause a temporary quasar. Scientists speculate that this could occur when the Milky Way collides with the Andromeda galaxy in about four to ten billion years.

*—Tyler Biscontini*

**Further Reading**

"Black Holes." *NASA Science: Astrophysics.* NASA, 30 Mar. 2015. Web. 12 May 2015.

Cain, Fraser. "Could the Milky Way Become a Quasar?" *Universe Today.* Universe Today, 26 Feb. 2015. Web. 11 May 2015.

Cain, Fraser. "Never a Star: Did Supermassive Black Holes Form Directly?" *Universe Today.* Universe Today, 7 Sept. 2007. Web. 11 May 2015.

GaBany, R. Jay. "A Singular Place." *Cosmotography.* R. Jay GaBany, 2015. Web. 11 May 2015.

Hall, Shannon. "How Do Black Holes Get Super Massive?" *Universe Today.* Universe Today, 13 Aug. 2013. Web. 11 May 2015.

Marshall, Michael. "Introduction: Black Holes." *New Scientist.* Reed Business Information, 6 Jan 2010. Web. 11 May 2015.

Miralda-Escudé, Jordi. "How do Quasars Really Work?" *Department of Astronomy and Meteorology.* University of Barcelona, n.d. Web. 11 May, 2015.

"1963: Maarten Schmidt Discovers Quasars." *Everyday Cosmology.* Observatories of the Carnegie Institution for Science, n.d. Web. 11 May 2015.

Redd, Nola Taylor. "Quasars: Brightest Objects in the Universe." *Space.com.* Purch, 23 Aug. 2012. Web. 11 May, 2015.

# R

## RED GIANT

### FIELDS OF STUDY

Astronomy; Cosmology; Stellar Astronomy

### SUMMARY

A red giant star has consumed all the hydrogen contained in its core, is past its peak, and is slowly dying. Helium has built up inside the core, and hydrogen has fused in the outer shells. This type of star has greatly expanded in size, and the outer layers burn cooler than the core, causing the star to appear red.

### PRINCIPAL TERMS

- **black hole:** a region of space with a gravitational pull so strong that nothing, not even light, can escape it.
- **core:** the dense, hot center of a star, where heat is normally generated.
- **fusion:** the joining together of two or more atomic nuclei to form a new, heavier nucleus of a different element.
- **giant:** a star larger than a dwarf star.
- **gravity:** a force of mutual attraction between two masses that increases with increased proximity.
- **luminosity:** the essential brightness of a celestial object; the amount of energy it emits.
- **luminosity classification:** a means of categorizing stars by their relative sizes, as derived from their spectral emissions, atmospheric pressure, distance from Earth, and loss of light to interstellar matter; also called the Morgan-Keenan classification.
- **main sequence:** in a Hertzsprung-Russell diagram, which plots the temperature/color of stars against their luminosity/brightness, the band of stars that crosses from the upper left to the bottom right, representing the evolution of a "normal" star over the course of its lifetime.
- **neutron star:** a dense, rapidly spinning star made of the material that remains after a star becomes a supernova. It is formed from the neutrons created by the reaction of the star's remaining protons and electrons.
- **spectral classification:** a method of classifying stars based on various elements of their emission spectra, as determined by such characteristics as surface temperature, luminosity, and size.
- **supernova:** a massive explosion that results when a dying star exhausts its fuel. The star's core collapses and explodes, releasing large amounts of energy and matter into the universe.
- **white dwarf:** a small, very dense star that has exhausted nearly all of its fuel and is nearing the end of its life cycle.

### RED GIANT PROCESS

Red stars start out just like any other star, where gravity causes hydrogen to fuse into helium, or performing what astronomers call "the main sequence." This is the process that makes stars appear to shine. In medium and large stars, eventually the hydrogen is depleted, the core of the star can no longer hold out against gravity, the outer layers collapse, and temperature and pressure in the core increase. The star becomes tighter and smaller, contracting into its own core. But with compression comes heat. More and more mass getting packed into a smaller and smaller space builds the temperature up to where the helium atoms begin to fuse into carbon atoms and oxygen atoms. Astronomers estimate that our sun has been performing this fusion for about 4.5 billion years so far.

This helium fusion can happen either gradually and slowly or happen suddenly and explosively, depending on the mass of the star. This fusion reaction again releases energy, causing the star to swell and grow, reaching massive diameters of 100 million to 1 billion kilometers, 100 to 1,000 times their original size. For example, our sun's diameter is currently about 0.01 AU. When it becomes a red giant, in approximately 5 billion years, its diameter will be

Principles of Astronomy

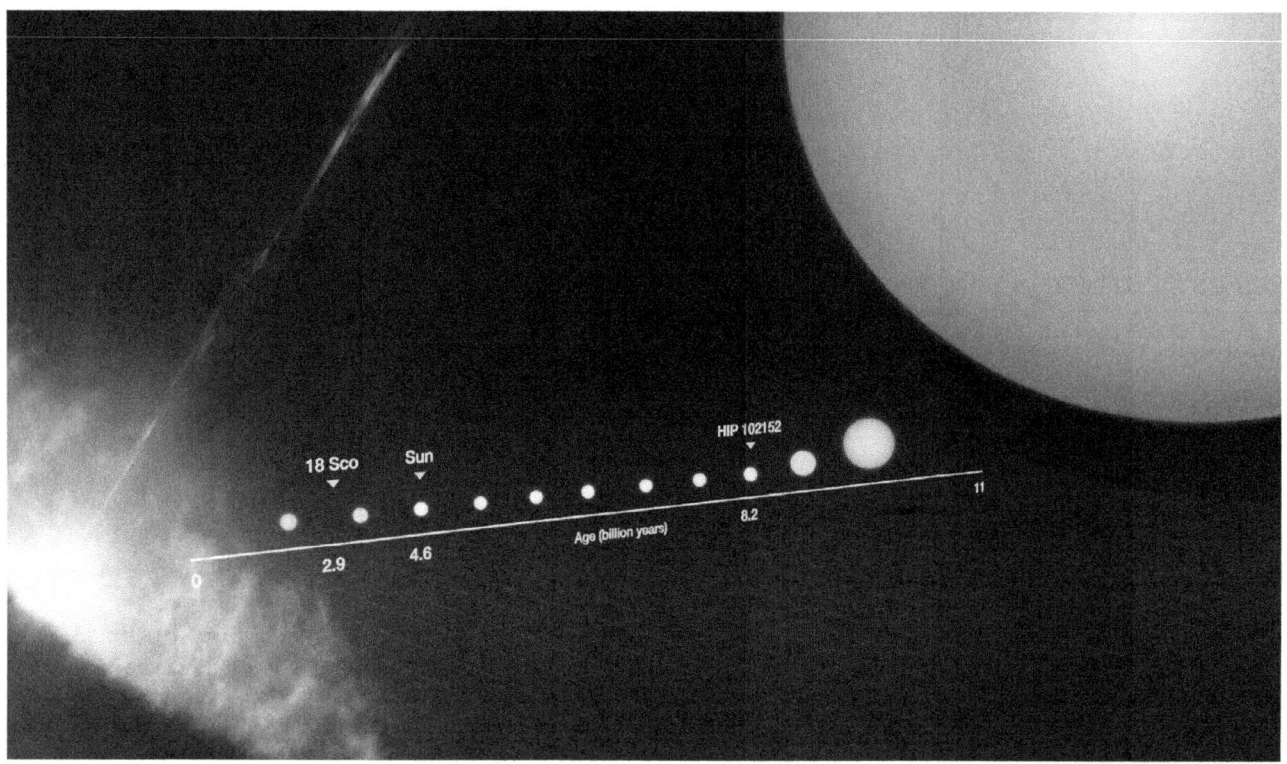

This image tracks the life of a Sun-like star, from its birth on the left side of the frame to its evolution into a red giant on the right after billions of years. By ESO/M. Kornmesser - Wikimedia Commons.

approximately 2 AU. At that point, it will have grown to encompass the orbits of Mercury, Venus, and Earth. During this expansion process, the luminosity of the star can greatly increase.

The energy created by the helium fusing into carbon is now dissipated over a much larger surface area, so the surface temperature of a red giant is much cooler than it was when the star was still fusing hydrogen into helium—about half the original temperature or a surface temperature ranging between 2,200C and 3,200C. In this temperature range, the stars appear anywhere from yellow-orange to bright red, leading to the name "red giant." Many stars, despite their current color (such as our sun, which is currently yellow-white) wll change the wavelength of light they put out and will appear to us to be red.

Medium stars (anywhere from one third the size of our Sun to about 8 times its size) go through this process and meet the spectral classification and luminosity classification of a red giant toward the end of their lives. Over the total life span of a star, only about 10 percent is spent as a red giant. However, not every star becomes a red giant at some point in its life. Stars that don't have a mass in this range transfer energy from their core to their surface by convection, and thus the process of fusion ends with helium. These types of stars don't get hot enough to become a red giant and become stagnant.

Some red giants still have hydrogen fusing into helium in the outer shell while the helium in the core continues to fuse into carbon. This process can go on for around another 1 billion years. But when there is no more helium left, the star will begin to shrink again, and helium from the outer shell will move toward the core and ignite. Because of the star's mass and its constant increase in temperature, its outer layers blow off as gas clouds, creating planetary nebulae. While in this phase, the star is known as a variable star, becoming brighter and dimmer, sometimes becoming bluer and smaller as its life progresses. But the core continues to collapse upon itself, causing this type of red giant to eventually become a white dwarf.

For very large red giants, the final collapse leads to the core exploding in a fiery and violent death, a supernova. This explosion leaves behind either a neutron star or a black hole.

### Red Giants Near Earth

Only 88 light years away, Gamma Crucis is the red giant nearest to the Earth. Arcturus is closer to Earth, only 36 light years away, but it is sometimes classified as an orange giant. Betelgeuse is an example of a red supergiant.

Usually, the color of a star is not visible with the naked eye. However, if you find the very bright red giants, Betelgeuse (the first star that was not our sun directly imaged), Aldebaran, Arcturus, or Antares in the night sky, you may be able to discern that they have a reddish color compared to the other stars in the area. Some of these stars are a million times brighter than the sun and are getting close to entering a supernova stage.

### Red Giants and Habitable Zones

When a star balloons up into a red giant, the zone around it in which liquid water can exist also changes. This means that when our sun eventually becomes a red giant, life may actually be able to exist near it for a billion years longer, and life may even be able to exist in places that have not had the light of the sun before but will be able to be heated by the sun when it expands. However, when our sun begins collapsing again after its life as a red giant to become a white dwarf, the habitable zone will change—the light of the sun will no longer reach as far as it does now, turning all water into ice and leaving no chance of life in its range.

—*Marianne M. Madsen, MSc*

### Further Reading

Cox, Brian and Andrew Cohen. *Wonders of the Universe.* Harper Design, 2011.

Dinwiddie, Robert, Will Gater, Giles Sparrow, and Carole Stott. *Nature Guide: Stars and Planets.* DK, 2012.

Eicher, David J. *The New Cosmos: Answering Astronomy's Big Questions.* Cambridge University Press, 2016.

Frebel, Anna. *Searching for the Oldest Stars: Ancient Relics from the Early Universe.* Princeton University Press, 2015.

King, Andrew. *Stars: A Very Short Introduction.* Oxford University Press, 2012.

Lang, Kenneth R. *The Life and Death of Stars.* Cambridge University Press, 2013.

Miglio, Andrea, Josephina Montelban, and Arlette Noels, eds. *Red Giants and Probes of the Structure and Evolution of the Milky Way.* Springer, 2012.

Owens, Steve. *Star Gazing for Dummies.* For Dummies, 2013.

Tirion, Wil. *The Cambridge Star Atlas.* Cambridge University Press, 2011.

Tyson, Neil deGrasse. *Death by Black Hole and Other Cosmic Quandaries.* W. W. Norton & Company, 2014.

# ROCKET ENGINES AND ROCKETRY

### FIELDS OF STUDY

Aerospace Engineering; Spacecraft Propulsion; Space Technology

### SUMMARY

Rockets are the vehicles that make space exploration possible. These powerful vehicles are launched from Earth and travel through the sky and into outer space. A rocket's engine is the component that propels the rocket upward. Newton's laws of motion play an important role in rocketry, as does mass fraction. Rockets have a rich history and continue to evolve.

### PRINCIPAL TERMS

- **escape velocity:** the speed needed for an object to escape the gravitational pull of a planet or moon. If the object reaches this speed, it will be able to enter outer space.
- **Newton's laws of motion:** three scientific laws developed by Isaac Newton that describe an object's response to a force that acts upon it. These laws relate force, acceleration, mass, and velocity.

- **nozzle:** an opening at the rear of a rocket that allows gases to be released and accelerates them so that thrust can be maximized.
- **payload:** the carrying capacity of a rocket, which can include anything from astronauts to a space probe.
- **propellant:** a mixture of fuel and oxidizer (usually oxygen) that propels a rocket. Both solid and liquid propellants are used in rockets.
- **thrust:** in rocketry, the force produced by a rocket engine that pushes a rocket upward.

#### FUNDAMENTALS OF ROCKETS

A rocket is a vehicle that typically travels to outer space. The term "rocket" can also refer to the engine that propels the vehicle. Besides space travel, rockets are often used in war. Both major types of rocket use a propellant, which is the chemical substance that launches the rocket. A propellant can be either solid or liquid and is usually a mixture of fuel and oxidizer. The oxidizer allows the fuel to burn, which is a key component to propelling the rocket. Oxygen is the most common oxidizer. As their names imply, a solid-propellant rocket uses propellant that is solid, and a liquid-propellant rocket uses propellant that is liquid. The propellant provides the engine with thrust, which is the force that actually drives the rocket. The thrust must be greater than the rocket's weight to propel the rocket upward.

A rocket contains a nozzle at its rear, which allows gases to be released. The nozzle also provides the rocket with the maximum amount of thrust. A nozzle has two components: a throat, which is the narrow part, and an exit cone.

Rockets also have a payload, which is a component that carries cargo. The payload can include astronauts, equipment, a spacecraft, a satellite, and/or a space probe. To launch the payload into either space or orbit, the rocket must first break free from the gravitational pull of Earth. This requires a very high speed, known as escape velocity. To escape the Earth's gravitational pull, a rocket (or any other spacecraft) must achieve an escape velocity of more than 40,000 kilometers per hour (25,000 miles per hour). Each planet and moon has its own escape velocity.

#### EARLY HISTORY OF ROCKETS

Rocketry dates back more than two thousand years. Perhaps the first rocket-like device was built around 400 BCE by a Greek named Archytas (ca. 400–350 BCE), who lived in a city that is now in Italy. According to the writings of the Roman Aulus Gellius (ca. 125– ca. 180 CE), Archytas flew a wooden bird that was propelled by steam. Several hundred years later, a Greek named Heron of Alexandria (ca. 10–85 CE) invented his own device, which was also propelled by steam. In the first century CE, the Chinese then began building gunpowder-filled devices that had rocket-like qualities. After experimenting with these devices, the Chinese built devices known as "fire-arrows," which were launched by escaping gas. In 1232, they used these fire-arrows in battle against the Mongols.

Experiments with rockets continued for centuries, particularly in Europe. Rockets eventually began being used for fireworks displays. For example, a sixteenth-century German fireworks maker named Johann Schmidlap invented a rocket that could

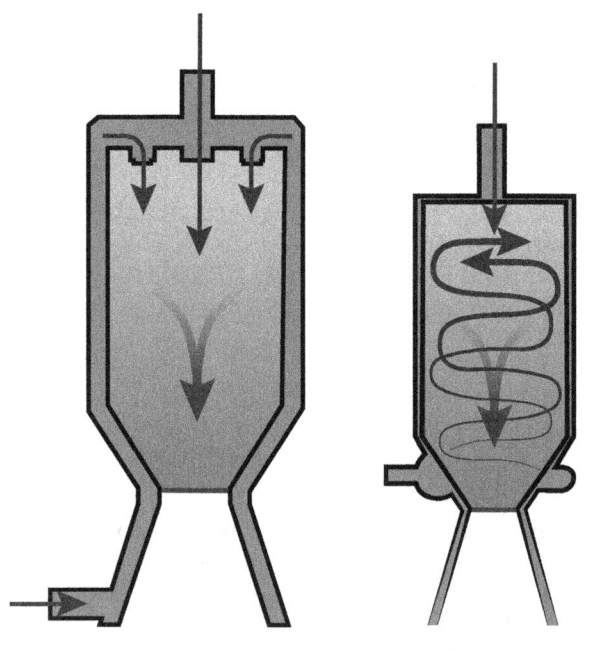

**Liquid Cooled Rocket Engine**    **Vortex Rocket Engine**

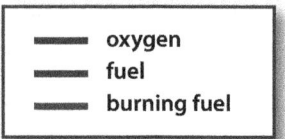

Diagram comparing a liquid-cooled rocket engine and a vortex rocket engine.

launch fireworks to higher altitudes than ever before. Known as the "step rocket," Schmidlap's invention carried two rockets of different sizes into the sky. The larger rocket lifted the device off the ground, and when it was exhausted, the smaller rocket took over, carrying the device to a higher altitude. The step rocket would become the basis for modern-day rockets used for space travel.

## NEWTON'S LAWS OF MOTION

In the late seventeenth century, English scientist Sir Isaac Newton (1642–1727) developed three scientific laws that would eventually have an impact on rocketry. Commonly referred to as Newton's laws of motion, the laws were compiled in Newton's *Principia Mathematica Philosophiae Naturalis* (1687; trans. 1729). The first law involves an object that is either at rest or moving uniformly. The law states that the object will remain at rest or in uniform motion as long as an external force does not act on the object. This state is known as "inertia."

Newton's second law addresses the velocity of an object. It states that an external force will change an object's velocity and that this change is dependent on the force applied ($F$), the mass of the object (m), and the acceleration of the object ($a$). The law is often expressed by the following equation:

$$F = ma$$

The third law involves action and reaction. The law explains that when one object exerts a force on a second object, the second object will exert an equal but opposite force on the first object. The following statement is commonly used to articulate the law: "For every action, there is an equal and opposite reaction."

## MODERN ROCKETRY

Newton's laws of motion paved the way for modern rocketry. The laws helped improve rocket design and also allowed for larger and more powerful rockets than previously existed. For example, German and Russian rocket experimenters developed rockets that weighed more than forty-five kilograms (one hundred pounds). Rockets were also extremely successful in war. In fact, the lyrics "the rockets' red glare" in the "Star-Spangled Banner," the national anthem of the United States, were inspired by the rockets used by British ships during the War of 1812.

However, it was not until 1898 that the true potential of rockets was realized. That year, Russian schoolteacher Konstantin Tsiolkovsky (1857–1935) set forth the idea that rockets could be used for space exploration. He published a report in 1903 outlining this idea. Furthermore, rocket experiments by American engineer Robert H. Goddard (1882–1945) in the early twentieth century helped guide rocketry closer to space exploration. It was not long before rocket societies were established throughout the world. Then in 1957, the Soviet Union launched *Sputnik I*, the world's first Earth-orbiting artificial satellite. The United States soon followed suit, launching the satellite *Explorer I* just a few months later. Since then, countless rockets from around the world have been launched into space, and rocketry has continued to evolve.

A rocket's effectiveness is generally determined by its mass fraction (MF), which is the mass of the rocket's propellants divided by the rocket's total mass. The mass fraction can be calculated by using the following equation

$$MF = \text{mass of propellants} / \text{total mass}.$$

The ideal rocket has a mass fraction of 0.91, or 91 percent. This means that 91 percent of the rocket's total mass should be propellants. The rocket's remaining mass should be distributed among the payload, engines, tanks, fins, and so on.

Determine the mass fraction of a rocket that has a propellant mass of 867,300 kilograms and a total mass of 1,029,600 kilograms.

To find the mass fraction, enter the rocket's propellant mass and total mass into the MF equation and solve:

$$MF = 867{,}300 \text{ kg} / 1{,}029{,}600 \text{ kg}$$

$$MF = 0.84$$

The rocket has a mass fraction of 0.84, or 84 percent, which is lower than that of the ideal rocket.

## The Future of Rocket Engines

A new type of rocket engine has been developed that scientists believe may change the way rockets are built. The Orbital Technologies Corporation (ORBITEC), a company based in Madison, Wisconsin, created a new vortex liquid-fuel rocket engine. A traditional liquid rocket engine must have a double wall to protect the combustion chamber from the extreme heat of the burning fuel. Unlike these traditional rocket engines, the vortex liquid-fuel rocket engine does not need a double wall because a different approach is used for the combustion chamber. In a vortex engine, liquid oxygen enters near the base of the chamber, forming a vortex of cold oxygen. This vortex then spirals up the walls of the chamber. Fuel enters the top of the chamber and mixes with the vortex of oxygen at the top, where it combusts, or burns. The burning mixture of oxygen and fuel forms a second vortex that then moves down the center of the chamber, traveling inside the first vortex. Essentially, the first vortex of oxygen is the outer vortex, and the second vortex of the burning mixture is the inner vortex. The reason for this approach is to prevent the walls of the chamber from getting too hot and potentially melting. Being cold oxygen, the outer vortex protects the chamber walls from the direct heat of the burning inner vortex. Therefore, the walls are exposed to only radiant heat, and a double wall is not required. Because these engines do not need a double wall, they are lighter, simpler, and less costly to build than engines that do require a double wall.

In October 2012, ORBITEC patented and successfully flight-tested a vortex liquid-fuel rocket engine. The rocket engine was placed in a launch vehicle made by the Garvey Spacecraft Corporation of Long Beach, California, and then was flight-tested at the Friends of Amateur Rocketry facility, located near the Edwards Air Force Base in California. The rocket engine supplied the 360-kilogram (800-pound), 7.6-meter (25-foot) launch vehicle with full thrust for approximately ten seconds. About twenty seconds later, the rocket reached its peak altitude. The rocket's top speed was approximately 270 meters per second (600 miles per hour). The successful test flight moved ORBITEC closer to its ultimate goal: a 13,600-kilogram (30,000-pound) thrust vortex engine that the company was developing for the National Aeronautics and Space Administration's Space Launch System and for the US Air Force Advanced Upper Stage Engine Program.

—*Michael Mazzei*

## Further Reading

"Brief History of Rockets." *NASA*. Natl. Aeronautics and Space Administration, 12 June 2014. Web. 28 May 2015.

"Escape Velocity: Fun and Games." *NASA*. Natl. Aeronautics and Space Administration, 10 Apr. 2009. Web. 28 May 2015.

Hague, Robin. "Vortex." *Celestial Mechanics*. Robin Hague, 2008. Web. 28 May 2015.

"Newton's Laws of Motion." *NASA*. Natl. Aeronautics and Space Administration, 5 May 2015. Web. 28 May 2015.

"Practical Rocketry." *NASA Quest*. Natl. Aeronautics and Space Administration, n.d. Web. 28 May 2015.

"What Is a Rocket?" *NASA*. Natl. Aeronautics and Space Administration, 21 Sept. 2010. Web. 28 May 2015

# S

# SAGITTARIUS

### FIELDS OF STUDY

Stellar Astronomy; Observational Astronomy

### SUMMARY

The constellation Sagittarius is a group of stars best seen in the Southern Hemisphere. The stars in this constellation form a shape ancient people thought resembled an archer. Some modern people think part of Sagittarius looks like a teapot. Ancient people studied the constellations for practical and spiritual reasons, and modern astronomers study them to gain more scientific information about the universe. Sagittarius is a large, bright constellation with diverse types of stars as well as galaxies, clouds called nebulae, and possibly a black hole.

### PRINCIPAL TERMS

- **celestial equator:** the imaginary line above Earth's equator that halves the celestial sphere; it is equally distant from the celestial poles.
- **constellation:** a region of space defined by a pattern of stars that can be seen in the night sky from Earth.
- **declination:** a space object's angular distance north or south of the celestial equator.
- **International Astronomical Union:** an association of professional astronomers from all over the world who define astronomical constants while promoting research, education, and discussion on important astronomical topics.
- **right ascension:** a space object's longitudinal arc along the celestial equator, measured eastward from the vernal equinox.

### ARCHER IN THE STARS

Sagittarius is a constellation, a region of space defined by a group of stars that resembles a picture or pattern. In ancient times, people studied the constellations for spiritual reasons or to help keep track of the changing seasons. Since 1930, the International Astronomical Union has recognized eighty-eight constellations, including Sagittarius, which according to many ancient astronomers represented an archer. Many myths even held that this archer was a half-horse, half-human hybrid. Sagittarius is one of about twelve or thirteen constellations that comprise the zodiac, the group of constellations that appear to intersect the path of the sun. Ancient people believed the zodiac constellations were especially important.

### ATTRIBUTES OF SAGITTARIUS

Sagittarius is a constellation most visible in the Southern Hemisphere between the latitudes of 55 degrees north and 90 degrees south. Its coordinates on the celestial sphere are a right ascension of nineteen hours and a declination of −25 degrees from the celestial equator. It is the largest constellation in the Southern Hemisphere, both in size and in number of planets encompassed. Southern stargazers get the best view of the constellation around August or September in the evenings around 9 p.m. Sagittarius is also visible in the Northern Hemisphere, but it does not rise as high into the sky and is harder to see. In very dark areas, observers can sometimes see the vast reach of the Milky Way, which appears like a whitish haze. Sagittarius is visible in the widest part of this band of stars.

Sagittarius is the fifteenth largest constellation acknowledged by modern astronomers. It is also noteworthy for its inclusion in the zodiac and its position in the center of the Milky Way galaxy, the enormous mass of stars, planets, and other space materials that includes Earth. During the winter solstice that occurs around December 21–22 each year in the Northern Hemisphere, the sun can be found in Sagittarius. That fact gave the constellation special importance for many groups of ancient people.

There are also many interesting and important stars and other space objects within this constellation.

These objects include binary stars, or pairs of stars that orbit a common center of mass, and variable stars, stars whose brightness increases or decreases. Many of these stars are named with letters of the Greek alphabet, such as alpha, beta, gamma, and so on, paired with the name Sagittarii.

Often the brightest star in a constellation is given the designation alpha. In the case of Sagittarius, however, the brightest star is epsilon Sagittarii. Epsilon Sagittarii is the thirty-sixth brightest star anywhere in the night sky. Even though it is 125 light-years from Earth, epsilon Sagittarii is about 375 times brighter than the sun and therefore still casts a powerful brightness.

Epsilon Sagittarii is often paired with three other stars, gamma, delta, and lambda Sagittarii. Together, these stars form the shape of the archer's bow and arrow. Appropriately, they all have names that reflect parts of the weapon. Epsilon is called Kaus Australis, meaning "southern end of the bow." Lambda is Kaus Borealis, "northern end of the bow." Delta is best known as Kaus Media or Kaus Meridionalis, "middle of the bow." Meanwhile, gamma is known as Nash, Nushaba, Al Nasl, or Alnasl, meaning "tip of the arrow." These stars, although appearing to be near each other, actually range greatly in distance from about 77 to 300 light-years away from Earth.

Other star names also reflect the theme of an archer in the sky. Alpha Sagittarii is also known as Rukbat, "knee of the archer," and beta 1 and beta 2 Sagittarii are called Arkab Prior and Arkab Posterior, which refer to leg parts. The third brightest star in Sagittarius, zeta Sagittarii, is called Ascella, or "armpit of the archer."

One of the most distinguishing features of Sagittarius involves several of the constellation's brightest stars. The stars delta, epsilon, zeta, phi, lambda, gamma 2, sigma, and tau Sagittarii take on a special shape unrelated to the archer. Many people consider this shape an asterism, or a small pattern that is not quite a constellation, that resembles a teapot with a spout, lid, and handle. Some modern stargazers find the Teapot asterism more appealing and easier to envision than the ancient archer interpretation.

In addition to stars, Sagittarius hosts an array of deep-sky objects. These include a star cloud, star clusters, different kinds of nebulae (clouds of gas and dust in which new stars form), and an irregularly shaped galaxy. Some of the most notable nebulae include the Lagoon Nebula, the Trifid Nebula, and the Omega Nebula (also known as the Swan, Horseshoe, or Lobster Nebula), which features a looped shape and twinkling stars. Some of these objects, though extremely far away, are bright enough to be detected with the naked eye in some areas.

Finally, one of the most mysterious and potentially important features of Sagittarius is known as Sagittarius A*. Located in the center of the Milky Way galaxy, Sagittarius A* is an area of space that emits radio waves. The very

Star chart of the Northern and Southern Hemispheres in June. Lines connect the stars of the constellation Sagittarius.

unusual behavior of Sagittarius A* has led some astronomers to believe that it may contain a black hole. A black hole is a little-understood space object with such powerful gravity that nothing can escape it.

**MONSTER MYTHS**

Stargazers first discovered the stars of the Sagittarius constellation thousands of years ago. The first ancient astronomers to identify the stars included Sumerians, Babylonians, and Arabs. Each of these peoples created their own interpretations and mythologies relating to the star patterns. In Babylonia, astronomers thought the stars represented the god Pabilsaĝ, a creature with a bull-like body, colorful legs, one human head and one panther head, and wings.

Ancient Greek astronomers also studied Sagittarius with great interest. Their mythical interpretations, first known to have been recorded around the second century CE, have become the basis for the modern view of Sagittarius. According to the Greeks, Sagittarius represented a creature with a human head and torso on the body of a horse. In Greek mythology, there were two monsters with these kinds of features: wild, aggressive centaurs and peaceful, playful satyrs.

Different versions of the Greek myths associate Sagittarius with centaurs or satyrs. In some myths, the constellation represents Crotus. Crotus was a wise and joyful satyr whose benevolence pleased the gods and earned him honor in the stars. In other interpretations, Sagittarius is a more aggressive centaur that is aiming its bow threateningly at its constellation neighbor Scorpio (scorpion). Centaurs were so common in myths that another constellation, Centaurus, was also associated with them.

**MODERN ASTRONOMICAL IMPORTANCE**

Today, amateur stargazers enjoy the beauty of stars and planets in constellations. Astronomers study the constellations to gain more scientific knowledge about the changing universe. Sagittarius is a popular constellation for both amateurs and professionals because of its diversity of stars, planets, and other space objects.

New features of Sagittarius are still being discovered. On March 15, 2015, an Australian astronomer discovered a likely nova, a star explosion, within the Teapot of Sagittarius. Experts feel this space object may have originally been a dim, distant star. It only became noticeable when its explosion created a flash of brightness.

—*Mark Dziak*

**FURTHER READING**

King, Bob. "New Binocular Nova Discovered in Sagittarius." *Universe Today*. Universe Today, 16 Mar. 2015. Web. 23 Apr. 2015.

Lomb, Nick. "Know Your Constellations: Sagittarius the Archer—Also Known as the Teapot." *Sydney Observatory*. Sydney Observatory, 25 Aug. 2011. Web. 23 Apr. 2015.

McClure, Bruce. "Sagittarius? Here's Your Constellation." *EarthSky*. Earthsky Communications, 7 Aug. 2014. Web. 23 Apr. 2015.

"Sagittarius (Archer)." Chandra X-Ray Observatory. *NASA*, 2 Dec. 2013. Web. 23 Apr. 2015.

"Sagittarius, the Archer." *StarDate*. University of Texas McDonald Observatory, n.d. Web. 30 Apr. 2015.

Zimmermann, Kim Ann. "Sagittarius Constellation: Facts about the Archer." *Space.com*. Purch, 20 June 2013. Web. 23 Apr. 2015.

# SCATTERED DISK OBJECTS (SDOS)

## FIELDS OF STUDY

Astronomy; Observational Astronomy; Sub-planet Astronomy

## SUMMARY

Scattered disk objects (SDOs) are small celestial bodies of icy rock with very eccentric orbits. The points in their orbits closest to the sun, or perihelia, are in the Kuiper Belt region between Neptune and Jupiter. Scattered disk objects are important because they are thought to date from the origin of the solar system.

## PRINCIPAL TERMS

- **detached object:** a celestial body in the solar system that lies beyond Neptune's orbit and mostly outside the gravitational influence of the planets.
- **eccentric orbit:** an orbit that is not a perfect circle, causing the distance between the orbiting object and the body it orbits to vary.
- **Edgeworth-Kuiper Belt:** a ring of icy rock objects located beyond the orbit of Neptune; also known as the Kuiper Belt.
- **periodic comet:** a comet that takes less than two hundred years to complete an orbit; also known as a short-period comet.

### ORIGINS OF SCATTERED DISK OBJECTS

Scattered disk objects are a type of Kuiper Belt object (KBO). KBOs are small, icy objects found in the Edgeworth-Kuiper Belt, a section of the solar system beyond Neptune's orbit. The Edgeworth-Kuiper Belt is also commonly known as the Kuiper Belt. It is named after the men who first theorized its existence, Kenneth Edgeworth (1880–1972) and Gerard Kuiper (1905–73). KBOs are classified into three groups based on their orbits. Classical KBOs and resonance KBOs orbit fully within the belt. Scattered KBOs, or scattered disk objects (SDOs), only partly orbit within the belt. All KBOs are part of the larger group known as trans-Neptunian objects, or TNOs.

SDOs have very eccentric orbits, meaning their distance from the sun varies. They travel within the range of the other objects in the Kuiper Belt but also sometimes extend much farther. An SDO's closest position to the sun, or perihelion, ranges from about 30 to 48 astronomical units (AU). However, their extremely elliptical orbits can take them beyond 60 AU from the sun at their farthest point, or aphelion. This means that SDOs spend very little time in an area where they are likely to be observed by Earth- or space-based telescopes. For this reason, it is likely that many SDOs have not yet been discovered. The distant region these objects inhabit is called the scattered disk.

Like the other objects found in the Kuiper Belt, SDOs are thought to have been formed outside Neptune's orbit in the earliest days of the solar system, 4.6 million years ago. Scientists have different theories about where and how they were formed. Some think that they were formed in their current location. Others believe that Neptune's migration into its current location may have moved the SDOs into the Kuiper Belt. In either case, the centers of SDOs likely contain untouched material from the origins of the solar system.

### OUTSIDE INFLUENCES ON SCATTERED DISK OBJECTS

Scientists think that the SDOs now orbiting within the Kuiper Belt represent just a fraction—possibly as little as 1 percent—of those that were there when the solar system formed. They believe that Neptune's gravitational force may have affected the distance between these objects and the sun. Many SDOs may have been essentially expelled from the solar system, while other SDOs may have been pulled deeper into the solar system, where they became centaurs. Centaurs are small space bodies in orbit between Neptune and Jupiter that exhibit characteristics of both comets and asteroids. Still other SDOs may have become what are known as detached objects. These are bodies that are outside of Neptune's influence. Scientists also theorize that some periodic comets may have once been SDOs. These are comets that complete their orbits in less than two hundred years. This includes the Jupiter-family comets, with orbit periods of less than twenty years, and the Halley-type comets, with periods of twenty to two hundred years.

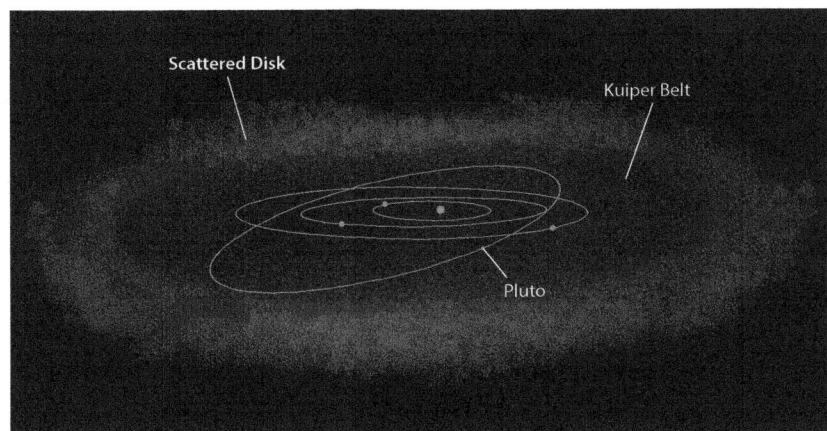

Model of the outer solar system. The scattered disk overlaps with the Kuiper Belt. Scattered disk objects are distinguishable from Kuiper Belt objects in that their orbit is more eccentric.

#### THE QUESTION OF SEDNA

In 2004, a team of researchers from the California Institute of Technology discovered a planet-like body at the far edges of the solar system. At approximately 1,700 kilometers (1,000 miles) in diameter, the object is larger than an asteroid but smaller than the dwarf planet Pluto. It is 76 AU from the sun at its perihelion and 1,000 AU from the sun at its aphelion. It is so far away that from its perspective, the sun is merely a bright star.

Scientists dubbed the planet-like body Sedna, after the Inuit goddess of the sea. They are not certain what it is. Some think it may be an object from the Oort cloud, a cloud of icy objects that surrounds the solar system. Many others believe that it is an SDO. In either case, it has a wildly eccentric orbit. While Pluto takes about 248 years to travel around the sun, Sedna takes more than 10,500 years.

—Janine Ungvarsky

#### FURTHER READING

Clavin, Whitney. "Planet-Like Body Discovered at Fringes of Our Solar System." *NASA.* NASA, 15 Mar. 2004. Web. 5 Mar. 2015.

"Comet." *Cosmos: The SAO Encyclopedia of Astronomy.* Swinburne University of Technology, n.d.. Web. 5 Mar. 2015.

Dick, Steven J. *Discovery and Classification in Astronomy: Controversy and Consensus.* New York: Cambridge University Press, 2013 Print.

"Kuiper Belt." *Cosmos: The SAO Encyclopedia of Astronomy.* Swinburne University of Technology, n.d. Web. 5 Mar. 2015.

"The Kuiper Belt." *New Horizons.* Johns Hopkins U Applied Physics Laboratory, n.d. Web. 5 Mar. 2015.

"Scattered Disk Objects." *Cosmos: The SAO Encyclopedia of Astronomy.* Swinburne University of Technology, n.d. Web. 5 Mar. 2015.

# SCORPIUS

### FIELDS OF STUDY

Stellar Astronomy; Observational Astronomy

### SUMMARY

Scorpius is a constellation, or grouping of stars, whose shape resembles a scorpion. Part of the zodiac, Scorpius is visible from the Northern and Southern Hemispheres and contains many bright stars as well as binary stars, star clusters, and a recurrent nova (exploding star). Over thousands of years, cultures around the world have associated various myths and legends with this constellation. Astronomers continue to study Scorpius and its many space objects to learn more about galaxies and the universe.

### PRINCIPAL TERMS

- **celestial equator:** the imaginary line above Earth's equator that halves the celestial sphere.
- **constellation:** a pattern of stars, identified by humans, that can be seen in the night sky from Earth.
- **declination:** the apparent north-south position of a celestial body relative to the celestial equator, expressed in degrees of arc.
- **International Astronomical Union:** an association of professional astronomers from all over the world who define astronomical constants while promoting research, education, and discussion on important astronomical topics.
- **right ascension:** the apparent east-west position of a celestial body relative to the location of the sun

at the vernal equinox, expressed in hours and minutes.

**A SCORPION IN SPACE**

Scorpius is a constellation, a grouping of stars that appear to form pictures or patterns in the night sky. Since constellations were first identified thousands of years ago, people around the world have been studying these star patterns. The ancient Greeks were responsible for describing more than half of the eighty-eight constellations now recognized by the International Astronomical Union (IAU). To make the designations more formal, Eugène Delporte assigned each constellation official boundaries on behalf of the IAU. One of these is Scorpius, whose fifteen main stars form a pattern that resembles a scorpion poised to strike.

Scorpius is also one of the twelve or thirteen constellations included in the zodiac, a collection of constellations through which the sun appears to travel on a path called the ecliptic. However, as the ecliptic is an illusion created by Earth's orbit around the sun, Scorpius's location is instead given with reference to the celestial equator. Scorpius has a declination of –40 degrees and a right ascension of about seventeen hours.

Scorpius is one of the brightest constellations in the night sky. It can be seen from either the Northern or the Southern Hemisphere. In the Northern Hemisphere, the constellation sits low in the sky, near the horizon. In the Southern Hemisphere, Scorpius is visible high in the sky, near the center of the Milky Way galaxy. Stargazers on Earth can get the best views of Scorpius during July evenings around 9:00 p.m. local standard time.

Scorpius is the thirty-third largest of the eighty-eight constellations. Its above-average size and exceptional brightness make it one of the most popular constellations among stargazers. It is also well known for its distinct and striking shape, which looks somewhat like a letter J with a curled tail.

The "head" of the scorpion is made up of three bright stars: beta Scorpii (also known as Graffias or

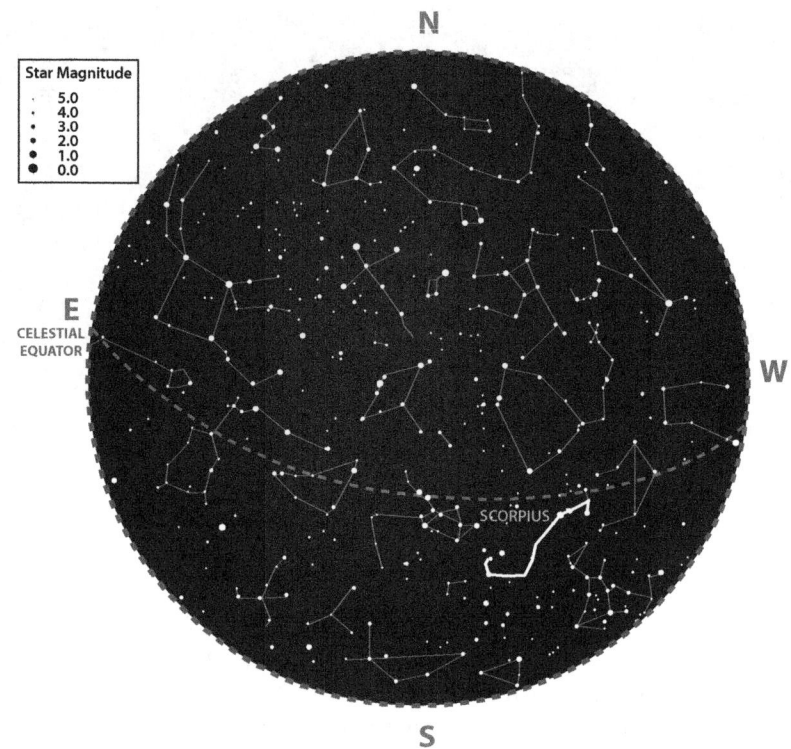

Star chart of the Northern and Southern Hemispheres in June. Lines connect the stars of the constellation Scorpius.

Acrab), delta Scorpii (Dschubba), and pi Scorpii. These three stars connect to another, even brighter star, a red supergiant called alpha Scorpii (Antares, or "rival of Ares"). In this case, "Ares" refers to the planet Mars. Early astronomers thought that Antares and Mars looked alike since they were both bright red space objects.

Antares is likely the most noteworthy star of the Scorpius constellation. It is the sixteenth-brightest star in the night sky. Often visible with the naked eye, it is about seven hundred times greater in diameter than the sun. A star as large as Antares is extremely powerful and contains enormous amounts of heat in its core. Scientists believe that one day Antares will explode into a supernova, releasing huge amounts of gas, dust, and other stellar materials that could eventually form new stars and planets.

In addition, Antares is a binary star, which means it is actually a pair of stars that orbit closely together. Scorpius contains several other multiple star systems, including beta Scorpii, xi Scorpii, and sigma Scorpii. Further bright and notable stars in the constellation include theta Scorpii (Sargas), lambda Scorpii (Shaula), nu Scorpii (Jabbah), and tau Scorpii (Al Niyat).

Many other space objects of interest to astronomers can be found in Scorpius. Most are star clusters, or large collections of stars. Two globular clusters, which are clusters that are roughly spherical in shape, are designated Messier 4 (M4) and Messier 80 (M80). M4 is noteworthy for being one of the closest globular clusters to Earth. Two open clusters, whose stars formed around the same time, include Messier 6 (M6) and Messier 7 (M7). M6 resembles the shape of a butterfly, and M7 is unusually bright due to the great age and size of its stars.

Finally, Scorpius also hosts a rare recurrent nova, or a star that experiences regular explosions, named U Scorpii. This nova is normally quite dim to Earth observers, but it brightens considerably during its eruptions. Astronomers have recorded at least seven major U Scorpii eruptions between 1863 and 2010.

## Myths from Many Lands

Scorpius was one of the first constellations identified by ancient astronomers. The Greek astronomer Ptolemy (ca. 100–ca. 170 CE) listed it among the forty-eight constellations he had studied, and many others noted it and its position within the zodiac. In ancient times, many people believed the zodiac stars had special powers because they appeared to be touched by the path of the sun. People created elaborate tales and belief systems based on the zodiac. Scorpius, like the other constellations known since ancient times, has a rich, colorful, and varied mythology.

The most famous lore involving Scorpius originated among the ancient Greeks. According to Greek myth, Orion, the hunter, bragged that he could kill any animal on Earth. This boast angered the gods and goddesses, including Gaia, the earth goddess, who sent a mighty scorpion to battle Orion. In most versions of the myth, the scorpion ultimately defeats the arrogant hunter. The myth concludes with the gods taking mercy on both combatants and honoring them with spaces in the night sky—at opposite ends, so that they appear in different seasons and cannot fight again.

Although the Greek interpretations of Scorpius are the best known today, many other ancient cultures created their own lore based on the star patterns. In North America, the Navajos saw Scorpius as two separate figures, an old man with a walking stick (known as "The First Big One") and a set of rabbit tracks, while the Skidi Pawnees saw it as a snake in the sky stalking a pair of swimming ducks. To the Maori of New Zealand, the constellation represented a fishhook that the trickster god Maui had used to pull islands out of the ocean. In various parts of Indonesia, the stars were seen to represent a swan or a coconut tree. Ancient Chinese myths saw the constellation as part of a great sky monster, the Azure Dragon.

## The Importance of Scorpius Today

In ancient times, people looked to the constellations for spiritual purposes and to help keep track of the changing seasons. In modern times, the constellations are still of interest, but for different reasons. Stargazers enjoy their awesome beauty, and astronomers use them to gain new scientific insights into the workings of the universe.

Like the other constellations, Scorpius has much to offer those who study it. Scientists often look to its binary stars and star clusters to study the behaviors of these stellar objects. In some cases, scientists can even gather clues about the workings of Earth and the solar system from these distant bodies. The recurrent nova U Scorpii, one of only about ten known examples, also promises a wealth of information about star behavior.

The bright stars of Scorpius are some of the most easily seen in the night sky. In dark areas, many people can see these stars without any visual aid. In cities, or on cloudier days, binoculars or different sizes of telescopes might be necessary. Amateur stargazers should have little trouble finding the three bright stars that make the front of the scorpion and then locating the distinctive "hook" shape. From there, stargazers can turn their attention to neighboring constellations, which include Sagittarius and Ophiuchus.

In addition, many telescopes turn to Scorpius in April, May, and June to watch two annual meteor showers that pass through the constellation. The Alpha Scorpiids begin in April and peak around May 3 each year. The June Scorpiids peak around June 5, with an average of about twenty meteor falls each hour.

—*Mark Dziak*

### Further Reading

Eratosthenes, Hyginus, and Aratus. *Constellation Myths: With Aratus's* Phaenomena. Trans. Robin Hard. Oxford: Oxford University Press, 2015. Print.

McClure, Bruce. "Scorpius? Here's Your Constellation." *EarthSky*. Earthsky Communications, 1 July 2014. Web. 30 Apr. 2015.

Plotner, Tammy. "Scorpius." *Universe Today*. Universe Today, 13 Jan. 2009. Web. 30 Apr. 2015. "Scorpius (the Scorpion)." *Chandra X-Ray Observatory*. NASA, n.d. Web. 30 Apr. 2015.

"Scorpius, the Scorpion." *StarDate*. University of Texas McDonald Observatory, n.d. Web. 30 Apr. 2015.

Zimmermann, Kim Ann. "Scorpius Constellation: Facts about the Scorpion." *Space.com*. Purch, 6 Aug. 2012. Web. 30 Apr. 2015.

# 90377 SEDNA

### FIELDS OF STUDY

Sub-planetary Astronomy; Cosmology

### SUMMARY

90377 Sedna is a large, icy trans-Neptunian object (TNO) that orbits the sun. It is located in the inner Oort cloud far beyond Neptune. Sedna is a rounded object likely about 1,600 kilometers (about 1,000 miles) in diameter. Therefore, it is probably a dwarf planet. However, its distant location makes it difficult for scientists to officially classify it as such. Sedna's location and unique orbit raise questions about the origin and nature of objects beyond the planets of the solar system.

### PRINCIPAL TERMS

- **aphelion:** the point farthest from the sun along an object's orbital path.
- **dwarf planet:** a celestial body with enough mass to attain a nearly round shape but lacking the gravity to keep its neighborhood free of other space objects. Dwarf planets orbit the sun and do not act as satellites to other space objects.
- **Oort cloud:** a vast, spherical area of space far beyond Neptune that is filled with trillions of icy objects and comets.
- **perihelion:** the point closest to the sun along an object's orbital path.
- **trans-Neptunian object (TNO):** a space object that orbits beyond Neptune at a distance more than 30 astronomical units (AU)—about 4.5 billion kilometers, or 2.8 billion miles—from the sun.

### FIRST LOOK INTO THE OORT CLOUD

Far beyond the orbit of Neptune lies a vast, spherical region of space that surrounds the solar system. This area, called the Oort cloud, is located at the very edge of the solar system. Most experts place the Oort cloud at a distance about 50,000 to 100,000 AU from the sun. The Oort cloud is more distant in space than even the Kuiper Belt, a disk-shaped region of space just beyond the orbit of Neptune. The Oort cloud is named for famed Dutch astronomer Jan Hendrik Oort (1900–92), who first hypothesized its existence.

Scientists believe the Oort cloud is filled with trillions of icy objects and comets. These are celestial remains from the formation of the solar system about 4.6 billion years ago. Because these objects are located at such an extreme distance, the sun's gravitational influence on them is significantly less than on closer objects. In fact, astronomers believe the sun's influence is quite weak in the Oort cloud. Thus, objects there are influenced by gravity from passing space objects.

Moving objects such as stars, star clusters, and even yet-to-be-discovered planets are believed to create gravitational disturbances that can change the orbits of Oort cloud objects. Some are ejected into interstellar space. Others are pushed into the inner solar system as long-period comets (comets that take at least two hundred years to orbit the sun).

Although astronomers had speculated about the Oort cloud since the early 1950s, its existence was not confirmed until 2003. At that time, astronomers at Palomar Observatory near San Diego, California, discovered 90377 Sedna.

The Palomar team was conducting an ambitious five-year project to systematically image the night sky using the forty-eight-inch Samuel Oschin Schmidt Telescope and the Palomar QUEST large-area CCD camera. The team was searching for previously undiscovered trans-Neptunian objects (TNOs). These are objects other than comets that orbit the sun at a greater average distance than the planet Neptune.

In November 2004, the equipment produced images of a large, rounded object orbiting at a distance nearly ninety times greater than that between Earth and the sun. The team studied the images carefully and referred to archived images from prior years. After performing a variety of calculations, they determined that the object was about 1,800 kilometers (1,118 miles) in diameter. They also found that it was situated at a distance greater than 100 AU (15 billion kilometers or 9.3 billion miles) from the sun.

Because this distance is far beyond the planetary region of the solar system, which ends at about 50 AU from the sun, the team realized they had identified the most distant object yet observed in the solar system. It was also the first observable "planetoid," or minor planet, in the inner Oort cloud.

The team named the icy object Sedna, after the Inuit goddess who rules over the frigid Arctic Ocean. Later, the International Astronomical Union (IAU) gave the object the official designation 2003 VB12.

## Characteristics of Sedna

In the years since Sedna's discovery, astronomers have been able to improve estimates of Sedna's size and characteristics. Some of this information has been derived from observations by the Herschel Space Observatory, a space telescope developed and managed by the European Space Agency (ESA). Herschel uses infrared and other invisible radiation to learn about distant stars and galaxies.

Information from Herschel and other research revealed that Sedna is smaller than originally estimated, with a diameter no larger than about 1,000 kilometers (620 miles). Thus, it is smaller than Pluto. Sedna's surface area is bright and quite red. In fact, it is nearly as red in color as the planet Mars. This suggests it has long been exposed to ultraviolet radiation. Its consistent coloration also suggests that it has not experienced impacts by other space bodies, which typically causes pits and craters that reveal more about the inner composition of a space object.

Sedna's surface temperatures are believed to be extremely icy. Its estimated average temperature is –240 degrees Celsius (–400 degrees Fahrenheit). Scientists believe its surface is composed of both methane ice and water ice.

## Unique Orbital Path

One of the most interesting characteristics of Sedna is its eccentric orbit. Sedna travels completely in the outer solar system. It orbits the sun along a long, oval path that takes about 11,400 years to complete. Its aphelion, or farthest point from the sun, is 937 AU, while its perihelion, or closest point, is 76.36 AU. One unresolved question about Sedna is if it classifies as a dwarf planet. The IAU created the dwarf planet class of space objects in 2006 to differentiate small, rounded planetary bodies from the eight regular planets of the solar system. IAU Resolution

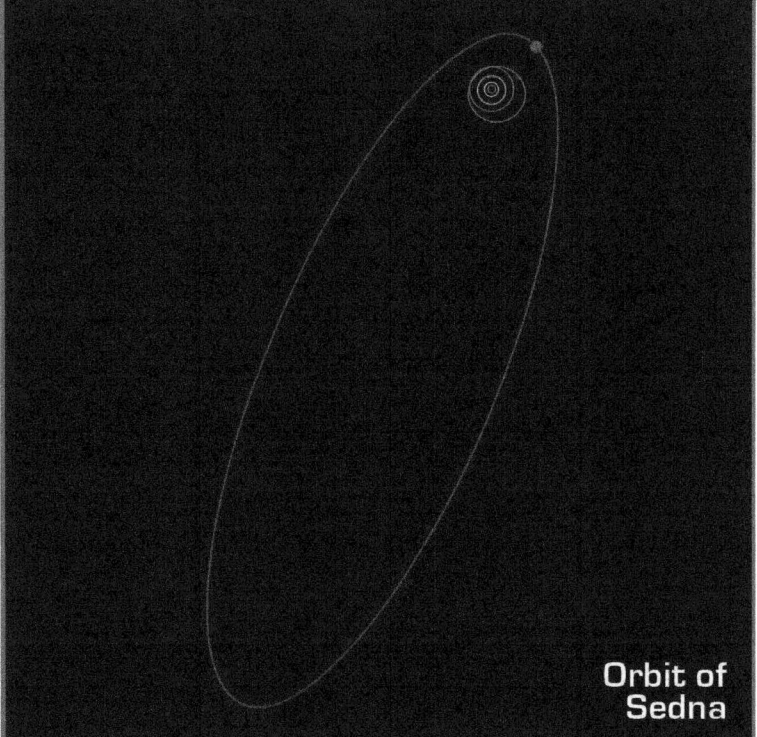

The orbit of dwarf-planet candidate 90377 Sedna compared to the outer Solar System. Sedna's full orbit is illustrated along with the object's current location. Sedna is nearing its closest approach to the Sun. Its 10,000-year orbit typically takes it to far greater distances. By NASA/Caltech [Public domain], via Wikimedia Commons

B5 defined dwarf planets as space objects that orbit the sun, have enough mass and gravity to attain a mostly round shape, and lack the gravitational power to clear other space objects out of their orbit. Additionally, dwarf planets are not satellites (moons) to other space objects.

Dwarf planets include Pluto, Eris, and Ceres. Although some researchers state with "near certainty" that Sedna should be classified as a dwarf planet, its extreme distance makes it difficult to officially classify it as such.

### ORIGINS AND EVOLUTION OF OORT CLOUD OBJECTS

Sedna's extreme distance and unique path of travel raise questions about its origin and evolution. Experts believe that Oort cloud objects are remnants from the formation of the solar system 4.6 billion years ago. Asteroids traveling in the asteroid belt are known to collide with each other and break into smaller pieces over time. However, objects in the Oort cloud seem relatively untouched and intact since their formation. This offers astronomers a unique look at the forces that shaped the early solar system. For example, because Sedna and similar objects are located so far from Neptune, most researchers do not believe that they were pushed into space by the gravity around Neptune. One theory is that an as-yet-to-be-identified planet exerted gravitational influence on Sedna and other objects in the Oort cloud. Other researchers have theorized that passing stars or star clusters may have played a role. Regardless of its origins, Sedna raises many questions about the origin and history of the early solar system.

—*Marie Keenan, MS*

### FURTHER READING

Barucci, M. A., et al. "(90377) Sedna: Investigation of Surface Compositional Variation." *Astronomical Journal* 140.6 (2010): 2095–2100. PDF file.

Brown, Michael. "The Largest Kuiper Belt Objects." *Solar System beyond Neptune.* Tucson: University of Arizona Press, 2008. PDF file.

Brown, Michael, et al. "Discovery of a Candidate Inner Oort Cloud Planetoid." *Astrophysical Journal* 617 (2004): 645–49. Web. 7 Apr. 2015.

"Dwarf Planets: Overview." *Solar System Exploration.* NASA, 2 Feb. 2015. Web. 24 Mar. 2015.

"Kuiper Belt and Oort Cloud: Overview." *Solar System Exploration.* NASA, 19 Nov. 2014. Web. 7 Apr. 2015.

Pál, A., et al. "TNOs Are Cool: A Survey of the Trans-Neptunian Region—VII. Size and Surface Characteristics of (90377) Sedna and 2010EK139." *Astronomy & Astrophysics* 541 (2012): n. pag. Web. 7 Apr. 2015.

Vergano, Dan. "Dwarf Planet Discovery Hints at Hidden World Orbiting Solar System." *National Geographic.* Natl. Geographic Soc., 26 Mar. 2014. Web. 7 Apr. 2015.

"What Is the Herschel Space Observatory?" *JPL.* NASA, n.d. Web. 7 Apr. 2015.

# SPACE NAVIGATION SYSTEMS

### FIELDS OF STUDY

Aerospace Engineering; Space Technology

### SUMMARY

Many factors influence space travel and must be considered before spacecraft are launched. A global navigation satellite system (GNSS) is a network of artificial satellites (unmanned spacecraft that orbit celestial objects) that have been launched into orbit around Earth to help guide spacecraft (as well as provide location information to electronic devices such as smartphones). The global satellite system, which consists of the Global Positioning System (GPS) in the United States and GLONASS in Russia as well as other developing systems around the world, is used in conjunction with other technology by teams of scientists and engineers on the ground and by those aboard space vessels or by the vessels themselves.

### PRINCIPAL TERMS

- **astronomical unit:** a unit of length equal to the average distance between the Earth and the sun, about 150 million kilometers (93 million miles); often used to measure distances within the solar system.

- **GPS:** Global Positioning System, a system of US-owned navigation satellites in medium Earth orbit.
- **inertia:** the resistance a physical object displays to a change in speed or direction.
- **satellite navigation:** navigation, on Earth or in space, with the aid of satellites that pinpoint an object's position using radio signals.
- **trajectory:** the path a spacecraft follows in space.

## NAVIGATING SPACE

Since ancient times, people have looked to the stars to help them navigate from place to place on Earth. In the twenty-first century, much technology has been developed to aid navigation both on Earth and in space. Teams of scientists, engineers, and astronauts use sophisticated networks of systems that include antennas, computers, radios, satellites, and more to navigate vessels through space.

In addition, mathematical calculations are used to determine any potential conditions that would affect the course of a spacecraft as it travels through space. Mathematical models account for things such as interferences, gravity, and solar radiation pressure (radiation from objects in space). Even the smallest of forces can greatly alter a spacecraft's mapped path, or trajectory. Accurate mathematical models are crucial for accounting for every possibility that could occur while navigating.

On manned spacecraft, navigation is done both by astronauts on the craft and teams controlling navigation systems on the ground. For unmanned spacecraft, navigation happens solely on the ground. Timing is key for space navigation. Radio signals are sent back and forth between the spacecraft and Earth. On Earth, a team of scientists and engineers use these signals to track the location and speed of the spacecraft and make adjustments to the flight course as necessary. In space, astronauts are responsible for communicating with these teams on the ground, sometimes referred to as ground control. Some unmanned space vessels use software that enables them to navigate through space and transmit data such as images and trajectory information back to Earth, eliminating much of work on the part of ground control.

The National Aeronautics and Space Administration (NASA) Christopher C. Kraft, Jr. Mission Control Center, also known as Mission Control, Houston, is in charge of manned space missions and missions to the International Space Station (ISS). Since NASA's Space Shuttle program ended in 2011, Mission Control focuses much of its attention on ISS missions. Teams at Mission Control check on crews aboard the ISS and monitor spacecraft to ensure everything operates as planned. The teams can respond to any issues that develop over the course of missions.

Teams at NASA's Jet Propulsion Laboratory, which is located near Pasadena, California, are responsible for the direction of unmanned space voyages. Each team has specific roles in the navigation of spacecraft and can respond to any problems that arise. Before missions, the teams plot navigation courses, calculating the positions of space objects that could interfere with the spacecraft's trajectory. During the mission, the teams use large dish antennas, radio transmissions, and images to determine the location of a spacecraft. They also ensure the spacecraft is on the correct trajectory.

## OBSTACLES TO NAVIGATING SPACE

Many obstacles exist that affect space navigation. First, everything in space is in motion, which affects a vessel moving through space. Earth rotates, and it orbits around the sun. Destinations such as planets, moons, asteroids, and stars are constantly moving. This motion affects calculating and planning missions.

Distance is another factor that affects navigation. Spacecraft must often travel far from Earth to reach targets that are of scientific interest. This distance is measured in astronomical units (AU); 1 AU is 150 million kilometers (93 million miles). Mars is only about 0.52 AU from Earth on average, but Jupiter is 4.2 AU away and Neptune 29.09 AU. Distance must be precisely calculated and considered when planning space missions, as it affects factors such as fuel use and communications.

While the development of advanced technology has helped to make communication to and from Earth easier, communicating while in space still has limitations. Spacecraft do not have much power available for radio communication. Most vessels have solar panels that pull energy from the sun. However, spacecraft that travel far from the sun cannot rely on its energy and therefore do not have as much power. This means any radio signals sent to Earth may take as long as a few hours to reach those on the ground.

The most important factor that affects space navigation is gravity. Gravity from the sun, planets, moons, and other celestial objects has a pull on the trajectory of a spacecraft. While it can be used to propel a spacecraft to save fuel, it must be calculated and timed perfectly to avoid potential problems.

### Inertial Navigation Systems (INS)

Early spacecraft, such as those used in the Apollo missions in the 1960s, used inertial navigation systems (INS) to keep track of their location. INS uses accelerometers and gyroscopes, instruments that use the degree of inertia they display to a change in speed or direction to measure the amount of that change. A computer then uses these measurements to calculate the spacecraft's location relative to a known starting point. These systems required very little computing power. The Apollo Navigation and Guidance System had less computing power than a twenty-first century pocket calculator.

In the twenty-first century, INS is still used in spacecraft, though often in combination with other forms of navigation, such as satellite navigation. INS is particularly useful for satellites in low Earth orbit, for which satellite navigation cannot be used.

### Global Navigation Satellite System (GNSS)

People can use the stars to determine their position on Earth, but this system has its limits. It is not exact and depends on clear skies. Scientists instead developed satellites that can nearly pinpoint an object's precise position. The satellites send signals to a receiver, which determines the receiver's distance in relation to the satellite by measuring the time it takes for the signal to reach the receiver. The more satellites used, the better a position can be pinpointed.

Satellite navigation quickly became an important tool, especially in space. Scientists have placed navigational satellites in medium Earth orbit (MEO), which gave them stability and afforded exact orbit predictions. These systems, known as global navigation satellite systems (GNSS), relay their locations in space and time to networks of ground control stations, which have teams of receivers that calculate their exact positions. These systems are not always accurate, however, and signals can be influenced by a variety of factors, such as atmospheric changes and even tall buildings. GNSSs are not only used in space but also on the ground by drivers, law enforcement officials, energy response teams, telecommunication companies, surveyors, and more. These satellites can control air traffic, computer networks, and power grids.

As of 2015 there are two fully functional GNSSs: the Global Positioning System (GPS) of the United States and the GLObal NAvigation Satellite System (GLONASS) of Russia. GPS, also known as Navstar Global Positioning System, has thirty-two satellites in orbit, while GLONASS has twenty-four satellites in orbit. These satellites interact with a network of ground stations to provide information on position, time, and velocity anywhere in the world regardless of weather conditions.

Other satellite systems are in development and operate regionally, with plans to be globally operational in the future. These include Europe's Galileo, China's BeiDou (BDS), India's Indian Regional Navigation Satellite System (IRNSS), and Japan's Quasi-Zenith Satellite System. Once these systems become fully functional, they will give access to signals from more than one hundred satellites around the globe.

—*Angela Harmon*

### Further Reading

"About Satellite Navigation." *European Space Agency*. European Space Agency, 14 Jan. 2013. Web. 14 May 2015.

"GNSS (Global Navigation Satellite System) Definition." *TechTarget*. TechTarget, n.d. Web. 14 May 2015.

Groves, Paul D. *Principles of GNSS, Inertial, and Multisensor Integrated Navigation Systems*. 2nd ed. Norwood: Artech House, 2013. Print.

Iannotta, Ben. "More Nations Crave Independent Satellite Navigation Systems." *Space.com*. Purch, 16 May 2007. Web. 14 May 2015.

"Navigating in Space." *National Air and Space Museum*. Smithsonian Institution, n.d. Web. 14 May 2015.

Rao, Joe. "Navigating by the Stars." *Space.com*. Purch, 19 Sept. 2008. Web. 14 May 2015.

"Satellite Navigation—Global Positioning System (GPS)." *Federal Aviation Administration*. FAA, 13 Nov. 2014. Web. 14 May 2015.

United Nations. *Education Curriculum: Global Navigation Satellite Systems*. Vienna: Author, 2012. PDF file.

# SPACE ROBOTICS

## FIELDS OF STUDY

Aerospace Engineering; Orbital Mechanics; Space Technology

## SUMMARY

Human curiosity about outer space has led to the development of robotic technology that provides information and images of places far from Earth. Two types of space robotics are robotic spacecraft, such as satellites, and planetary rovers, which land on a planet and explore the terrain. Scientists are testing humanoid robots that can work in space and service robots that can refuel or repair existing spacecraft in orbit. This technology is important, as it allows humans to explore and work in space with less risk, including in areas beyond human capacity.

## PRINCIPAL TERMS

- **command and control interface:** communication link allowing humans at a ground station to remotely control robots.
- **machine vision and compensation:** computerized imaging system used in robotics to allow independent navigation and tracking of objects.
- **orbital robotics:** field that focuses on using robotic technology such as satellites, space probes, and service spacecraft and equipment in space.
- **planetary rover:** mobile robotic science lab sent to explore and study the surface of other planets.
- **power sources and recharging:** batteries and cells needed to run space robots, typically using solar power for electricity and recharging.
- **radiation hardening:** process of making electronic components resistant to radiation in space.
- **teleoperation:** remote operation and control of robotics.
- **thermal considerations:** how exposure to heat or cold affects components in spacecraft or robotics.

## ROBOTIC STRUCTURES AND SYSTEMS

Space robotics involves designing and building machines that can work in the adverse conditions and remote locations of outer space. To work effectively, all robots need a body, power, mobility, control, sensors, and whatever tools they require for their purpose. Robots can go where it is too dangerous or too far away for humans to travel. Robotics was a natural solution to the challenges of exploring the moon, the sun, and other planets.

Structures and systems that work on Earth were not designed for the extreme conditions of outer space. The harsh temperatures and other punishing environmental conditions beyond Earth's atmosphere require numerous adjustments to robotic systems. Scientists have to allow for size and weight limitations in addition to finding materials and processes that can survive the difficult conditions. Space robots require special power sources and recharging abilities. They usually run on batteries and solar power. Electronic parts must be prepared through a process called radiation hardening to protect them during exposure to radiation. Robotic structures and systems are tested and retested for durability and a long lifespan. Once a space robot is launched, it usually cannot be repaired.

The field of orbital robotics involves using unmanned technology and spacecraft, such as satellites, space probes, and service equipment, to get information about space. Planetary rovers land and explore where humans have yet to venture. They roll across rough terrain taking samples of soil and gases, testing them, and sending the results back to Earth. The National Aeronautics and Space Administration (NASA) has landed several rovers on Mars. The shape of the rovers was designed to move through the terrain and protect the systems so that the rovers can complete the work they were sent to do.

Space robotics includes the use of mechanical arms that can help astronauts complete tasks or make repairs. NASA has also designed humanoid robots called robonauts. The scientists made the first successful model in 2000 and they continue to improve them. Another project has involved building robotic spacecraft that could service orbiting craft and satellites, such as the International Space Station (ISS).

## EARLY ROBOTICS

Robots have a long history in the imagination of mankind, if not in reality. Leonardo da Vinci was among the early thinkers who imagined some sort of mechanical man. Yet until the middle of the twentieth century, the field of robotics was not advanced enough to provide machines for practical use. By 1975, mechanical arms were in use on industrial assembly lines. In the 1980s, computer-controlled walkers were tested in hazardous terrain. Technology continued to advance, and in the 1990s robots were designed to do more varied and independent tasks.

Since that time, NASA has been building and field testing what it calls "exploration systems" technologies. NASA's Computational Sciences Division at the Ames Research Center in Mountain View, California, builds and tests a variety of robots. Its testing sites use terrain similar to what might be found on the moon or Mars. The scientists can work with rovers and human-robot teams. They test such features as the command and control interface as well as teleoperation, through which the human controls the robot. Machine vision and compensation uses camera "eyes" and computer software to allow the rover to track interest points and choose a route. Machine vision was used on the K9 rover to navigate various types of terrain and avoid hazards in its path. Flying robots, such as Personal Satellite Assistants and the Yamaha rotorcraft *RMAX*, use machine vision for navigation and control in flight.

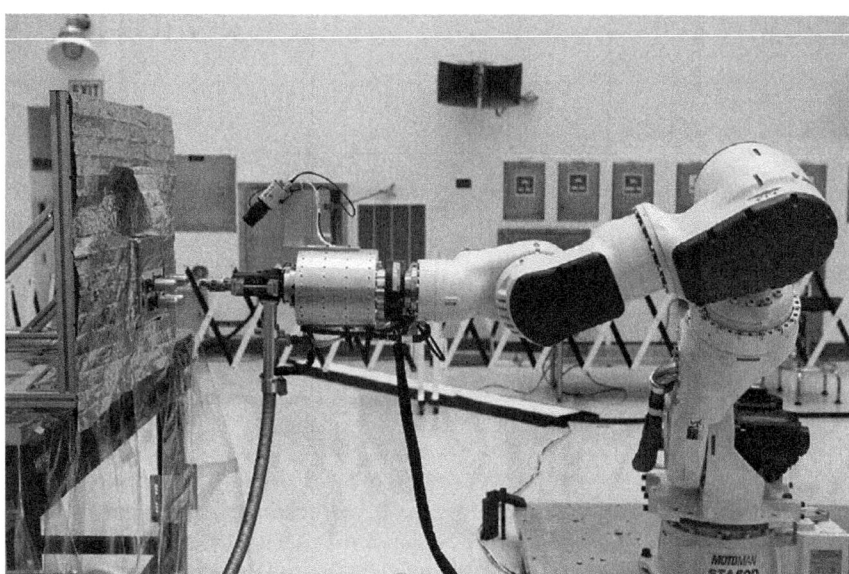

The RROxiTT industrial robot located at Kennedy Space Center in Florida, during a test run of remote operations conducted by a roboticist in Maryland. NASA/KSC.

## UNMANNED SPACE EXPLORATION

Once humans succeeded in sending astronauts into simple orbits around the Earth, they began to look farther into the galaxy. The moon was the first target for exploration. However, before scientists could plan a manned mission, they had to have more information. NASA sent robotic spacecraft, such as the *Surveyor* missions, to see whether it was feasible to send humans. *Surveyor 1* made a successful soft landing on the moon in 1966. It sent more than ten thousand photographs back from a remote-controlled camera. Later *Surveyor* missions included a robotic arm, which was able to pick up soil samples from the surface and analyze them for chemical elements. While the *Surveyor* robotics allowed for numerous important discoveries, the long-term purpose of the missions was to prepare to land a human on the moon. The *Surveyor* studies included testing the ability of the lunar surface to support the weight of a manned spacecraft. More than two years after its soft landing, *Surveyor 3* received a visit from American astronauts who had arrived on the *Apollo 12* lunar module.

## ORBITAL ROBOTICS

The initial use of robotics in space exploration was through satellites designed to fly by or orbit planets and send pictures back to Earth. To ensure that such missions reach their targets, scientists must allow for a variety of conditions and forces in space. They use mathematical formulas and algorithms, or predicted patterns, to design the robotics and each detail needed to put the satellite into the desired orbit. While scientists initially had some experience in launching rockets to deploy satellites, robotic technology was not well developed when the first orbiters were sent up. There were many challenges in working in the space environment. Scientists had to design electromechanical and control systems that would allow them to operate the machines at great distances. Thermal considerations affected the technology, which had to be made to withstand extremely high and low temperatures. Radiation hardening was needed to protect electronic components as the spacecraft circled distant objects in the solar system.

The *Ulysses* mission was launched from the space shuttle *Discovery* in 1990 to orbit the sun and gather information from positions that could not be seen from Earth. The robotic spacecraft carried ten instruments to measure and characterize cosmic radiation, energy particles, and the solar wind. Because scientists had foreseen how hostile conditions in space would be and designed the orbital robot accordingly, *Ulysses* orbited the sun and continued to send information to Earth for nearly twenty years. NASA scientists could observe and measure the solar wind for a long period of time, revealing that it was gradually dying down.

Humanoid, or human-like robots, are popular in books, films, and video games, but are rarely used in space. However, NASA has developed a humanoid model called Robonaut 2 to help astronauts. R2, as it is called, has a helmet-like head with cameras that function as eyes. It also has moving arms and hands with fingers that can hold and use tools like a person does. It can be attached to a stand for stationary work or mounted on wheels or legs for mobility. The scientists can control R2 remotely using teleoperation, or they can program it to handle simple tasks by itself. NASA sent one to the ISS to assist the astronauts and see if it would perform well in space. The robot has been capable of helping the astronauts with repairs, but it is more likely to be used for routine cleaning and upkeep. This can give the astronauts more time to complete their scientific studies. Because it is in orbit on the ISS, R2 experiences weightlessness, or microgravity. Microgravity locomotion—moving in low gravity—is a challenge for both humans and robots. NASA later sent a pair of climbing legs for R2 so the astronauts could see how well the robot could navigate the spacecraft in such conditions. With climbing legs, R2 could leave the spacecraft and work on the outside of the ISS. Astronauts would not have to risk their lives to make repairs.

**PLANETARY ROVERS**

The planet Mars has always attracted the interest of humans. Scientists imagine it to be a place that might support life in some form. After exploring the planet with orbiting robots, NASA expanded the Mars mission to include planetary rovers. The ultimate goal has been to send astronauts to Mars. However, until that becomes possible, robots on its surface have taken thousands of images and completed a variety of experiments there. This long-term project has three goals in addition to future exploration by humans. They are to learn about the climate, to examine the geology of the planet, and to determine whether there ever was life on Mars.

NASA has sent four unmanned robotic vehicles to the red planet to gather data from various areas. The first efforts at using planetary rovers were not very successful, but they helped scientists improve the design and increase the abilities and efficiency of the rovers. The first *Mars* rover, *Sojourner*, landed on the planet's surface in July of 1997. It was followed by *Spirit* and *Opportunity*, which both reached Mars in the winter of 2004. More than ten years later, *Opportunity* was still operating and had traveled more than 40 kilometers (25 miles). Later, *Curiosity* was able measure the relative humidity in the atmosphere of Mars, which ranged from 5 to 100 percent, depending on the season. Using its built-in testing instrument, the robot confirmed the identity of one of the minerals that had been observed through orbital robotics. It also made progress in the hunt for signs of life. *Curiosity* found evidence of a lake bed where the environment would have been favorable for the development of a simple life form.

Although Venus is the closest planet to Earth, its extremely hot temperature has made it difficult to land robotic spacecraft on its surface. Russia has landed several *Venera* probes on Venus. However, they have only been able to transmit for a few hours at most. As of early 2015, NASA was working on a new type of rover that would hopefully be able to withstand the harsh conditions for longer.

—*Nancy Comstock*

**FURTHER READING**

"A Foundation for Robotics." *National Science Foundation.* Natl. Science Foundation, n.d. Web. 22 May 2015.

"Machine Vision for Robotics." *NASA.* NASA, 29 Mar. 2008. Web. 22 May 2015.

"NASA's Curiosity Mars Rover Finds Mineral Match." *NASA.* NASA, 4 Nov. 2014. Web. 22 May 2015.

Paat-Dahlstrom, Emeline. "A Giant Leap for Curiosity, a Small Step for Robot-Kind." *Forbes.* Forbes.com, 25 July 2012. Web. 22 May 2015.

"Robot Systems." *Rover Ranch*. NASA, 3 Apr. 2003. Web. 22 May 2015.

"The Surveyor Program." *Lunar and Planetary Institute*. Lunar and Planetary Inst., n.d. Web. 22 May 2015.

"Ulysses." *NASA*. NASA, n.d. Web. 22 May 2015.

# SPACE TRANSPORTATION SYSTEMS

### FIELDS OF STUDY

Aerospace Engineering; Spacecraft Propulsion; Space Technology

### SUMMARY

Space Transportation System program focused on advancing space exploration through a piloted vehicle that would allow reliable, low-cost, and regular travel into space. This goal led to the development of the space shuttle, a partly reusable orbital spacecraft. Over the course of many decades, various space shuttles successfully performed numerous significant space missions. However, the overall success of the program has been debated.

### PRINCIPAL TERMS

- **external tank:** a container that held fuel in the form of liquid hydrogen and an oxidizer in the form of liquid oxygen. The tank supplied the pressurized fuel and oxidizer to the three main engines of the space shuttle during lift-off and ascent.
- **Orbital Maneuvering System:** a system that provided the thrust needed for orbit insertion, transfer, and deorbit via two smaller engines.
- **propellant:** a chemical mixture contained in the space shuttle's solid rocket boosters. When ignited, it helped thrust the shuttle into space.
- **solid rocket boosters:** two solid-propellant motors that provided the space shuttle with the main thrust needed to lift it off the launch pad and into space against the pull of Earth's gravity.

### A Brief History

In 1969, with Project Apollo in full swing, US president Richard Nixon set up the Space Task Group to study the United States' future in space exploration. Sending a human to the moon was only the beginning. The group envisioned a space program that continued the exploration of space with piloted flights. This would benefit humankind through advances in communications, science, technology, navigation, and more. The team felt that by expanding humans' understanding of the universe, the US space program had the potential to improve the quality of life on Earth.

Because of the success of Project Apollo, the Space Task Group found that the National Aeronautics and Space Administration (NASA) should accept a piloted mission to Mars as a key goal for the space program. They believed that NASA would be able to carry out such a program within fifteen years. Reliable, long-lived, yet low-cost and simpler operational space systems with flexible flight configurations would make this possible. The Space Task Group supported continuing any future robotic space exploration. However, the group proposed that these activities be carried out with the piloted Mars mission in mind for optimal benefit.

The Space Transportation System (STS) was the official name for NASA's space shuttle program. The name was taken from the Space Task Group's plan for a system of interrelated, reusable, multipurpose spacecraft. The objective of the STS was to offer NASA an efficient, reusable method of sending astronauts to a permanently piloted space station. Following the Apollo lunar landings, it was important to guarantee a permanent US presence in space. Space shuttles could also be used as satellite delivery vehicles. As a partly reusable, low-Earth orbital spacecraft system, the space shuttle was the only item in the plan that was eventually funded for development. Nixon announced the beginning of NASA's development of the space shuttle system in January 1972.

### Developing a Space Airplane

NASA's space shuttle system differed from the rocket of the *Apollo* missions. More technologically advanced, the space shuttle combined attributes of the earlier, rocket-based spacecraft with the transportation qualities of a conventional airplane. The launch configuration and thrust mechanism of the shuttle were also different and more controlled. It was the first winged US spacecraft that could make a horizontal landing on a runway upon its return to Earth.

NASA had initially envisioned a reusable spacecraft that would be operated in two stages. Two piloted winged crafts would have been launched together. The smaller craft would have served as a booster to bring the second craft into orbit. The smaller craft would have then disconnected and landed back on Earth. Due to budget issues, however, NASA ultimately designed the three-part shuttle system used in the missions that would span several decades. Its main components were an orbiter, which held the crew; an external tank (ET), which contained fuel for the main engines; and two solid rocket boosters (SRBs), which gave the shuttle most of its thrust during the beginning of flight. With the exception of the ET, which was disposed of after the launch and burned up in Earth's atmosphere, all of the components were reused.

The shuttle was the first operational orbital spacecraft that was designed to be reused. It carried payloads (cargo) into orbit, provided the International Space Station (ISS) with crew rotation and supplies, and performed service and repair on satellites. The space shuttle also recovered payloads such as satellites from orbit and carried them back to Earth.

## Attributes of the Space Shuttle

The space shuttle was designed to transport cargo into a low orbit about 304 to 528 kilometers (190 to 330 miles) above Earth. The payload was held in a cargo bay measuring 4.6 meters (15 feet) wide and 18.3 meters (60 feet) long. The orbiter could carry a crew of eight people, with the capability of transporting a total of ten people under emergency conditions.

A pair of SRBs provided about 71.4 percent of the thrust the space shuttle needed to lift off the pad. The SRBs were the largest solid-propellant motors to ever be flown. Each booster measured 45.5 meters (149 feet) long and 3.7 meters (12 feet) in diameter. They were also the first solid-propellant motors designed for reuse. The propellant in the SRBs was a mixture of ammonium perchlorate (an oxidizer), aluminum (as fuel), iron oxide (a catalyst), and a polymer to bind the mixture together. This mixture provided a high thrust upon ignition, followed by a reduction in thrust after lift-off. This prevented excessive stress on the vehicle under dynamic pressure. The SRBs

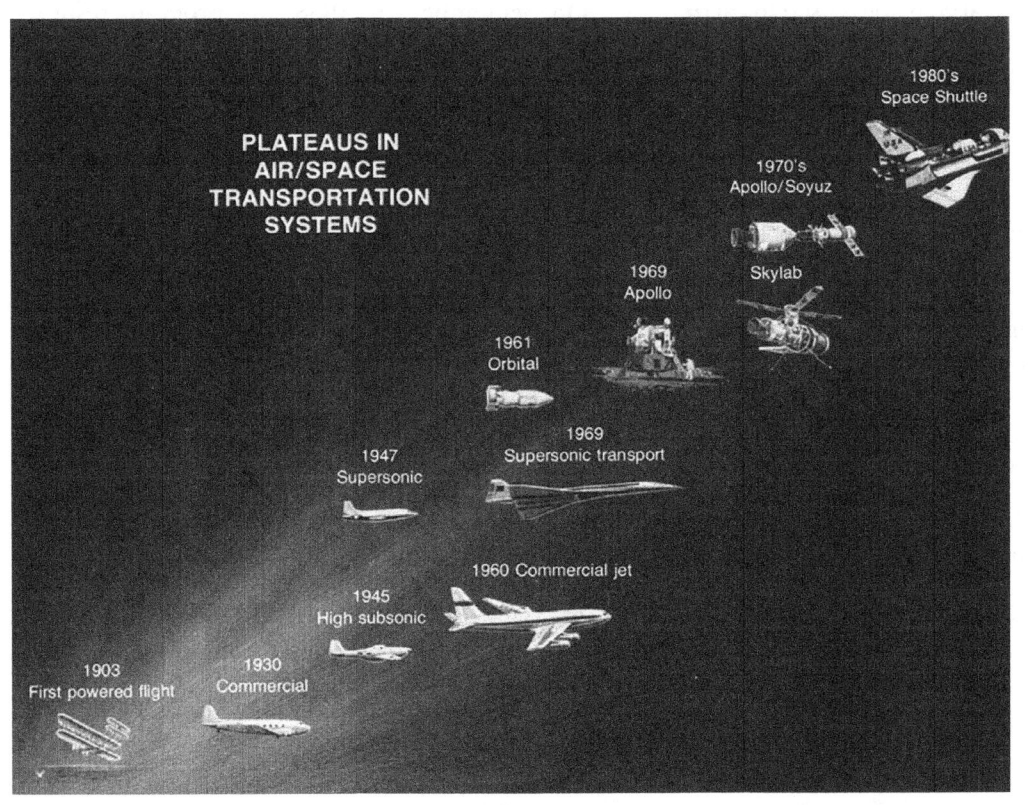

Plateaus in air/space transportation systems. An illustrated timeline of space transportation systems. NASA.

were ignited after the verification of the thrust level of the three main engines. The two SRBs also supported the weight of the ET and the orbiter. The boosters were spent in about two minutes and separated from the ET.

The ET contained liquid hydrogen fuel and liquid oxygen oxidizer. This pressurized propellant was supplied to the three main engines of the space shuttle inside the orbiter during lift-off and ascent. The SRBs were later recovered from predetermined points in the ocean so they could be reused on future space flights. Right before the spacecraft was inserted into orbit, the space shuttle's main engines shut down and the ET was discarded into Earth's atmosphere. It broke apart upon reentering Earth's atmosphere and was not recovered. Once the orbiter was clear of the ET, the Orbital Maneuvering System (OMS) engines were initiated. The OMS provided the thrust needed to insert the orbiter into orbit, maneuver once in orbit, and reduce speed upon reentry.

Another unique feature of the space shuttle was a large cargo bay that had doors that opened at the top of the spacecraft. This cargo bay was used to carry the orbiter's payload. The cargo door configuration allowed for the deployment of large satellites, including the Hubble Space Telescope. The cargo door also allowed for payloads to be captured and returned to Earth. The orbiter had an average payload capacity of 22,700 kilograms (50,045 pounds). This capacity could vary depending on the launch configuration.

### Successes and Tragedies

After years of delays, the first orbital test mission began in 1981 with the historic launch of the shuttle *Columbia* on April 12 from Kennedy Space Center (KSC). It returned safely fifty-four hours later at Edwards Air Force Base. This successful mission proved that the shuttle could fly into orbit, remain in orbit while conducting operations, and return to Earth safely. In 1981 and 1982, this shuttle engaged in three more test flights. The final orbital test flight ended on July 4, 1982, with *Columbia* completing 95 percent of its objectives and the first scientific experiment of the shuttle program. US president Ronald Reagan declared that *Columbia* would be fully operational for its next flight. The space shuttle program's first operational flight began at KSC on November 11, 1982.

From 1982 to 1985, additional space shuttles were added to the program's fleet—*Challenger*, *Atlantis*, and *Discovery*. The shuttles ascended from the launch pad at KSC an average of four or five times annually. Most of the missions ended at Edwards Air Force Base. The program seemed to be a huge success.

On January 28, 1986, *Challenger* broke apart seventy-three seconds after launch. Americans watched in horror as the shuttle exploded. There were no survivors. The program was put on hold for more than two years while a commission formed by President Reagan investigated the accident. It was determined that the failure of an O ring in the joint between two segments of the SRBs had caused the disaster. The commission also found numerous management failures within NASA. Many managers were aware of the issue but did not take proper precautions. A review revealed that there was pressure to announce that the shuttle was operational and that this had led to system resources being stretched too far. The tragedy of the Challenger accident led to many tangible changes within NASA and the shuttle program.

On September 29, 1988, the program resumed flights with the launch of *Discovery*. For the next several years, the program was successful. Flights were more productive in terms of payload, while the number of missions increased to an average of six per year. On July 30, 1991, US president George H. W. Bush and Mikhail Gorbachev, leader of the falling Soviet Union, agreed that their space agencies would work together. In February of 1994, a joint program commenced. The first Russian flew on a space shuttle that month, the first American went aboard the Russian Mir space station in 1995, and the space shuttle *Atlantis* docked at the station that same year. Millions of people around the world breathed a collective sigh of relief—the Cold War was officially over. Russia and the United States would go on to collaborate on other important projects that involved the space shuttles, including the International Space Station (ISS).

The shuttle program lost its second vehicle in February 2003. Sixteen minutes before its scheduled landing, *Columbia* disintegrated over Texas; all seven crew members died. In 2004, US president George W. Bush declared that all shuttles in NASA's STS would be retired. *Atlantis*, *Discovery*, and *Endeavor* (the replacement for *Challenger*) flew regularly from 2006 to 2011, especially to finish putting the ISS together.

Atlantis touched down for the final time at KSC on July 21, 2011, marking the end of the program. *Discovery* was installed at the Smithsonian National Air and Space Museum in 2012.

#### SIGNIFICANCE

Piloted flights using the shuttle led to many advances in science and technology. Astronauts on *Discovery* placed the Hubble Space Telescope into orbit. Satellites placed into orbit by shuttles have enhanced national security. In addition, the shuttles were responsible for carrying remaining pieces and necessary equipment to the ISS. This station has fostered international cooperation and allowed humans to live and work in space for lengthy periods of time. With the entire program reportedly costing billions of dollars, however, some critics saw the shuttle program as an unnecessary loss of life and a disappointing, limiting distraction. They also argued that it was a poor use of resources based on a design that was unrealistic.

—Michael Mazzei

#### FURTHER READING

Archaeological Consultants. *NASA-Wide Survey and Evaluation of Historic Facilities in the Context of the US Space Shuttle Program: Roll-Up Report.* Sarasota: Archaeological Consultants, 2008. PDF file.

Dowling, Stephen. "What Caused the Space Shuttle Columbia Disaster?" *BBC Future.* BBC, 30 Jan. 2015. Web. 1 July 2015.

Mosher, Dave. "Space Shuttle in Extreme Detail: Exclusive New Pictures." *National Geographic.* Natl. Geographic Soc., 16 Apr. 2012. Web. 1 July 2015.

"Overview." *NASA.* NASA, 7 Apr. 2002. Web. 29 May 2015.

"Report of the Space Task Group, 1969." *NASA Headquarters.* NASA, n.d. Web. 26 June 2015.

"Space Shuttle Basics: Solid Rocket Boosters." *Human Space Flight.* NASA, 22 Oct. 2002. Web. 26 June 2015.

"Space Shuttle Basics: Space Shuttle History." *Human Space Flight.* NASA, 27 Feb. 2008. Web. 26 June 2015.

"Space Shuttle Program." *National Geographic.* Natl. Geographic Soc., 2015. Web. 26 June 2015.

"Space Shuttle Program Fast Facts." *CNN.* Cable News Network, 4 Aug. 2014. Web. 26 June 2015.

Zeeberg, Amos. "How to Avoid Repeating the Debacle That Was the Space Shuttle." *Discover.* Kalmbach, 22 July 2011. Web. 26 June 2015.

# SPACE-TIME PROPERTIES

#### FIELDS OF STUDY

Astrophysics; Theoretical Astronomy

#### SUMMARY

Although time and space are thought of as being the same for everyone, science has proven that they are not. Each is dependent on the perspective of the observer, a concept known as the theory of relativity. This theory claims that only space-time—the sum of all events—is an absolute, not the individual parts. This is important because all the laws of physics that relate to time and space are subject to this relativity.

#### PRINCIPAL TERMS

- **general relativity:** the theory that gravity is the result of matter causing space-time to curve; one of Albert Einstein's two theories of relativity, the other being special relativity.
- **gravity:** the fundamental force that causes objects with mass to attract one another.

#### MAKING SPACE AND TIME RELATIVE

Until the early twentieth century, it was generally believed that time and space are the same for everyone. In 1905, Albert Einstein (1879–1955) proposed the theory of special relativity. This theory states that the laws of physics remain the same in all frames of reference that are not experiencing acceleration, whether

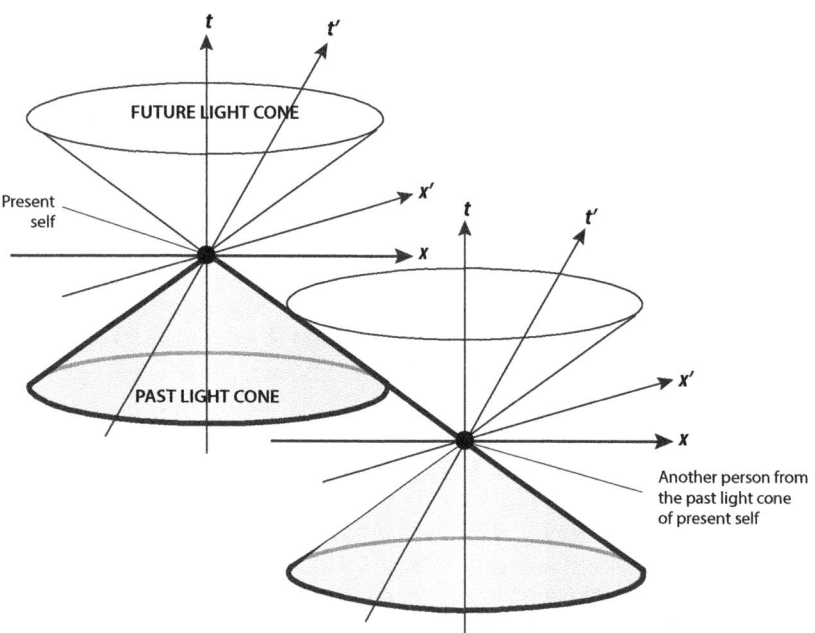

A model of space-time shows the relationship between present self and the space-time of another person from one's present self's past. The cone surfaces defined in space (x, x ') and time (t, t ') dimensions shows the extension of light, or existence, in the past and future, extending from the pres-ent self's existence. Another person will have their own future and past light surfaces.

is constant, at 299,792,458 meters per second (about 983,571,088 feet per second). Light appears the same to all observers, regardless of their individual speeds. Nothing travels faster than the speed of light.

The constancy of light speed provides a way to determine whether any one event could be the cause of another. To conclude that an event at space-time point A caused an event at space-time point B, one must prove that light could have traveled from point A to point B in the time that elapsed between the two events. Because nothing travels faster than the speed of light, if the two points are too far apart or the events happened too quickly for light from point A to have reached point B in time, it is impossible for event A to have caused event B.

Another property of space-time is that time passes more slowly for moving objects or objects of greater mass than it does for motionless objects and those of lesser mass. For instance, a moving clock will measure time more slowly than a clock that is standing still. At everyday speeds, however, the difference is so infinitesimal as to be nearly nonexistent. It only truly becomes evident at speeds approaching the speed of light. This phenomenon is often illustrated with the so-called twin paradox. If twins were separated so that one twin traveled in a spaceship at near light speed while the other remained on Earth, both would feel time passing normally. However, when the first twin returned from space, he or she would be much younger than the second twin, because time moved more slowly on the spaceship than on Earth.

they are stationary or moving at a constant speed, while the speed of light remains constant in any and all frames of reference. For this to be true, the passage of time must depend on the position of the observer with respect to the observed. Put simply, the faster somebody is moving, the slower time passes for him or her. Thus, while space-time—the continuum of actual events—is an absolute, time and space themselves are relative and dependent on the observer.

In 1915, Einstein published his theory of general relativity, which is based on his theory of special relativity. General relativity states that gravity is caused by massive objects distorting space-time and causing it to curve. According to this theory, space-time is basically flat unless energy or mass disrupts it.

## Properties of Space-Time

Under Einstein's theory, space-time is not only absolute, it also operates under a causal structure. By examining space-time, it is possible to determine what events could or could not cause or influence others. The determining factor for this would be the speed of light, or the speed at which electromagnetic radiation can move through a vacuum. The speed of light

## Additional Space-Time Concepts

Other scientists, both before and after Einstein, have proposed theories tied to the concepts of relativity and space-time. In 1898, French mathematician Henri Poincaré (1854–1912) proposed that time intervals might depend on the observer. In 1904, one year before Einstein proposed his theory of special relativity, Poincaré suggested that the speed of light could be unsurpassable.

In 1908, one of Einstein's former professors, German mathematician Hermann Minkowski (1864–1909), described space and time as constituting a four-dimensional space-time fabric. Minkowski's four-dimensional fabric is often portrayed as two light cones stacked vertically, tip to tip. The top cone represents future events; the bottom cone, the past. The point at which the tips of the two cones meet is the present moment. A horizontal plane at this meeting point represents space, and a vertical line drawn through the point represents time. The cones' walls represent light and provide a reference point for the speed of light. These cones are used to conceptualize and plot past and future events and determine whether other events can be causal or not.

## Proof of Einstein's General Relativity Theory

In the years since Einstein proposed his theory of relativity, scientists have continued to test and prove its validity. In 2011, a National Aeronautics and Space Administration (NASA) mission called Gravity Probe B (GP-B) provided additional proof.

Einstein had theorized that Earth was sitting on top of a plane of space-time, pressing down on it much like a person sitting on the middle of a bed. He predicted that gravity would make objects follow the lines formed by the Earth pressing down into the space-time fabric. Because the Earth rotates, it should cause corresponding twists in space-time, creating a space-time vortex.

To test this, NASA launched the GP-B spacecraft with a gyroscope aimed at a fixed point in the universe. In theory, without gravity or other factors to affect it, the gyroscope would continue to point in the same direction indefinitely. If Einstein's prediction of a space-time fabric twisted by the pressure of the Earth were accurate, however, the gyroscope would slowly change direction. The gyroscope did change direction, proving Einstein's theory. This also meant that scientists can measure the twist in space-time by measuring the change in direction.

NASA experiments proved Einstein right in another way as well. He had predicted that space-time would be smooth except for where it was interrupted by energy and matter. Other scientists argued that space-time consisted of many very small particles that would create a foamy or frothy surface. In 2009, NASA's Fermi Gamma-Ray Space Telescope captured three photons from a gamma-ray burst more than seven billion light-years away. Despite their long journey, the three photons arrived just millionths of a second apart. Scientists think this near-simultaneous arrival is additional proof of Einstein's theory that space-time is smooth. Photons are so tiny that even the smallest bits of matter in their path would have been enough to jostle them off course and delay them.

## Effects of Space-Time

The accuracy of the space-time theory is proved not only in the depths of space but on Earth as well. The Global Positioning System (GPS) that provides directions and locations for so many things on Earth, from military operations to cell phones, also experiences the effects of time dilation. GPS works by sending out synchronized signals to various receivers on Earth. The amount of time it takes the signal to reach Earth pinpoints the location of the object, so the accuracy of the timepieces in these GPS satellites is maintained to the nanosecond (one billionth of a second). But these orbiters are located about 20,300 kilometers (about 12,600 miles) above Earth and travel at about 10,000 kilometers (about 6,213 miles) per hour. At that distance and speed, and with gravity factored in, they lose about seven thousand nanoseconds per day to time dilation—enough to cause an error of 8 kilometers (5 miles) in GPS directions after only one day. Thus, the satellites must correct for time dilation in order to remain accurate.

Space-time properties are part of everyday life, whether one is aware of them or not. Future study of space-time is expected to provide new ways to not just adapt to space-time properties but use them to the advantage of humankind, in space travel and in everyday life.

—*Janine Ungvarsky*

## Further Reading

Dambeck, Susanne. "Einstein: Travelling through Spacetime." *Lindau Nobel Laureate Meetings*. Lindau Nobel Laureate Meetings, 17 Apr. 2015. Web. 5 May 2015.

"Einstein Was Right: Space-Time Is Smooth, Not Foamy." *Space.com*. Purch, 10 Jan. 2013. Web. 5 May 2015.

Emspak, Jesse. "Eight Ways You Can See Einstein's Theory of Relativity in Real Life." *LiveScience.* Purch, 26 Nov. 2014. Web. 5 May 2015.

Overduin, James. "Einstein's Spacetime." *Gravity Probe B: Testing Einstein's Universe.* Stanford U, Nov. 2007. Web. 5 May 2015.

Phillips, Tony. "NASA Announces Results of Epic Space-Time Experiment." *NASA Science.* NASA, 4 May 2011. Web. 5 May 2015.

Pössel, Markus. "Elementary Einstein." *Einstein Online.* Max Planck Inst. for Gravitational Physics, n.d. Web. 5 May 2015.

"Speed of Light." *Cosmos: The SAO Encyclopedia of Astronomy.* Swinburne University of Technology, n.d. Web. 25 Apr. 2015.

# SPUTNIK SPACE PROGRAM

## FIELDS OF STUDY

Space Technology; Aerospace Engineering; Orbital Mechanics

## SUMMARY

The Sputnik space program was the crucial first space program of the Soviet Union. It successfully launched *Sputnik I*, the first man-made satellite to orbit Earth. The program gathered important data about Earth's atmosphere and played a significant role in the Cold War.

## PRINCIPAL TERMS

- **Object D:** early designation of *Sputnik 3*, designed by the Soviets in 1956 and intended to be the first satellite sent into space. Because of its complicated design, it was not finished in time for the first launch.
- **R-7 rocket:** the Soviet-designed rocket that launched Sputnik 1 into space. Versions of the rocket have continued to be used to send craft into space.
- **satellite:** technically, any small object, either natural or artificial, that orbits around a larger object. However, the term is most often applied to man-made objects that are placed in orbit around Earth or another celestial body for research, monitoring, or communication purposes.
- **Soviet Union:** also known as the Union of Soviet Socialist Republics (USSR), established in 1922 in the aftermath of the Russian Revolution in 1917. By its end in 1991, the union was made up of fifteen republics.

## First Step into Space

On October 4, 1957, a small metal object was shot into space and into history. *Sputnik 1*, a silver, beach-ball-sized satellite with four long, thin antennae on one side, was launched into space by the Soviet Union. Launched from the Baikonur Cosmodrome with the help of an R-7 rocket, the 83.6-kilogram (about 184 pounds) *Sputnik 1* satellite orbited Earth once every ninety-eight minutes. Its distinctive signal—a beep about three-tenths of a second long, followed by an equal silence in a repeating cycle—was transmitted for about three weeks before *Sputnik 1's* chemical batteries ran out and the satellite lapsed into silence. It remained in orbit for 1,440 passes around Earth before its orbit began to decay ninety-two days after it was launched. The satellite burned up in the atmosphere on January 4, 1958. *Sputnik 1* was the first artificial satellite to orbit Earth, but it was not the last. A man-made object has been in geocentric orbit almost consistently since its launch.

In addition to making the Soviets the first into space, *Sputnik 1* had other scientific purposes as well. As it orbited, the satellite gathered information about the thickness of the upper layers of Earth's atmosphere and the effect of the ionosphere on the transmission of radio waves.

After the end of World War II and the beginning of the Cold War, the United States and the Soviet Union engaged in a parallel battle to prove superiority in space. The *Sputnik* program and the launch of *Sputnik 1* were spurred by this tense incentive as well as by an initiative from the International Council of Scientific Unions issued in 1952. The union knew that solar activity cycles would be at a peak between July 1, 1957, and December 31, 1958, and it designated that time frame as the International Geophysical Year (IGY).

A related resolution in October 1954 encouraged the launch of artificial satellites during the IGY for Earth-mapping purposes. The United States moved forward with its *Vanguard* satellite program in 1955. However, the Soviets surprised much of the world when *Sputnik 1* won the race into space.

During the time it orbited Earth, *Sputnik 1* was observable with the naked eye as a blip of light in the sky. Its signal could also be heard indirectly on radio and television broadcasts. Ham-radio operators were able to hear it live while it was in orbit overhead.

### The Sputnik Program Continues

*Sputnik 1* was the first in a series of *Sputnik* missions launched by the Soviet Union between 1957 and 1961. While *Sputnik 1* was still orbiting and the world was continuing to absorb the significance of this feat, the Soviets followed up on November 3, 1957, with the launch of *Sputnik 2*. This was a heavier, larger unit. It also carried the first animal into space, a dog named Laika. Along with the equipment necessary for data transmission and orbiting, *Sputnik 2* included food, water, a camera and other monitoring equipment, and a waste-collection system. Neither the craft nor its passenger returned to Earth. However, the mission provided information on the radiation experienced in space and helped scientists learn the effects of space travel on a life-form.

The next eight *Sputnik* missions were launched between 1958 and 1961. Design of the heavy satellite that flew as *Sputnik 3* actually began in 1956 with the goal of launching it as the first satellite in space. However, the building process for the craft, known during planning phases as Object D, took longer than planned. To ensure victory over the United States, the Soviets launched *Sputnik 1*, a simpler satellite, instead. The next missions included instruments aimed at determining the parameters of a space vehicle and equipment that would be needed for manned space flight. Because of this focus, the Soviets began naming the crafts *Korabl-Sputnik* ("spaceship satellite") to distinguish them as tests for future manned vehicles. *Korabl-Sputnik 2* carried two dogs, Belka and Strelka, as well as the usual scientific instruments. It was the first space vehicle to successfully return to Earth. *Korabl-Sputnik 4* carried both a canine passenger named Chernushka and a dummy astronaut. It, too, made a successful reentry. All three dogs were recovered unharmed.

### Sputnik Leads to Another First for the Soviet Union

The information gathered from the Sputnik missions paved the way for the manned missions on the various *Vostok* spacecraft. *Vostok 1* carried the first human being into space. Yuri A. Gagarin (1934–68), a twenty-seven-year-old fighter pilot, was launched into space on April 12, 1961. His spherical craft was equipped with external radio antennae. It had two main compartments. One was a service module that housed the support systems, orientation rockets, and chemical batteries. The other was the cabin that housed Gagarin. This section included communications equipment, instruments for data review and recording, a life-support system, and an ejection seat. The two sections were set to separate before the

*Sputnik 1* replica. By NSSDC, NASA, public domain via Wikimedia Commons.

manned capsule returned to Earth. Because scientists had no way to know for sure how Gagarin would be affected by weightlessness, *Vostok 1* was designed to be piloted from Earth. The onboard flight controls could only be operated with the use of a key kept in a sealed envelope for emergencies.

After one orbit lasting one hour and forty-eight minutes, the body of the spacecraft returned to Earth. Gagarin ejected when he was about seven kilometers (a little more than four miles) above Earth and returned via a parachute. The mission was captured by both radio and television transmissions from space. It preceded the first flight to put an American into space by less than one month.

### Significance of the Sputnik Program

The Sputnik program accomplished a number of scientific goals, including the first orbits of Earth by both a machine and a human being. The various missions gathered vast amounts of information about Earth and its atmosphere. These were the first observations and studies of space to be undertaken from space itself.

The importance of the program went beyond its scientific accomplishments. The Soviets took the early lead with *Sputnik 1* and the first manned flight into space. Driven by both the desire to achieve technological superiority and concerns about the effects of Soviet satellites in Earth orbit while relations between the two countries remained tense, the United States created the National Aeronautics and Space Administration (NASA) in 1958. A short time later, President John F. Kennedy (1917–63) issued a challenge to the fledgling space agency when he promised that the United States would put a man on the moon by the end of the decade.

The Soviets struggled with their lunar program while the American program experienced many successes, including the first manned moon landing on July 20, 1969. The period between the launch of *Sputnik 1* in 1957 and the United States' *Apollo 11* lunar landing in 1969 is often referred to as the "space race."

As the Cold War died down, interest began to grow in pooling knowledge and resources between the two superpowers. The joint *Soyuz-Apollo* mission in 1975 involved spacecraft from the two countries docking with each other. The crews of each shook hands in space. This was a precursor to the joint missions that would occur once the long-term Mir space station was launched.

While the Soviets' early success in space caught much of the world off guard and created concern at a time when tensions were high between the two superpowers, some identified another important factor of the accomplishment. Dwight D. Eisenhower (1890–1969), president of the United Sates at the time of the *Sputnik 1* launch, knew that the Soviets' efforts were establishing the idea of "freedom of space." Air space over a country is considered to belong to that country and is monitored, patrolled, and protected by that country. Outer space, however, belonged to everyone and no one. Satellites of one country could freely orbit over another country. Without widespread acceptance of this concept, the orbit of satellites would have been restricted and made less effective or even impossible.

In the years that followed, the United States and the country formerly known as the Soviet Union collaborated on a number of projects in space, including the International Space Station. The technology that made these twenty-first-century accomplishments possible began with the first tentative steps into space that led to the Sputnik 1 launch.

—*Janine Ungvarsky*

### Further Reading

Dickson, Paul. "Sputnik's Impact on America." *Nova.* WGBH, 6 Nov. 2007. Web. 4 June 2015.

Hall, Rex, and David J. Shayler. *The Rocket Men: Vostok & Voskhod, the First Soviet Manned Spaceflights.* New York: Springer, 2001. Print.

Kluger, Jeffrey. "Gagarin's Golden Anniversary: The High Price Paid by the First Man in Space." *Time.* Time, 12 Apr. 2011. Web. 4 June 2015.

Launius, Roger D. "It All Started with Sputnik." *Air & Space.* Smithsonian Inst., July 2007. Web. 4 June 2015.

"The Space Race." *History.com.* A&E Television Networks, n.d. Web. 4 June 2015.

"Sputnik and the Dawn of the Space Age." *NASA History Program Office.* NASA, 10 Oct. 2007. Web. 4 June 2015.

"Sputnik 1." *National Space Science Data Coordinated Archive.* NASA, 26 Aug. 2014. Web. 4 June 2015.

# STAR STRUCTURE

## FIELDS OF STUDY

Stellar Astronomy; Astronomy; Cosmology

## SUMMARY

The structure of a star is dependent on its stage of life. Stars are formed from compressed clouds of gas in outer space. Early stars, called protostars, are simpler than their older counterparts. Most stars contain a hydrogen or helium core, surrounded by a radiative zone and a convective zone. Above the convective zone are the photosphere, or surface of the star; the chromosphere, or the star's atmosphere; and the corona, an area of plasma found around the outside of the star.

## PRINCIPAL TERMS

- **chromosphere:** the middle layer of a star's atmosphere.
- **convection:** the transfer of heat via the movement of molecules in a fluid.
- **core:** the dense, hot center of a star, where heat is normally generated.
- **corona:** the outermost layer of a star's atmosphere, a zone of extremely low-density plasma that is several million degrees hotter than the surface of the star.
- **main sequence:** in a Hertzsprung-Russell diagram, which plots the temperature/color of stars against their luminosity/brightness, the band of stars that crosses from the upper left to the bottom right, representing the evolution of a "normal" star over the course of its lifetime.
- **photosphere:** the effective surface of a star and the innermost layer of its atmosphere.
- **radiative zone:** the layer of a star's interior in which heat is transferred from the core toward the surface via radiative diffusion, or the absorption and reemission of photons.

## YOUNG STARS

Stars begin their lives as molecular clouds, or giant interstellar clouds of dust and gas, usually helium or hydrogen. The cloud, which accumulates ever more matter, eventually becomes too dense and massive to resist its own strong gravity. At that point, the cloud collapses and breaks into a number of smaller, though still massive, clumps. These clumps continue to collapse, becoming even smaller and denser. As their densities increase, they heat up, becoming protostars.

Protostars are one of the simplest and most common types of star. They are relatively small and normally glow shades of red. Protostars are surrounded by disks of gas and other material pulled in by their strong gravity; this influx of matter causes the fledgling stars to grow. Unlike most stars, protostars do not undergo nuclear reactions in their core. Instead, they use convection to move heat from the core to the photosphere, the surface of the star. Some protostars never progress beyond this stage. If they fail to accumulate enough matter to begin a nuclear reaction in their core, they become brown dwarfs, a type of failed star. Brown dwarfs do not produce energy on their own. Instead, they slowly radiate away their energy as light and heat. When all of their energy has been expended, they become dead stars. However, sometimes protostars manage to generate so much heat that they begin fusing the hydrogen atoms in their core into helium. This creates a main-sequence star, which has a layered internal structure.

## THE MAIN SEQUENCE

The main sequence is one of the most stable periods of a star's life. Main-sequence stars are perfectly balanced in a very important way. The gravity generated by their mass does not cause them to collapse because it is countered by thermal energy produced by the nuclear reactions in their cores. In a main-sequence star, these two forces—gravity and thermal energy—are very close to equal. This balance allows main-sequence stars to stay the same size for as long as their cores continue to burn. Because smaller main-sequence stars do not burn through their fuel as fast as larger ones, the smaller stars survive longer. Earth's sun is a relatively average main-sequence star.

The structure of a main-sequence star is more complex than that of a protostar. All main-sequence stars undergo nuclear reactions in their cores, constantly fusing hydrogen into helium to generate thermal energy. Beyond the core, the star's interior structure varies depending on its mass. If the star is between about 0.5 and 1.5 solar masses (that is, 0.5–1.5 times the mass of the sun), the core is surrounded by a tightly packed layer known as the radiative zone, in which thermal energy is transported toward the surface via radiative diffusion (the absorption and reemission of photons). That layer is surrounded by a less dense layer called the convective zone, in which thermal energy is transported via convection. If the star is less than 0.5 solar masses, however, it has no radiative zone; its interior is entirely convective. If it is more than 1.5 solar masses, the two layers are reversed, with the convective zone inside the radiative zone.

Next is the star's outer layer, the photosphere, where the thermal energy from the core is emitted in the form of light. The photosphere is considered the lowest layer of the star's atmosphere. A stellar atmosphere, by definition, comprises the layers from which photons (that is, light) can emerge. The next atmospheric layer is the chromosphere, followed by the corona, a distant, volatile region full of dramatic solar winds.

Despite being able to study Earth's sun as an example, astronomers still do not fully understand some aspects of main-sequence stars. For example, the temperature of the sun's photosphere is roughly 5,500 kelvins (5,227 degrees Celsius, or 9,440 degrees Fahrenheit). However, even though it is farther away from the extremely hot core, the corona has a temperature of more than one million kelvins. Astronomers are unsure why the corona is so much hotter than the photosphere.

**DIVERGENT EVOLUTIONARY TRACKS**
Eventually, all main-sequence stars convert all their hydrogen to helium. First they burn up all the hydrogen in the core. When this is exhausted, their cores contract and their outer layers expand. They start burning hydrogen in these outer layers instead. As each layer burns, the star collapses a little more. Every time it contracts, gravitational potential energy is converted into thermal energy, causing the temperature in the core to increase.

The next stage in a star's evolution depends on its mass. In theory, very low-mass stars continue to collapse until they are almost entirely helium. Then they shed their outer layers, leaving only their white-hot cores. These cores slowly radiate away all the heat from the star's time in the main sequence. During this time, the star is called a white dwarf. After all the remaining heat is gone, the star stops glowing and becomes a black dwarf. However, because low-mass stars have such long life spans, this process has never been directly observed. It would take many billions of years for such a star to use up all its hydrogen and become a white dwarf—longer than the estimated 13.8 billion years since the

The sun's surface and internal structure. The interior of the sun consists of a core, a radiative layer, and a convective layer. Surrounding the sun, beyond its surface, is the corona.

universe began—and trillions more for it to radiate away all its heat.

Larger main-sequence stars take a different path. Their greater mass means that more gravitational energy is converted into thermal energy, raising the temperature of the core to reach more than 100 million kelvins. This allows the core to begin fusing helium into carbon, stabilizing the star. Because helium fusion releases more energy than hydrogen fusion, the star expands far beyond its original size. Such stars are called red giants or red supergiants, depending on their luminosity.

When smaller red giants run out of helium to burn, they shed their outer layers and become white dwarfs. Billions of years from now, Earth's sun will take this path. However, larger red giants have enough mass that their cores eventually heat up to more than 600 million kelvins. At this temperature, they begin fusing carbon, a multistep conversion process that ends with the production of iron. The star gradually forms onion-like layers of progressively heavier elements, ranging from the outer hydrogen and helium layers to the iron core. Other layered elements include neon, oxygen, and silicon.

Once all of the carbon in a large red giant has been converted, there is nothing left for it to burn. While hydrogen, helium, and carbon fusion all generate energy, iron fusion consumes it. At this point, the star finally succumbs to its own gravity and begins its last collapse. What happens next depends on the mass of the core.

If the core is less than about three solar masses, the core contracts to just a fraction of its original size—typically only about twenty kilometers (twelve miles) in diameter. With such a massive amount of matter compressed into such a small space, the core becomes so dense that its atoms are crushed together, something not normally possible. The protons and electrons in these atoms are smashed together to create neutrons. Eventually, due to a phenomenon known as neutron degeneracy pressure, the core can contract no further. The collapsing matter rebounds, creating a massive shock wave that blows away the red giant's outer layers. This explosion is called a supernova. The remaining core is called a neutron star.

Because of their extreme density, neutron stars are incredibly heavy. On Earth, one teaspoon of neutron star matter would weigh roughly ten million tons. Because neutron stars have no fuel to burn, they do not glow like conventional stars. Instead, their highly compressed nature gives them an incredibly powerful magnetic field. Neutron stars also rotate extremely quickly. In some neutron stars, known as pulsars, this causes their magnetic fields to produce powerful beams of radiation that, if properly oriented, can be detected from Earth.

If the core of the collapsing red giant is more than three solar masses or so, even neutron degeneracy pressure is not sufficient to halt its collapse. In this case, the core continues to contract, compressing more and more matter into a smaller and smaller space, until its gravitational pull is so strong that even light cannot escape it. At this point, the star has become a black hole.

**THEORETICAL EXOTIC STARS**
Astronomers and physicists theorize that several more types of stars may exist. However, none of these stars have ever been observed. These rare stellar structures are called exotic stars.

One of the most commonly mentioned types of exotic star is the quark star. Quarks are elementary particles that combine to make up hadrons, a class of particles that includes protons and neutrons. They are among the smallest particles in existence. In theory, a quark star is created when a neutron star is compressed beyond the neutron degeneracy limit, but not so much that it becomes a black hole. This would cause some or all of the neutrons break down into their constituent quarks. Though scientists are unsure whether the laws of physics allow for quark stars, many astronomers believe they do. Several known ultradense neutron stars have been proposed as potential quark stars, including XTE J1739-285, in the Ophiuchus constellation, and the pulsar 3C 58.

Another theoretical type of exotic star is the electroweak star. Scientists have proposed that as a neutron star is about to collapse into a black hole, the extreme temperatures generated by the collapse may be hot enough to merge two fundamental forces of the universe, the electromagnetic force and the weak nuclear force, into a single force known as the electroweak force. This force would convert the quarks of the neutron star into leptons, elementary particles that have considerably less mass than quarks. The mass difference would be converted into energy, which could halt the star's collapse for ten million years or more. Such stars, if they exist, would only emit a small amount of visible light. The majority of their energy emissions would be in the form of neutrinos, a kind of lepton, which are very difficult to detect.

—*Tyler Biscontini*

**FURTHER READING**

Brainerd, Jerome James. "Protostars." *Astrophysics Spectator*. Astrophysics Spectator, 7 May 2008. Web. 21 May 2015.

"Energy Transport in Stars." *Astronomy 162: Stars, Galaxies, and Cosmology*. University of Tennessee Knoxville, n.d. Web. 21 May 2015.

Fraknoi, Andrew. "The Lives of Stars." *Seeing in the Dark*. PBS, Mar. 2008. Web. 21 May 2015.

Hunter, Matt. "Have Quark Stars Been Discovered?" *From Quarks to Quasars*. FQTQ, 26 Mar. 2014. Web. 29 May 2015.

"Life Cycle of a Star." *National Schools' Observatory*. Natl. Schools' Observatory, 2004–15. Web. 21 May 2015.

Pevtsov, Alex. "Explain the Coronal Heating Problem, Please." *Stanford Solar Center*. Stanford U, n.d. Web. 29 May 2015.

"Post–Main Sequence Stars." *Australia Telescope National Facility*. Commonwealth Scientific and Industrial Research Org., n.d. Web. 21 May 2015.

Schombert, James. "Star Formation." *Astronomy 122: Birth and Death of Stars*. University of Oregon, n.d. Web. 21 May 2015.

Schombert, James. "Stellar Structure." *Astronomy 122: Birth and Death of Stars*. University of Oregon, n.d. Web. 21 May 2015.

"Theorists Propose a New Way to Shine—and a New Kind of Star." *Astronomy Magazine*. Kalmbach, 15 Dec. 2009. Web. 29 May 2015.

# STAR-FORMING REGIONS

**FIELDS OF STUDY**

Stellar Astronomy; Astrophysics

**SUMMARY**

Astronomers have determined that certain circumstances are needed for stars to form. These conditions exist in places where molecular clouds—gatherings of very dense, relatively cold dust and gas—are affected by gravity. Scientists study these star-forming regions to learn how existing stars came into being and to determine where new stars and planetary systems will form in future.

**PRINCIPAL TERMS**

- **Bok globule:** the smallest detectable dark nebula, associated with the formation of protostars.
- **Herbig-Haro (HH) object:** a bright object formed when the hot, high-velocity gas ejected from a protostar interacts with and ionizes the cooler material surrounding it. HH objects are often associated with Bok globules.
- **infrared:** long-wavelength light that is just below visible light on the electromagnetic spectrum.
- **protostar:** an infant star that forms when a cold, dense molecular cloud of dust and gas collapses, its gravity causing part of it to become hotter and denser until it eventually produces the core of the new star.
- **T Tauri star:** a low-mass star that formed less than ten million years ago and is not hot enough for nuclear fusion to begin. In terms of stellar evolution, T Tauri stars come between protostars and low-mass main-sequence stars such as Earth's sun.

**STAR NURSERIES**

Certain areas of space contain very dense, relatively cold clouds of hydrogen gas and dust known as molecular clouds. When gravity from nearby objects affects part of this cloud, the affected region can begin to collapse inward. As gravity continues to act on the region, it begins to form a cloud fragment both denser and hotter than the surrounding gas cloud. This fragment draws more dust and debris toward itself, forming an accretion disk that is surrounded by even more dust and gas. Astronomers call this newly formed star a protostar.

Some parts of the universe feature conditions that are favorable to the creation of protostars. These are known as star-forming regions, sometimes called "star

Giant stellar nursery. Young stars form within a large cloud of gas in the galaxy M33. Many of these stars are visible using the Hubble Space Telescope. NASA.

nurseries." The protostars in these nurseries have yet to reach a mass and temperature that would allow for fusion and the creation of energy and light. For this reason, they are very dim, making them difficult to detect and observe.

Astronomers have discovered that telescopes and other tools that detect infrared light can be very helpful in finding and watching these young stars as they continue to form. Infrared light is not visible to the human eye. However, it gives off a distinct signature that can be detected by instruments even when the target objects are obscured by dense clouds of dust and gas, as in the case of protostars.

### Characteristics of a New Star

Astronomers have discovered that protostars not only collect matter in their core and accretion disk but also eject material from both poles. This outflow is expelled in jets, typically following the protostar's axis of rotation. It can reach speeds of hundreds of kilometers per second. As it streams out, the flow interacts with the material around it. The resulting hot, ionized gas creates glowing balls of matter known as Herbig-Haro (HH) objects. HH objects are typically brightest at the point of first contact between the emissions and the surrounding cloud.

HH objects are often found near another space formation known as a Bok globule. These small, dark nebulas surround the newborn star. They are named for Dutch American astronomer Bart Bok (1906–83), who first theorized that they were related to protostars. Bok globules may be associated with the formation of individual stars, as opposed to star clusters.

As these new stars age, they begin to change. Lower-mass stars less than ten millions years old have not yet reached a temperature high enough for fusion to take place in their cores. However, they still sometimes release radiation in variable bursts. They also tend to be brighter than main-sequence stars such as Earth's sun. These stars represent an intermediate stage between a protostar and a main-sequence star. Such stars are called T Tauri stars, after T Tauri, the first such star discovered. Because T Tauri stars retain accretion disks, they are likely candidates to play a role in the formation of planets and even entire planetary systems.

### Nearby Star Formation

Astronomers from the National Aeronautics and Space Administration (NASA) have determined that the constellation Orion contains a huge active star-formation region about fourteen hundred light-years away from Earth. Using the Spitzer Space Telescope, NASA researchers have identified at least three hundred protostars in this region.

In March 2015, NASA scientists announced that they had observed one of these protostars, known as HOPS 383, releasing outbursts of light and heat. They believe that the brightness of these outbursts was caused by a sudden influx of gas from the protostar's accretion disk. As material from the accretion disk neared the protostar's core, the temperatures of both rose, causing their brightness to increase.

By studying HOPS 383 and other protostars, astronomers get a firsthand look at how Earth's sun and other stars and planetary systems may have formed.

—*Janine Ungvarsky*

**FURTHER READING**

"Bok Globule." *Cosmos: The SAO Encyclopedia of Astronomy*. Swinburne University of Technology, n.d. Web. 1 Apr. 2015.

Fazekas, Andrew. "Multiple-Star Birth Revealed in Stellar Nursery." *National Geographic News*. Natl. Geographic Soc., 11 Feb. 2015. Web. 1 Apr. 2015.

"Herbig-Haro Object." *Cosmos: The SAO Encyclopedia of Astronomy*. Swinburne University of Technology, n.d. Web. 1 Apr. 2015.

"Infrared Waves." *Mission:Science*. NASA, 13 Aug. 2014. Web. 1 Apr. 2015.

"Protostar." *Space Book*. Las Cumbres Observatory Global Telescope Network, 2012. Web. 1 Apr. 2015.

Reddy, Francis. "NASA Satellites Catch 'Growth Spurt' from Newborn Protostar." *WISE: Wide-Field Infrared Survey Explorer*. NASA, 23 Mar. 2015. Web. 1 Apr. 2015.

# STELLAR CLASSIFICATION

**FIELDS OF STUDY**

Stellar Astronomy; Observational Astronomy; Historical Astronomy

**SUMMARY**

Stellar classification measures the characteristics of stars using information from a variety of sources. Most commonly, the light that stars give off are used to measure location, trajectory, movement speed, mass, size, and composition. Stars are usually classified into one of seven temperature-based types: O, B, A, F, G, K, or M. Following that initial classification, stars are given a second classification to denote their current life-cycle phase.

**PRINCIPAL TERMS**

- **Harvard scheme:** a spectral classification system created at Harvard University in the late nineteenth and early twentieth century that assigned successive letters to stars based on their spectra and later rearranged to reflect their surface temperatures.
- **luminosity classification:** a means of categorizing stars by their relative sizes, as derived from their spectral emissions, atmospheric pressure, distance from Earth, and loss of light to interstellar matter; also called the Morgan-Keenan classification.
- **spectral classification:** a means of categorizing stars by color/temperature, which is derived from the light spectra emitted by their elements.

**TYPES OF STARS**

Astronomers usually classify stars according to their spectra and temperature. The spectral classification categories are signified by the first letter of the star's name. The categories include O, B, A, F, G, K, and M. Astronomers have also added the L, T, and Wolf-Rayet (WR) categories.

Type O stars are extremely large and hot. They project bright blue light and burn at more than 25,000 kelvins (K). The average type O star has more than sixty times the mass, roughly fifteen times the size, and more than one million times brightness of the sun. However, because type O stars burn fuel very quickly, they have very short lives. Type O stars are some of the brightest stars in the sky. The star 10 Lacertae is a type O star.

Type B stars also burn a bright blue but are significantly smaller than type O stars. The average type B star burns at 11,000 to 25,000 K. They have roughly eighteen times the mass and are roughly seven times the size of the sun. Type B are also roughly twenty thousand times brighter than the sun. They have longer lives than type O stars but shorter ones than most other stars. The famous star Rigel is a type B.

Type A stars are much smaller than type O and type B stars. They have only 3.2 times the mass of the sun and are only 2.5 times its size. They burn between 7,500 and 11,000 K and are eighty times as bright as the sun. Sirius, the Dog Star, is a type A star.

Type F stars are only slightly larger than the sun. They have 1.7 times its mass and are 1.3 times its size. Type F stars burn between 16,000 and 7,500 K and are roughly six times the size of the sun. Procyon, the dimmer star in the constellation Canis Minoris, is a type F star.

Type G stars radiate white or yellow light and burn between 5,000 and 6,000 K. The sun is a relatively average-size type G star. The sun is vastly closer than any other star, making it relatively easy to study; thus, astronomers know the most about type G stars.

Type K stars glow orange or red. They burn between 3,500 and 5,000 K. The average type K star has 0.8 times the mass of the sun, 0.9 times its size, and only 0.4 times its brightness. Arcturus, found in the constellation Boötes, is a type K star. Though Arcturus is a relatively dim star, its close proximity to Earth makes it one of the brightest stars in the night sky.

Type M stars glow a dark red color and burn at under 3,500 K. They are very small, boasting only 0.3 times the mass of the sun and 40 percent of its size. Type M stars are usually less than half as bright as the sun. The star Betelgeuse, found in the famous constellation Orion, is a type M star.

Type L and Type T stars are called dwarf stars. Some of these are stars that never formed. Others are stars that have burned all their fuel and collapsed. Type L and T stars do not generate heat; instead, they radiate away their heat over millions or billions of years.

Wolf-Rayet (WR) stars are extremely massive stars, thought to be giving off giant bursts of energy at the end of their lives. WR stars live extremely short lives, but they may grow large enough to become black holes.

The second half of a star's classification, its luminosity classification, is an indicator of its place in the stellar life cycle. Ia stars are extremely bright supergiants. Ib stars are dimmer supergiants. II stars are extremely bright giants. III stars are average giants. IV stars are subgiants. V stars are main sequence stars and dwarfs. VI stars are subdwarf stars. Lastly, VII are white dwarf stars. The sun is classified as a GV star. That means the sun is a main-sequence, yellow glowing star burning between 5,000 and 6,000 K.

## HISTORY OF STELLAR CLASSIFICATION

The history of stellar classification is often traced to the amateur astronomer Henry Draper (1837–82). Draper was a medical doctor but spent most of his spare time studying and practicing astronomy. He pioneered astrophotography and was the first person ever to photograph the moon and the Orion Nebula. After Draper's death, his wife dedicated most of his fortune to advancing his research.

One of the prominent astronomers who continued Draper's research was Edward Charles Pickering (1846–1919). Pickering taught physics at MIT for nine years, after Harvard hired him as a professor of astronomy. Pickering helped to combine the two fields of study and is considered one of the founders of astrophysics.

Most of Pickering's astronomical work focused on using astrophotography, stellar spectroscopy, and precise measurements of stars' relative brightness to categorize and study them. This required huge amounts of time-consuming, tedious cataloging and computing of stellar characteristics. In order to lighten his workload, Pickering hired his secretary to help with the astronomical cataloguing and computations. However, Pickering was not satisfied with the secretary's work ethic. He fired the man, replacing him with his maid, Williamina Fleming (1857–1911).

Fleming's talent and work ethic quickly impressed Pickering. As his telescopes began to photograph more work than the pair could hope to catalog, Pickering decided to hire more help. Many women were available with equivalent education to men. At the time, women were paid much less than men for equivalent work. For this reason, Pickering could afford to hire many more women in his observatory. These women came to be known as the Harvard Computers. Working together, Pickering and the Harvard Computers published the *Henry Draper Catalogue*. The book was the largest star catalog of its time, containing more than ten thousand stars classified according to the Harvard scheme.

Several of the most notable members of Harvard Computers, such as Fleming and the now-famous Annie Jump Cannon (1863–1941), later became full-fledged astronomers. While the Harvard Computers received little credit in their own time, they are now recognized as important contributors to the field of astronomy.

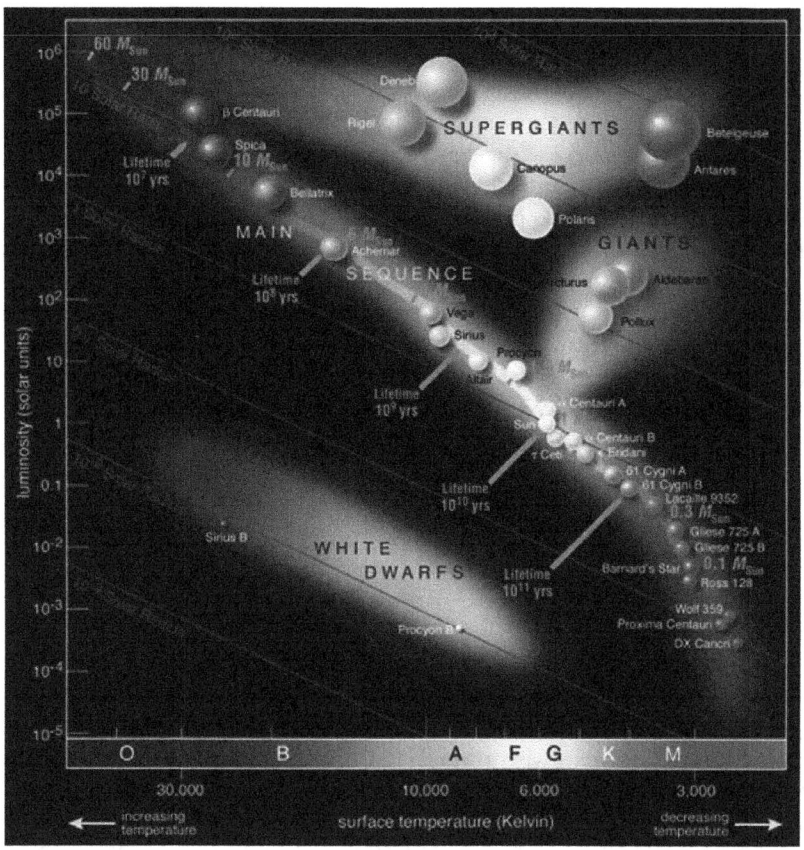

Hertzsprung-Russell diagram showing well-known stars by luminosity, surface temperature, and Morgan-Keenan classification. European Southern Observatory (eso.org), via Wikimedia Commons.

## Makeup of Stars

Astronomers measure the composition of specific stars through a variety of methods. Most commonly, they use spectroscopy. This is the process of measuring the light spectrum emitted by the star. Radiation leaving the inside of a star, such as visible and infrared light, form a continuous spectrum of colors. When viewed, this spectrum looks like a rainbow. With specialized, high-quality equipment, astronomers can see that spectra are sometimes interrupted by black lines. These lines are called "absorption lines."

Absorption lines are caused by cold elements placed in the path of hot elements. When any element is superheated, it gives off a unique light spectrum. However, because the element does have color, it also absorbs small, specific frequencies of light. When these elements are present in the cool atmosphere of a star, they absorb some of the light emitted from the surface of the star. When the star's spectrum is interpreted on Earth, the missing colors appear as absorption lines.

Stellar spectra are not visible with the naked eye. When stellar spectra were discovered, astronomers used prisms to view them. Prisms are precisely cut, triangular glass shapes. Starlight would be captured by a telescope and filtered through a narrow slit to create a thin beam of visible light. Then that light would be aimed at one of the corners of the prism, forcing the light to split. The prism would then project the stellar spectrum onto a backdrop for photographing and measuring.

Modern astronomers no longer use prisms because they are delicate, difficult to make, and require extremely high-quality glass for a clear reading. Instead, astronomers use diffraction gratings. These are glass plates with precisely-cut small lines etched onto them. Diffraction gratings are less delicate and more adjustable, making them suitable for a wider variety of astronomical spectroscopic purposes.

—*Tyler Biscontini*

### Further Reading

"Annie Jump Cannon." *San Diego Supercomputer Center.* SDSC, n.d. Web. 28 May 2015.

"Edward Charles Pickering." *Encyclopaedia Britannica.* Encyclopaedia Britannica, n.d. Web. 3 June 2015.

Geiling, Natasha. "The Women Who Mapped the Universe and Still Couldn't Get Any Respect." *Smithsonian Magazine.* Smithsonian, 18 Sept. 2013. Web. 28 May 2015.

Gibson, Steven J. "Draper, Henry." *The Biographical Encyclopedia of Astronomers.* Ed. Thomas A. Hockey, Virginia Trimble, and Katherine Bracher. New York: Springer, 2007. Print.

"Howell, Elizabeth. "What Is a Wolf-Rayet Star?" *Universe Today.* Universe Today, 5 Feb. 2015. Web. 28 May 2015.

Rothstein, Dave. "What Can We Learn from the Color of a Star?" *Ask an Astronomer.* Astronomy Dept., Cornell U, 17 Mar. 2002. Web. 28 May 2015.

"Spectral Classification of Stars." *University of Nebraska-Lincoln.* Astronomy Education, University of Nebraska-Lincoln, n.d. Web. 28 May 2015.

# STELLAR POPULATION MAPPING

### FIELDS OF STUDY

Astronomy; Astrophysics; Stellar Astronomy

### SUMMARY

Stellar populations are collections of stars with similar chemical content, age, or distribution in space. Scientists have determined that the Milky Way Galaxy has three key stellar populations: disk, halo, and bulge. By studying these groupings, and the ways their stars have changed, astronomers can learn more about the way stars evolved to form the galaxy.

### PRINCIPAL TERMS

- **extinction:** the dimming of stellar objects because of the presence of dust between the object and observer.
- **infrared:** light just beyond that which is visible on the electromagnetic spectrum at the red end.
- **metallicity:** in astronomy, elements such as iron and silicon that are heavier than the hydrogen and helium that are fused in star cores.
- **spectra:** plural of "spectrum"; a range or distribution of light or wavelengths.
- **spectroscopy:** the study of the light frequency emitted by atoms and molecules.

### Stellar Mapping

Researchers have determined that the stars of the Milky Way Galaxy can be grouped into categories based on location, composition, and age. This is known as stellar mapping. The characteristics of stars are not necessarily fixed, so the way a star changes over time is also part of the mapping process. Studying, categorizing, and tracking star groupings provides a wealth of information about the origins of the galaxy and how it has evolved over time.

Scientists map the stellar population of areas by first determining the attributes of individual stars. The key factors under consideration include spatial distribution (the number of stars in relation to the area occupied), kinematics (the speed or velocity ranges of these stellar objects), star chemistry (the kinds and amounts of chemicals present), and the age of stars in the area. Scientists then look for similarities in these traits among the star population. By first identifying the attributes of individual stars and then comparing them to others nearby, researchers can determine patterns of star distribution. They can identify "sibling" stars that may have come from the same source. They can also develop theories about how these stars, and others like them, formed and how they will act in the future.

Scientists have a number of methods and tools for studying and categorizing stellar populations. One of the most powerful tools is spectroscopy. Both light and radiation are organized in spectra by color, intensity, or wavelength. Astronomers use spectroscopy to examine the light frequencies emitted by different parts of a celestial body. This allows them to determine the body's light wavelengths; its velocity, age, and degree of extinction; and the temperature and pressure of the body's atmosphere.

Spectroscopy can also reveal the composition of a stellar body. An object's metallicity is its chemical components that are heavier than the helium and hydrogen commonly fused into energy in a star's core. This includes oxygen and carbon, which are not normally considered metals. Astronomers have determined that the higher the metallicity of a stellar object—that is, the more oxygen, silicon, carbon, or iron it has—the more likely it is to survive. They have also learned that objects with heavier metals are more likely than are those with lighter metals to have the materials necessary to develop a planetary system.

When an object is too dim to be seen with a conventional telescope or satellite-mounted camera, astronomers turn to other tools. Sometimes a star is dimmed by a process known as extinction. Stellar extinction refers to the visual obstruction of a star by a cloud of dust. The dust blocks the blue waves that are most easily seen by observers, making the object

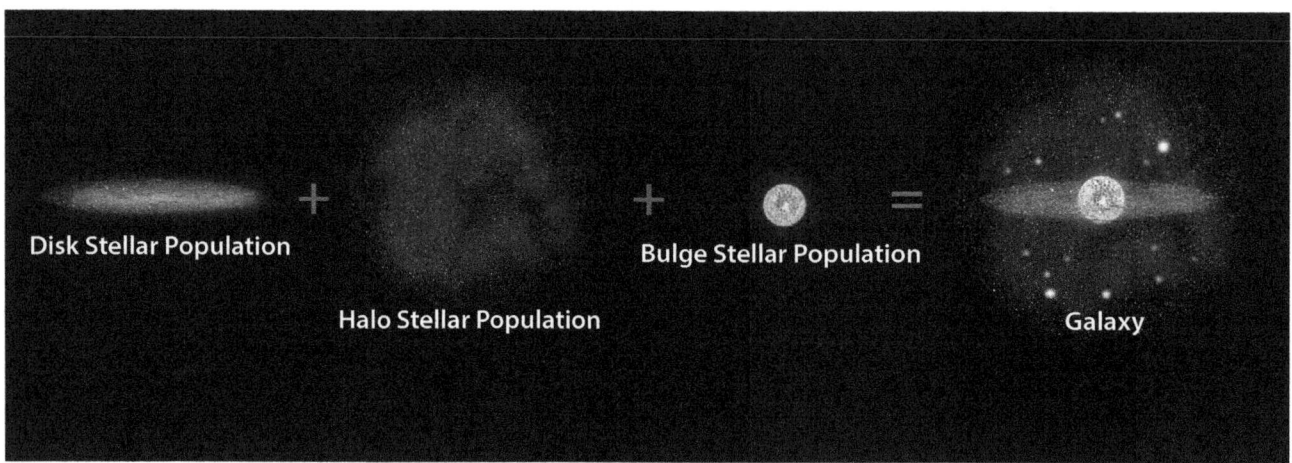

Three stellar populations: the disk population, the bulge population, and the halo population, combine to make up the stars in the galaxy. Each population of stars has different physical, chemical, and kinematic properties.

appear both less bright and redder. In this case, scientists turn to infrared technologies to reveal the red light that is emitted by all celestial bodies. Infrared allows scientists to see, study, and categorize stars that would otherwise not be visible.

**EXAMPLES OF STELLAR POPULATION MAPPING**
Using these tools and methods, astronomers group the stars in Earth's galaxy into disk, halo, and bulge components. The isotropic halo component (which has equal properties in all directions) includes the oldest stars and globular clusters (spherical groups of stars with similar origins). It is located at the outer edge of the galaxy. The bulge component rotates in the center of the galaxy and includes some old stars as well as a violently active core. The disk population includes gas and mostly young stars. It makes up the rotating, flat part of the Milky Way Galaxy.

Uses of Stellar Population Mapping
Stellar population mapping enables astronomers to theorize about the origins of the galaxy. The presence of enough heavy elements to enrich Earth and its sun to their current levels indicates that other, earlier stars experienced supernovas (massive explosions) and expelled these elements into the universe.

By studying and mapping the bodies of the Milky Way Galaxy, scientists can explain why stars with a higher metallicity and relatively slow dispersing cloud of dust and gas are able to form planetary systems. Stellar mapping gives scientists an understanding of the processes that led to the formation of the Milky Way and other galaxies. Understanding how and why stars and planets form provides insight not only into how conditions evolved to allow life to develop on Earth, but also how and where conditions might exist for life in other parts of the universe.

—*Janine Ungvarsky*

**FURTHER READING**
"Extinction." *Cosmos: The SAO Encyclopedia of Astronomy*. Centre for Astrophysics and Supercomputing, Swinburne University of Technology, n.d. Web. 29 Mar. 2015.

"Infrared Waves." *NASA: Mission Science*. NASA, 13 Aug. 2014. Web. 28 Mar. 2015.

Majewski, Steven R. "Stellar Populations and the Formation of the Milky Way." *Globular Clusters*. Ed. C. Martinez Roger, I. Pérez-Fournon, and F. Sanchez. New York: Cambridge University Press, 1999. Print.

Sanders, Ray. "When Stellar Metallicity Sparks Planet Formation." *Astrobiology Magazine*. Astrobio.net, 9 Apr. 2012. Web. 29 Mar. 2015.

Schultheis, M., et al. "Mapping the Milky Way Bulge at High Resolution: The 3D Dust Extinction, CO, and X Factor Maps." *Astronomy & Astrophysics* 566 (2014): n. pag. Web. 10 Apr. 2015.

"Spectroscopy." *Cosmos: The SAO Encyclopedia of Astronomy*. Centre for Astrophysics and Supercomputing, Swinburne University of Technology, n.d. Web. 29 Mar. 2015.

"Stellar Populations." *Astrophysics*. Amer. Museum of Natural History, n.d. Web. 29 Mar. 2015.

# STELLAR REMNANTS

## FIELDS OF STUDY

Astronomy; Astrophysics; Stellar Astronomy

## SUMMARY

As stars burn energy, they change, often dramatically. What is left behind after the final change is called a "stellar remnant." The type of remnant is determined by the nature of the original star. By studying stellar remnants, scientists can learn more about how stars are created, how long they last, and how they interact with other forces in space.

## PRINCIPAL TERMS

- **black hole:** a region of space with a gravitational pull so strong that even light cannot escape it.
- **electromagnetic (EM) radiation:** energy created by interactions between electrically charged particles and transmitted in the form of waves; includes the visible light spectrum, as well as radio waves, microwaves, infrared, ultraviolet, x-rays, and gamma rays.
- **neutron star:** an incredibly dense, rapidly spinning star, consisting mainly of closely compacted neutrons left behind after the gravitational collapse of a massive star.
- **nova:** a sudden, intense, ultimately short-lived increase in the brightness of a white dwarf star, caused by an accumulation of hydrogen that triggers a runaway thermonuclear reaction.
- **supernova:** the immensely forceful explosion produced when a massive star reaches the end of its life cycle and collapses under its own mass.
- **white dwarf:** a small, very dense star that has exhausted nearly all of its fuel and is nearing the end of its life cycle.

## LIFE CYCLES OF STARS

Stars are not living things, but they do have a life cycle. They are created by the force of gravity acting on molecular clouds and stellar gases. The energy that sustains them is produced in their core, where various elements undergo constant nuclear fusion. As these elements are depleted, the stars change and eventually die. This cycle happens over the course of millions or even billions of years, so it is not possible to observe it directly. However, astronomers learn about the process by studying the remnants left behind by dying stars.

One of the ways scientists learn about stars is by studying the various kinds of electromagnetic (EM) radiation they emit. The visible light emitted by a star provides information about its temperature, with brighter stars typically being hotter. The presence of a considerable amount of ultraviolet light reveals that

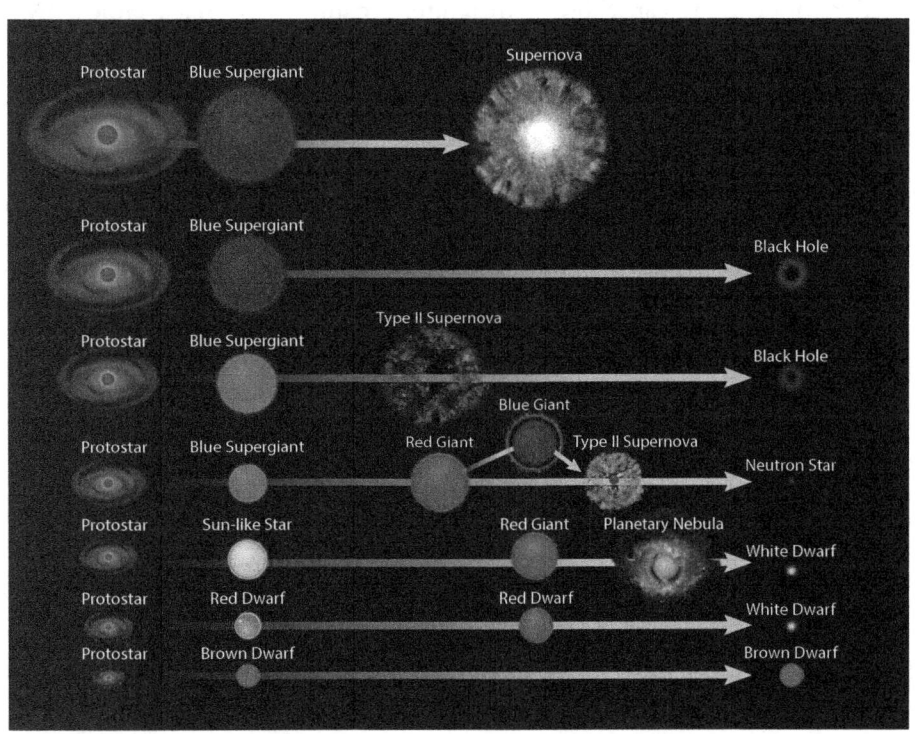

Timeline showing the evolution of stars, beginning with protostars and progressing to the various stellar remnants left by different types of stars.

a star is very active and new. X-rays are associated with neutron stars and black holes, and white dwarfs emit both x-rays and ultraviolet (UV) radiation. Gamma rays, which have the shortest wavelength of all EM radiation, are emitted by both black holes and the supernovas that often precede them.

**HOW STARS EVOLVE**

Stars evolve in different ways, depending on their type. Astronomers believe that most stars in the universe, including Earth's sun, will eventually become white dwarfs. The core of a white dwarf is formed by fusing helium into carbon. White dwarfs do not get hot enough to burn carbon, so this core will contract until its electrons occupy the smallest space allowed by quantum mechanics. While this happens, the temperature of the core increases to about 100,000 kelvins. This residual heat allows the white dwarf to continue glowing, even though its light will be dim.

Sometimes, if a white dwarf is part of a binary system and its companion star is close enough to be affected by its gravity, the white dwarf will attract matter from the atmosphere of the companion star. When enough matter has accreted, the hot core of the white dwarf can cause it to ignite, creating an explosive flare called a nova. Astronomers once thought that novas were caused by the birth of new stars. It is now known that they are, in fact, signposts of dying white dwarfs.

When a star is more than eight times the mass of Earth's sun, it ends its life in a different way. As a massive star's core loses energy, it can no longer resist the force of its own gravity. The star collapses, then explodes in a supernova. Much of the star's matter is lost in the explosion. The remaining core collapses once again, this time shrinking to such an extent that its protons and electrons are forced to combine and form neutrons. The result is a small, incredibly dense object called a neutron star.

However, if the core remaining after a supernova is massive enough—about three times the mass of the sun or greater—its gravity is so great that it simply does not stop collapsing. Eventually, all of the core's mass is concentrated in a single point, known as a singularity. The region surrounding it is called a black hole.

If, after the supernova, the star retains a mass greater than two-and-a-half times that of Earth's sun but does not collapse into a black hole, the star may become a quark star instead. However, no quark stars have yet been observed.

**STUDYING STELLAR REMNANTS**

Stellar remnants, such as white dwarfs, neutron stars, and black holes, offer astronomers a look into both the past and the future of the universe. Learning how stars change and die provides insight into star formation and composition, as well as how they may have affected Earth in the past and how they may affect it in the future.

—*Janine Ungvarsky*

**FURTHER READING**

Bernoskie, Brandi, Heather Deiss, and Denise Miller. "What Is a Supernova?" *NASA Education*. NASA, 4 Sept. 2013. Web. 31 Mar. 2015.

"Black Holes." *NASA Science*. NASA, 30 Mar. 2015. Web. 31 Mar. 2015.

Keeton, Charles. *Principles of Astrophysics: Using Gravity and Stellar Physics to Explore the Cosmos*. New York: Springer, 2014. Print.

Kornreich, Dave. "What Is a Nova?" *Ask an Astronomer*. Cornell U Astronomy, 24 Jan. 1999. Web. 31 Mar. 2015.

"Neutron Stars: Incomprehensible Density." *Science and Space*. Natl. Geographic Soc., n.d. Web. 31 Mar. 2015.

O'Neill, Ian. "Why Are Quark Stars So Strange?" *Discovery News*. Discovery Communications, 19 Jan. 2010. Web. 6 Apr. 2015.

"White Dwarf." *Cosmos: The SAO Encyclopedia of Astronomy*. Swinburne University of Technology, n.d. Web. 31 Mar. 2015.

"White Dwarfs: Aging Stars." *Science and Space*. Natl. Geographic Soc., n.d. Web. 31 Mar. 2015.

# SUB-BROWN DWARFS

## FIELDS OF STUDY

Astronomy; Observational Astronomy

## SUMMARY

Sub-brown dwarfs (SBDs) are celestial bodies that have attributes similar to stars and planets. SBDs are similar to brown dwarfs (BDs) and extrasolar planets, but they have some important differences. SBDs form in the same way stars form, but SBDs are smaller and do not produce thermonuclear fusion. Although scientists believe that BDs and SBDs are common in the galaxy, these objects are difficult to identify and confirm. Scientists study SBDs in part because they believe that SBDs could possibly be hospitable for extraterrestrial life.

## PRINCIPAL TERMS

- **brown dwarf:** a celestial object that forms like a star but does not have enough mass to produce thermonuclear fusion like a star.
- **thermal energy:** energy that is in the form of heat.

## Brown Dwarfs and Sub-Brown Dwarfs

Sub-brown dwarfs (SBDs) are celestial bodies that are generally smaller than stars and larger than planets. They formed in the same way stars form, but they are too small to create the thermonuclear reaction that stars have. Brown dwarfs (BDs) are similar low-mass celestial objects that cannot burn hydrogen but can at least burn deuterium, a heavy form of hydrogen. SBDs do not have enough mass for this fusion, either. Sometimes SBDs are surrounded by circumstellar disks. Circumstellar disks are made up of gas and dust, and they surround stars and other objects. Scientists believe that circumstellar disks produce planets.

SBDs and BDs are sometimes called "failed stars." This is because their mass is too small, and they failed to produce the thermonuclear fusion that powers other stars. However, these bodies do give off some thermal energy. When SBDs are first created, they have about the same temperature as a very cool star, about 1,500 to 2,000 kelvins (1227 to 1727 degrees Celsius, or 2,240 to 3,140 degrees Fahrenheit). SBDs radiate their thermal energy in the form of heat and dim light. As SBDs age, they become cooler and cooler.

SBDs and BDs are mostly observed through infrared imaging. This is because these failed stars are too faint to see with traditional telescopes. BDs that are up to about one hundred light-years away can be detected through infrared imaging. This is not a very far cosmic distance. Scientists have not been able to observe many of these objects. The imaging problem becomes even more complex when researchers try to observe SBDs. These objects are even smaller than BDs. Scientists cannot see many of the BDs and SBDs that they believe exist in the universe. Some scientists believe that the Milky Way Galaxy has about as many BDs and SBDs as it has stars. However, this cannot yet be proven.

## SBDs and Other Celestial Bodies

SBDs are related to, but not the same as, both BDs and extrasolar planets. BDs are celestial objects that formed in a similar way as stars. Stars, BDs, and SBDs all formed when a ball of gas and dust collapsed under its own weight. Scientists believe that planets formed from circumstellar disks of gas and dust that surround stars, BDs, and SBDs.

Stars have to be quite large so that they can create a thermonuclear reaction that causes them to produce light and heat. Stars usually have to be over seventy-five times as large as Jupiter to create fusion. Objects between thirteen and seventy-five times the size of Jupiter are BDs.

Some objects form in the same way as BDs, but they are smaller than thirteen times the size of Jupiter. Some of these objects are extrasolar planets. Extrasolar planets orbit stars. Objects that are thirteen times the size of Jupiter and smaller but do not orbit stars are SBDs.

Mass, not radius, distinguishes the different types of objects. This phenomenon is an important part of why stars achieve thermonuclear fusion and less massive objects do not. An object compresses as it becomes more massive, and the gases inside the object become hotter and hotter. Once the gases become

Against the background of the solar system is an ultra-cool brown dwarf, glowing green in this infrared im-age. The brown dwarf is cooler than the surrounding stars. NASA/JPL-Caltech/UCLA.

very hot, hydrogen molecules can fuse together to make helium, causing thermonuclear fusion.

The differences among stars, BDs, SBDs, extrasolar planets, and planets are still being discussed among scientists. It is clear that these objects are different, but their exact definitions are being refined. As scientists learn more about these objects, they will be able to further refine the definitions.

## Studying SBDs

One reason to study SBDs that scientists believe these objects might be capable of supporting life. The temperature of an SBD is hottest when the object is first created, but scientists believe that SBDs can have fairly stable temperatures for long periods of time. Some SBDs may have atmospheres that hold in their heat. Because SBDs can capture their own heat, they might have liquid on them, which could allow for life to form. Earth has liquid water that allows life to exist, but scientists believe life could develop in places where liquids other than water are present. However, life that evolved on an SBD would most likely be much different from life that evolved on Earth.

—*Elizabeth Mohn*

### Further Reading

"Circumstellar Disks." *Formation and Evolution of Planetary Systems.* University of Arizona, n.d. Web. 27 Mar. 2015.

Comins, Neil F. *Discovering the Universe: From the Stars to the Planets.* New York: Freeman, 2009. Print.

Moskowitz, Clara. "Searching for Alien Life? Try Failed Stars." *Space.com.* Purch, 31 Mar. 2011. Web. 27 Mar. 2015.

Mullen, Leslie. "Star Bright . . . and the Sub-Brown Dwarfs." *Astrobiology Magazine.* Astrobio.net, 4–6 Aug. 2003. Web. 27 Mar. 2015.

O'Neill, Ian. "Clouds of Water Possibly Found in Brown Dwarf Atmosphere." *Discovery News.* Discovery Communications, 26 Aug. 2014. Web. 30 Mar. 2015.

Perryman, Michael. *The Exoplanet Handbook.* New York: Cambridge University Press, 2011. Print.

Plait, Phil. "The Upper Limit to a Planet." *Discover.* Kalmbach, 7 Sept. 2006. Web. 27 Mar. 2015.

"Super-Cool Brown Dwarf Helps Distinguish Dwarfs and Giant Planets." *Marie Curie Actions.* European Commission, 28 Apr. 2014. Web. 30 Mar. 2015.

# SUPERNOVA REMNANTS

## FIELDS OF STUDY

Astronomy; Observational Astronomy; Stellar Astronomy

## SUMMARY

Supernova remnants are the material that is left after a massive star has exploded into a supernova. The explosion of the supernova ejects material and rips a hole in the interstellar medium, causing a shockwave. The remaining material is the supernova remnant. Different types of supernova remnants include plerions, shell-type remnants, plerionic composites, and thermal composites. Supernova remnants are responsible for creating supercharged particles and cosmic rays. Scientists study supernova remnants in order to learn more about how galaxies evolve.

## PRINCIPAL TERMS

- **cosmic rays:** highly energized subatomic particles traveling at near the speed of light.
- **electromagnetic spectrum:** the range of all possible frequencies and wavelengths of electromagnetic radiation, including radio waves, microwaves, infrared, visible light, ultraviolet light, x-rays, and gamma rays.
- **interstellar medium:** gas and dust found in the spaces between stars; the matter from which new stars are formed.
- **plerion:** a type of supernova remnant, also called a crab-type remnant, that emits radiation from a pulsar in its center.
- **plerionic composite:** a type of supernova remnant that appears like a plerion at both x-ray and radio wavelengths but also exhibits the shell of a shell-type remnant.
- **pulsar:** a rapidly rotating neutron star that emits pulses of electromagnetic radiation at regular, extremely short intervals.
- **supernova:** the immensely forceful explosion produced when a massive star reaches the end of its life cycle and collapses under its own mass.
- **synchrotron radiation:** radiation produced when charged particles move in a curved path.
- **thermal composite:** a type of supernova remnant that appears shell-like at radio wavelengths but crab-like at x-ray wavelengths.

### SUPERNOVAE AND SUPERNOVA REMNANTS

Supernova remnants are the materials that are left after a massive star or, in some cases, a white dwarf explodes, turning into a supernova. Supernovae are the most powerful events in the universe. They are so bright that when they first explode, they can briefly outshine the galaxies they reside in.

When stars are first born, they are made mostly of hydrogen and some helium, which are the lightest elements in the universe. However, after many years of nuclear fusion, the makeup of a star's interior begins to change, until eventually the core is made up of heavy elements, including iron and nickel. These heavy elements cause the star to collapse in on itself and explode into a supernova.

When a star explodes into a supernova, material is ejected from the star's center, ripping a hole through the interstellar medium that surrounds it. The explosion creates an expanding shock wave that forms a type of bubble. Inside the bubble, the material ejected from the star combines with the interstellar

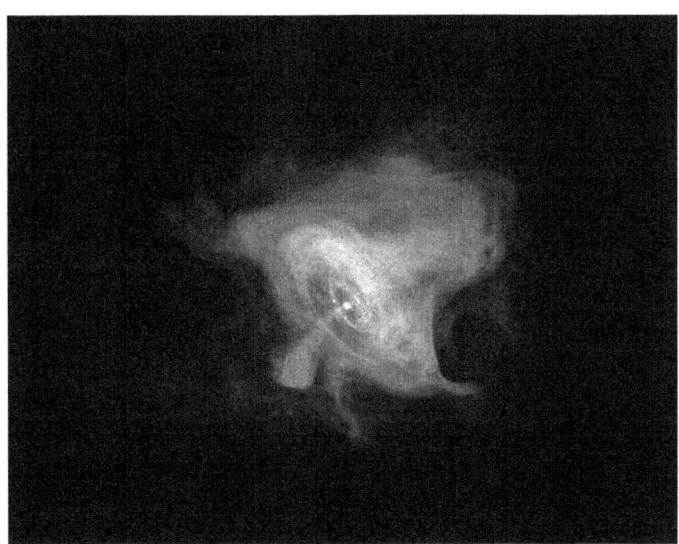

X-ray image of Crab Nebula from Chandra X-ray Observatory's Advanced CCD Imaging Spectrometer. Crab Nebula is an example of a crab-like supernova remnant. NASA/CXC/SAO/F. Seward et al.

medium it displaces. This combined material is the supernova remnant. It can reach up to several million degrees Celsius.

**FORMATION OF SUPERNOVA REMNANTS**
Scientists study the energy emitted from supernova remnants at different wavelengths of the electromagnetic spectrum. At first, the heat from the supernova explosion is so great that it creates thermal x-rays and radiation. As the remnants cool, they emit more ultraviolet radiation and less x-rays and synchrotron radiation. The remnants will continue to cool and disperse into the surrounding interstellar medium for up to about ten thousand years after the initial explosion.

Supernova remnants are classified into three main groups: shell-type remnants, crab-type remnants, and composite remnants. Shell-type remnants are surrounded by ring-like shells of material disturbed by the shock wave. They emit most of their radiation from these shells. Crab-type remnants, also called plerions, have pulsars at their centers. They emit most of their radiation from inside their expanding shock waves. Composite remnants have characteristics of both shell-type and crab-type remnants, depending on the wavelength at which they are observed. Thermal composites appear shell-like at radio wavelengths but crab-like at x-ray wavelengths. Plerionic composites appear crab-like at both radio and x-ray wavelengths, but they have a shell typical of shell-type remnants.

**DISCOVERIES FROM SUPERNOVA REMNANTS**
In the 2010s, scientists discovered that cosmic rays are produced by supernova remnants. For nearly a century, scientists had tried to determine the source of cosmic rays. This was difficult because magnetic fields throughout the solar system cause the particles that make up cosmic rays to change course as they travel.

Even though scientists could not be sure, they had assumed that supernova remnants cause particles to accelerate and create cosmic rays. Eventually, they learned that the magnetic fields in the remnants cause these particles to be bounced back and forth. Sometimes the particles are pushed through the shock waves surrounding the remnants, which energizes them. They escape the remnants and travel through the universe as cosmic rays. Some of these particles eventually reach Earth.

**STUDYING SUPERNOVAE AND SUPERNOVA REMNANTS**
Studying supernovae and supernova remnants is important for a number of reasons. One reason is that scientists believe supernovae seed the universe with many of the elements that help create planets and make life possible, including carbon, oxygen, nitrogen, and iron. Supernovae are responsible for creating all elements in the universe that are heavier than lead. Therefore, metals such as copper, gold, and iron all exist because of supernovae.

Scientists are also interested in studying supernova remnants because these objects give scientists more information about the Milky Way and the evolution of galaxies in general. Supernovae play an important role in the galaxy, and studying their remnants could help scientists better understand the formation of the Milky Way and its current composition.

—*Elizabeth Mohn*

**FURTHER READING**
Burrows, David N., et al. "Supernova Remnants." *Penn State University Department of Astronomy and Astrophysics.* Pennsylvania State U, 18 Oct. 2004. Web. 20 Mar. 2015.

Butcher, Ginger. *Tour of the Electromagnetic Spectrum.* N.p.: NASA, 2010. *Mission:Science.* Web. 20 Mar. 2015.

Hobbs, Maryam. "An Introduction to Pulsars." *Australia Telescope National Facility.* CSIRO, n.d. Web. 20 Mar. 2015.

"Introduction to Supernova Remnants." *NASA's HEASARC: High Energy Astrophysics Science Archive Research Center.* NASA, 11 May 2011. Web. 20 Mar. 2015.

McKee, Maggie. "Cosmic Rays Originate from Supernova Shockwaves." *Nature.* Nature, 15 Feb. 2013. Web. 20 Mar. 2015.

"Supernova Remnant Type." *COSMOS: The SAO Encyclopedia of Astronomy.* Swinburne University of Technology, n.d. Web. 20 Mar. 2015.

"Supernovas & Supernova Remnants." *Chandra X-ray Observatory.* NASA / Smithsonian Inst., 6 May 2013. Web. 20 Mar. 2015.

# TAIL (COMET ANATOMY)

## FIELDS OF STUDY

Astrometry; Astrophysics; Cosmology

## SUMMARY

Comets are large objects that travel through space. Comet tails are the long projections often seen behind comets. Tails can be made from a variety of substances and appear to be different lengths depending on the comet's position in relation to the sun. Historically, comet tails have been poorly understood. People have considered comet tails powerful omens or curses.

## PRINCIPAL TERMS

- **comet:** a celestial body mostly made of carbon dust and ice.

### WHAT IS A COMET?

Sometimes called the "dirty snowballs" of space, comets are balls of ice, dust, and dirt found in outer space. Some scientists think comets are made of undisturbed material dating to the formation of the universe. Comets have large orbits that occasionally pass through the solar system. They come in a variety of shapes and sizes. The largest comet observed was forty kilometers across. Because their orbits are elliptical, comets will return at predictable intervals. Some comets return quickly, while some will not return for thousands of years.

When comets pass close to a star, some of the material on the surface of the comet evaporates. The evaporated material trails behind the comet, forming a long tail often visible from Earth. Comet tails sometimes reach millions of kilometers long.

### COMET TAILS

All comets have two types of tails: the dust tail and the ion tail. The dust tail is composed of solid matter that the sun's heat caused to radiate off the comet. The ion tail is composed of various gases, including methane, ammonia, water, carbon dioxide, and carbon monoxide, that evaporated from the surface of the comet. Sunlight ionizes these gases, causing them to glow.

Comet tails always point away from the sun. The ion tail points in a straight line, directly away from the sun. The dust tail, however, curves toward that path of the comet's orbit. Ion tails are significantly thinner, longer, and brighter than dust tails, and they tend to glow a pale blue color.

Comet Hale-Bopp with two tails, a blue gas tail and a white dust tail. By E. Kolmhofer and H. Raab, Johannes-Kepler-Observatory, Linz, Austria, via Wikimedia Commons.

When a comet passes by Earth, its tail leaves a trail of tiny, leftover pieces of the comet. Sometimes, Earth passes through these trails. The pieces of comet burn brightly when they enter Earth's atmosphere, causing meteor showers.

### Comet Tails in History

Comet tails visible from Earth are large, dramatic displays. Some even look like giant balls of fire in the sky. Consequentially, ancient people sometimes considered their appearance a significant, magical event. In one instance, when Halley's comet appeared dramatically in 1066 CE, it was believed to be a bad sign for the Anglo-Saxon king Harold II but a good sign for William the Conqueror, duke of Normandy.

—*Tyler Biscontini*

### Further Reading

"About Comets." *Lunar and Planetary Institute.* USRA, 31 Oct. 2012. Web. 9 Mar. 2015.

"Anatomy of a Comet." *Rosetta.* Jet Propulsion Laboratory, Caltech, n.d. Web. 19 Feb. 2015.

Choi, Charles Q. "Comets: Facts about the 'Dirty Snowballs' of Space." *Space.com.* Purch, 15 Nov. 2014. Web. 9 Mar. 2015.

"Comets." *Solar System Exploration.* NASA, 3 Feb. 2014. Web. 12 Mar. 2015. "Comet Facts for Students." *JPL Education.* NASA, 2 Nov. 2010. Web. 19 Feb. 2015.

Koberlein, Brian. "What Are Comet Tails?" *Universe Today.* Universe Today, 21 July 2014. Web. 9 Mar. 2015.

"What Are the Types of Comet Tails?." *HubbleSite.* Space Telescope Science Inst., n.d. Web. 12 Mar. 2015.

# TAURUS

### FIELDS OF STUDY

Astronomy; Observational Astronomy

### SUMMARY

The constellation Taurus is a group of stars visible in the Northern and Southern Hemispheres. The main stars take on a shape that some people believe looks like a bull. In ancient times, people studied the stars for practical and religious purposes. Today, astronomers still look to the constellations to learn more about science. Taurus contains bright stars, paired stars, and important star clusters called the Hyades and Pleiades. In addition, the Crab Nebula and two meteor showers within the constellation attract the attention of Earth observers.

### PRINCIPAL TERMS

- **celestial equator:** the imaginary line in the night sky that follows Earth's equator, dividing the celestial sphere in half.
- **constellation:** a group of stars that can be seen in the night sky from Earth and is officially recognized by the International Astronomical Union.
- **declination:** the north-south position of a celestial body relative to the celestial equator, expressed in degrees of arc.
- **International Astronomical Union:** the authoritative international organization of astronomers.
- **right ascension:** the east-west position of a celestial body when viewed from the Earth's equator, defined in relation to the vernal equinox (one of two points at which the ecliptic intersects the celestial equator) and expressed in hours and minutes.

### The Bull Constellation

Taurus is a constellation, a section of the night sky as designated by the International Astronomical Union (IAU). Constellations are based on asterisms, or groups of stars that are interpreted as a picture or pattern. People have observed asterisms for thousands of years. In ancient times, people believed these star shapes had spiritual importance. They also helped people keep track of the changing seasons. Today, the IAU recognizes eighty-eight official constellations.

Taurus, one of the most famous of all constellations, looks like a line with V shapes on either end. Many ancient stargazers believed this shape represented a bull. These astronomers included Taurus in the zodiac, a circle of twelve or thirteen constellations through which the sun appears to travel over the course of a year. This path that the sun follows through the zodiac is called the ecliptic.

Modern astronomers locate constellations by measuring their position relative to the celestial equator, the projection of Earth's equator onto the night sky. Taurus has a declination of approximately +15 degrees and a right ascension of four hours. Declination and right ascension correspond to latitude and longitude on Earth.

### ATTRIBUTES OF TAURUS

The Taurus constellation can be seen from the Northern and Southern Hemispheres between the latitudes of 90 degrees north (the North Pole) and 65 degrees south. Although it is visible in the night sky from November until March, it is best viewed during January evenings around 9:00 p.m. Taurus is one of the most easily found and visually striking of the constellations.

The most distinctive feature of Taurus is a *V*-shaped asterism that makes up the bull's horns and head. Toward the base of the *V* is the bright red giant star alpha Tauri, better known as Aldebaran, which is more than forty times the size of the sun. Although it is about sixty-five light-years away from Earth, Aldebaran is still about the fourteenth brightest star in the night sky. Astronomers think of it as the eye of the bull, both for its powerful brightness and its blazing, seemingly bloodshot appearance.

At one end of the V is beta Tauri, better known as Elnath, a blue-white star that represents the tip of the bull's horn. Some of the other named stars that make up the bull's head and body include gamma Tauri (Hyadum I), epsilon Tauri (Ain), and eta Tauri (Alcyone). Lambda Tauri, in the middle of the bull's body, is an eclipsing binary star, which means it is actually a pair of stars that revolve around each other. To Earth observers, this motion results in regular eclipses in which one star blocks the light of the other. Theta Tauri, kappa Tauri, sigma Tauri, phi Tauri, and chi Tauri are other examples of double stars.

Taurus also contains several important deep-space objects. One of these is a diffuse nebula, a vast cloud of dust and gases floating in space. This cloud is variously designated as Messier 1 (M1), New General Catalog 1952 (NGC 1952), and Taurus A and is popularly known as the Crab Nebula due to its crab-like shape. It is the remnant of a supernova.

Taurus also contains two major star clusters, the Hyades and the Pleiades. The Hyades cluster, which contains about two hundred stars, helps form the "head" of the bull constellation. The clustered stars appear to encircle Aldebaran, but in fact they about 150 light-years away, compared to Aldebaran's 65 light-years. The cluster of stars and dust known as Pleiades, sometimes called the Seven Sisters, is located above the body of the bull.

### MYTHOLOGY

Historians today recognize Taurus as one of the first constellations to be identified. Stories about the constellation date back to the early Bronze Age, and it may have been recognized even earlier than that. Some scientists believe that humans depicted a bull in the stars in cave paintings dating back to distant prehistoric times.

Taurus was also one of the most important constellations. Its position in the zodiac marked the spring equinox to stargazers in ancient Greece and the Middle East. To many cultures, the appearance of the bull represented the beginning of a bright new year and hopes for a productive harvest season.

Cultures around the world developed their own myths and lore about the star pattern. The most detailed myths that have survived to the present day, and those most influential to the modern view of Taurus, originated in ancient Greece. Many ancient Greek myths dealt with bulls, which were considered at the time to be symbols of strength and fertility.

In one myth, the king of the gods, Zeus, fell in love with a beautiful princess named Europa. Zeus appeared before Europa disguised a white bull with golden horns and knelt down. The princess climbed on its back and the bull went racing across the seas to a faraway island, where Zeus finally revealed his true identity.

Another myth involves a bull sent to King Minos of Crete. The king was supposed to sacrifice the animal to the sea god, Poseidon, but neglected to do so. Poseidon was angered by the apparent slight and arranged for Minos's wife, Pasiphaë, to fall in love with the bull. Pasiphaë then gave birth to the Minotaur, a half-bull, half-human monster that would later be killed by the hero Theseus.

Other Greek myths involve heroes hunting bulls. One of the greatest Greek mythical heroes, Heracles (romanized as Hercules), was tasked with subduing the bull of Crete. Other legends tell of the great hunter Orion stalking bulls for prey. This latter tale is reflected in the constellations, as Taurus and Orion seem to be eye-to-eye, forever locked in combat.

Although the Greek myths are most widely recognized today, many other cultures around the world developed their own interpretations of the star patterns. Early Arabian peoples also imagined a bull in the stars. Some Egyptian astronomers drew a connection to their god Apis, who took the shape of a bull. Other Egyptians instead made a connection to the god Osiris, who had a human form but was usually depicted with a large, two-peaked hat that resembled horns. Chinese astronomers thought of the constellation as a tiger or a bridge in the stars.

Other myths relate to the formations within Taurus. The Greeks had related the stars of the Pleiades cluster to the seven mythical daughters of the Titan Atlas and the nymph Pleione. The Zuni people of New Mexico considered the Pleiades a symbol of seeds. When these "star seeds" appeared in the sky, it was a clear reminder that the time had come to plant crops for the new harvest season.

## Studying Taurus Today

Ancient people considered the constellations extremely important. These star patterns offered spiritual meanings as well as a calendar of the changing seasons. Today, most people look to the stars for other reasons. Amateur stargazers enjoy the beautiful marvels of the heavens. Professional astronomers study the constellations to learn more about their various features and gain scientific insights into the workings of the universe.

Taurus is popular with stargazers because some of its most impressive features, such as Aldebaran and the Pleiades, are often visible to the naked eye. Other main parts of the constellation can be seen with binoculars or small telescopes. Experienced stargazers have developed several techniques for quickly locating Taurus in the night sky. One of the most

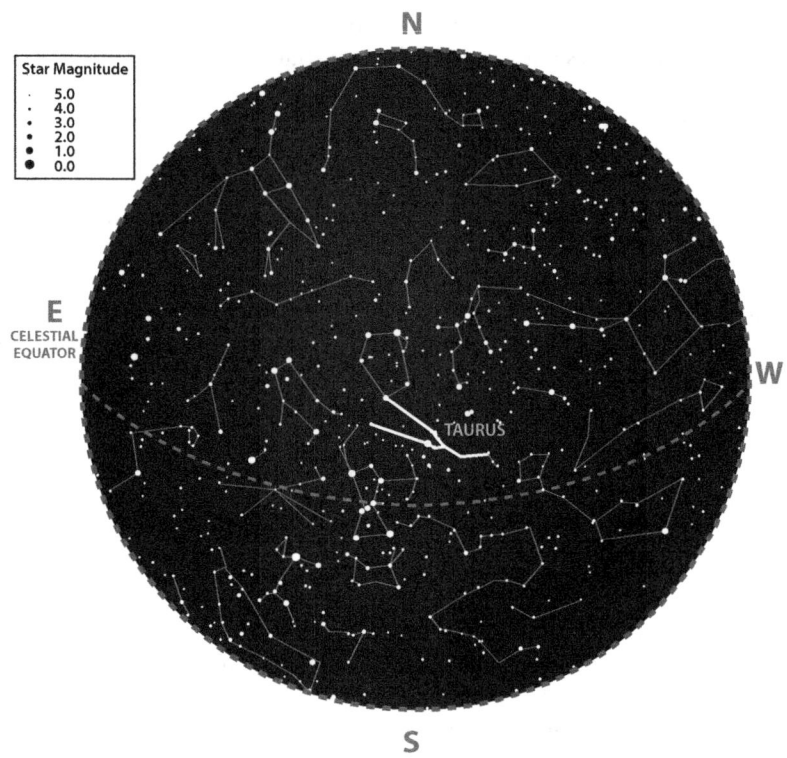

Star chart of the Northern and Southern Hemispheres in November. Lines connect the stars of the constellation Taurus.

popular is finding Orion's belt, which will lead the observer's eye to Aldebaran and the Hyades cluster. From there, observers can locate the rest of the constellation with ease.

Astronomers often look to Taurus to study developments among its stars, clusters, and nebula. In addition, each year many telescopes turn to this constellation in search of meteors. From September through December, the Taurid meteor showers takes place. These showers cover both the Northern and Southern Hemisphere, peaking at different times for each hemisphere. In June and July, daylight meteor showers called the Beta Taurids can send about twenty-five meteors streaking through the sky every hour.

—*Mark Dziak*

## Further Reading

Kronberg, Christine. "Taurus." *The Munich Astro Archive*. U Observatory Munich / Leibniz Supercomputing Centre, 21 Mar. 1997. Web. 6 May 2015.

McClure, Bruce. "Taurus? Here's Your Constellation." *EarthSky*. Earthsky Communications, 5 Jan. 2015. Web. 6 May 2015.

Sessions, Larry, and Deborah Byrd. "Aldebaran Is the Bull's Fiery Eye." *EarthSky*. Earthsky Communications, 29 Dec. 2014. Web. 7 May 2015.

"Taurus (the Bull)." *Chandra X-Ray Observatory*. NASA, 2 Dec. 2013. Web. 6 May 2015.

"Taurus, the Bull." *StarDate*. University of Texas McDonald Observatory, 2015. Web. 6 May 2015.

Zimmermann, Kim Ann. "Taurus Constellation: Facts about the Bull." *Space.com*. Purch, 14 Aug. 2012. Web. 6 May 2015.

# THEORETICAL ASTROPHYSICS

## FIELDS OF STUDY

Astrophysics; Astronomy; Cosmology

## SUMMARY

Theoretical astrophysics involves the study of the physics behind celestial objects, such as stars and planets, and other phenomena, including black holes and dark energy. It includes observation and analysis of their structures, properties, processes, and behaviors. Theoretical astrophysics is also applied to the formation of galaxies and the expansion of the universe. Among the concepts that have come from theorists in this branch of science are the big bang theory, cosmic inflation, and relativity.

## PRINCIPAL TERMS

- **big bang:** the theoretical initial explosion that caused the universe to expand outward from a single, infinitely dense point.
- **dark energy:** energy of an unknown nature that is believed to be accelerating the expansion of the universe.
- **dark matter:** a hypothetical type of matter that cannot be directly observed but is believed to constitute most of the matter in the universe.
- **general relativity:** the theory, developed by Albert Einstein, that gravity is the result of matter causing space-time to curve.
- **cosmic inflation:** the theory that a sudden, exponential expansion of the universe occurred within the first second of the big bang.
- **polytrope:** a solution to a certain equation that can be used as an approximate model of a star.
- **wormhole:** a hypothetical tunnel through time and space.

## THE UNIVERSE IN THEORY

Theoretical astrophysics is a branch of astronomy that is concerned with the study of the universe. It involves not only the physical aspects of stars and planets but also some of the mysteries of outer space, such as dark matter, dark energy, and possible wormholes. Astrophysicists try to understand planets' movements and their effects on each other, how light travels in space, and the causes and effects of gravity. They rarely can make close-up examinations or measurements, and some of the phenomena they study are invisible. Using computers, astrophysicists construct models or simulations that allow them to analyze and test their theories.

## PIONEERS IN THEORETICAL ASTROPHYSICS

While the science of astronomy dates back thousands of years, astrophysics began to emerge as a subfield only in the nineteenth century. Early astrophysicists included Swiss astronomer and meteorologist Robert Emden (1862–1940), who studied the relationship between the pressure and density within stars. He developed an early mathematical model called a polytrope. A polytrope is a solution to an equation known as the Lane-Emden equation, named for Emden and American astrophysicist Jonathan Homer Lane (1819–80). This equation describes the structure of a gaseous sphere whose gravity is balanced by its internal pressure. A polytrope can serve as a simplified model for analyzing the structure of stars.

More early contributions to the field of astrophysics came from American astronomer Edwin Hubble (1889–1953). In the 1920s, astronomers were still debating whether the Milky Way galaxy constituted the whole of the universe. Hubble was the first to demonstrate conclusively that what were previously thought to be nebulae within the Milky Way were in fact other galaxies outside of it. He also helped discover the expansion of the universe and calculate its age.

The work of German-born physicist Albert Einstein (1879–1955) in the early twentieth century closed some longstanding gaps in theoretical physics. Einstein was the first to determine that the speed of light is always constant. He also developed the theory of general relativity, which states that a large object, such as a planet, can warp space-time, just as a heavy ball placed on an outstretched rubber sheet will warp its shape. The indentation caused by the object creates a gravity well, resulting in gravitational pull, which in turn can cause light to bend.

Russian-born physicist George Gamow (1904–68) helped develop the big bang theory, which posits that the universe began with the explosive expansion of a singularity, or a single point of infinite density. Gamow believed that if this theory were correct, scientists should be able to find the radiation left over from this explosion throughout the universe. This radiation, later dubbed cosmic microwave background radiation (CMB), was discovered by American physicists Robert Woodrow Wilson (b. 1936) and Arno Allan Penzias (b. 1933) in 1965, nearly twenty years after Gamow predicted its existence. Wilson and Penzias shared the 1978 Nobel Prize in Physics for their discovery.

American astronomer Carl Sagan (1934–96) was another pioneer in astrophysics. Sagan was the director of the Laboratory for Planetary Studies at Cornell University. In addition to astrophysics, he specialized in many other subfields of astronomy, including cosmology, astrobiology, and planetary science. Sagan was well known both as a popular-science writer and for presenting the award-winning documentary series *Cosmos: A Personal Voyage* (1980). He also shared his fascination with exobiology, or the search for and study of extraterrestrial life, with many people who respected his ideas on the subject.

More recently, theoretical physicists such as Stephen Hawking (b. 1942) have brought more new ideas and evidence to the field. Hawking focused particularly on the study of black holes. In 1974, he proposed that, contrary to prior scientific understanding, some energy can in fact escape from black holes. Based on the nature of this hypothetical energy, later named Hawking radiation, Hawking suggested in 1981 that physical information can be irretrievably lost within a black hole. This suggestion came to be known as the "information paradox," so called because it violates the tenets of quantum theory. Since then, Hawking has revised his conclusions. In 2014, he proposed that instead of having an event horizon from which nothing can escape, a black hole has an "apparent horizon" that can change.

The big bang theory was developed to explain the physics of the universe. According to this theory, within the first second after the big bang, the universe expanded extremely rapidly during a period called cosmic inflation. Following cosmic inflation, the universe continued to expand, though at a slower rate.

## THEORIES IN ASTROPHYSICS

Astrophysicists have continually developed and improved on important theories related to the creation of the universe. In some cases, new theories and knowledge support earlier ideas. In others, assumptions are proved wrong and new theories must be developed. For example, the theory of cosmic inflation was developed in the 1980s to resolve outstanding problems with the big bang theory. One such problem was the uniform temperature of the universe in all directions. Without a period of rapid inflation in the earliest moments of the universe, there simply would not have been enough time since the big bang for the universe to reach thermal equilibrium.

Another such problem also involves the expansion of the universe. Astrophysicists had long supposed that the rate of expansion was gradually slowing due to the effects of gravitational force. However, in 1998, data from the Hubble Space Telescope revealed that it is actually speeding up. Astrophysicists scrambled to explain how this could happen. While the cause has yet to be conclusively identified, the most widely accepted hypothesis is that some strange force in the universe is causing the acceleration. This force, dubbed dark energy, is an as-yet-unknown form of energy that is believed to account for about 68.3 percent of the total mass and energy of the universe. In an effort to learn more, the National Aeronautics and Space Administration (NASA) assisted the European Space Agency (ESA) in developing the Planck mission. In 2009, the ESA launched the *Planck* spacecraft, an unmanned observatory that carried instruments capable of measuring variations in the oldest light in the universe. NASA also helped analyze the data.

*Planck* collected evidence of both dark energy and dark matter, a hypothetical form of matter that makes up an estimated 84.5 percent of all matter in the universe, or 26.8 percent of the total mass-energy content. (Ordinary matter accounts for only 4.9 percent of the universe's mass-energy.) Dark matter is so called because it does not emit, reflect, or absorb light and thus cannot be directly observed. Its existence was first proposed in the 1930s to account for the discrepancy between the observable masses of galaxies and calculations of their masses based on their gravitational effects. The percentage of dark matter in the universe was similarly calculated from the gravitational force exerted on observable matter.

The Planck mission lasted until 2013. At that point it had collected enough information for scientists to process the data through a supercomputer. They were then able to revise prior theories on the age of the universe, which is now believed to be 13.8 billion years old.

Another popular subject in theoretical astrophysics is the possible existence of wormholes. Wormholes are hypothetical tunnels through time and space that could allow for shortcuts in space travel—and, in science-fiction stories, often do. Although there is no physical evidence that they exist, Einstein's theory of general relativity means that they are mathematically possible. Some theorists believe that if large, stable examples of wormholes could be found, they might make space travel and even time travel possible.

—*Nancy Comstock*

**FURTHER READING**

"Astronomy." *Astrophysical.org*. Istituto Scientia, 2004–14. Web. 4 May 2015.

Clavin, Whitney. "Supercomputer Helps Planck Mission Expose Ancient Light." *Planck*. California Inst. of Technology, 21 Mar. 2013. Web. 4 May 2015.

Durham, Ian T. "Emden, Robert." *Biographical Encyclopedia of Astronomers*. Ed. Thomas Hockey et al. 2nd ed. Vol. 2. New York: Springer, 2014. Print.

Magrass, Yale R. "Stephen Hawking." *Salem Press Biographical Encyclopedia* (2014): n. pag. *Research Starters*. Web. 3 June 2015.

Redd, Nola Taylor. "What Is a Wormhole?" *Space.com*. Purch, 13 Apr. 2015. Web. 4 May 2015.

Tate, Karl. "Einstein's Theory of Relativity Explained (Infographic)." *Space.com*. Purch, 5 Mar. 2015. Web. 4 May 2015.

Wanjek, Christopher. "Ringside Seat to the Universe's First Split Second." *NASA*. NASA, 16 Mar. 2006. Web. 4 May 2015.

"What Is Dark Matter?" *NASA Education*. NASA, 22 Feb. 2012. Web. 4 May 2015.

# THREE-BODY PROBLEM

### FIELDS OF STUDY
Astrometry; Orbital Mechanics; Theoretical Astrophysics

### SUMMARY
The three-body problem is a concept that involves the orbital motions of celestial bodies in a three-body system. The problem arises when attempting to predict how three given bodies will orbit one another around the common center of mass. Sir Isaac Newton addressed orbital motions in the 1680s, but could not predict the way three bodies would orbit each other. Jules Henri Poincaré developed the three-body problem in the 1880s. Other scientists, including Joseph-Louis Lagrange and Leonhard Euler, Roger Broucke and Michel Hénon, and Milovan Šuvakov and Veljko Dmitrašinović, discovered solutions to the three-body problem. Each solution is generally grouped into one of sixteen families.

### PRINCIPAL TERMS

- **common center of mass:** balance point of a system of orbiting celestial bodies; also known as the barycenter.

### BRIEF HISTORY
The three-body problem has roots in the work of English physicist Sir Isaac Newton (1643–1727) in the 1680s. Newton's law of gravity proved that the orbits of two celestial bodies could always be accurately predicted, and that they almost always orbit each other in an elliptical path. Newton, however, was unable to predict the way in which three celestial bodies in a three-body system would repeatedly orbit one another. Scientists had struggled with this problem for about two centuries by the time Jules Henri Poincaré (1854–1912) developed his work on it in the late 1880s.

Poincaré claimed that a three-body system is a chaotic deterministic system. The term "deterministic" suggests that the positions and speeds of the three bodies are fixed, but Poincaré believed that the system is also chaotic. This means that a disturbance in the original state of the bodies could cause differences in a later state, and the final state would not be able to be predicted. An example of such a disturbance is a small change in the initial position of one of the bodies. Poincaré's work on the three-body problem contained the first description of chaotic behavior in a dynamic system. His paper on the subject won an international competition sponsored by King Oscar II of Sweden and Norway in 1889 and was published in *Acta Mathematica* in 1890.

### BASICS OF THE THREE-BODY PROBLEM
The three-body problem involves predicting the way the three objects will repeatedly orbit one another in a three-body system. To grasp the concept, an understanding of the common center of mass is generally required. Every celestial body has a center of mass, or balance point. This balance point may be located in the middle of the body or somewhere else. For example, if a celestial body was thought of as a ruler, its center of mass would be exactly in the center of the ruler because a ruler is uniform, with both ends being equally heavy. Placing one's finger right in the middle of the ruler would balance the ruler. On the other hand, comparing a celestial body to a nonuniform object such as a hammer would place its center of mass somewhere other than its center. Trying to balance a hammer by placing a finger at its center would cause the hammer to tip over. This is because one end of the hammer is heavier than the other end. Therefore, the hammer's center of mass is closer to the heavier end.

The center of mass also applies to the orbit of a celestial body. The body will orbit another body based on the common center of mass of the system, which is also called the barycenter. One celestial body does not technically orbit another celestial body; the two bodies actually orbit each other. The reason for this is gravity. Gravity causes the bodies to pull on each other, and there is a balance point. The balance point is the common center of mass. The common center of mass is closer to the body with the larger mass. The two bodies, therefore, orbit each other around this balance point.

A system containing more than two celestial bodies also has a common center of mass. Three celestial

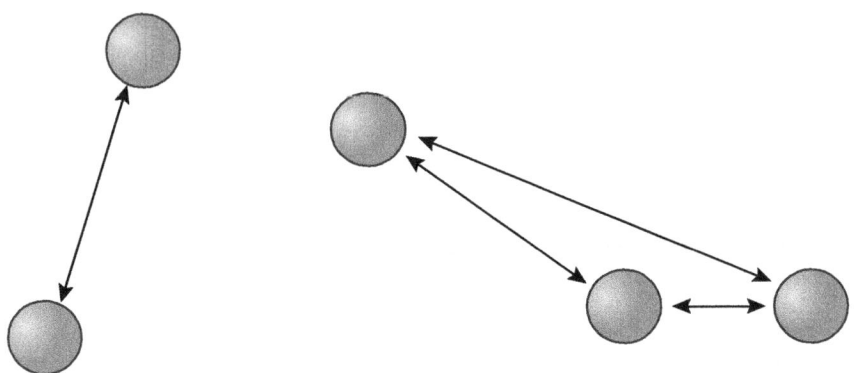

A two-body diagram has only one line of interaction between the objects. A three-body diagram has three lines of interaction among the objects. The number of interactions among objects increases faster than the number of objects.

bodies will orbit one another around the common center of mass of all three bodies. A good example is the system of the earth, moon, and sun. The earth has more mass than the moon, so the common center of mass between the two bodies is inside the earth. The earth and moon orbit each other around this common center of mass. Similarly, the sun has more mass than the earth, so the two bodies' common center of mass is inside the sun, and they orbit each other around this balance point.

## Solutions to the Three-Body Problem

German mathematician Heinrich Bruns claimed that a general solution to the three-body problem was impossible. He further argued that the only possible solutions were those that could be used under certain conditions. Various solutions to the three-body problem have been discovered, however. Typically, the solutions fall into one of several families of solutions, including the Lagrange-Euler family, the Broucke-Hénon family, and the figure-eight family.

In the eighteenth century, mathematicians Joseph-Louis Lagrange (1736–1813) and Leonhard Euler (1707–83) identified the solutions that eventually fell under the Lagrange-Euler family. The solutions in this family feature three equally spaced bodies that rotate around in a circle. The solutions in the Broucke-Hénon family were discovered in the 1970s by American mathematician Roger Broucke and French astronomer Michel Hénon. These solutions have a complex configuration, as they contain two bodies that orbit on the inside and one body that orbits on the outside. The solutions in the figure-eight family were first alluded to in a paper by American mathematician Richard Moeckel in 1988, but were discovered by American physicist Cristopher Moore in 1993. The figure-eight solutions feature three bodies that orbit one another in a shape that resembles an eight. These solutions were later proven by French mathematician Alain Chenciner and American mathematician Richard Montgomery. Spanish mathematician Carles Simó further investigated the solutions.

In 2013, thirteen new families of solutions to the three-body problem were discovered. Two physicists from the Institute of Physics Belgrade, Milovan Šuvakov and Veljko Dmitrašinović, made the discoveries. They used a computer simulation to come up with the new solutions. They also created a new classification system for the solutions and also used a "shape-sphere" to illustrate the shape of the bodies' orbits. One of the new solutions has the appearance of a ball of yarn.

Counting the original families of solutions along with Šuvakov and Dmitrašinović's new solutions, there are a total of sixteen families of solutions to the three-body problem. However, only one of these families is accepted as actually existing. The Lagrange-Euler family can be seen in the system that consists of Jupiter, the sun, and a nearby asteroid.

—*Michael Mazzei*

## Further Reading

Barrow-Green, June. "Oscar II's Prize Competition and the Error in Poincaré's Memoir on the Three Body Problem." *Archive for History of Exact Sciences* 48.2 (1994): 107–31. Print.

Cartwright, Jon. "Physicists Discover a Whopping 13 New Solutions to Three-Body Problem." *Science*. Amer. Assoc. for the Advancement of Science, 8 Mar. 2013. Web. 15 May 2015.

Casselman, Bill. "A New Solution to the Three Body Problem." *Feature Column*. American Mathematical Society. Web. 15 May 2015.

Gray, Jeremy John. "Poincaré, Henri." *Britannica Biographies* 1 Mar. 2012: 1–2. *Biography Reference Center.* Web. 9 June 2015.

Murzi, Mauro. "Jules Henri Poincaré (1854–1912)." *Internet Encyclopedia of Philosophy.* Internet Encyclopedia of Philosophy. Web. 15 May 2015.

Valtonen, M. J., and Hannu Karttunen. *The Three-Body Problem.* Cambridge: Cambridge University Press, 2005. *eBook Collection (EBSCOhost).* Web. 9 June 2015.

# TRANS-NEPTUNIAN OBJECTS

## FIELDS OF STUDY

Observational Astronomy; Sub-planet Astronomy

## SUMMARY

Trans-Neptunian objects (TNO), including Kuiper Belt objects (KBO), are small, icy celestial bodies that are in solar orbit beyond the orbit of Neptune. The first TNO to be discovered was Pluto in 1930. The orbital behavior and characteristics of TNOs provide clues to what lies beyond the boundaries of the known solar system.

## PRINCIPAL TERMS

- **detached objects (resonant):** small icy bodies in the Kuiper Belt that are in elongated orbits between 35 and 100 astronomical units (AU) from the sun and are far enough away from Neptune that they are not affected by its gravity.
- **Kuiper Belt:** a ring-shaped section of the solar system populated with small ice and rock objects; also known as the Edgeworth-Kuiper Belt.
- **Kuiper Belt objects (classical and resonant):** small, icy space bodies in the Kuiper Belt, also known as trans-Neptunian objects; most KBOs are classical, with orbits of low eccentricity and inclination; resonant KBOs orbit the sun in a consistent ratio to Neptune's orbit.
- **minor planets:** a celestial object that does not meet the criteria to be a planet or a comet but orbits around the sun.
- **Oort cloud:** a theoretical vast collection of icy space bodies surrounding the solar system beyond the known planets and the Kuiper Belt.
- **plutinos:** KBOs found in the inner edges of the Kuiper Belt that have the same orbital resonance as Pluto, or a resonance of 3:2.
- **scattered disk objects:** also known as scattered Kuiper Belt objects; space bodies made of ice and rock with very eccentric, or flattened oval, orbits that can go both closer and farther from the sun than the other types of KBOs.

## ORIGIN AND DISCOVERY OF TRANS-NEPTUNIAN OBJECTS

Trans-Neptunian objects (TNOs) are thought to have been formed outside Neptune's orbit in the earliest days of the solar system about 4.6 billion years ago. Many TNOs are likely to be found within one of two regions outside of Neptune's orbit: the Kuiper Belt or the scattered disk. There are differing theories on whether these objects were formed where they are or whether the effects of Neptune's migration into its current position in the solar system moved them to where they are. In either case, the centers of these rocky bodies, which are also known as Kuiper Belt objects (KBOs), likely contain untouched material from the origins of the solar system.

The Kuiper Belt, the innermost trans-Neptunian region, is so far away from Earth that its existence was only a theory for many years. Clyde Tombaugh discovered Pluto, then considered a planet, beyond the orbit of Neptune in 1930. It was thought that other objects near planet size might be out there, but none were found until 1992. In that year, astronomers David Jewitt (b. 1958) and Jane X. Luu (b. 1963) confirmed the existence of the first TNO, 1992 QB1.

## TYPES OF TNOs

This first discovery led to the identification of other objects in this region of space and the realization that Pluto had more in common with those objects than with the eight primary planets. Pluto is larger than most TNOs and relatively reflective, making it easier to see from Earth. This is why it was designated a planet in 1930. However, with the discovery of other TNOs of similar size and orbit, Pluto was reclassified as a dwarf or minor planet in 2006 by the International Astronomical Union. Minor planets are round and orbit the sun like the primary planets do but are not large enough to knock other objects out of their orbits. They are smaller than the primary planets. They are not thought to be capable of maintaining any known form of life.

Plutinos are another type of TNO. They have the same 3:2 orbital resonance as Pluto, meaning they make three trips around the sun for every two made by Neptune. Because Pluto shares this resonance, it is also considered to be a plutino. The first plutino other than Pluto was discovered in 1993. Plutinos make up the largest class of resonant TNOs, or objects that orbit in resonance with Neptune. The orbital paths of many plutinos cross Neptune's orbit, but because they orbit in resonance, they will never collide. Plutinos make up the inner edge of the Kuiper Belt. This part of the belt is closest to the sun.

Scattered disk objects (SDO) are also known as scattered Kuiper Belt objects. These bodies of ice and rock collect in the scattered disk, a flat band encircling the solar system. They have very eccentric, or thin and long, orbits that bring them as close as 35 AU to the sun or more than 100 AU away. With such eccentric orbital paths, SDOs can cross through the Kuiper Belt but also travel well beyond it through the Oort cloud and toward the outer regions of the solar system. Scientists theorize that some objects in the Oort cloud may have originated in the scattered disk, and that comets can originate in both areas.

Detached objects are another type of TNO. Even when their orbit is at its perihelion, or closest point to the sun, detached objects are far enough away from Neptune that it and the other planets can exert only minimal gravitational forces against them. As a result, they seem to be detached from the rest of the solar system. It is theorized that the long elliptical orbits of detached objects are caused by passing contact from a star other than Earth's sun or by the influences of planets other than the gas giant planets such as Neptune and Jupiter. This makes them different from the other TNOs, all of which have orbits affected by the influence of Neptune or the other large planets. The most commonly known detached object is Sedna. Discovered in 2003, Sedna is a planet-like body at the very edge of the solar system. An estimated 1,300 to 1,700 kilometers (about 800 to 1,000 miles) in diameter, Sedna is larger than an asteroid but smaller than Pluto. Its orbit ranges between about 76 and 972 AU from the sun and takes about 12,000 years to complete. From Sedna the distant sun looks like a bright star.

## SIGNIFICANCE OF TNOs

It might seem that TNOs are so far from Earth that they could be of little significance for study or discovery, but scientists find them to be very important. TNOs are believed to be made up of the material left over from the beginnings of the universe. They have been untouched in the deep regions of space for billions of years. This makes them valuable sources of information about the formation of the universe. It is also thought that the region where they are found could be populated with additional planets.

Using archived Hubble Space Telescope photos, scientists have identified and named a number of

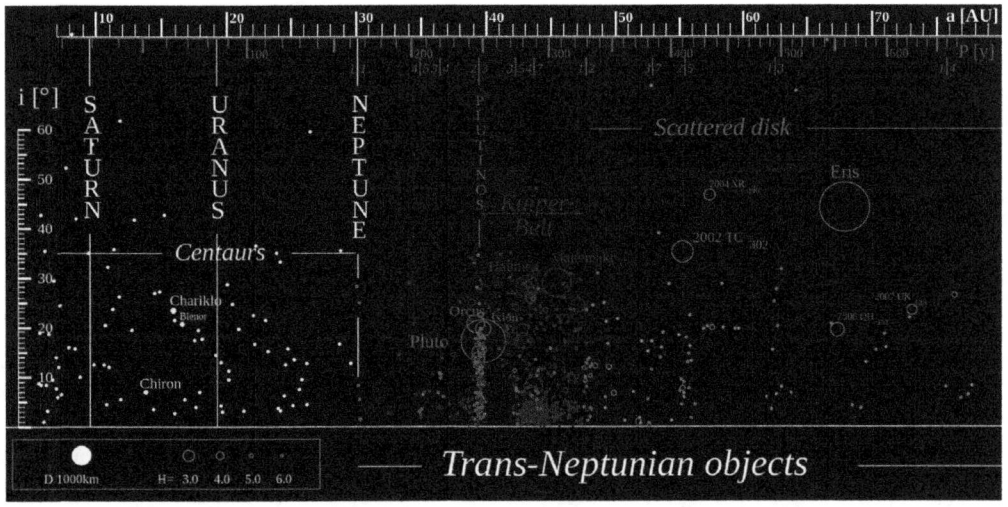

Trans-Neptunian objects are found within the Trans-Neptunian Asteroid Belt, which falls outside of Neptune's orbit (includes the Kuiper Belt and some scattered disk objects). By Eurocommuter via Wikimedia Commons.

TNOs. The Hubble Telescope uses long-exposure photography to capture images as it peers into space. This means the camera lens stays open and captures light for a period of time. When the telescope is pointed at one object and something else moves between the object and the lens, its movement creates a streak of light. Scientists determined that when Hubble is pointed something at the far edge of the solar system and a streak is seen, it is a TNO. More than a dozen new TNOs were identified in early 2015 by this technique.

Scientists have also used mathematical calculations to determine that there could be two or more TNOs that qualify as planets beyond the orbit of Pluto. In observing a number of objects in this region of space, scientists noticed that several showed characteristics that cannot be accounted for unless they are under the influence of a planetary body. Since these TNOs are outside the reach of the influence of Neptune, the logical deduction is that there are other planets beyond the known solar system bodies that are affecting the orbit of these objects.

Study will continue into the TNOs, and an additional tool has been deployed to help with the efforts. The National Aeronautics and Space Administration launched the New Horizons mission to Pluto in 2006. By mid-2015, the probe was more than halfway to its destination. It will spend between five and ten years gathering more information about trans-Neptunian objects.

—*Janine Ungvarsky*

**FURTHER READING**

Chambers, John and Jacqueline Mitton. *From Dust to Life: The Origin and Evolution of Our Solar System.* Princeton: Princeton University Press, 2013. Print.

Jaggard, Victoria. "Pluto Gets 14 New Neighbors." *National Geographic: Voices.* Natl. Geographic Soc., 15 Sept. 2010. Web. 10 May 2015.

"Kuiper Belt." *Cosmos: The SAO Encyclopedia of Astronomy.* Swinburne Centre for Astrophysics and Supercomputing, Swinburne University of Technology, n.d. Web. 10 May 2015.

"Kuiper Belt and Oort Cloud." *Solar System Exploration.* National Aeronautics and Space Administration, 19 Nov. 2014. Web. 10 May 2015.

"Kuiper Belt Objects." *Cosmos: The SAO Encyclopedia of Astronomy.* Swinburne Centre for Astrophysics and Supercomputing, Swinburne University of Technology, n.d. Web. 10 May 2015.

Plataforma SINC. "Trans-Neptunian Objects Suggest That There Are More Dwarf Planets in Our Solar System." *ScienceDaily.* ScienceDaily, 15 Jan. 2015. Web. 10 May 2015.

"Scattered Disk Objects." *Cosmos: The SAO Encyclopedia of Astronomy.* Swinburne Centre for Astrophysics and Supercomputing, Swinburne University of Technology, n.d. Web. 10 May 2015.

# TROJAN ASTEROIDS

## FIELDS OF STUDY

Astronomy; Observational Astronomy

## SUMMARY

Trojan asteroids are small, rocky objects that share an orbital period with a planet and maintain a consistent distance from the planet. This makes collision between the planet and a Trojan asteroid unlikely. Trojan asteroids' stable paths make them good candidates for future space study missions, and their composition could tell scientists much about the early solar system.

## PRINCIPAL TERMS

- **asteroid:** a small, rocky space object that orbits the sun and is categorized by its composition.
- **Lagrangian points:** a set of five points found in relation to two large objects, such as a planet and the sun, where gravitational forces allow a third, smaller object to maintain a consistent distance from the other two; also known as libration points or L-points.
- **libration points:** another name for Lagrangian points, or the five points shared by two large objects where gravitational forces allow a third ob-

ject to maintain a consistent distance from the other two.
- **Trojan asteroid:** an asteroid that shares an orbital period with a planet and maintains a consistent distance from it at the Lagrangian points.

## ORBITAL PATHS OF TROJANS

Some asteroids cross paths with planets and could collide with them. Trojan asteroids share an orbital period with a planet and orbit at a consistent distance from that planet. This reduces the potential for collisions. This stable path is possible because the asteroids are located at the libration points. Libration points are places where the gravitational forces of the sun and a planet are such that a third, small body is able to remain steady. Trojan asteroids hold a consistent orbit alongside, ahead of, or behind a planet as it orbits the sun.

These libration points are also called Lagrangian points after Joseph-Louis Lagrange (1736–1813). Lagrange was a mathematician who tried to calculate a solution to the "three-body problem." Scientists and mathematicians had long tried to figure out the relative motion of three space bodies based on their mass, velocity, and position. Lagrange determined that if one body is much smaller than the other two, it is possible to identify five points in relation to the two larger objects where the smaller object could maintain a consistent distance from the others. At the Lagrangian points, the gravitational force of the planet is equal to that of the sun. Although the sun is much more massive than the planet, the sun is so far from the Trojan asteroid that it exerts a gravitational pull equal to the planet's. These points of equilibrium became known as the Lagrangian points, or L-points, and are labeled L1 through L5. Asteroids located in the more stable L4 or L5 positions form one point of an equilateral or equal-sided triangle. The sun and the planet make up the other two points.

## WHERE TROJANS ARE FOUND

Most of the known Trojan asteroids share an orbital period with Jupiter. The first Trojan asteroid to be discovered, 588 Achilles, was first identified in orbit around Jupiter by astronomer Max Wolf (1863–1932) in 1906. About 65 percent of Jupiter's Trojans are in orbit ahead of the planet in the L4 position, and the rest trail behind the planet in the L5 spot. The gravitational forces of other planets—most notably Saturn—cause slight oscillations in the orbit of those asteroids. However, Jupiter's gravitational force is sufficient to keep them within a small range of variation. These asteroids are named after warriors from the Trojan War. L4 asteroids have Greek names, and L5 asteroids have Trojan names. There are two exceptions to this rule. One of Jupiter's L4 asteroids was named after Hektor, a Trojan who spied on the Greeks, and an L5 asteroid was named Patroclus, after a Greek warrior who tricked the Trojans.

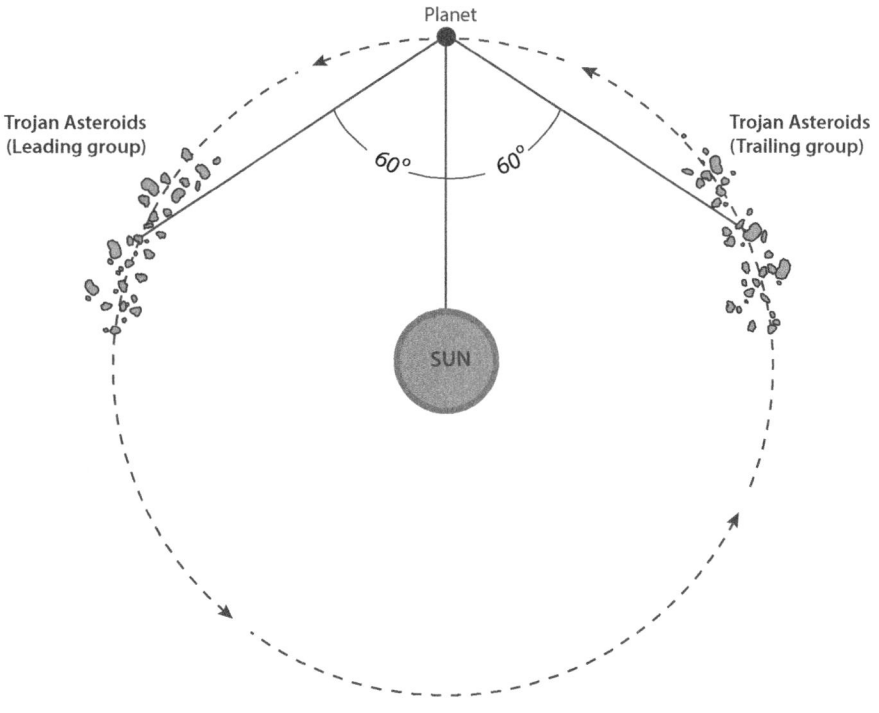

Trojan asteroids orbit the sun in the same path as a particular planet, either in front of or following the planet. The angles from the sun to the planet to the asteroids are 60 degrees.

While most of the known Trojan asteroids are near Jupiter, they have been found orbiting several other planets as well, including Mars and Neptune. In 2011, scientists using the National Aeronautic and Space Administration's Wide-Field Infrared Survey Explorer (WISE) space telescope discovered a Trojan asteroid orbiting with Earth. Scientists had long theorized that a Trojan asteroid would be found near Earth. However, the asteroid's small size and tendency to be obscured by the sun's brightness made finding it difficult. Earth's Trojan asteroid is about 300 meters (1,000 feet) across and 20 million kilometers (12.4 million miles) ahead of Earth in its orbital path. Its name is 2010 TK7.

**FUTURE TROJAN SPACE BASES**
Asteroids are valuable to scientists because they are made up of materials left over from the formation of the solar system about 4.6 billion years ago. Because of their stable paths, Trojan asteroids are less likely than other asteroids to have collided with other objects. Trojan asteroids are of interest to astronomers because they have likely gone unchanged for billions of years.

Their consistent presence near other planets gives Trojan asteroids further potential for mining sites and as space bases and observation posts. The possibility of traveling to a Trojan asteroid could provide valuable data about the origins of the solar system. It could also provide a base for observation of other parts of the galaxy.

—*Janine Ungvarsky*

**FURTHER READING**
"Asteroids: Overview." *Solar System Exploration*. NASA, 13 Jan. 2015. Web. 6 Mar. 2015.

Celletti, Alessandra, and Ettore Perozzi. *Celestial Mechanics: The Waltz of the Planets*. Chichester: Praxis, 2007. Print.

Clavin, Whitney, and Trent J. Perrotto. "NASA's WISE Mission Finds First Trojan Sharing Earth's Orbit." *Mission News*. NASA, 27 July 2011. Web. 6 Mar. 2015.

Cornish, Neil J. "The Lagrange Points." *Wilkinson Microwave Anisotropy Probe*. NASA, July 2012. Web. 6 Mar. 2015.

Lang, Kenneth R. *The Cambridge Guide to the Solar System*. 2nd ed. Cambridge: Cambridge University Press, 2011. Print.

Than, Ker. "Trojan Asteroid Found Sharing Earth's Orbit—A First." *National Geographic News*. Natl. Geographic Soc., 28 July 2011. Web. 6 Mar. 2015.

# U

## URSA MAJOR

### FIELDS OF STUDY

Stellar Astronomy; Observational Astronomy

### SUMMARY

Ursa Major is a constellation best seen in the Northern Hemisphere. The main stars in Ursa Major form a shape that many observers believe resembles a bear. The name Ursa Major means "great bear." Some of the brightest stars in the constellation form a scoop-like shape, which is called the Big Dipper. In addition, Ursa Major includes interesting varieties of stars, many galaxies (groups of stars and planets), and a nebula (a huge cloud of gas and dust). People have been aware of this constellation for thousands of years. Ursa Major has a very rich, colorful history in myth and lore from cultures around the world.

### PRINCIPAL TERMS

- **celestial equator:** the imaginary line above Earth's equator that halves the celestial sphere; it is equally distant from the celestial poles.
- **constellation:** a region of space defined by a pattern of stars that can be seen in the night sky from Earth.
- **declination:** a space object's angular distance north or south of the celestial equator, expressed in degrees of arc.
- **International Astronomical Union:** an association of professional astronomers from all over the world who define astronomical constants while promoting research, education, and discussion on important astronomical topics.
- **right ascension:** a space object's longitudinal arc along the celestial equator, measured eastward from the vernal equinox and expressed in hours.

### THE GREAT BEAR CONSTELLATION

Ursa Major is a constellation. A constellation is a region of space defined by a group of stars that resembles a pattern or picture when seen from Earth. For thousands of years, people have searched for such patterns among the stars. People used these patterns to track the changing seasons. Some believed the images held spiritual significance as well. Many constellations have been identified by world cultures, and astronomers recognize eighty-eight constellations today.

Possibly the best-known of all constellations is Ursa Major, also known as the Great Bear. The stars of this constellation form a quadrangle with extensions on each corner that many observers believe resembles a bear. In some regions, people are most familiar with one portion of this constellation, which they call the Big Dipper.

### ATTRIBUTES OF URSA MAJOR

Ursa Major is a prominent Northern Hemisphere constellation best seen between the latitudes of 90 degrees north and 30 degrees south. Its right ascension is 9 hours, 46 minutes, and 31.7 seconds. Its declination is about +57 degrees from the celestial equator. Although it is visible to stargazers throughout the year, this constellation is clearest during its culmination, or highest point, in April. Ursa Major is the third largest of all eighty-eight constellations recognized by the International Astronomical Union.

Although the image of the bear is the definitive symbol of Ursa Major, the constellation is better known in many regions as the Big Dipper. This name comes from the distinctive shape of one asterism, or star pattern, within Ursa Major. This asterism resembles a large scoop or spoon as might be dipped into food or water. The Big Dipper shares many stars of the larger constellation.

The stars of the Big Dipper are designated by letters of the Greek alphabet as well as individual names. These stars include alpha (Dubhe), beta

(Merak), gamma (Phad or Phecda), delta (Megrez), epsilon (Alioth), zeta (Mizar), and eta (Alkaid or Benetnash) Ursae Majoris. These stars, particularly Alioth, Dubhe, Alkaid, Mizar, and Merak, are among the brightest in the entire constellation. These bright stars make the Big Dipper highly distinct in the night sky.

Many of the stars of Ursa Major are notable for their variations. Mizar, for example, is considered a double star because it appears to be right next to another star, 80 Ursae Majoris (Alcor). Alcor itself is a binary star. A binary star is a pair of stars that orbit around a common center of mass. Xi Ursae Majoris (Alula Australis) is another binary star. Some other named stars from Ursa Major include iota (Talitha), lambda (Tania Borealis), mu (Tania Australis), and omicron (Muscida) Ursae Majoris.

Aside from stars, deep-space objects exist within the Ursa Major constellation. These include a vast assortment of galaxies, collections of stars, and other space materials bound together by the force of gravity. The galaxies take on various forms. Some are irregular galaxies that do not take on a specific shape. Among the notable spiral galaxies within Ursa Major are Messier 81 (M81), also called Bode's galaxy; M82, Cigar galaxy; and M101, Pinwheel galaxy.

Finally, Ursa Major also contains a planetary nebula, M97. A planetary nebula is a vast cloud of gases surrounding an expanding star. In 1848, astronomer William Parsons, the third earl of Rosse, named M97 the Owl Nebula for the two distinct holes in the cloud, which reminded him of an owl's eyes.

### TALES FROM THE PAST

The first clearly recorded references to Ursa Major date back to ancient times. The Greek Egyptian astronomer Ptolemy (ca. 100–170 CE) listed it among the constellations he recognized. Although cultures around the world speculated on the meaning of this constellation, Greek myths are the most well-known today. The most famous Greek myth about Ursa

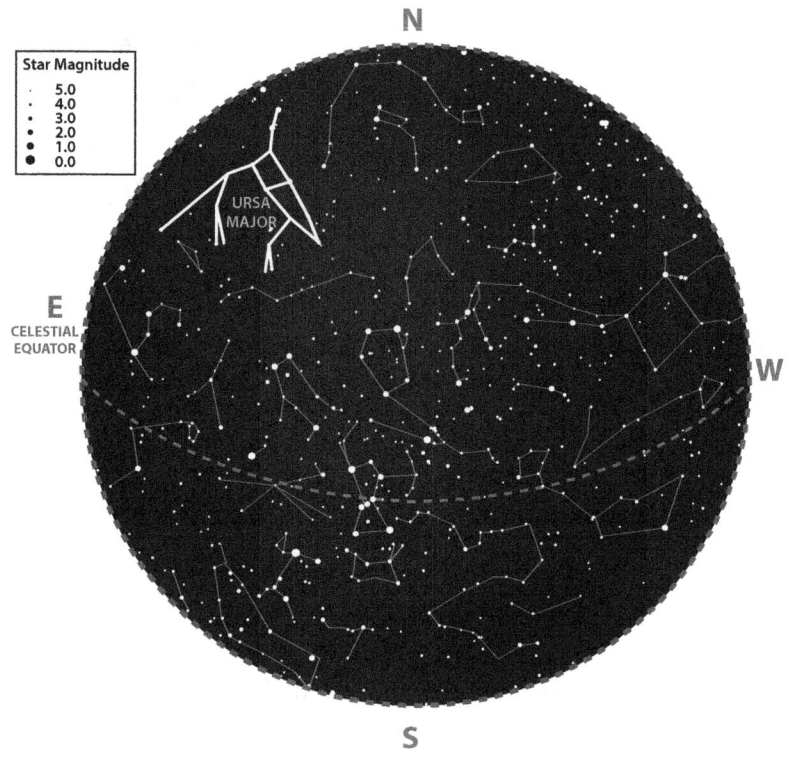

Star chart of the Northern and Southern Hemispheres in November. Lines connect the stars of the constellation Ursa Major.

Major has several minor variations. The main idea of the myth, however, deals with the chief god, Zeus.

In this myth, Zeus fell in love with a beautiful nymph, or spirit, named Callisto. Zeus's wife, the goddess Hera, noticed his interest and became jealous of the nymph. Hera used her power to transform Callisto into a bear. Shortly afterward, Callisto's son Actas was hunting when he spotted the bear. Without realizing it was actually his mother, he prepared to shoot it with his bow. Zeus intervened by turning Actas into a bear as well and then hurling them both into the stars. There, they formed the Great Bear (Ursa Major) and Small Bear (Ursa Minor) constellations.

Some American Indian cultures also had bear stories for Ursa Major. According to Iroquois and Mi'kmaq lore, the constellation represents human hunters who are forever pursuing the bear, hoping to kill and eat it. Similarly, the Zuni people believed that the stars were a magic bear that moderated the seasons. Every spring the bear would appear and chase away the cold. In the winter, the bear would fade into hibernation, allowing the cold to return.

Other cultures did not see a bear at all. In the Basque region of Spain, people told stories of the constellation representing a group of people searching for stolen oxen. In some Arabian lands, people interpreted the stars as a funeral procession. Four stars of the quadrangle represented the coffin and other nearby stars stood for mourners. Many groups saw vehicles in the sky. North Pacific peoples in the Marshall Islands believed the stars formed a canoe. In German mythology, the stars were a large wagon. English people thought that the legendary King Arthur lived in the sky near the constellation, which was later known as King Arthur's Chariot, while some people in Ireland believed it represented a chariot belonging to King David.

Other interpretations dealt with scoops or drinking vessels. In ancient China, farmers thought these star patterns looked like containers for grain. In North America, many people saw a ladle or pan, thus leading to the name Big Dipper. (Ursa Minor, the companion constellation, is often called the Little Dipper.) In the early nineteenth century, when slavery was still legal in the American South, enslaved people sang songs about the stars forming a "drinking gourd" that offered happiness. These songs had a hidden meaning. If escaped slaves walked toward the "gourd" in the northern sky, they would eventually find themselves in the northern states, where slavery was outlawed.

### Studying Ursa Minor Today

Astronomers search the constellations to gather scientific information about their diverse stars and planets. Constellations are a handy way of mapping features of space so they can be located easily. For both amateur and professional stargazers, Ursa Major is a good starting point for a visual exploration of the night sky. From it, stargazers can locate the constellations Ursa Minor, Auriga, and Gemini.

Ursa Major contains double and binary stars, a nebula, and a striking number of galaxies within its boundaries. Scientists have noted that the star called Groombridge 1830, unlike the vast majority of stars, never moves. In addition, Earth observers often look to Ursa Major to see meteor showers, including the Alpha Ursa-Majorids, Ursids, and Leonids-Ursids.

—*Mark Dziak*

### Further Reading

"The Constellation of Ursa Major (Great Bear)." *Royal Museums Greenwich*. Natl. Maritime Museum, 2015. Web. 30 Apr. 2015.

Kronberg, Christine. "Ursa Major." *Munich Astro Archive*. C. Kronberg, 21 May 1997. Web. 30 Apr. 2015.

"The Myths of Ursa Major, the Great Bear." *AAVSO: American Association of Variable Star Observers*. AAVSO, 16 Apr. 2010. Web. 30 Apr. 2015.

"Ursa Major (Big Bear)." *Chandra X-Ray Observatory*. NASA, 2 Dec. 2013. Web. 30 Apr. 2015.

"Ursa Major." *Constellations*. Peoria Astronomical Soc., n.d. Web. 30 Apr. 2015.

"Ursa Major Myths." *Astronomy in Different Cultures*. Project ASTRO New Mexico, NF Observatory, n.d. Web. 30 Apr. 2015.

"Ursa Major, the Great Bear." *StarDate*. University of Texas McDonald Observatory, 4 May 2015. Web. 4 May 2015.

# URSA MINOR

### FIELDS OF STUDY

Astronomy; Observational Astronomy

### SUMMARY

Ursa Minor is a circumpolar constellation in the Northern Hemisphere. The constellation resembles a ladle and is often referred to as the Little Dipper. Its counterpart is Ursa Major, which contains the Big Dipper. Ursa Minor contains Polaris, which has long been an important star. Various myths exist about Ursa Minor, particularly Greek myths. Ursa Minor has several valuable uses in navigation and in the field of astronomy.

### PRINCIPAL TERMS

- **celestial equator:** the imaginary projection of Earth's equator onto the night sky.

- **circumpolar constellation:** a pattern of stars that, from the perspective of the observer, never sets and is always visible in the night sky, despite Earth's rotation.
- **constellation:** a pattern of stars that can be seen in the night sky from Earth.
- **declination:** a space object's angular distance north or south of the celestial equator.
- **International Astronomical Union (IAU):** a group founded in 1919 to foster cooperation between astronomers from around the world and promote the science of astronomy.
- **right ascension:** a space object's east-west position along the celestial equator, measured eastward from the vernal equinox.

## THE LITTLE DIPPER

Ursa Minor is one of the eighty-eight constellations that are officially recognized by the International Astronomical Union (IAU). Ursa Minor is a circumpolar constellation, because from the perspective of a viewer in the Northern Hemisphere, it never sets and is visible at all times in the night sky. It is more precisely classified as a northern circumpolar constellation, as compared to a southern circumpolar constellation, which is visible at all times in the Southern Hemisphere night sky. Besides Ursa Minor, other northern circumpolar constellations include Ursa Major, Draco, Perseus, Auriga, Lynx, Cepheus, Cassiopeia, and Camelopardalis. Ursa Minor is bordered by the constellations Draco, Chamaeleon, and Cepheus.

Ursa Minor means "little bear" in Latin. The larger constellation Ursa Major, also known as the Great Bear, is Ursa Minor's counterpart. Ursa Minor resembles a ladle, which is why it is often referred to as the Little Dipper. Ursa Major also contains a ladle-shaped pattern of stars, known as the Big Dipper. According to Tom Kerss, an astronomer at the Royal Observatory Greenwich, Ursa Minor's Little Dipper got its name from the Big Dipper.

At one time, Ursa Minor was part of the constellation Draco, forming an asterism called the Dragon's Wing. An asterism is a pattern of stars that is generally simpler than a constellation and is not recognized by the IAU. It may form part of an official constellation, or it may include stars from more than one constellation.

## KEY ATTRIBUTES

Ursa Minor is located in the Northern Hemisphere. It takes up an area of 256 square degrees and is the fifty-sixth largest of the eighty-eight constellations. Ursa Minor is visible between the latitudes of 90 degrees north (the North Pole) and 10 degrees south. It is best seen in June around 9:00 p.m. The constellation resembles a ladle, complete with a "bowl" and "handle." There are seven stars in the constellation that form this ladle: alpha Ursae Minoris (better known as Polaris), beta Ursae Minoris (Kochab), gamma Ursae Minoris (Pherkad), delta Ursae Minoris (Yildun), epsilon Ursae Minoris (Urodelus), zeta Ursae Minoris (Akhfa al Farkadain), and eta Ursae Minoris (Anwar al Farkdain). Polaris, Yildun, and Urodelus form the handle of the ladle, and Kochab, Pherkad, Akhfa al Farkadain, and Anwar al Farkdain make up the bowl.

Polaris is one of the principal features of Ursa Minor. Also known as the North Star or the Pole Star, Polaris is a yellow supergiant star and is the brightest star in Ursa Minor. It has a right ascension of 2 hours, 31 minutes, and 48.7 seconds. Right ascension is measured eastward from the vernal equinox, one of two points at which the celestial equator intersects with the ecliptic, or the apparent path of the sun through the sky. Polaris has a declination of about +89 degrees. It aligns with Earth's axis, which is also called the celestial pole.

Polaris is part of a multiple star system, meaning that it has several companion stars. One of the companion stars is hard to see because it is so close to Polaris. It was first observed in 2006 with the Hubble Space Telescope. Another of Polaris's companion stars was first discovered in 1780 by Sir William Herschel (1738–1822). Located at the end of Ursa Minor's handle, Polaris is about 430 light-years from Earth.

Kochab, the second-brightest star in Ursa Minor and the brightest in the bowl of the ladle, is an orange giant about 126 light-years from Earth. Pherkad is the third-brightest star in the constellation. It is about 480 light-years from Earth and about 1,100 times brighter than the sun. Kochab and Pherkad are commonly called the "Guardians of the Pole" because they were twin pole stars from about 1500 BCE to 500 CE, before Polaris became the pole star. Yildun, a white dwarf, is about 183 light-years away. Urodelus is a triple star system that is approximately

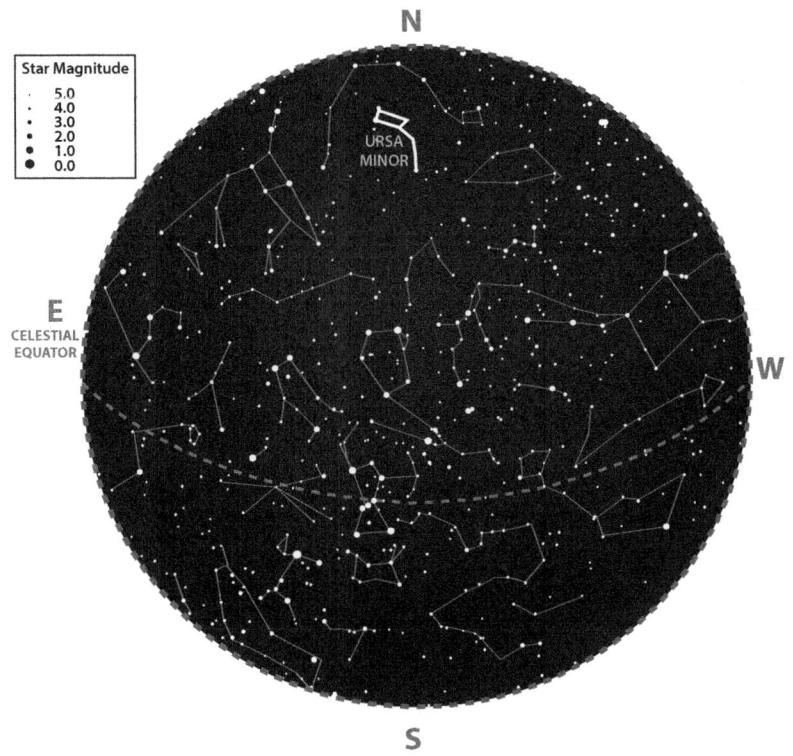

Star chart of the Northern and Southern Hemispheres in November. Lines connect the stars of the constellation Ursa Minor.

347 light-years from Earth. Its principal star is a yellow giant. Akhfa al Farkadain is a white main-sequence dwarf that will eventually become a giant star. It is about 380 light-years away. Anwar al Farkdain is a yellow-white main-sequence dwarf that lies about 97 light-years away.

One of the distinguishing features of Ursa Minor is the annual Ursid meteor shower. As seen from Earth's northern hemisphere, the Ursids radiate from near Ursa Minor and Ursa Major. The meteor shower begins around December 17 and continues for the remainder of the month. During the period of greatest activity, observers can see as many as ten meteors per hour shooting across the sky over a twelve-hour period.

Another key feature of Ursa Minor is the Ursa Minor Dwarf, which is a dwarf elliptical galaxy. Discovered in 1954 by Albert George Wilson (1918–2012) at the Lowell Observatory, the Ursa Minor Dwarf galaxy is a satellite galaxy to the Milky Way galaxy. Its stars began forming stars about eleven billion years ago.

## History

Although the discovery of Ursa Minor cannot be pinpointed for certain, the description and naming of the constellation is thought to have originated with the Greek philosopher Thales of Miletus (ca. 625 BCE–545 BCE), although he may have been describing a constellation already used by the Phoenicians, from whom his family descended. Regardless, Ursa Minor was catalogued by the Greek astronomer Ptolemy (ca. 100–ca. 170) in the second century CE. Ptolemy included Ursa Minor in his list of forty-eight constellations.

Numerous myths surround Ursa Minor. In ancient Greece, the constellation was known as Cynosura, Greek for "dog's tail." It was seen as part of the constellation Canis Major, "the great dog." The constellation was later described in relation to a Greek myth about the nymphs Ida and Adrasteia, who nursed Zeus on Mount Ida in Crete when he was young. According to the myth, Zeus honored his nymph nurses by placing them in the sky as Ursa Major and Ursa Minor. A different Greek myth contends that the seven stars of the Little Dipper represent the Hesperides, who were the seven daughters of Atlas.

## Navigation and Astronomy

Ursa Minor has played an important role in navigation for thousands of years. The Phoenicians often used Ursa Minor for navigation because it was a good guide to true north, leading the ancient Greeks to often refer to the constellation as the Phoenician. Polaris is especially significant, as it aligns with Earth's axis and stays stationary in the night sky over the North Pole, while the other stars in the Northern Hemisphere appear to rotate around it. This fact, along with its brightness, has long made Polaris a key star for navigation.

Polaris is a pulsating variable star, which means it has defined periods of brightness and pulsation. This attribute makes it a valuable star for measuring distances between objects in space. In 2014 astronomers discovered that Polaris is brighter than it was in the past. A team of astronomers determined that the star is approximately two and a half times brighter than it was about two hundred years ago. In fact, since astronomers began studying Polaris, the star has brightened by 15 percent. It began dimming in the 1990s but began to brighten again in 2000.

—*Michael Mazzei*

**FURTHER READING**

Howell, Elizabeth. "Reference: The Little Dipper; Host of the North Star." *Space.com*. Purch, 18 Dec. 2014. Web. 1 May 2015.

McClure, Bruce. "How Far Is a Light-Year?" *EarthSky*. Earthsky Communications, 29 Aug. 2014. Web. 1 May 2015.

"Northern Circumpolar Constellations." *Windows to the Universe*. Natl. Earth Science Teachers Assn., 2012. Web. 1 May 2015.

Plotner, Tammy. "Ursa Minor." *Universe Today*. Fraser Cain, 24 Jan. 2009. Web. 1 May 2015.

Redd, Nola Taylor. "The North Star Polaris Is Getting Brighter." *Space.com*. Purch, 28 Jan. 2014. Web. 8 May 2015.

"Ursa Minor." *International Astronomical Union*. IAU, n.d. Web. 8 May 2015.

"Ursa Minor (Little Bear)." *Windows to the Universe*. Natl. Earth Science Teachers Assn., 2012. Web. 1 May 2015.

# VENUS

## FIELDS OF STUDY

Planetary Science; Planetary Geology; Orbital Mechanics

## SUMMARY

Venus is the second of the terrestrial planets, following a nearly circular orbit around the Sun. It is surrounded by dense clouds of sulfuric acid, and has surface temperatures of 460°C or more, due to a runaway greenhouse effect from its atmosphere of 96.5 percentcarbon dioxide. Atmospheric pressure at the surface of Venus is 90 times that of Earth. There is neither a magnetic field nor tectonic or volcanic activity. Surface features are known primarily from satellite-based radar scans, although there have been a number of actual probe landings on the surface since 1962.

## PRINCIPLE TERMS

- **astrochemistry:** the scientific study of chemical elements in space and how they interact with each other and with radiation.
- **convection:** the transfer of heat via the movement of molecules in a fluid.
- **eccentricity:** the extent to which a celestial body's orbit deviates from a perfect circle.
- **magnetic field:** an area around an object that has a magnetic influence.
- **opacity:** the degree to which a substance or object lets various forms of electromagnetic radiation pass through it.
- **reflection:** the return of a light or sound wave to its source. The wave bounces back after interacting with a surface.

## VENUS AND ITS PLACE

The planet Venus is one of the four terrestrial planets, along with Mercury, Earth and Mars. Like the other terrestrial planets Venus is a solid, rocky planet. It is most comparable in size to Earth, and has long been thought of as Earth's 'sister planet'. Venus follows a very nearly circular orbit about the sun, with an average diameter of $1.082 \times 10^8$ kilometers. The eccentricity of the Venusian orbit is just 0.0068, as compared to 0.0000 for a perfect circle. At 12,104 kilometers, the diameter of Venus is similar to Earth's diameter of 12,741 kilometers. The mass of the planet, however, is just 82 percent of the mass of Earth, indicating that there is a significantly higher percentage of lighter elements in the make-up of Venus. The atmosphere of Venus is 96.5 percent carbon dioxide, 3 percent nitrogen and small amounts of water and other gases. The average density of Venus is 5.24 g.cm-3, as compared to the average density of Earth of 5.52 g.cm-3. Like the planet Mercury, Venus has no moons in orbit around it.

### QUALITIES

Venus occupies the most circular orbit of all of the planets, varying from a perfect circle by just 0.68 percent. The period of rotation of Venus is 243 Earth days, and its orbital period is just 224.7 days. In comparison, Earth has a period of rotation of 1 day and an orbital period of 365.245 days. The rotation of Venus is so slow that it actually rotates backwards, and is just one of three Solar system planets having retrograde rotation.

The orbital tilt of the planet is designated as 177 degrees, and not 3 degrees as might seem logical. A planet having an orbital tilt of 3 degrees would still rotate in a counterclockwise direction, like Earth and other planets in the Solar system. But because the rotation of Venus is clockwise, in the opposite direction, it is designated as having an orbital tilt that is also the opposite of 3 degrees.

The slow rotation of Venus is responsible for the extreme atmospheric conditions on Venus. At the surface of the planet, the thick, carbon dioxide atmosphere of Venus has a pressure approximately 90 times that of Earth. Studies of the astrochemistry of Venus have found that at altitudes of from about 50 to 70 kilometers, the planet is shrouded in thick clouds consisting primarily of sulfuric acid droplets. The high opacity of the clouds explains the high ability of Venus for the reflection of light, giving it an albedo of 0.75, the same as that of Earth. Light that does reach the surface through the clouds and haze has a distinct yellow color. A constant haze extends down to an altitude of about 30 kilometers. From there to the ground level, the atmosphere is quite clear.

In the upper atmosphere, the winds are some 350 kilometers per hour in the direction of the planet's rotation, but they decrease steadily as altitude decreases until there is effectively no wind at ground level. The atmosphere of Venus is thus effectively static and rotates in unison with the planet, with no areas of high or low pressure nor convection to drive winds in different directions.

The composition of the atmosphere is believed to be a result of the amount of carbon dioxide emitted by volcanic activity in Venus's past. There is essentially no water on Venus to absorb the carbon dioxide, whereas on Earth the vast majority of carbon dioxide that was emitted in the past has been trapped by water and reacted to produce carbonic acid, which has been converted to carbonate-based rock. This effect being absent on Venus, the atmospheric carbon dioxide accumulation has produced a run-away greenhouse effect such that the surface temperature of Venus exceeds 460°C (860°F), greater than the melting point of some metals and minerals.

The internal structure of Venus is believed to be very similar to that of Earth as well, with a nickel-iron core, mantle and silicate crust. The surface of Venus is known to be older than that of Earth, but younger than the surface of Earth's Moon, and the crust is relatively thin and flexible. There has been tectonic and volcanic activity on Venus in the distant past, but there are neither active volcanoes nor tectonic plates on Venus today.

Venus has no detectable magnetic field, which is a great puzzle given the similarity its size and composition compared to Earth's. One of the possible explanations for this is that, since it is known that the magnetic field of Earth reverses polarity fairly regularly, the magnetic field of Venus may be in just now in the middle of a magnetic field reversal cycle.

### VENUS IN HISTORY AND MYTHOLOGY

Like Mercury, the planet Venus has been known since ancient times, due to its ready visibility. It has long been known as "the morning star" and is considered to be the most beautiful of celestial objects seen from Earth. Accordingly, the Greeks named it Aphrodite after the goddess of beauty. When the Roman Empire superseded Greek culture, the "star" was renamed Venus, for the Roman goddess of beauty. Like Mercury, Venus is seen in the sky only after sunset and is the last "star" to disappear before sunrise, hence its common name of "the morning star." This is because the planet is never seen more than 47 degrees away from the Sun, just as the planet Mercury is never seen beyond about 21 degrees. Accordingly, Venus is the first bright "star" to be seen in the night sky and the last before sunrise.

Galileo and other astronomers since have been able to view the planet through telescopes. Direct observation of Venus by satellite and landing probe began in 1962 with the Russian *Venera* probes that first landed on the surface of Venus. The inhospitable conditions of the planet were

The internal structure of Venus – the crust (outer layer), the mantle (middle layer) and the core (yellow inner layer) CC BY-SA 3.0, Wikimedia Commons.

very surprising, and the extreme pressure and temperature destroyed the ability of early probes to function after as little as fifteen minutes. Pictures were obtainable only with cameras that had small lenses made of diamond. Since then, most of the information about the surface features of Venus has been obtained by radar scans from orbiting satellites. Radar easily penetrates the thick cloud cover, and the scans have revealed approximately one thousand large craters over the surface of the planet, as well as a number of ancient volcanic peaks. Most of the surface features on Venus are named for women.

### MEASURING VENUS

Venus is the second of the four terrestrial planets, and since its orbit is between Earth and the Sun, its proximity to the Sun determines the maximum angle at which it can be seen from Earth. For Venus, the maximum viewing angle is 47 degrees to either side of a straight line from Earth to the Sun. The orbit of Venus is very nearly a perfect circle around the Sun, and has an eccentricity of just 0.0068. All other planetary orbits are more eccentric ellipses. Understanding the geometry of ellipses is an important aspect of orbital mechanics. Every ellipse has two points called a focus situated on its major axis, the distance between the two points that are farthest apart. Half of that distance is called the semimajor axis, and is typically quoted as the average distance of a planet from the Sun. The eccentricity of an ellipse, or of an elliptical orbit, is calculated as the ratio of the distance between the two focal points to the distance of the major axis.

—*Richard M. Renneboog M.Sc.*

### FURTHER READING

Bengtsson, Lennart, Roger-Maurice Bonnet, David Grinspoon, Symeon Koumoutsaris, Sebastien Lebonnois, and Dmitri Titov, eds. *Towards Understanding the Climate of Venus*. New York, NY: Springer Science+Business Media, 2014. Print.

Elkins-Tanton, Linda T. *The Sun, Mercury, and Venus*. New York, NY: Chelsea House, 2006. Print.

Esposito, Larry W., Ellen R. Stofan, and Thomas E. Cravens, eds. *Exploring Venus as a Terrestrial Planet*. Washington, DC: American Geophysical Union, 2007. Print.

Grego, Peter. *Venus and Mercury and How to Observe Them*. New York, NY: Springer Science+Business Media, 2008. Print.

Taylor, Frederic W. *The Scientific Exploration of Venus*. New York, NY: Cambridge University Press, 2014. Print.

# 4 VESTA

### FIELDS OF STUDY

Observational Astronomy; Cosmology; Sub-planetary Astronomy

### SUMMARY

4 Vesta is the second-largest asteroid orbiting in the asteroid belt of the solar system. Though rocky, Vesta is unique in that it appears to have evolved more like a terrestrial planet. Vesta was studied at length in 2011 and 2012 as part of NASA's Dawn mission.

### PRINCIPAL TERMS

- **asteroid belt:** an oval-shaped ring in which millions of asteroids orbit, located between the orbits of Mars and Jupiter, about 2 to 4 astronomical units (AU; 186 to 370 million miles) from the sun.
- **Dawn mission (NASA):** NASA-funded mission to study the asteroids Vesta and Ceres to learn more about their composition, history, and evolution. The mission launched in 2007 and reached Vesta in July 2011 and Ceres in March 2015.
- **dwarf planet:** a celestial body that has enough mass to attain a nearly round shape, lacks the gravity to keep its orbit free of other space objects, orbits the sun, and does not act as a satellite.
- **meteorites:** small particles that break off comets or asteroids and survive the journey through Earth's atmosphere to reach the planet's surface.

## A Massive Asteroid

4 Vesta is an asteroid. An asteroid is an irregularly shaped, rocky space object that orbits the sun. Vesta is found in the asteroid belt, an oval-shaped ring between the orbits of Mars and Jupiter, about 2 to 4 AU (186 to 370 million miles) from the sun.

Scientists believe that planets formed during the early stages of the solar system when space objects accreted, or came together to create single bodies. However, once the gas giant Jupiter formed, it began interfering with this process. Its size and gravitational force prevented a single planet from coming together from space debris in the outer parts of the solar system. Then, its gravity caused these smaller bodies to collide. These events produced the tens of thousands of rocky, orbiting space objects known as asteroids.

4 Vesta, asteroid. NASA/JPL–Caltech/UCAL/MPS/DLR/IDA.

As space objects go, most asteroids are quite small. In fact, many of them are less than a kilometer across. However, Vesta is much larger and more massive. It is 569 kilometers (359 miles) in diameter at its widest point. This makes Vesta the second-largest asteroid in the solar system after 2 Pallas. It is also the second most massive body in the asteroid belt after Ceres. Ceres, once thought to be an asteroid, is large enough to be considered a dwarf planet. Dwarf planets, though nearly round, are not the same as regular planets. They are smaller and do not have the gravity to clear their orbits of other space objects.

## Vesta's Characteristics

Vesta's cold, rocky surface at first appears similar to most asteroids in the asteroid belt. However, extensive study by NASA's Dawn mission has revealed a unique object that is more like a planet than an asteroid. It is also one of the few large, intact space objects to have survived since the early days of the solar system.

The Dawn mission helped researchers to understand that Vesta most likely formed like the terrestrial planets Earth and Mars. Like them, Vesta has a separate crust, mantle, and iron core. Vesta's core is large for its size, with a radius of 110 kilometers (approximately 68 miles). However, unlike Earth's core, Vesta's core melted early on. This produced basaltic rock, or solidified lava, over the surface.

Volcanic melting and collisions with other space objects resulted in Vesta's varied surface features. These include mountains, boulders, and steep cliffs as well as deep craters and troughs. One of the most notable surface features of Vesta is the enormous impact basin Rheasilvia near the asteroid's southern pole. Researchers believe this crater formed from an impact with a large space object about one billion years ago. This collision most likely launched a large number of rock fragments into space. Some of these objects are believed to have made impact with Earth in the form of meteorites. Meteorites are rocky space particles that survive the journey through the Earth's atmosphere to land on Earth. The chemical makeup of these meteorites point to their origins on Vesta.

## NASA Dawn Mission

The *Dawn* spacecraft launched from Cape Canaveral, Florida, in September 2007. The mission's goal was to reach and orbit Vesta and Ceres to learn about their makeup, history, and evolution. These space objects are of particular interest because they are the largest relatively intact bodies remaining from the early days of the solar system.

The *Dawn* spacecraft reached Vesta in July 2011 and orbited around the asteroid until September 2012. Images and data sent back to Earth helped scientists to gain a deeper understanding of its unique nature. The mission revealed Vesta's planet-like qualities and confirmed it as the origin of many meteorites found on Earth. The *Dawn* spacecraft reached Ceres's orbit on March 6, 2015. This marked the first time a spacecraft had entered the orbit of two space bodies.

—Marie Keenan, MS

### FURTHER READING

Cook, Jia-Rui. "It's Complicated: Dawn Spurs Rewrite of Vesta's Story." *Jet Propulsion Laboratory*. Caltech, 6 Nov. 2013. Web. 16 Mar. 2015.

Coulter, Dauna. "Space Mountain Produces Terrestrial Meteorites." *NASA Science: Science News*. NASA, 30 Dec. 2011. Web. 16 Mar. 2015.

"Dawn's Targets—Vesta and Ceres." NASA. 28 July 2013. Web. 16 Mar. 2015.

International Astronomical Union. *IAU Resolution B5: Definition of a Planet in the Solar System*. N.p.: IAU, 24 Aug. 2006. PDF file.

"NASA Dawn Mission Reveals Secrets of Large Asteroid." *SSERVI: Solar System Exploration Research Virtual Institute*. NASA, n.d. Web. 16 Mar. 2015.

"NASA Spacecraft Becomes First to Orbit a Dwarf Planet." *NASA News*. NASA, 6 Mar. 2015. Web. 16 Mar. 2015.

Rayman, Marc D., et al. "Dawn: A Mission in Development for Exploration of Main Belt Asteroids Vesta and Ceres." *Acta Astronomica* 58 (2006): 605–16. Print.

Russell, C. T., et al. "Dawn: A Journey to the Beginning of the Solar System." *Proceedings of the Conference on Asteroids, Comets and Meteors, July 29–August 2, 2002*. Berlin: Technical University of Berlin, 2002. Web. 16 Mar. 2015.

# VIRGO

### FIELDS OF STUDY

Stellar Astronomy; Observational Astronomy

### SUMMARY

Virgo is the largest of the zodiac constellations and the second largest of all the constellations. It consists of twenty-six stars and thirty-three exoplanets, which are planets that are outside of the solar system and orbit a star other than the sun. It also contains numerous deep-sky objects. The Virgo constellation can be viewed in the night sky between the months of April and September. In ancient times, people studied the stars, especially those of the zodiac, for practical and religious purposes. In modern times, scientists look to constellations such as Virgo to learn more about the universe.

### PRINCIPAL TERMS

- **celestial equator:** the imaginary line above Earth's equator that halves the celestial sphere; it is equally distant from the celestial poles.
- **constellation:** a region of space defined by a pattern of stars that can be seen in the night sky from Earth.
- **declination:** a space object's angular distance north or south of the celestial equator, expressed in degrees of arc.
- **International Astronomical Union:** an association of professional astronomers from all over the world who define astronomical constants while promoting research, education, and discussion on important astronomical topics.
- **right ascension:** a space object's longitudinal arc along the celestial equator, measured eastward from the vernal equinox and expressed in hours, minutes, and seconds.

### THE VIRGIN CONSTELLATION

A constellation is an area of space defined by a group of stars in the night sky that resembles a picture or pattern. People have searched for constellations for thousands of years. In ancient times, people believed these star shapes had spiritual importance. They also helped show the changing of the seasons. In modern times, astronomers study constellations to learn more about the universe. In 1930, the International

Astronomical Union (IAU) recognized eighty-eight official constellations and set their boundaries in the night sky.

The constellation Virgo, which means "virgin" in Latin, is tied to agriculture, fertility, fortune, and justice. It is said to look like a woman with angel wings and a bunch of wheat in her left hand. Virgo is one of the constellations in the zodiac, a group of twelve or thirteen constellations through which the sun appears to pass over the course of a year. The sun's path through the zodiac is called the ecliptic.

**ATTRIBUTES OF VIRGO**
Virgo is visible in both the Northern and the Southern Hemispheres. The second-largest constellation, it covers 1,294 square degrees and is visible between latitudes 80 degrees north and 80 degrees south. The best time to view it is during the month of May around 10:00 p.m. local time. Virgo is below the constellations Boötes and Coma Berenices and sits west of Leo and east of Libra. Corvus and Crater are just below it and the ecliptic. Its declination is −3.73 degrees from the celestial equator, and its right ascension is 13.2 hours.

Many of the stars in the Virgo constellation are dim, but its brightest star, alpha Virginis, also known as Spica, is easy to find in night sky. Spica represents the bundle of wheat in Virgo's left hand. Using the Big Dipper in the Ursa Major constellation as a guide, viewers can follow its curved handle to the southeast until they see the bright star alpha Boötis (Arcturus) in Boötes. The next brightest star down is Spica.

Spica is considered the sixteenth-brightest star in the sky, with a visual magnitude of 0.98. It is actually a binary star, or two stars that revolve around a common center of mass. This pair of giant blue stars is located about 260 light-years from Earth. Spica A is ten times as massive and more than 12,000 times as luminous as of the sun. Its companion, Spica B, is only six times as massive as the sun and 1,500 times as luminous.

Gamma Virginis, also known as Porrima or Arich, is the next-brightest star of Virgo. It is also a binary star. The third-brightest star of Virgo is the yellow giant star epsilon Virginis, also known as Vindemiatrix, the grape gatherer. Zeta Virginis (Heze) is a dwarf star that rotates in less than half a day. The star 70 Virginis has one of the first confirmed exoplanets, which was found in 1996.

In addition to its twenty-six stars and thirty-three exoplanets, Virgo contains eleven Messier objects as well as other deep-sky objects. Early astronomers searching for comets with telescopes saw objects that were later determined to be nebulae, which are clouds of gases or space dust, and deep-sky objects. During the eighteenth century, French astronomer Charles Messier (1730–1817) catalogued these objects, which were named after him. Among the constellations, Virgo has the second-largest number of Messier objects; Sagittarius has the most, at fifteen. Some of Virgo's Messier objects (abbreviated M) include Messier 49 (M49), Messier 58 (M58), Messier 60 (M60), Messier 87 (M87), and Messier 104 (M104), the Sombrero galaxy. It also contains New General Catalog 4435 (NGC 4435) and New General Catalog 4438 (NGC 4438), the Eyes galaxies; New General Catalog 4567 (NGC 4567), the Siamese Twins; and the quasar 3C273. A quasar is a very bright object in space that gives off electromagnetic radiation and may get its energy from a massive, central black hole in its galaxy. In addition, Virgo and the constellation Coma Berenices contain the galaxy cluster known as the Virgo Cluster, which has at least two thousand galaxies.

**THE VIRGIN MYTH**
Greek Egyptian astronomer Ptolemy (ca. 100–ca. 170) identified and classified the constellation Virgo in the second century CE. Virgo, the virgin, is associated with several myths. Some link her to agriculture, fertility, fortune, and justice. Virgo is mostly associated with the season of spring. In addition, Virgo is located next to the constellation Libra, the scales of justice, which links Virgo to justice.

In some Greek myths, Virgo is associated with Demeter, also known as the corn goddess of fertility. According to one myth, long ago Earth experienced eternal spring. Then, one day, the god of the underworld, Hades, abducted Demeter's daughter, Persephone, and Demeter turned her back on her role as goddess of fruitfulness. As a result, Earth saw frigid winters and scorching summers, which deterred growth and new life. Zeus, the king of gods, demanded that Persephone be returned. Persephone was reunited with her mother, but because she ate Hades's pomegranate, she was obligated to return to the underworld for four months a year. The time she is with Demeter represents spring and summer in

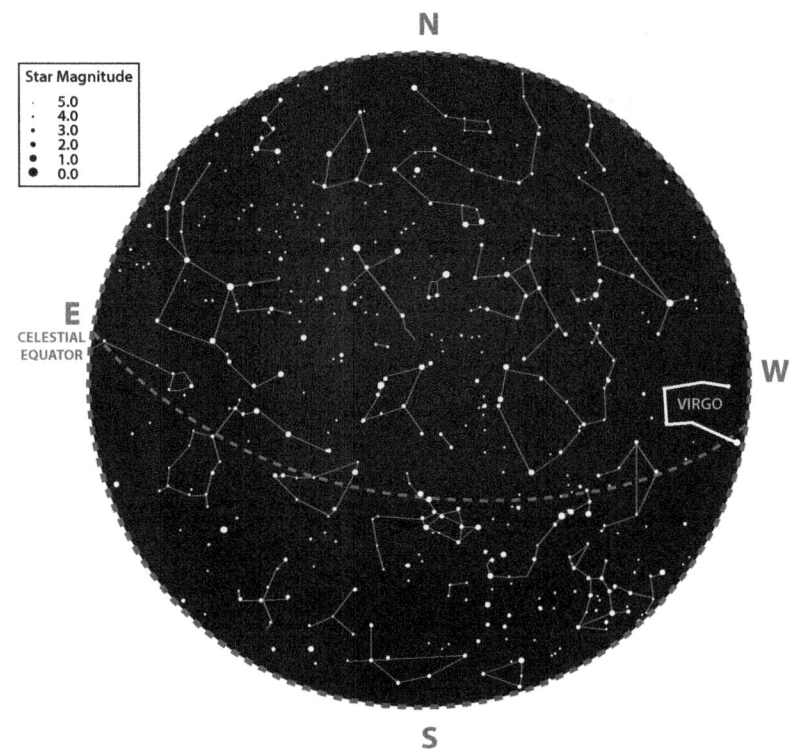

Star chart of the Northern and Southern Hemispheres in June. Lines connect the stars of the constellation Virgo.

the Northern Hemisphere, which is why Virgo can be seen in the sky then. In a similar Babylonian myth, the goddess Ishtar, associated with Virgo, descends to the underworld to retrieve her brother-lover Tammuz.

In another Greek myth, Virgo is associated with Astraia, or Dike, the goddess of justice. According to the myth, Dike lived on Earth during the Golden Age, a time of peace, abundance, and spring. In time, the Bronze Age began and mortals became lawless. Dike tried to reason with the humans, but she eventually fled Earth for the sky.

## Modern Astronomical Importance

In ancient times, people used the Virgo constellation to track the changing seasons. Because Virgo was visible in the spring months in the Northern Hemisphere, it helped farmers determine when they could start planting. In modern times, scientists study the stars to determine much about outer space and in turn learn more about Earth.

Virgo made headlines in November 2012 when light from an exploding star, or supernova, in the Virgo Cluster reached Earth. Scientists determined that the star exploded 230 million years ago, during the Triassic period on Earth. They classified the event as supernova 2012ha, or Sherpa. The scientists learned that the supernova was the result of an exploding white dwarf star as large as the sun. They believe that when the star died, its core shrunk to about the size of Earth.

Studying supernovas provides astronomers information about dark energy. Dark energy is a force that works opposite to that of gravity. It accelerates the expansion of the universe. By studying supernovas, scientists can calculate the rate of expansion.

—*Angela Harmon*

### Further Reading

Frommert, Hartmut. "Charles Messier (June 26, 1730–April 12, 1817)." *SEDS*. Students for the Exploration and Development of Space, n.d. Web. 6 May 2015.

Kaler, Jim. "Spica (Alpha Virginis)." *STARS*. Author, 3 July 2009. Web. 6 May 2015.

McClure, Bruce. "Virgo? Here's Your Constellation." *EarthSky*. Earthsky Communications, 11 May 2014. Web. 4 May 2015.

O'Meara, Stephen James. *Deep-Sky Companions: The Messier Objects*. 2nd ed. Cambridge: Cambridge University Press, 2014. Print.

"Virgo." *StarDate*. University of Texas McDonald Observatory, n.d. Web. 6 May 2015.

"White Dwarf Supernovae Are Discovered in Virgo Cluster Galaxy and in Sky Area 'Anonymous.'" *SMU Research*. Southern Methodist U, 26 Feb. 2013. Web. 6 May 2015.

# WORMHOLES

## FIELDS OF STUDY

Theoretical Astronomy; Astrophysics; Cosmology

## SUMMARY

A wormhole is a hypothetical passage or tunnel through space-time that could theoretically allow travel from one place in time and space to another. The theory of general relativity allows the warping of space-time in a way that could allow for the formation of wormholes. However, no direct evidence of their existence has been discovered. The study of wormholes remains important because investigating the properties of space-time leads to greater understanding of its nature.

## PRINCIPAL TERMS

- **black hole:** a region of space that exerts such strong gravitational force that nothing can escape, not even light.
- **exotic matter:** any type of matter, whether real or hypothetical, whose physical properties violate the known laws of physics; for example, matter with a mass or energy density of less than zero.
- **general relativity:** Albert Einstein's theory that gravity is the result of matter and energy causing space-time to curve, which in turn affects the movement and distribution of matter and energy in space.
- **singularity:** a theoretical point or region of infinite density where the curvature of space-time, and thus the gravitational force, likewise becomes infinite.
- **time travel:** the theoretical or imagined ability to travel between two points in time.

### Tunneling through Space-Time

In 1915, German-born physicist Albert Einstein (1879–1955) proposed his theory of general relativity. According to this theory, the presence of matter in the universe causes time and space to curve around it. This curvature gives rise to the phenomenon of gravity, which in turn affects how matter moves through space. Rather than being independent and progressing in a straight line, time and space are affected by each other and by any matter within them.

One consequence of Einstein's theory is the idea that gravity can bend light. According to classical physics, this would be impossible, because gravity is the result of one body with mass acting on another body with mass. However, if gravity is instead the result of space-time being curved, any particle in the universe would still have to follow the curve. This is true even of massless particles, such as photons (light particles).

One example of gravity exerting overwhelming force on everything around it is a black hole. The singularity at the center of a black hole is created when the remnant of a dying star is overwhelmed by its own gravity and collapses. As all of the star's matter falls toward its center, it is condensed past the point where even the neutrons in its atomic nuclei are crushed. The end result is a dimensionless point in space with a mass many times that of the sun—in essence, a point of infinite density. A singularity can swallow vast amounts of gas and debris. It is surrounded by an event horizon, a boundary in space-time that defines the radius of a black hole. Within the event horizon, the gravitational pull of the singularity is so strong that not even light can escape.

In the 1930s, Einstein and his assistant Nathan Rosen (1909–95) reviewed a paper that had been published by Vienna-based physicist Ludwig Flamm (1885–1964) in 1916. Flamm's paper summarized German physicist Karl Schwarzschild's (1873–1916) solutions to Einstein's general relativity theory. These solutions described some characteristics of black holes, which at the time were not yet proved to exist. In his summary, Flamm noted that Einstein's theory allowed for the possibility of a second solution mathematically connected to the first. This second solution described what was later called a "white hole." A white hole is a hypothetical region of space with the reverse properties of a black hole, unable to be entered but capable of emitting light and matter. Einstein and Rosen proposed that a black hole could combine with a white hole to form a bridge between two areas of space-time. Matter would be drawn into the black hole on one side of the bridge and then expelled through the white hole on the other side. This theoretical model came to be called an Einstein-Rosen bridge.

In the 1950s, Princeton University–based theoretical physicist John Archibald Wheeler (1911–2008) began to study Einstein-Rosen bridges. It was Wheeler who first coined the term "wormhole." He noted that the path wormholes describe is much like the hole a worm makes through an apple, if the surface of the apple represents conventional space-time. Although Wheeler determined that the Einstein-Rosen model would be too unstable for any matter to travel through it, the idea of a possible shortcut through space-time remained.

### The Challenges of Wormholes

While general relativity allows for the existence of wormholes, research strongly indicates that even if they exist, they would not be viable for human travel. An Einstein-Rosen bridge between two singularities would be too unstable, collapsing almost immediately after it formed. Anything attempting to travel through the bridge would be destroyed.

In the 1980s, American astronomer and popular science writer Carl Sagan (1934–96) worked with theoretical physicist Kip Thorne (b. 1940), then a professor at the California Institute of Technology, to devise a plausible method of faster-than-light travel for Sagan's 1985 novel *Contact*. Thorne and his graduate students proposed that a wormhole could be stabilized using exotic matter. This hypothetical matter would have negative energy, meaning that its energy per unit volume would be less than zero. Such matter would be repelled by gravity rather than attracted by it.

No exotic matter with negative energy has been found to exist naturally in the universe, but very small amounts can be created in laboratory conditions. As a result, Thorne and other scientists agree that such stable wormholes could not form naturally. Even if a wormhole could be stabilized in this manner, the introduction of nonexotic matter—for example, a person or a spaceship—could still cause it to collapse.

### Other Potential Portals

While wormholes may be unlikely, scientists working with the United States National Aeronautics and Space Administration (NASA) have discovered another type of portal. Data from NASA's Time History of Events and Macroscale Interactions during Substorms (THEMIS) satellites and the European Space Agency (ESA) Cluster II probes have revealed points in Earth's magnetosphere where the planet's magnetic field connects directly to the magnetic field of the sun, a phenomenon known as "magnetic reconnection." This connection allows high-energy

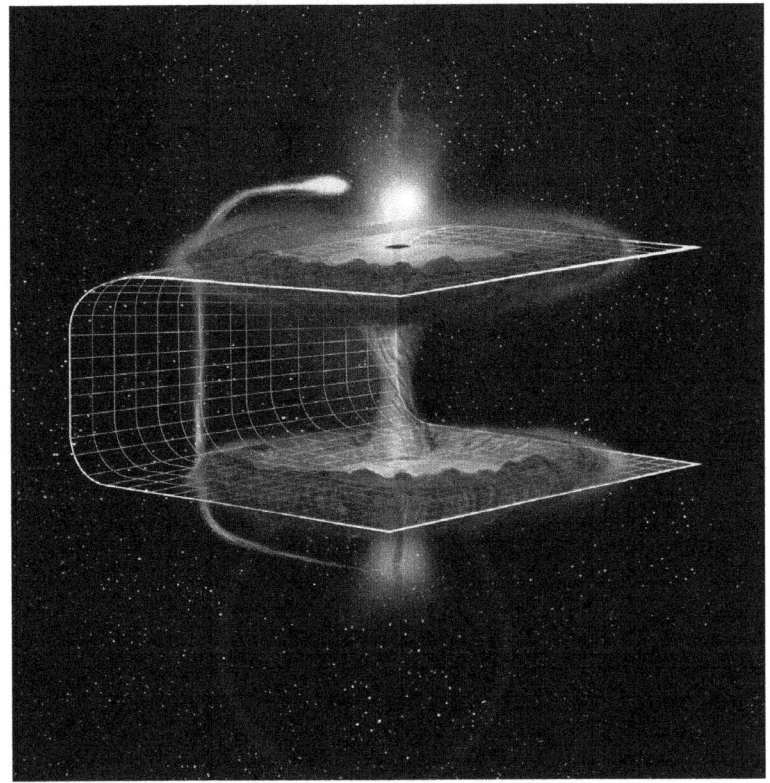

A model of a wormhole bridge formed between two different points in space-time. Theoretically, wormholes are two spheroidal openings connected by a tunnel through space-time, creating a path that reduces time and distance traveled.

particles to pass from the sun into Earth's upper atmosphere. Such transfers are called "flux transfer events." The interaction of these particles with Earth's magnetic field causes geomagnetic storms and auroras, among other phenomena.

These magnetic portals can be located by identifying points where the magnetic fields of Earth and the sun cross each other. NASA scientists call these points "X-points." Once opened, the portals form a direct, uninterrupted path from Earth to the sun, more than ninety-three million miles away. In March 2015, NASA launched the Magnetospheric Multiscale Mission (MMS) to further study these portals, how they work, and their potential impact. Their existence also raises the possibility that other as-yet-unidentified portals might be waiting to be found.

**TINY WORMHOLES AND TIME TRAVEL**
In 1955, Wheeler proposed that if the universe could be viewed on the smallest possible scale (the Planck scale, about 10–35 meters in length), virtual particles, black holes, and wormholes would appear to be constantly popping in and out of existence. This idea is called the "quantum foam" hypothesis because space-time at this scale would resemble a mass of churning foam.

In a 2010 article for the *Daily Mail*, renowned British theoretical physicist Stephen Hawking (b. 1942) wrote that the laws of physics allow for the possibility of trapping one of these short-lived, infinitesimal wormholes and expanding it enough to travel through. He also claimed that time travel through such wormholes may be theoretically possible, although that possibility raises many other issues. Traveling back in time would violate the universal law that says cause must happen before effect. It would leave the universe vulnerable to paradox—an event that could change the past in such a way as to prevent itself from happening. Further, if a wormhole were opened to a point in the past, the natural radiation passing through it would create a feedback loop that would ultimately destroy it. For these reasons, Hawking believes that a wormhole to the past could never exist.

However, it is known that time passes at different speeds in different parts of the universe. For the Global Positioning System (GPS) satellites that orbit Earth, time passes slightly faster than it does on Earth's surface, while in the vicinity of a black hole, time slows nearly to a standstill. The faster an object is moving, the slower the rate at which it experiences time. In this sense, time travel to the future is already possible, although the technology does not yet exist to travel more than a short "distance." To go farther forward in time would require a form of transportation that could travel at near light speed.

Wormholes may seem to be too theoretical and improbable to be of any serious interest, but the universe is full of the unknown and undiscovered. While scientists concede that any real wormholes may bear little to no relation to those envisioned in science fiction, there is still a good chance that they exist, and researchers continue to investigate the possibilities.

—*Janine Ungvarsky and Randa Tantawi, PhD*

**FURTHER READING**
Anderson, David Lewis. "Wormholes." *The Anderson Institute*. Anderson Multinatl., 2012. Web. 10 July 2015.

"Elementary Einstein." *Einstein Online*. Max Planck Inst. for Gravitational Physics, n.d. Web. 10 July 2015.

Grush, Loren. "Will Wormhole Travel Ever Be Possible?" *Popular Science*. Bonnier, 26 Oct. 2014. Web. 10 July 2015.

Hawking, Stephen. "How to Build a Time Machine." *Daily Mail*. Assoc. Newspapers, 27 Apr. 2010. Web. 10 July 2015.

Jaggard, Victoria. "Would Astronauts Survive an Interstellar Trip through a Wormhole?" *Smithsonian.com*. Smithsonian Inst., 7 Nov. 2014. Web. 10 July 2015.

Overduin, James. "Einstein's Spacetime." *Gravity Probe B: Testing Einstein's Universe*. Stanford U, Nov. 2007. Web. 10 July 2015.

Phillips, Tony. "Hidden Portals in Earth's Magnetic Field." *NASA Science*. NASA, 29 June 2012. Web. 10 July 2015.

Thorne, Kip. "The Man Who Imagined Wormholes and Schooled Hawking." Interview by Susan Kruglinski. *Discover*. Kalmbach, 9 Nov. 2007. Web. 10 July 2015.

Woo, Marcus. "Will We Ever…Travel in Wormholes?" *BBC Future*. BBC, 26 Mar. 2015. Web. 10 July 2015.

# XMM-NEWTON

### FIELDS OF STUDY

Space Technology; Observational Astronomy; Aerospace Engineering

### SUMMARY

*XMM-Newton*, or the X-Ray Multi-Mirror Mission, was launched in 1999 by the European Space Agency (ESA). The unmanned satellite carries three x-ray telescopes and an optical monitor. X-rays produced in space do not reach Earth's surface and can only be studied from space. The special equipment aboard *XMM-Newton* was designed to detect, observe, and record data needed by scientists to study x-rays and the areas of space where they are found.

### PRINCIPAL TERMS

- **black hole:** a region of space with such strong gravitational pull that not even light can escape.
- **European Photon Imaging Camera (EPIC):** a camera that uses charge-couple devices (CCDs) to detect and capture images of x-rays from aboard *XMM-Newton*. Two metal oxide semiconductors (MOS CCDs) and a pn-CCD allow each of two EPICs to cover a large expanse of space to identify and map locations of x-ray emissions that help scientists learn about a variety of space objects and phenomena.
- **European Space Agency:** an intergovernmental organization made up of twenty-two European countries in charge of Europe's space program and research since 1975.
- **satellite:** any small object, natural or artificial, that orbits around a larger object; one example is Earth orbiting the sun. Artificial satellites are generally deployed for research, monitoring, communication, or weather prediction purposes, and all include at least one antenna for sending and receiving data and a power source such as a battery or solar panels.
- **x-ray:** a form of electromagnetic radiation with short wavelengths and high energy that is produced in space by sources including supernovas, active galactic nuclei, quasars, and the superheating of gases being drawn into black holes.

### MISSION HISTORY

When *XMM-Newton* was launched from Kourou, French Guiana, on December 10, 1999, it was the largest satellite built and launched by the European Space Agency. It contained some of the most sensitive and sophisticated scientific instruments ever made. Initially the mission was known simply as *XMM*. It was later renamed *XMM-Newton* after English physicist Isaac Newton (1642–1727). Newton was the inventor of spectroscopy, which provided a way to measure the wavelengths of light and energy. The most state-of-the-art spectroscopy then available was incorporated into the *XMM-Newton* to facilitate its mission to detect and study x-rays.

X-rays offer specific challenges to researchers. They have small wavelengths and high energy. This makes them act more like a photon, or individual particle of light, than like a wave of light. Earth's atmosphere filters out most x-rays before they reach the surface, and the small particles of x-rays also go through some seemingly solid surfaces. This is what allows a doctor to use x-ray technology to see dense objects like bone through soft tissue. "Seeing" x-rays in space requires both special equipment and a space-based platform beyond Earth's atmosphere.

X-rays, when detected, reveal a great deal of information about the objects and areas of space where they originate. By collecting individual photons of x-ray light and studying how many are collected, their energy, and their speed, scientists can learn about the objects that released the x-rays. This helps scientists study such emitters as the sun, stars, neutron stars, binary stars, supernova remnants, and comets.

Black holes also release x-rays by superheating the gases drawn toward their event horizons. Many of these objects are difficult or impossible to see, so detecting and studying the x-rays they produce provides valuable insight into their origin and properties. *XMM-Newton* established a new pinnacle in research into x-rays.

### XMM-NEWTON DESIGN

The *XMM-Newton* satellite unit is 10 meters (32.8 feet) from end to end and 16 meters (52.5 feet) across its solar panel array. A true international effort, the satellite includes parts created by forty-seven companies from fourteen European countries and the United States.

Three main parts make up *XMM-Newton*: the operating systems (propulsion and electrical systems), the telescope tube and assembly for the instruments, and the x-ray mirror module. During its forty-eight-hour orbit, the satellite uses three main devices and several secondary ones to gather x-ray data.

The European Photon Imaging Camera (EPIC) is aligned to the short wavelength and small particle size of x-rays. In addition to capturing images of the x-ray photons, EPIC can also reveal the time the x-ray hit the camera. Using what is known about various types of stars, supernovas, and so forth, researchers can apply data about the arrival time to deduce physical characteristics of the object that emitted the x-rays.

*XMM-Newton* is also equipped with the Reflection Grating Spectrometer (RGS). The RGS measures wavelengths of radiation and is enhanced with 364 mirrors each etched with 645 grooves per millimeter. The grooves disperse the x-rays at different angles and provide an alternative way to gather data on the x-rays. The Optical Monitor (OM) telescope is aligned with the main x-ray detecting telescope to provide images in the optical and ultraviolet ranges. The OM and RGS work with EPIC to capture a full range of images of a wide section of sky over time. This provides the raw data scientists need to map the section in several wavelengths, giving a full picture of the x-ray patterns emitted by the objects in that area.

Other tools aboard *XMM-Newton* include optical equipment for observations and a particle detector known as EPIC Radiation Monitor System (ERMS). ERMS detects and measures the radiation from Earth's radiation belt and solar flares so that it can be factored out of the radiation data collected from space.

Satellite operations are monitored and controlled from the European Space Operations Centre (ESOC) in Darmstadt, Germany. There are also stations in French Guiana, Australia, and several other backup locations.

When *XMM-Newton* was launched, scientists hoped that it would remain viable in orbit for at least two years. As the satellite remained serviceable beyond that period and continued to provide groundbreaking data that changed how scientists viewed space, the mission was extended several times. *XMM-Newton* could remain in orbit until at least December 2018.

### DISCOVERIES FROM THE XMM-NEWTON

The *XMM-Newton* mission has achieved a great deal of success. From its earliest days, the satellite gathered and returned data from stellar nurseries and graveyards, shedding light on how stars come into being and how they fade out and die. X-ray emissions

Replica of the XMM observatory system. By Kucharek (own photograph/assembled panorama), public domain via Wikimedia Commons.

also provide a way to spot and study objects such as distant pulsars, binary stars, and black holes.

In 2010, after eleven years of gathering data, the *XMM-Newton* captured something that took scientists by surprise. A neutron star is a dense, rapidly spinning star made of leftover matter from a supernova, or powerful stellar explosion. They are normally faint and nearly impossible to see. The *XMM-Newton* detected a large burst of light from one tiny neutron star that was part of a binary pair in Supergiant Fast X-Ray Transient IGR J18410-0535. Using data from EPIC, astronomers determined that the dense gravity of the smaller star pulled in a huge discharge of superheated gas from its companion blue giant. This created a flare that lasted four hours. Scientists had long suspected such star cannibalism events, in which one star captures a burst of gas or matter from another to fuel itself. However, the bursts are infrequent and brief, making it difficult to capture images of one from end to end. This was the first direct evidence of the phenomenon.

The x-ray studies have also produced a wealth of information about black holes, which cannot be seen but are identified by observing conditions around them. A black hole leads to the formation of x-rays when its massive gravity causes gases entering its event horizon to superheat. While other telescopes and devices can detect evidence of black holes, *XMM-Newton* provides much more detailed information. In 2013, researchers working with *XMM-Newton* discovered a black hole of three thousand or more solar masses consuming material in a small dwarf galaxy. The black hole was undetectable with conventional telescopes and appeared much smaller when viewed with the Wide- Field Infrared Survey Explorer (WISE) telescope. *XMM-Newton* detected x-ray emissions more than one hundred times as strong as expected from such a small galaxy. This raised new questions as to the origins and development of black holes.

## Importance of *XMM-Newton* Research

Being able to see things in space that are otherwise invisible and undetectable provides scientists with glimpses into not only how the universe is shaped but also what it is made of and how it came to be. The extensive, detailed observations of *XMM-Newton* and its ability to peer into the inner workings of celestial objects and phenomena give researchers unprecedented insight into the universe.

—Janine Ungvarsky

## Further Reading

"The EPIC Instrument." *XMM-Newton*. Astrophysics Science Division, NASA Goddard Space Flight Center, 18 June 2012. Web. 18 May 2015.

"European Photon Imaging Camera (EPIC)." *XMM-Newton Users Handbook*. XMM-Newton Science Operations Center, European Space Agency, 3 Mar. 2003. Web. 18 May 2015.

"The European Photon Imaging Camera (EPIC) Onboard XMM-Newton." *XMM-Newton*. European Space Agency, 29 Sept. 2009. Web. 18 May 2015.

Fazekas, Andrew. "Star Caught Eating Another Star, X-Ray Flare Shows." *National Geographic News*. Natl. Geographic Soc., 6 July 2011. Web. 18 May 2015.

Stillman, Dan. "What Is a Satellite?" *NASA Education*. NASA, 12 Feb. 2014. Web. 18 May 2015.

"Summary." *XMM-Neuton*. European Space Agency, 19 July 2011. Web. 18 May 2015.

"X-Ray Detectors." *Imagine the Universe*. NASA Goddard Space Flight Center, Oct. 2013. Web. 18 May 2015.

"XMM-Newton Reveals Origin of Elements in Galaxy Clusters." *XMM-Newton*. European Space Agency, 11 May 2006. Web. 18 May 2015.

# Appendices

# NOBEL NOTES

While there is no Nobel Prize designated specifically for astronomy, quite a few of the Physics Prizes and at least one of the Chemistry prizes have some connection with one of the following areas: observational astronomy, the big bang theory, or the ultimate fate of the universe.

The Nobel Prize for Physics in 1936 was split equally between Carl D. Andersen, for his discovery of the positron, and Victor F. Hess, for the discovery of cosmic rays.

In 1967 the prize went to Hans Albrecht Bethe for his analysis of the hydrogen fusion reaction that powers most stars.

Mention should be made here of the 1971 Chemistry prize to Gerhard Herzberg, a spectroscopist, who studied the electronic structure of highly reactive species, many of which could only exist in the near vacuum of space. Using a variety of innovative techniques he was able to determine the absorption spectra of these highly reactive species.

The Nobel Physics prize was shared in 1974 by Sir Martin Ryle and Anthony Hewish for their investigation of radio sources.

In 1978, the Nobel Prize for Physics went to Arno Penzias and Robert Wilson for identifying the cosmic ray background, the isotropic radiation background reflecting a black body temperature of 2.7 K, confirming the large scale expansion of the universe as proposed by Hubble. This was the first Nobel Award for cosmology, the study of the universe as a whole.

In 1983 the Nobel Prize in Physics was split between Subramanyan Chandrasekhar and William H. Fowler. Chandrasekhar had established the maximum size possible for a white dwarf star and contributed to the theory of colliding black holes. Fowler, an experimental nuclear physicist, had clarified the sequence of reactions that powered the sun and other stars.

In 1993 the Prize was awarded jointly to Russell A. Hulse and Joseph H. Taylor Jr. for the discovery of a binary neutron star which appeared to be losing energy due to gravitational radiation, a prediction of Albert Einstein' general theory of relativity.

The Nobel Prize in 1995 honored two investigators in neutrino physics. Martin Perl was honored for the discovery of the tau lepton, which completed the set of 6 leptons in parallel to 6 quark flavors while Frederick Reines was honored for the actual detection of neutrinos in a water tank near a nuclear reactor.

In 2002, the Nobel Prize for physics was shared by three investigators: Raymond Davis, Jr. and Masatoshi Koshiba, for the detection of neutrinos from cosmic sources, and Riccardo Ginacconi for identifying cosmic x-ray sources which are most likely the result of matter falling into a black holes.

The Nobel committee returned to cosmology in 2006 with an award to John C. Mather and George F. Smoot for confirming that the cosmic background radiation was of the form of a black body spectrum and for the first determination of the anisotropy of the radiation. Physicists can now look to the development of that isotropy is the earliest moments of the universe to determine how it gave rise to the current distribution of galaxies arose.

The Nobel Prize in Physics for 2015 was awarded to Takaaki Kajita and Arthur B Macdonald for the discovery of neutrino oscillations. If the three known families of neutrinos are allowed to have a slightly different amounts of rest mass, then it could be argued that the flux of electron neutrinos from the sun would in the eight minutes it took to reach the earth turned in to a roughly equal mixture of electron, muon, and tauon neutrinos, accounting for the fact that only one third of the neutrinos expected are found among the solar neutrinos.

If we examine the Nobel Prizes that touch on Astronomy and Cosmology, we can readily identify a number of important trends. The establishment of the big bang cosmology is certainly one. Tying together advances in particle physics with their astrophysical consequences is certainly another. Enrico Fermi's "little neutral one," the *neutrino*, is continuing to surprise us. The next few chapters in the history of astronomy are still to be written.

# GLOSSARY

**ablation shield:** the material on spacecraft that is meant to ablate to protect the craft during reentry.

**absorption:** the process by which the energy of an electromagnetic wave passing through matter is retained and transferred to a molecule or atom.

**accretion disk:** a relatively flat, rotating collection of debris and gas that forms around large celestial objects with great gravitational force, such as black holes and young stars.

**active galactic nucleus (AGN):** a relatively compact region at the center of a galaxy that releases immense amounts of energy and electromagnetic radiation, believed to be the result of mass accumulation by a supermassive black hole.

**adaptive optics:** a system installed on astronomical telescopes that allows astronomers to see celestial bodies that are very bright.

**Aeolis Palus:** a plain that stretches between the northern rim of Gale Crater and the mountain Aeolis Mons on the planet Mars.

**aerodynamics:** the study of the way in which air or gases act on an object; often used when designing aircraft, automobiles, buildings, and bridges.

**Amor asteroid group:** a group of near-Earth asteroids that come within 0.3 astronomical units (AU) of Earth but do not cross its orbit, although most do cross Mars's orbit.

**angular momentum:** the momentum, or force of motion, of a rotating object or system, determined by the object's mass, speed, and distance from the point about which it is rotating.

**anomaly:** the angular distance between the position of a celestial body and its point in orbit nearest to the body it revolves around.

**aphelion:** the point in a space object's orbital path that is farthest from the sun.

**Apollo asteroid group:** the group of asteroids whose perihelion lies within Earth's orbit and whose aphelion lies outside it, causing them to cross Earth's orbit as they travel around the sun.

**asteroid:** a small, irregularly shaped celestial body that orbits the sun and is composed of rock, metal, and silicate.

**asteroid belt:** an oval-shaped ring located between the orbits of Mars and Jupiter, containing millions of asteroids in orbit around the sun.

**astrochemistry:** the interdisciplinary field that studies the chemical composition and evolution of the universe, including the chemical reactions that occur between the elements of interstellar dust.

**astronomical unit:** a unit of measure used to estimate long distances in space; equal to equal to about 149.6 million kilometers (about 93 million miles).

**astrophysics:** the branch of astronomy that deals with the physics of the universe, covering such areas as physical properties, composition, processes, and other phenomena.

**Aten objects:** asteroids that orbit between Earth and the sun and are classified as near-Earth asteroids. Aten asteroids have a semimajor axis, or half the greatest diameter of its orbit, of less than 1 astronomical unit (AU).

**attitude control:** the process of obtaining and sustaining the proper orientation in space.

**axion:** an as-yet undetected type of low-mass WIMP particle that could explain the absence of a measure of polarity for neutrons and that may have been made in large quantities during the big bang. Axions are thought to be one possible component of dark matter.

**big bang nucleosynthesis (BBN):** the coming together of free neutrons and protons to create hydrogen isotopes, and the subsequent fusion of hydrogen nuclei to form other light elements, within the first three minutes of the big bang.

**big bang theory:** a widely accepted explanation of the origin of the universe that says the universe was formed by the sudden expansion of a very hot, dense primordial fireball about 13.7 billion years ago.

**binary centaur:** two small solar system bodies that orbit around a common center of mass while also following a larger, nonresonant, unstable orbit around the sun that crosses the path of one or more outer planets.

**binary star:** two stars that orbit around a common center of mass.

**binary trans-Neptunian object:** two minor planets that orbit around a common center of mass while also following a larger orbit around the sun at a greater average distance than Neptune.

**black hole:** a region of space with a gravitational pull so strong that even light cannot escape it.

**blackbody spectrum:** the range of wavelengths of thermal radiation given off by a blackbody, an object that neither reflects nor transmits received light but gives off some light at all wavelengths, with shorter wavelengths for hotter objects.

**Bok globule:** the smallest detectable dark nebula, associated with the formation of protostars.

**bolide:** a fireball that explodes in Earth's atmosphere and creates a sonic boom.

**Bragg spectrometer:** an instrument that uses x-rays to examine and measure variations in the scattering angles of crystals; originally invented by William Henry Bragg.

**brown dwarf:** a celestial body that is between 15 and 75 times the mass of Jupiter and cannot maintain the hydrogen fusion needed to become a star.

**bulge:** a dense, spheroidal group of stars found in the center of many spiral and lenticular galaxies.

**Canadian Space Agency:** Canada's government agency established in 1989 to manage Canadian space activities and research.

**carbon-nitrogen-oxygen (CNO) cycle:** a six-step process that converts hydrogen into helium within the cores of high-mass stars.

**celestial equator:** the imaginary line above Earth's equator that halves the celestial sphere; it is equally distant from the celestial poles.

**Ceres:** a dwarf planet in the main asteroid belt, an oval-shaped ring of objects between the orbits of Mars and Jupiter. Ceres is being studied on the Dawn Mission.

**chemical propulsion:** the use of a chemical reaction to propel an object.

**chromosphere:** the middle layer of a star's atmosphere.

**circumpolar constellation:** a constellation that is always visible in the night sky; the Northern and Southern Hemispheres have different sets of circumpolar stars.

**classical mechanics:** the study of the motion of bodies, rooted in Isaac Newton's physical and mathematical principles; also called Newtonian mechanics.

**coma:** the atmosphere that forms around a comet when it passes near the sun.

**combustion:** a chemical process in which a fuel source combined with an oxidizer reacts to create heat.

**Combustion Integrated Rack (CIR):** a combustion chamber on the International Space Station (ISS) that allows scientists to remotely conduct combustion experiments in the extreme low-gravity conditions of space.

**comet:** an icy small solar system body that heats up and releases gas when it passes near the sun.

**command and control interface:** communication link allowing humans at a ground station to remotely control robots.

**common center of mass:** balance point of a system of orbiting celestial bodies; also known as the barycenter.

**conservation of energy:** a fundamental law of physics that states that the amount of energy in a domain remains constant over time. Although the energy in the domain can be converted from one form to another, it cannot be created or destroyed.

**conservation of momentum:** a fundamental law of physics that states that the amount of momentum in a domain remains constant over time. Although momentum can be changed through the action of forces, it cannot be created or destroyed.

**constellation:** an apparent pattern of stars, identified by humans, that can be seen in the night sky from Earth.

**continuous spectrum:** electromagnetic radiation that is emitted at all frequencies and wavelengths within a given range, with no apparent gaps.

**convection:** the transfer of heat via the movement of molecules in a fluid.

**core:** the dense, hot center of a star, where heat is normally generated.

**corona:** the outermost layer of a star's atmosphere, a zone of extremely low-density plasma that is several million degrees hotter than the surface of the star.

**cosmic microwave background radiation:** thermal radiation that is present in nearly uniform quantities throughout the observable universe and produces a blackbody spectrum characteristic of a temperature of approximately 2.7 kelvins. It believed to be left over from an early stage of universe expansion..

**cosmic rays:** extremely high-energy charged particles, primarily atomic nuclei, that travel through space at near light speed.

**cosmological constant:** developed by Einstein, the concept that empty space has sufficient density and pressure to prevent the universe, once believed to be static, from collapsing under the pressure of its own gravity.

**cosmology:** the astrophysical science that seeks to understand why the universe is the way it is by looking at how it formed, how it has changed, how it is structured now, and how it may change in future.

**crater:** a bowl-shaped depression on the surface of a planet, moon, or other astronomical body, usually caused by either volcanic activity or a high-speed impact with another, smaller body.

**C-type asteroid:** one of three broad classes of asteroids. C-type asteroids are mainly composed of black carbon.

**dark halo:** a hypothetical massive cloud of dark matter that surrounds a galaxy's disk.

**Dawn mission:** a NASA-funded mission launched in 2007 to study the two largest objects in the asteroid belt, the asteroid Vesta and the dwarf planet Ceres.

**declination:** the north-south position of a celestial body relative to the celestial equator expressed in degrees of arc.

**Deep Space Network:** a collection of large radio antennae positioned in three locations around the world that provide a communications system that can reach into space.

**deep-sky object:** a space object that exists beyond the solar system and cannot be seen with the naked eye.

**density:** mass per unit volume.

**detached objects (resonant):** small icy bodies in the Kuiper Belt that are in elongated orbits between 35 and 100 astronomical units (AU) from the sun and are far enough away from Neptune that they are not affected by its gravity.

**disk:** the roughly flat, circular component of a spiral or lenticular galaxy that contains the majority of the galaxy's stars, most of which orbit in the same direction; also spelled "disc."

**duct propulsion:** a form of propulsion that uses a duct to direct air or fluid into an engine, increase its momentum, and expel it to generate thrust.

**dust coma:** a cloud of dust that surrounds the nucleus of a comet when it approaches the sun.

**dust tail:** a visible stream of dust that flows off a comet's nucleus and forms an arc behind it as the comet approaches the sun.

**dust-grain model:** a model of molecular hydrogen formation in which the reaction takes place on the surfaces of interstellar dust grains.

**dwarf planet:** a celestial body that orbits the sun and has enough mass to attain a nearly round shape but lacks the gravity necessary to keep its neighborhood free of other space objects.

**dwarf star:** a low- to medium-mass star that becomes a white dwarf when it burns out.

**eccentric orbit:** an orbit that is not a perfect circle, causing the distance between the orbiting object and the body it orbits to vary.

**eccentricity:** the extent to which a celestial body's orbit deviates from a perfect circle.

**Edgeworth-Kuiper Belt:** a ring of icy rock objects located beyond the orbit of Neptune; also known as the Kuiper Belt.

**electric field:** a field around an object that is created by electric charges or magnetic fields.

**electromagnetic (EM) radiation:** energy created by interactions between electrically charged particles and transmitted in the form of waves; includes the visible light spectrum, as well as radio waves, microwaves, infrared, ultraviolet, x-rays, and gamma rays.

**electromagnetic spectrum:** the standard range of all possible frequencies of electromagnetic radiation that classifies the colors of light (red, orange, yellow, green, blue, and violet) by their wavelength and energy. Beginning with red, each color emits more energy than the previous, while the wavelength shortens as it moves from red to violet.

**electromagnetic waves:** the classical form of electromagnetic radiation, produced when electric and magnetic fields come together and interact; can be in the form of radio waves, microwaves, infrared, optical, ultraviolet, x-rays, or gamma rays, depending on their frequency, energy, and wavelength.

**electron:** a negatively charged subatomic particle.

**Eley-Rideal mechanism:** a model of molecular hydrogen formation in which one hydrogen atom adsorbs to a grain of dust, then reacts directly with a second hydrogen atom as it passes.

**ellipse:** a shape that resembles an elongated circle; mathematically speaking, a closed conic section.

**elliptical galaxy:** a roughly spherical or ellipsoid galaxy with no bulge or disk, consisting mainly of old stars that follow random orbits.

**Eris:** a dwarf planet on the edge of our solar system.

**Eros:** a very large S-type asteroid, the first NEA discovered, and the first asteroid to be landed on by a spacecraft.

**escape velocity:** the speed needed for an object to escape the gravitational pull of a planet or moon. If the object reaches this speed, it will be able to enter outer space.

**European Photon Imaging Camera (EPIC):** a camera that uses charge-couple devices (CCDs) to detect and capture images of x-rays from aboard *XMM-Newton*. Two metal oxide semiconductors (MOS CCDs) and a pn-CCD allow each of two EPICs to cover a large expanse of space to identify and map locations of x-ray emissions that help scientists learn about a variety of space objects and phenomena.

**European Space Agency:** an intergovernmental organization made up of twenty-two European countries in charge of Europe's space program and research since 1975.

**event horizon:** the boundary of a black hole beyond which nothing can escape.

**exotic matter:** any type of matter, whether real or hypothetical, whose physical properties violate the known laws of physics; for example, matter with a mass or energy density of less than zero.

**external tank:** a container that held fuel in the form of liquid hydrogen and an oxidizer in the form of liquid oxygen. The tank supplied the pressurized fuel and oxidizer to the three main engines of the space shuttle during lift-off and ascent.

**extinction:** the dimming of stellar objects because of the presence of dust between the object and observer.

**extravehicular activity (EVA):** all phases of a mission that take place while the crew member is in a specially designed pressurized suit in an unpressurized environment. These are commonly called space walks when occurring outside of a spacecraft. EVA can be a planned part of a mission or an unplanned addition to deal with a problem or urgent need.

**fireball:** a meteor that is unusually bright.

**Flame Extinguishment Experiments (FLEX):** a series of experiments conducted in the CIR from 2009 to 2013 that studied the burning of fuel droplets in the microgravity conditions of space and their response to various gases intended to suppress burning.

**flatness:** a point of near-critical density, or the point at which the universe stops expanding.

**flux:** the amount of electromagnetic radiation from an object passing through a detector

**free-free radiation:** radiation produced from electrons that are not associated with atoms or ions before or after bremsstrahlung.

**fuel injection:** a computer-controlled process in a vehicle's fuel system in which the fuel and oxygen are mixed in the ratio needed to power the vehicle's engine

**fusion:** the joining together of two or more atomic nuclei to form a new, heavier nucleus of a different element.

**galactic nucleus:** the center of a galaxy, relatively small but with a high concentration of stars.

**galaxy:** a huge collection of stars, gas, dust, and dark matter that is held together by gravity.

**gamma ray:** a form of electromagnetic radiation with high energy and a very short wavelength.

**gas coma:** the cloud of gas surrounding a comet's nucleus.

**gas-grain model:** the theory that grains of space dust interact with gas to form stars.

**general relativity:** Albert Einstein's theory that gravity is the result of matter causing space-time to curve. Among other things, this implies that gravity can bend light.

**geospatial technology:** the equipment used to see, measure, and analyze Earth and its geography, weather, and other features. Examples include global positioning systems (GPS), geographical information systems (GIS), and remote sensing (RS).

**giant:** a star larger than a dwarf star.

**Global Positioning System (GPS):** a satellite navigation system developed in the United States; uses satellite data to determine the position, navigation, and timing of objects on Earth.

**globular cluster:** a dense, spheroidal collection of old stars of the same age and origin, bound together by gravity and orbiting within the halo of a large galaxy.

**grand unification theory:** a vision or goal in physics that involves unifying the fundamental forces such as strong, weak, and electromagnetic forces. This unification would allow for a better understanding of the way the universe is organized.

**gravitational field:** a force field surrounding an object that has a gravitational influence.

**gravitational force:** the attractive force or pull between objects due to their mass.

**gravitational lens:** a phenomenon that occurs when space is distorted by a large object in the foreground of the field of vision, such as a black hole, a sun, or a galaxy. The light from distant objects behind these larger objects appears brighter and larger as it is warped by the gravity of the larger object, much as a drop of water magnifies an image under it.

**gravity:** a fundamental force that attracts two or more bodies to each other.

**ground-based navigation:** a grounded communication unit that controls the navigation of a spacecraft.

**halo:** a roughly spherical region that extends beyond the visible reaches of a galaxy, containing old, dim stars, dark matter, gas, and globular clusters.

**harmonic standing spherical wave:** a wave that is created when moving waves crash into one another.

**Harvard scheme:** a spectral classification system created at Harvard University in the late nineteenth and early twentieth century that assigned successive letters to stars based on their spectra and later rearranged to reflect their surface temperatures.

**Herbig-Haro (HH) object:** a bright object formed when the hot, high-velocity gas ejected from a protostar interacts with and ionizes the cooler material surrounding it. HH objects are often associated with Bok globules.

**Hertzsprung-Russell diagram:** a tool used by scientists to classify and chart stars according to their temperature and luminosity.

**high-resolution spectroscopy:** advanced spectroscopic technology capable of analyzing spectra in minute detail, giving greater information about the structure of the material that emitted it.

**HII region:** an interstellar region formed when gas clouds consisting primarily of atomic hydrogen are ionized by the ultraviolet radiation given off by OB stars.

**horizon:** the limit of the observable universe, or the farthest region of the universe from which Earth can receive information (i.e., radiation), located about forty-five billion light-years away.

**hot core region:** an area of space where stars form.

**Hubble classification:** a galactic classification system based on shape, invented by Edwin Hubble; also called the Hubble sequence.

**hydrogen cloud:** an invisible layer of hydrogen that surrounds a comet's coma.

**Ida:** a heavily cratered asteroid that orbits the sun in the asteroid belt. It is likely at least a billion years old.

**inertia:** the principle that states that an object remains at rest or in motion unless an outside force acts upon it.

**infrared:** electromagnetic wavelengths that are slightly longer than red visible light waves but shorter than microwaves.

**International Astronomical Union:** an association of professional astronomers from all over the world who define astronomical constants while promoting research, education, and discussion on important astronomical topics.

**interstellar dust:** solid particles of matter that exist in the vast regions of space between star systems.

**interstellar medium:** a mix of gas, dust, and cosmic rays that fills the space between star systems in a galaxy.

**ion propulsion:** a system found on certain spacecraft that gives xenon atoms a negative electrical charge, accelerates them, and then emits them to propel the vehicle through space.

**ion tail:** a stream of ions that flows out behind a comet when it approaches the sun and its gas coma interacts with the solar wind.

**irregular galaxy:** a massive group of stars and other space objects that has an unusual shape and does not share characteristics with other kinds of galaxies.

**Japan Aerospace Exploration Agency:** Japan's government agency formed in 2003 to head the country's civilian space activities and research.

**jet:** a plume of electrons and positrons that is emitted from a black hole when its magnetic field interacts with other space objects.

**jet propulsion:** the means of propelling an object by discharging a jet of ejected matter in the opposite direction of the intended motion.

**Kleopatra:** an asteroid that is a rubble pile with two moons, Cleoselene and Alexhelios.

**Koronis asteroid:** a main-belt asteroid with an irregular shape that spins at about the same rate as the others in its family.

**Kuiper Belt:** a disk-shaped region of space beyond the orbit of Neptune, about thirty to fifty-five astronomical units (AU) from the sun. The Kuiper Belt contains a collection of more than a thousand icy space objects that may be untouched material from the formation of the solar system.

**Kuiper Belt objects (classical and resonant):** small, icy space bodies in the Kuiper Belt, also known as trans-Neptunian objects; most KBOs are classical, with orbits of low eccentricity and inclination; resonant KBOs orbit the sun in a consistent ratio to Neptune's orbit.

**Lagrangian points:** a set of five points found in relation to two large objects, such as a planet and the sun, where gravitational forces allow a third, smaller object to maintain a consistent distance from the other two; also known as libration points or L-points.

**laser:** stands for "light amplification by stimulated emission of radiation"; a device that creates a steady beam of visible radiation by stimulating electronic, ionic, or molecular particles to a higher energy level, causing them to emit some of that energy.

**lenticular galaxy:** a disk galaxy with a prominent central bulge, no arms, and little interstellar medium; so named because its shape is similar to that of a lens.

**libration points:** another name for Lagrangian points, or the five points shared by two large objects where gravitational forces allow a third object to maintain a consistent distance from the other two.

**light element:** one of the first three elements of the periodic table—hydrogen (in the form of the isotope deuterium), helium, and lithium—which are believed to have been formed within the first few minutes of the big bang.

**luminescence:** the light given off by matter that does not get its energy from the creation of heat.

**luminosity:** the essential brightness of a celestial object; the amount of energy it emits.

**luminosity classification:** a means of categorizing stars by their relative sizes, as derived from their spectral emissions, atmospheric pressure, distance from Earth, and loss of light to interstellar matter; also called the Morgan-Keenan classification.

**machine vision and compensation:** computerized imaging system used in robotics to allow independent navigation and tracking of objects.

**magnetic field:** an area around an object that has a magnetic influence.

**magnetic monopole:** a hypothetical magnetic particle that has only a north pole, unlike ordinary magnets, which have both a north and a south pole.

**main sequence:** in a Hertzsprung-Russell diagram, which plots the temperature/color of stars against their luminosity/brightness, the band of stars that crosses from the upper left to the bottom right, representing the evolution of a "normal" star over the course of its lifetime.

**main-belt asteroid:** an asteroid that orbits the sun in the region between Mars and Jupiter.

**massive compact halo objects (MACHOs):** a category of objects considered likely possibilities to be dark matter. MACHOs are objects made of ordinary matter— electrons, neutrons, and protons—and include neutron stars, brown dwarfs, and black holes, all of which emit no light and are large enough to cause gravitational lensing.

**Messier object:** an astronomical deep-sky object cataloged by French astronomer Charles Messier.

**metallicity:** in astronomy, elements such as iron and silicon that are heavier than the hydrogen and helium that are fused in star cores.

**meteor:** a piece of rock or metal that enters Earth's atmosphere and becomes visible as a burning, glowing streak in the sky. The streak of light is also called a meteor.

**meteorites:** small particles that break off comets or asteroids and survive the journey through Earth's atmosphere to reach the planet's surface.

**meteoroid:** a small body of rock or metal that travels through space.

**minor planets:** a celestial object that does not meet the criteria to be a planet or a comet but orbits around the sun.

**mirror bee:** a small spacecraft equipped with mirrors intended to reflect and focus the sun's light on a specific spot on an asteroid, which would superheat and explode. This would release a plume of gas that would act much like a propulsion system to redirect the asteroid. The mirror bee idea was the precursor to laser bees.

**mission payload:** the extra equipment carried by a craft for a specific mission. For a launch vehicle, payload usually refers to scientific instruments, satellites, probes, and spacecraft attached to the launcher.

**molecular radioastronomy:** the use of radio telescopes and remote sensing to detect signals from molecules in space.

**moving wave:** a wave that travels along the medium; also known as a traveling wave.

**National Aeronautics and Space Administration (NASA):** the government agency established in 1958 by an act of Congress to run and oversee the United States' space program and research.

**n-body problem:** a mathematical model used to determine how gravity affects the motions and interactions of a group of celestial bodies.

**near-Earth asteroid (NEA):** a small, irregularly shaped celestial body with an orbital path that brings it close to Earth. NEAs travel within 0.3 astronomical units (AU) of Earth's orbit and within 1.3 AU of the sun. An astronomical unit is equal to about 149.6 million kilometers (about 93 million miles).

**near-Earth object:** a celestial object that orbits the sun within 1.3 astronomical units (AU) and that potentially poses a collision hazard to Earth.

**neutralino:** an as-yet undetected type of very large, light WIMP particle that may be a kind of dark matter and that is part of supersymmetry, a physics theory that endeavors to combine all the known forces of physics.

**neutrino:** a type of fundamental particle created in large amounts by nuclear reactions in stars, and one possible type of WIMP that could contribute to dark matter. Neutrinos pass right through Earth without causing any effect, making them difficult to detect.

**neutron star:** a dense, rapidly spinning star made of the material that remains after a star becomes a supernova. It is formed from the neutrons created by the reaction of the star's remaining protons and electrons.

**Newton's laws of motion:** three scientific laws developed by Isaac Newton that describe an object's response to a force that acts upon it. These laws relate force, acceleration, mass, and velocity.

**nova:** a sudden, intense, ultimately short-lived increase in the brightness of a white dwarf star, caused by an accumulation of hydrogen that triggers a runaway thermonuclear reaction.

**nozzle:** an opening at the rear of a rocket that allows gases to be released and accelerates them so that thrust can be maximized.

**nuclear propulsion:** the use of a nuclear reactor to propel an object.

**nucleus:** the body of a comet, made up of ice, dust, and carbon.

**OB star association:** a loose collection of O and B stars. One of the best-known OB star associations is Orion B1 in the Milky Way galaxy.

**Object D:** early designation of *Sputnik 3*, designed by the Soviets in 1956 and intended to be the first satellite sent into space. Because of its complicated design, it was not finished in time for the first launch.

**Oort cloud:** a vast, spherical area of space far beyond Neptune that is filled with trillions of icy objects and comets.

**opacity:** the degree to which a substance or object lets various forms of electromagnetic radiation pass through it.

**Opportunity:** a rover that, as of 2015, has spent the longest amount of time and traveled the greatest distance on the surface of Mars to date. It landed in 2004 and has provided crucial data about the planet.

**optical navigation system (ONS):** a navigation system that uses optical physics to measure the degree of the relative motion (both speed and magnitude) between a navigation device and the surface being navigated. ONS processes the reflection and scattering of light to aid in the navigation of surfaces.

**optics:** science of the properties and behaviors of light

**Orbital Maneuvering System:** a system that provided the thrust needed for orbit insertion, transfer, and deorbit via two smaller engines.

**orbital mechanics:** the study of the movements of artificial satellites and spacecraft; also called flight mechanics or astrodynamics.

**orbital robotics:** field that focuses on using robotic technology such as satellites, space probes, and service spacecraft and equipment in space.

**particle accelerator:** a machine designed to increase the speed and energy of subatomic particles to extremely high levels.

**payload:** the carrying capacity of an aircraft or spacecraft. With spacecraft, the payload may include astronauts, equipment, a satellite, and/or a space probe.

**perihelion:** the point in a space object's orbital path that is closest to the sun.

**periodic comet:** a comet that takes less than two hundred years to complete an orbit; also known as a short-period comet.

**perturbation:** a change in the orbit of a celestial object caused by the gravitational force of another object.

**photoelectric effect:** a process in which electrons are freed from a solid after the solid is exposed to electromagnetic radiation.

**photon:** an elementary particle of light that moves and has energy but lacks mass and electrical charge.

**photosphere:** the effective surface of a star and the innermost layer of its atmosphere.

**Planck relationship:** an equation that relates the energy of a moving particle to its frequency; first proposed by physicist Max Planck to explain the properties of radiation.

**planetary rover:** mobile robotic science lab sent to explore and study the surface of other planets.

**plasma:** a fluid-like state of matter consisting of highly ionized gas that contains positively charged ions and negatively charged electrons. Plasmas can propagate waves and are highly electrically conductive.

**platform:** all parts of a spacecraft that are not part of the payload; also known as the bus.

**plerion:** a type of supernova remnant, also called a crab-type remnant, that emits radiation from a pulsar in its center.

**plerionic composite:** a type of supernova remnant that appears like a plerion at both x-ray and radio wavelengths but also exhibits the shell of a shell-type remnant.

**plutinos:** KBOs found in the inner edges of the Kuiper Belt that have the same orbital resonance as Pluto, or a resonance of 3: 2.

**polarization:** a property in which the oscillations in a wave are perpendicular to the direction of travel. It can cause waves to produce other waves.

**potentially hazardous object:** a space object greater than 140 meters (about 460 feet) in diameter with an orbit that passes within 0.05 astronomical units (about 7.5 million kilometers, or 4.6 million miles) of Earth's orbit.

**power sources and recharging:** batteries and cells needed to run space robots, typically using solar power for electricity and recharging.

**propellant:** a mixture of fuel and oxidizer (usually oxygen) that propels a rocket. Both solid and liquid propellants are used in rockets.

**propulsion:** the way of creating force to move an object.

**protostar:** an infant star that forms when a cold, dense molecular cloud of dust and gas collapses, its gravity causing part of it to become hotter and denser until it eventually produces the core of the new star.

**pulsar:** a neutron star formed from the remains of a supernova explosion that gives off radio waves in intermittent bursts or pulses.

**Q-type asteroid:** an asteroid that is made up of olivine (a rare iron silicate), pyroxene, and metals.

**quantum physics:** the subfield of physics that studies the often contradictory and counterintuitive behavior of the smallest units of matter and energy, such as subatomic particles, atoms, and molecules.

**quark:** an elementary particle that is part of every hadron (proton or neutron). It is found in pairs or triplets.

**quark-nova:** an explosion believed to occur after a high-mass star experiences a supernova, resulting from a second collapse of its core.

**quasar:** short for "quasi-stellar"; an extremely bright celestial object that produces very large amounts of energy.

**R-7 rocket:** the Soviet-designed rocket that launched *Sputnik 1* into space. Versions of the rocket have continued to be used to send craft into space.

**radiation hardening:** process of making electronic components resistant to radiation in space.

**radiative zone:** the layer of a star's interior in which heat is transferred from the core toward the surface via radiative diffusion, or the absorption and reemission of photons.

**radiometric resolution:** the ability of an imaging system or sensor to capture small increments of energy in an image.

**reflection:** the return of a light or sound wave to its source. The wave bounces back after interacting with a surface.

**relativistic energy expression:** a physics expression that shows that in high-speed collisions, particles with mass can be created at the expense of kinetic energy.

**right ascension:** the east-west position of a celestial body defined in relation to the celestial equator and expressed in hours and minutes, not degrees of arc.

**rocket propulsion:** a class of jet propulsion in which the thrust is produced by ejecting a propellant such as fuel; used as the main form of spacecraft propulsion because it does not need air.

**rover:** an unmanned, self-propelled mobile research unit that can travel independently to gather information about a location, especially a planet, moon, or other celestial body.

**rubble-pile asteroid:** a nonsolid asteroid that is made up of rocks of various sizes held together by gravity.

**Russian Federal Space Agency:** the government agency created in 1992 to officially coordinate Russian space activities and research.

**satellite:** any small object, natural or artificial, that orbits around a larger object; one example is Earth orbiting the sun. Artificial satellites are generally deployed for research, monitoring, communication, or weather prediction purposes, and all include at least one antenna for sending and receiving data and a power source such as a battery or solar panels.

**satellite navigation:** a system of navigation in which location is determined based on information transmitted from satellites. A receiver records the time of regular transmissions from multiple satellites equipped with atomic clocks, allowing it to determine its longitude, latitude, altitude, and time.

**scattered disk objects:** space bodies made of ice and rock with very eccentric, or flattened oval, orbits. They are found in the area beyond Neptune in the outer area of the solar system.

**semimajor axis:** the distance from the center of an ellipse and the farthest point of its perimeter.

**singularity:** a theoretical point or region of infinite density where the curvature of space-time, and thus the gravitational force, likewise becomes infinite.

**small solar system body:** an astronomical object other than a planet, dwarf planet, or satellite that orbits the sun.

**Sojourner:** the first robotic rover to land on Mars. It landed in 1997 and began sending information from Mars's surface back to Earth.

**solar neutrino:** a neutrino created by nuclear reactions in the sun.

**solar propulsion:** propulsion using solar energy, whether collected by solar panels to power an electric engine (ion propulsion) or harnessed directly to push a solar sail.

**solid rocket boosters:** two solid-propellant motors that provided the space shuttle with the main thrust needed to lift it off the launch pad and into space against the pull of Earth's gravity.

**Soviet Union:** the Union of Soviet Socialist Republics (USSR), a communist state established in the aftermath of the Russian Revolution of 1917. Following World War II, the Soviet Union and the United States raced for supremacy in space as part of the Cold War. The Soviets were the first to put a satellite in orbit, the first to put a human in orbit around Earth, and the first to launch a space station.

**Soyuz:** the Russian spacecraft used to carry astronauts to and from the International Space Station.

**space station:** an orbiting space vehicle intended for long-term support of a human crew. Space stations include facilities for living and research in addition to the systems necessary to run the station. Crew members can enter and leave the space station while it remains in orbit by means of secondary craft used for transport.

**spatial resolution:** the number of pixels, or individual units, that make up a digital image. A greater number of pixels provides a higher spatial resolution and therefore greater clarity and detail.

**spectra:** plural of "spectrum"; a range or distribution of light or wavelengths.

**spectral classification:** a method of classifying stars based on various elements of their emission spectra, as determined by such characteristics as surface temperature, luminosity, and size.

**spectral patterns:** lines in the electromagnetic spectra of stars that represent the presence or absence of wavelengths.

**spectral resolution:** the ability of an astronomical device called a spectrograph to differentiate various wavelengths of light.

**spectroscopy:** the study of the spectrum of light wavelengths emitted by atoms and molecules. This tool helps researchers determine the wavelengths of an object's light. It can also verify the object's velocity, temperature, and makeup.

**spectrum:** the range of wavelengths that make up light, such as visible, infrared, and ultraviolet light.

**spin-orbit coupling:** the rotation rate of a celestial object in relation to the rate at which it orbits another body; also known as orbital resonance.

**spiral galaxy:** a disk galaxy with spiral arms that extend into the disk from either a central bulge or a bar that passes through the bulge.

**Spirit:** a rover that landed on Mars in 2004. It sent information about the planet back to Earth until it ceased transmission in 2011.

**staged combustion:** a chemical process in which fuel is ignited, reacts with oxygen, and begins burning; the resulting heat is forced into a second stage chamber, where it ignites additional fuel and completes the combustion process.

**standing wave:** a wave that stays in place as it vibrates up and down.

**star cluster:** a group of stars that share a common origin and are held together by gravity for a period of time.

**stellar halo:** a spherical mass of stars, star clusters, dust, and gas that surrounds certain types of galaxies.

**stellar neutrino:** a neutrino created by a star other than the sun.

**stellar nucleosynthesis:** the fusion of light elements into heavier elements, such as carbon, oxygen, and iron, within the interiors of stars.

**Strömgren sphere:** a spherical region of ionized atomic hydrogen (HII) surrounding a young OB star in an HII region.

**s-type asteroid:** fairly bright asteroids that are usually made of stone (silicate), nickel, and iron.

**sublimation:** the process of a frozen solid becoming a gas without passing through the liquid phase.

**sunspots:** cooler, darker areas of the sun that appear in pairs or groups and result from the sun's magnetic field pushing through its surface.

**Sunyaev-Zel'dovich effect:** an absence of microwaves in the cosmic background radiation, caused by x-rays interacting with hot gas.

**supergiant:** the largest type of star in the universe, and usually the hottest.

**supermassive black hole:** a black hole with a mass greater than one hundred thousand times that of the sun.

**supernova:** a massive explosion that results when a dying star exhausts its fuel. The star's core collapses and explodes, releasing large amounts of energy and matter into the universe.

**synchrotron radiation:** radiation produced when charged particles move in a curved path.

**T Tauri star:** a low-mass star that formed less than ten million years ago and is not hot enough for nuclear fusion to begin. In terms of stellar evolution, T Tauri stars come between protostars and low-mass main-sequence stars such as Earth's sun.

**tail:** ionized gas and dust that is pushed away from the head of a comet.

**telemetry:** the process of transmitting measurement data via radio to operators on the ground. Telemetry is used to improve spaceflight performance and accuracy. It provides important information about standard operational health and status of a craft as well as mission-specific payload data.

**teleoperation:** remote operation and control of robotics.

**temporal resolution:** the amount of time that it takes a remote sensing device to make a return trip to an exact location over Earth and acquire information. A shorter delay between revisits means the temporal resolution is high.

**theory:** an explanation of some aspect of the natural universe that has been developed and thoroughly tested by means of the scientific process of observation, development of a hypothesis or reasoned explanation, experimentation, and conclusion.

**thermal composite:** a type of supernova remnant that appears shell-like at radio wavelengths but crab-like at x-ray wavelengths.

**thermal considerations:** how exposure to heat or cold affects components in spacecraft or robotics.

**thermal control:** the system aboard a spacecraft that controls the temperature of various components to ensure safety and accuracy during a mission.

**thermal energy:** energy that is in the form of heat.

**thermal spectrum:** the continuous spectrum of electromagnetic radiation that an object emits in the form of heat, produced by the movement of the particles that make up the object.

**thick disk:** the part of a galactic disk that contains older stars, typically more than ten billion years old.

**thin disk:** the part of a galactic disk that contains younger stars, typically less than ten billion years old, as well as the dust and gas from which new stars are formed.

**thrust:** in rocketry, the force produced by a rocket engine that pushes a rocket upward.

**tidal evolution:** the change in the rise and fall of an ocean caused by the gravitational force of a nearby celestial object.

**tidal force:** the effect of the gravitational force of one celestial body on another.

**time travel:** the theoretical or imagined ability to travel between two points in time.

**tracking and commanding:** tracking takes account of a craft's position in relation to the ground base with transponders, radar, or other systems. Commanding refers to the ground station sending signals to a craft to change settings such as ascent and orbit paths.

**trajectory:** the curved path that an object, either natural or artificial, takes through space.

**trans-Neptunian object (TNO):** a space object that orbits beyond Neptune at a distance more than 30 astronomical units (AU)—about 4.5 billion kilometers, or 2.8 billion miles—from the sun.

**Trojan asteroid:** an asteroid that shares an orbital period with a planet and maintains a consistent distance from that planet.

**turbomachinery:** devices that use an engine with rotating blades to generate or transfer energy from a fluid.

**ultraviolet radiation:** short-wavelength electromagnetic radiation just beyond the visible light spectrum.

**US Explorer 6:** the sixth in a series of Earth-imaging satellites launched in the late 1950s and the first to send back a picture of Earth from orbit in 1959.

**vacuum:** an empty space devoid of air or any other gas; also sometimes defined as the absence of matter, although scientists now believe that even the vacuum of space, where no atomic matter is found, is occupied by dark energy and dark matter.

**vector:** a quantity having both magnitude and direction.

**Vesta:** the largest asteroid in the main asteroid belt, an oval-shaped ring of objects between the orbits of Mars and Jupiter. Vesta was studied on the Dawn Mission.

**visible light spectrum:** portion of the electromagnetic spectrum visible to the human eye

**wavelength:** the distance between two peaks of a light wave.

**weakly interacting massive particle (WIMP):** a category of objects considered likely possibilities to be dark matter. WIMPs are not made up of ordinary matter but do have mass, can be light or heavy, and can pass through ordinary matter without effect. Possible types include axions, neutrinos, and neutralinos.

**white dwarf:** a small, very dense star that has exhausted nearly all of its fuel and is nearing the end of its life cycle.

**Wide-field Infrared Survey Explorer telescope:** an instrument used to collect images and data from throughout the Milky Way.

**wind tunnel tests:** aerodynamics tests that involve moving air past an object in a tunnel to study the resulting effects.

**x-ray:** a form of electromagnetic radiation with short wavelengths and high energy that is produced in space by sources including supernovas, active galactic nuclei, quasars, and the superheating of gases being drawn into black holes.

**x-ray astronomy:** the branch of astronomy that is concerned with detecting x-rays in space and studying the phenomena that produce them.

**YORP effect:** a phenomenon in which objects in space change their rate of spin when exposed to rays from the sun.

**zodiac:** an imaginary band across the sky that follows the ecliptic (the path along which the sun appears to travel over one year when viewed from Earth) and is divided into twelve equal sections, each occupied by one of twelve constellations.

# BIBLIOGRAPHY

Achenbach, Joel. "The God Particle." *National Geographic*. Natl. Geographic Soc., Mar. 2008. Web. 12 June 2015.

Aleksandrov, Aleksandr Pavlovich, and Richard Fullerton. "Extravehicular Activity (EVA)." *Phase 1 Program Joint Report*. Ed. George C. Nield and Pavel Mikhailovich Vorobiev. NASA, Jan. 1999. *NASA*. Web. 4 June 2015.

Allain, Rhett. "Why Does a Meteor Explode in the Air?" *Wired*. Condé Nast, 18 Feb. 2013. Web. 11 Mar. 2015.

Allen, Jesse S. "The Classification of Stellar Spectra." *LHEA Team and Group Web Sites*. NASA, 1998. *Professor Paul Crowther, University of Sheffield*. Web. 8 Apr. 2015.

Amos, Jonathan. "Controllers Now Banking on Philae Wake-Up Call." *BBC*. BBC, 30 Jan. 2015. Web. 5 Mar. 2015.

Anderson, David Lewis. "Wormholes." *The Anderson Institute*. Anderson Multinatl., 2012. Web. 10 July 2015.

Anderson, Scott R. "Lecture 6: The Laws of Motion and Gravity." *Open Course Info*. Author, 10 Apr. 2003. Web. 6 May 2015.

Angelo, Joseph A. "Ablation." *Encyclopedia of Space and Astronomy*. New York: Facts On File, 2006. Print.

Archaeological Consultants. *NASA-Wide Survey and Evaluation of Historic Facilities in the Context of the US Space Shuttle Program: Roll-Up Report*. Sarasota: Archaeological Consultants, 2008. PDF file.

Bahcall, John N. "Solving the Mystery of the Missing Neutrinos." *Nobelprize.org*. Nobel Media, 28 Apr. 2004. Web. 15 Apr. 2015.

Baldwin, Emily. "Asteroid Kleopatra Spawned Twins." *Astronomy Now*. Pole Star, 23 Feb. 2011. Web. 16 Feb. 2015.

Balogh, André, Leonid Ksanfomality, and Rudolf von Steiger, eds. *Mercury*. New York, NY: Springer Science+Business Media BV, 2008. Print.

Barber, Elizabeth. "What Is a Space 'Centaur'? Scientists Now Know the Answer." *Christian Science Monitor*. Christian Science Monitor, 26 July 2013. Web. 20 Mar. 2015.

Barlow, Nadine G. *Mars: An Introduction to Its Interior, Surface and Atmosphere*. New York, NY: Cambridge University Press, 2008. Print.

Barrow-Green, June. "Oscar II's Prize Competition and the Error in Poincaré's Memoir on the Three Body Problem." *Archive for History of Exact Sciences* 48.2 (1994): 107–31. Print.

Barucci, M. A., et al. "(90377) Sedna: Investigation of Surface Compositional Variation." *Astronomical Journal* 140.6 (2010): 2095–2100. PDF file.

Beatty, Kelly. "Tour November's Sky: Cassiopeia." *Sky & Telescope*. F+W Media, 31 Oct. 2014. Web. 22 Apr. 2015.

Belfiore, Michael. "How to Mine an Asteroid." *Popular Mechanics*. Hearst Digital Media, 27 Oct. 2014. Web. 20 Mar. 2015.

Bengtsson, Lennart, Roger-Maurice Bonnet, David Grinspoon, Symeon Koumoutsaris, Sebastien Lebonnois, and Dmitri Titov, eds. *Towards Understanding the Climate of Venus*. New York, NY: Springer Science+Business Media, 2014. Print.

Beutel, Jacob, Harold L. Kundel, and Richard L. Van Metter. *Handbook of Medical Imaging: Physics and Psychophysics*. Bellingham: SPIE, 2000. Print.

Bhaskaran, Shyam. "Autonomous Navigation for Deep Space Missions." *Twelfth International Conference on Space Operations, June 11–15, 2012*. Pasadena: Jet Propulsion Laboratory, 2012. PDF file.

Bilger, Burkhard. "The Martian Chroniclers." *New Yorker*. Condé Nast, 22 Apr. 2013. Web. 5 June 2015.

"Binary Star." *Cosmos: The SAO Encyclopedia of Astronomy*. Swinburne University of Technology, n.d. Web. 30 Mar. 2015.

Bittencourt, J. A. *Fundamentals of Plasma Physics*. 3rd ed. New York: Springer, 2004. Print.

Bjornerud, Marcia. *Reading the Rocks: The Autobiography of the Earth*. New York, NY: Basic Books, 2005. Print.

Blevins, John A, et al. "An Overview of the Characterization of the Space Launch System Aerodynamic Environments." *Aerospace Research Central*. American Inst. of Aeronautics and Astronautics, 13 Jan 2014. Web. 8 June 2015.

Blunck, Jürgen. *Solar System Moons: Discovery and Mythology*. New York, NY: Springer, 2010. Print.

"Bode's Law." *Astro 2201: Our Home in the Universe*. Cornell U, n.d. Web. 11 May 2015.

Bond, Peter. *Exploring the Solar System*. Oxford: Wiley, 2012. Print.

Bond, Peter. "Galileo Spacecraft Crashes into Jupiter." *Spaceflight Now.* Spaceflight Now, 21 Sept. 2003. Web. 16 Mar. 2015.

Braeunig, Robert A. "Orbital Mechanics." *Rocket & Space Technology.* Author, 2013. Web. 4 June 2015.

Brainerd, Jerome James. "Protostars." *Astrophysics Spectator.* Astrophysics Spectator, 7 May 2008. Web. 21 May 2015.

Bratton, Mark. *The Complete Guide to the Herschel Objects: Sir William Herschel's Star Clusters, Nebulae and Galaxies.* New York: Cambridge University Press, 2011. Print.

"Breakthroughs in Cosmology: Taking the Universe's Baby Pictures." *HubbleSite.* NASA, n.d. Web. 7 Apr. 2015.

"Brief History of Rockets." *NASA.* Natl. Aeronautics and Space Administration, 12 June 2014. Web. 28 May 2015.

Brown, Michael E., and Emily L. Schaller. "The Mass of Dwarf Planet Eris." *Science* 316.5831 (2007): 1585. Print.

\_\_\_\_\_, et al. "Discovery of a Candidate Inner Oort Cloud Planetoid." *Astrophysical Journal* 617 (2004): 645–49. Web. 7 Apr. 2015.

\_\_\_\_\_. "The Largest Kuiper Belt Objects." *Solar System beyond Neptune.* Tucson: University of Arizona Press, 2008. PDF file.

\_\_\_\_\_, Al Conrad, and Linda Copman. "Planetary Astronomy: The New Solar System." *Cosmic Matters.* W. M. Keck Observatory, Fall 2007. Web. 27 Mar. 2015.

Burchfield, Joe D. *Lord Kelvin and the Age of the Earth.* Chicago: University of Chicago Press: U of Chicago, 1990. Print.

Burrows, David N., et al. "Supernova Remnants." *Penn State University Department of Astronomy and Astrophysics.* Pennsylvania State U, 18 Oct. 2004. Web. 20 Mar. 2015.

Buta, Ronald J., Harold G. Corwin Jr., and Stephen C. Odewahn. *The de Vaucouleurs Atlas of Galaxies.* New York: Cambridge UP, 2007. Print.

Byrd, Deborah. "Orion the Hunter Easy to Spot in January Night Sky." *EarthSky.* Earthsky Communications, 9 Jan. 2015. Web. 1 May 2015.

Byrne, Michael. "The Black Holes We See in Space Might Already Be White Holes." *Motherboard.* Vice Media, 21 July 2015. Web. 30 Apr. 2015.

Cammorata, Nicole. *Words You Should Know 2013: The 201 Words from Science, Politics, Technology, and Pop Culture that Will Change Your Life This Year.* Avon: Adams, 2013. Print.

Cancellieri, Brian M., and Vladimir G. Mamedov, eds. *Interstellar Medium: New Research.* Hauppage: Nova Science, 2012. Print.

Carr, Michael. *The Surface of Mars.* New York, NY: Cambridge University Press, 2007. Print.

Cartlidge, Edwin. "Neutrinos Point to Rare Stellar Fusion." *Physicsworld.com.* IOP, 9 Feb. 2012. Web. 6 Apr. 2015.

Cartwright, Jon. "Physicists Discover a Whopping 13 New Solutions to Three-Body Problem." *Science.* Amer. Assoc. for the Advancement of Science, 8 Mar. 2013. Web. 15 May 2015.

\_\_\_\_\_. "Reading between the Lines: Space Spectroscopy." *Chemistry World* Dec. 2009: 46–49. Print.

Casselman, Bill. "A New Solution to the Three Body Problem." *Feature Column.* American Mathematical Society. Web. 15 May 2015.

Cassidy, David C. "Focus: *Landmarks*: Photons Are Real." *Physics Rev. Focus* 13.8 (2004): n.pag. *Physics.* Web. 15 Apr. 2015.

Celletti, Alessandra, and Ettore Perozzi. *Celestial Mechanics: The Waltz of the Planets.* Chichester: Praxis, 2007. Print.

Cessna, Abby. "Planetoid." *Universe Today.* Fraser Cain, 10 Aug. 2009. Web. 6 May 2015.

Chaisson, Eric J. "Origin of Heavy Elements." *Cosmic Evolution.* Harvard-Smithsonian Center for Astrophysics, 2013. Web. 2 June 2015.

Chambers, John and Jacqueline Mitton. *From Dust to Life: The Origin and Evolution of Our Solar System.* Princeton: Princeton University Press, 2013. Print.

Chapman, Clark R. "The Hazard of Near-Earth Asteroid Impacts on Earth." *Earth and Planetary Science Letters* 222.1 (2004): 1–15. Print.

Chapman, Mary. *The Geology of Mars.: Evidence from Earth-Based Analogs.* New York, NY: Cambridge University Press, 2007. Print.

Chappell, Bill. "Meteorite Impact on Moon Sets Record as Brightest Ever Seen." *The Two-Way.* Natl. Public Radio, 24 Feb. 2014. Web. 9 Mar. 2015.

"Circumstellar Disks." *Formation and Evolution of Planetary Systems.* University of Arizona, n.d. Web. 27 Mar. 2015.

Cobb, Allan B. *Earth Chemistry*. New York, NY: Chelsea House, 2009. Print.

Cockell, Charles, ed. *An Introduction to the Earth-Life System*. New York, NY: Cambridge University Press, 2007. Print.

Comins, Neil F. *Discovering the Universe: From the Stars to the Planets*. New York: Freeman, 2009. Print.

Condon, J. J., and S. M. Ransom. "Free-Free Radio Emission from an HII Region." *Essential Radio Astronomy*. Natl. Radio Astronomy Observatory / Assoc. Univs., n.d. Web. 26 Mar. 2015.

Congiu, Sasha. "SLS Model 'Flies' through Langley Wind Tunnel Testing." *NASA*. NASA, 28 Nov. 2012. Web. 8 June 2015.

Connolly, Amy R. "Scientists: Evidence of Big Bang Theory Fails to Space Dust." *UPI. com*. United P Intl., 31 Jan. 2015. Web. 22 Apr. 2015.

"Constellations: Frequently Asked Questions." *Tools for Science*. Coll. of St. Benedict & St. John's U, n.d. Web. 27 Mar. 2015.

Cook, Jia-Rui. "It's Complicated: Dawn Spurs Rewrite of Vesta's Story." *Jet Propulsion Laboratory*. Caltech, 6 Nov. 2013. Web. 16 Mar. 2015.

Cooksey, Diana. "Understanding the Global Positioning System (GPS)." *MSU Global Positioning System (GPS) Laboratory*. Montana State U, n.d. Web. 10 July 2015.

Cornish, Neil J. "The Lagrange Points." *Wilkinson Microwave Anisotropy Probe*. NASA, July 2012. Web. 6 Mar. 2015.

Coulter, Dauna. "Space Mountain Produces Terrestrial Meteorites." *NASA Science: Science News*. NASA, 30 Dec. 2011. Web. 16 Mar. 2015.

Cox, Brian and Andrew Cohen. *Wonders of the Universe*. Harper Design, 2011.

Crockett, Christopher. "Surprising Rings Circle Comet-Asteroid Hybrid." *Science News for Students*. Soc. for Science & the Public, 30 Mar. 2014. Web. 3 Mar. 2015.

Croswell, Ken. "Star Struck." *National Geographic*. Natl. Geographic Soc., Dec. 2010. Web. 13 Mar. 2015.

Dambeck, Susanne. "Einstein: Travelling through Spacetime." *Lindau Nobel Laureate Meetings*. Lindau Nobel Laureate Meetings, 17 Apr. 2015. Web. 5 May 2015.

Dana, Peter H. "A Guide to the Global Positioning System." *Geographer's Craft Project*. U of Colorado at Boulder, 1999–2015. Web. 10 July 2015.

David, Leonard. "Russia Reignites Its Rocket Industry with New Angara Booster." *Space. com*. Purch, 19 Aug. 2014. Web. 8 June 2015.

Dearden, Lizzie. "One of the Most Earth-Like Planets in Our Galaxy Gliese 581d 'Really Does Exist,' Astronomers Say." *Independent*. Independent. co.uk, 7 Mar. 2015. Web. 30 Mar. 2015.

"Deep Space, Branching Molecules, and Life's Origins?" *Life, Unbounded*. Scientific Amer., 30 Sept. 2014. Web. 26 May 2015.

de Pater, Imke, and Jack J. Lissauer. *Planetary Sciences*. Updated 2nd ed. Cambridge: Cambridge University Press, 2015. Print.

Desain, John D. "Green Propulsion: Trends and Perspectives." *Aerospace*. Aerospace Corp., 2011. Web. 5 June 2015.

Dickinson, David. "Why This Weekend Is Perfect for a Messier Marathon." *Universe Today*. Fraser Cain, 5 Mar. 2013. Web. 5 May. 2015.

Dickson, Paul. "Sputnik's Impact on America." *Nova*. WGBH, 6 Nov. 2007. Web. 4 June 2015.

Dick, Steven J. *Discovery and Classification in Astronomy: Controversy and Consensus*. New York: Cambridge University Press, 2013 Print.

Dinwiddie, Robert, Will Gater, Giles Sparrow, and Carole Stott. *Nature Guide: Stars and Planets*. DK, 2012.

Dotson, Ben. "How Particle Accelerators Work." *Energy.gov*. US Dept. of Energy, 18 June 2014. Web. 12 June 2015.

Dowling, Stephen. "What Caused the Space Shuttle Columbia Disaster?" *BBC Future*. BBC, 30 Jan. 2015. Web. 1 July 2015.

Dunbar, Brian. "Mission to Mars: Mars Science Laboratory." *NASA*. NASA, 28 July 2013. Web. 15 Apr. 2015.

Dupraz, Sébastien, Trent Dupuy, and Sandrine Bottinelli. "Formation of Astrobiologically Important Molecules in Extraterrestial Environments." *Astrobiology Institute*. U of Hawaii NASA Astrobiology Inst., 2004. Web. 26 Mar. 2015.

Ďurech, J., et al. "New Photometric Observations of Asteroids (1862) Apollo and (25143) Itokawa: An Analysis of YORP Effect." *Astronomy & Astrophysics* 488.1 (2008): 345–50. Print.

Durham, Ian T. "Emden, Robert." *Biographical Encyclopedia of Astronomers*. Ed. Thomas Hockey et al. 2nd ed. Vol. 2. New York: Springer, 2014. Print.

Duxbury, T. C., et al. "Asteroid 5535 Annefrank Size, Shape, and Orientation: Stardust First Results." *Journal of Geophysical Research* 109.E2 (2004): n. pag. Web. 13 Mar. 2015.

"Dwarf Planets." *National Geographic*. Natl. Geographic Soc., 1996–2015. Web. 9 Mar. 2015.

Dye, Lee. "An Aging Pioneer Has Halley Comet Show All to Itself." *Los Angeles Times*. Los Angeles Times, 27 Feb. 1986. Web. 10 Mar. 2015.

Dymock, Roger. *Asteroids and Dwarf Planets and How to Observe Them*. New York: Springer, 2010. Print.

Eicher, David J. *The New Cosmos: Answering Astronomy's Big Questions*. Cambridge University Press, 2016.

"Einstein Was Right: Space-Time Is Smooth, Not Foamy." *Space.com*. Purch, 10 Jan. 2013. Web. 5 May 2015.

"Elementary Einstein." *Einstein Online*. Max Planck Inst. for Gravitational Physics, n.d. Web. 10 July 2015.

Elert, Glenn. "Standing Waves." *Physics Hypertextbook*. Glenn Elert, n.d. Web. 6 May 2015.

Elkins-Tanton, Linda T. *The Sun, Mercury, and Venus*. New York, NY: Chelsea House, 2006. Print.

_____. *Uranus, Neptune, Pluto and the Outer Solar System* New York, NY: Chelsea House, 2006. Print.

Emspak, Jesse. "Eight Ways You Can See Einstein's Theory of Relativity in Real Life." *LiveScience*. Purch, 26 Nov. 2014. Web. 5 May 2015.

"Energy Transport in Stars." *Astronomy 162: Stars, Galaxies, and Cosmology*. University of Tennessee Knoxville, n.d. Web. 21 May 2015.

"Engine Fuel System." *Glenn Research Center*. NASA, 12 June 2014. Web. 21 Apr. 2015.

Eratosthenes, Hyginus, and Aratus. *Constellation Myths: With Aratus's* Phaenomena. Trans. Robin Hard. Oxford: Oxford University Press, 2015. Print.

Esposito, Larry W., Ellen R. Stofan, and Thomas E. Cravens, eds. *Exploring Venus as a Terrestrial Planet*. Washington, DC: American Geophysical Union, 2007. Print.

"European Photon Imaging Camera (EPIC)." *XMM-Newton Users Handbook*. XMM-Newton Science Operations Center, European Space Agency, 3 Mar. 2003. Web. 18 May 2015.

*Europe's Satellite Navigation Systems*. European Space Agency, 2000–2010. Web. 22 Apr. 2015.

"Explorer 6." *NASA Space Science Data Center*. NASA, 26 Aug. 2014. Web. 24 Feb. 2015.

"Exploring Impact Craters with Google Earth." *Las Cumbres Observatory Global Telescope Network*. LCOGT, 2012. Web. 9 Mar. 2015.

Falkner, David E. *The Mythology of the Night Sky: An Amateur Astronomer's Guide to the Ancient Greek and Roman Legends*. New York: Springer, 2011. Print.

"Faraway Eris Is Pluto's Twin." *ESO*. European Southern Observatory, 26 Oct. 2011. Web. 27 Mar. 2015.

Faure, Gunter, and Teresa M. Mensing. *Introduction to Planetary Science: The Geological Perspective*. New York, NY: Springer, 2007. Print.

Fazekas, Andrew. "Multiple-Star Birth Revealed in Stellar Nursery." *National Geographic News*. Natl. Geographic Soc., 11 Feb. 2015. Web. 1 Apr. 2015.

_____. "Star Caught Eating Another Star, X-Ray Flare Shows." *National Geographic News*. Natl. Geographic Soc., 6 July 2011. Web. 18 May 2015.

Ferguson, Harry, and Andy Fruchter. "Drizzle." *Space Telescope Science Institute*. STScI, 1997. Web. 7 Apr. 2015.

Ferron, Karri. "What Are We Learning about the True Identity of Centaurs?" *Astronomy* Jan. 2014: 19. *Science Reference Center*. Web. 3 Mar. 2015.

Fishbine, Brian, et al. "Nuclear Rockets: To Mars and Beyond." *National Security Science*. Los Alamos Natl. Laboratory, 2011. Web. 21 Apr. 2015.

Fishman, Charles. "5,200 Days in Space." *Atlantic*. Atlantic Monthly, Jan.–Feb. 2015. Web. 13 May 2015.

Fitzpatrick, Richard. "Brief History of Plasma Physics." *Home Page for Richard Fitzpatrick*. University of Texas at Austin, 31 Mar. 2011. Web. 19 May 2015.

"Foci (Focus Points) of an Ellipse." *Math Open Reference*. Math Open Reference, 2009. Web. 4 June 2015.

"Forces and the Grand Unified Theory." *Particle Adventure*. Particle Data Group, n.d. Web. 6 May 2015.

Fowler, Michael. "The Photoelectric Effect." *Modern Physics*. U of Virginia, n.d. Web. 15 Apr. 2015.

Fraknoi, Andrew. "The Lives of Stars." *Seeing in the Dark*. PBS, Mar. 2008. Web. 21 May 2015.

Francis, Matthew. "Moonday: A Bite-Sized Moon." *Galileo's Pendulum*. Galileo's Pendulum, 5 Mar. 2012. Web. 16 Mar. 2015.

Frank, Adam. "The Violent, Mysterious Dynamics of Star Formation." *Discover*. Kalmbach, 26 Mar. 2009. Web. 4 May 2015.

Frebel, Anna. *Searching for the Oldest Stars: Ancient Relics from the Early Universe*. Princeton University Press, 2015.

Frommert, Hartmut. "Charles Messier (June 26, 1730–April 12, 1817)." *SEDS*. Students for the Exploration and Development of Space, n.d. Web. 6 May 2015.

Fuchs, Jim. "Constellation History." *Modern Constellations*. Author, n.d. Web. 17 Apr. 2015.

GaBany, R. Jay. "A Singular Place." *Cosmotography*. Author, n.d. Web. 30 Apr. 2015.

Gaherty, Geoff. "Doggie Constellations Are a Skywatcher's Best Friends This Week." *Space.com*. Purch, 16 Feb. 2012. Web. 17 Apr. 2015.

"Galileo Fact Sheet." *European Space Agency*. ESA, 15 Feb. 2013. Web. 26 June 2015.

Gannon, Megan. "Bizarre Asteroid with Six Tails Spotted by Hubble Telescope." *Space. com*. Purch, 7 Nov. 2013. Web. 6 May 2015.

Garg, Rahul. "Optical Navigation Systems: The Foundation of Modern Pointing Devices." *EDN Network*. UBM Canon, 12 Aug. 2013. Web. 7 May 2015.

Garino, Brian W., and Jeffrey D. Lanphear. "Spacecraft Design, Structure, and Operations." *AU-18 Space Primer*. Air U, 2009. Web. 9 June 2015.

Geiling, Natasha. "The Women Who Mapped the Universe and Still Couldn't Get Any Respect." *Smithsonian Magazine*. Smithsonian, 18 Sept. 2013. Web. 28 May 2015.

Gendler, Robert, Lars Lindberg Christensen, and David Malin. *Treasures of the Southern Sky*. New York: Springer, 2011. Print.

Geveden, Rex D. "DAWN Cancellation Reclama." Letter to director, Jet Propulsion Laboratory. 27 Mar. 2006. PDF file.

Gibson, Steven J. "Draper, Henry." *The Biographical Encyclopedia of Astronomers*. Ed. Thomas A. Hockey, Virginia Trimble, and Katherine Bracher. New York: Springer, 2007. Print.

Gilmour, Jess K. *The Practical Astronomer's Deep-Sky Companion*. London: Springer, 2003. Print.

Giunti, Carlo, and Chung W. Kim. *Fundamentals of Neutrino Physics and Astrophysics*. New York: Oxford University Press, 2007. Print.

Glanz, James. *The Pervasive Plasma State*. College Park: Amer. Physical Soc., 1996. *Division of Plasma Physics*. Web. 26 May 2015.

"Global Navigation Satellite System (GLONASS) Overview." *PosiTim*. PosiTim, May 2010. Web. 6 May 2015.

"Global Positioning Satellites." *HyperPhysics*. Georgia State U, n.d. Web. 26 June 2015.

"GNSS (Global Navigation Satellite System) Definition." *TechTarget*. TechTarget, n.d. Web. 14 May 2015.

Goebel, Greg. "Spaceflight Propulsion." *Vectors*. Greg Goebel, 1 Feb. 2014. Web. 5 May 2015.

Gold, Lauren. "Asteroids Spin at YORP Speed, Thanks to the Effects of Sunlight, Cornell and Belfast Astronomers Discover." *Cornell Chronicle*. Cornell U, 7 Mar. 2007. Web. 7 Apr. 2015.

"GPS Info." *Naval Oceanography Portal*. US Navy, n.d. Web. 10 July 2015.

Gray, Jeremy John. "Poincaré, Henri." *Britannica Biographies* 1 Mar. 2012: 1–2. *Biography Reference Center*. Web. 9 June 2015.

"Greek Mythology and the Constellations." *Skywise Unlimited*. Western Washington U, n.d. Web. 9 Apr. 2015.

Greene, Brian. "How the Higgs Boson Was Found." *Smithsonian*. Smithsonian Inst., July 2013. Web. 12 June 2015.

Gregersen, Erik, ed. *The Milky Way and Beyond: Stars, Nebulae, and Other Galaxies*. New York: Britannica, 2010. Print.

Grego, Peter. *Venus and Mercury and How to Observe Them*. New York, NY: Springer Science+Business Media, 2008. Print.

Griggs, Mary Beth. "Laser Bees Could Save Us from Asteroids." *Smart News*. Smithsonian Inst., 12 July 2013. Web. 4 June 2015.

Groves, Paul D. *Principles of GNSS, Inertial, and Multisensor Integrated Navigation Systems*. 2nd ed. Norwood: Artech House, 2013. Print.

Grundy, W. M., et al. "The Orbit, Mass, Size, Albedo, and Density of (65489) Ceto/Phorcys: A Tidally-Evolved Binary Centaur." *Icarus* 191.1 (2007): 286–97. Print.

Grush, Loren. "Will Wormhole Travel Ever Be Possible?" *Popular Science*. Bonnier, 26 Oct. 2014. Web. 10 July 2015.

Hague, Robin. "Vortex." *Celestial Mechanics*. Robin Hague, 2008. Web. 28 May 2015.

Hall, Geoffrey. "Astronomical Image Processing A1." *Workspace: Physics Undergraduate Laboratories.* Imperial Coll. London, 6 Mar. 2015. Web. 7 Apr. 2015.

Hall, R. Cargill. *Lunar Impact: The NASA History of Project Ranger.* Chelmsford: Courier, 2013. 7–12. Print.

Hall, Rex, and David J. Shayler. *The Rocket Men: Vostok & Voskhod, the First Soviet Manned Spaceflights.* New York: Springer, 2001. Print.

Hall, Shannon. "How Do Black Holes Get Super Massive?" *Universe Today.* Fraser Cain, 13 Aug. 2013. Web. 30 Apr. 2015.

Hannes, Alfvén, and Gustaf Arrhenius. "Evolution of the Solar System: 4. The Small Bodies." *NASA History Office.* NASA, 1976. Web. 9 Mar. 2015.

Harbison, Rebecca. "Is There Really a 10th Plant?" *Ask an Astronomer.* Curious Team, 1997–2015. Web. 27 Mar. 2015.

Harrington, J. D., and Whitney Clavin. "NASA's WISE Mission Sees Skies Ablaze with Blazars." *Wide-Field Infrared Survey Explorer.* NASA, 12 Apr. 2012. Web. 13 Mar. 2015.

Harris, Alan W., et al. "Near-Earth Objects." *Encyclopedia of the Solar System.* Ed. Tilman Spohn, Doris Breuer, and Torrence V. Johnson. 3rd ed. Waltham: Elsevier, 2014. 603–23. Print.

Haug, Eberhard, and Werner Nakel. *The Elementary Process of Bremsstrahlung.* River Edge: World Scientific, 2004. Print.

Hawking, Stephen. "How to Build a Time Machine." *Daily Mail.* Assoc. Newspapers, 27 Apr. 2010. Web. 10 July 2015.

Head, Marilyn. "Planetary Lineup Excites the Sun." *ABC Science.* ABC, 2 July 2008. Web. 24 Apr. 2015.

Hearnshaw, J. B. *The Analysis of Starlight: Two Centuries of Astronomical Spectroscopy.* New York: Cambridge UP, 2014. Print.

"Herbig-Haro Object." *Cosmos: The SAO Encyclopedia of Astronomy.* Swinburne University of Technology, n.d. Web. 1 Apr. 2015.

Hermans-Killam, Linda. "Infrared Astronomy." *Ask an Infrared Astronomer.* California Inst. of Technology, n.d. Web. 15 Apr. 2015.

"High Mass Star." *Space Book.* Las Cumbres Observatory Global Telescope Network, 2012. Web. 18 Mar. 2015.

"History of Photometry in Astronomical Observations." *McCormick Museum.* University of Virginia, 2009. Web. 10 June 2015.

Hobbs, Maryam. "An Introduction to Pulsars." *Australia Telescope National Facility.* CSIRO, n.d. Web. 20 Mar. 2015.

Hockey, Thomas, et al., eds. *Biographical Encyclopedia of Astronomers.* 2nd ed. New York: Springer, 2014. Print.

Holder, Jamie. "Gamma-Ray Astronomy." *Department of Physics and Astronomy.* University of Delaware, 2008. Web. 8 Apr. 2015.

Hollas, J. Michael. "Low and High Resolution Spectroscopy." *High Resolution Spectroscopy.* Oxford: Butterworth-Heinemann, 2013. 85–86. Print.

Holloway, April. "The Ancient Wonder and Veneration of the Dog Star Sirius." *Ancient Origins.* Ancient Origins, 1 Feb. 2014. Web. 17 Apr. 2015.

"How Do DS1 and Other Spacecraft Propel Themselves through Space?" *Qualitative Reasoning Group.* Northwestern U, n.d. Web. 5 June 2015.

"How Does a Comet Work?" *Rosetta.* California Inst. of Technology, n.d. Web. 5 Mar. 2014.

"How Do Objects Travel in Space?" *Qualitative Reasoning Group.* Northwestern U, n.d. Web. 4 June 2015.

Hrovat, Mary. "1905, Hertzsprung Notes Relationship between Star Color and Luminosity." *Great Events from History: The 20th Century, 1901–1940.* Ed. Robert F. Gorman. Vol. 1. Pasadena: Salem, 2007. *eBook Academic Collection.* Web. 8 Apr. 2015.

"Hubble Directly Observes the Disc around a Black Hole." *ESA Science & Technology.* European Space Agency, 4 Nov. 2011. Web. 31 Mar. 2015.

"Hubble Image Processors: Raiders of the Hubble Archive." *HubbleSite.* NASA, n.d. Web. 7 Apr. 2015.

"Hubble's Classification Scheme." *Center for Educational Resources (CERES) Project.* NASA/Montana State U–Bozeman, n.d. Web. 26 May 2015.

"Hubble Zooms in on Magnified Galaxy." *Hubble Space Telescope.* NASA, 2 Feb. 2012. Web. 5 June 2015.

Hughes, Stefan. *Catchers of the Light.* Cyprus: ArtDeCiel, 2013. Print.

Hunter, Matt. "Have Quark Stars Been Discovered?" *From Quarks to Quasars.* FQTQ, 26 Mar. 2014. Web. 29 May 2015.

Hutchinson, Ian H. "Introduction." *Introduction to Plasma Physics*. Massachusetts Inst. of Technology, 2001. Web. 19 May 2015.

"Identify the Winter Circle and Winter's Brightest Stars." *EarthSky*. Earthsky Communications, 30 Jan. 2015. Web. 17 Apr. 2015.

"Impact Cratering." *Explore!* Lunar and Planetary Inst., 3 Dec. 2013. Web. 9 Mar. 2015.

"Infrared Waves." *Mission:Science*. NASA, 13 Aug. 2014. Web. 1 Apr. 2015.

"Inside Blazars." *Astronomy Magazine*. Kalmbach, 23 Apr. 2008. Web. 13 Mar. 2015.

International Astronomical Union. *IAU Resolution B5: Definition of a Planet in the Solar System*. N.p.: IAU, 24 Aug. 2006. PDF file.

"Ionization and Plasmas." *Astronomy 162: Stars, Galaxies, and Cosmology*. University of Tennessee, Knoxville, n.d. Web. 19 May 2015.

"Irving Langmuir—Biographical." *Nobelprize.org*. Nobel Foundation, n.d. Web. 26 Mar. 2015.

"It's Not Over Yet." *Science News*. NASA, 14 Feb. 2001. Web. 5 May 2015.

Jaggard, Victoria. "Pluto Gets 14 New Neighbors." *National Geographic: Voices*. Natl. Geographic Soc., 15 Sept. 2010. Web. 10 May 2015.

———. "Would Astronauts Survive an Interstellar Trip through a Wormhole?" *Smithsonian.com*. Smithsonian Inst., 7 Nov. 2014. Web. 10 July 2015.

Jansen, Albert. *Star Maps for Southern Africa: An Easy Guide to the Night Skies*. Cape Town: Struik, 2004. Print.

Jones, Brian. "Skywatch Spring: Canis Major." *Cicerone Extra*. Cicerone, 10 Mar. 2015. Web. 17 Apr. 2015.

Jones, Daniel C., and Iwan P. Williams. "High Inclination Meteorite Streams Can Exist." *Advances in Meteoroid and Meteor Science*. Ed. J. M. Trigo-Rodríguez. New York: Springer, 2008. Print.

Jones, Jeremy. "How Do Space Probes Navigate Large Distances with Such Accuracy and How Do the Mission Controllers Know When They've Reached Their Target?" *Scientific American*. Scientific American, 27 Mar. 2006. Web. 7 May 2015.

"JPL History." *Jet Propulsion Laboratory California Institute of Technology*. NASA, n.d. Web. 5 June 2015.

Kambic, Bojan. *Viewing the Constellations with Binoculars: 250+ Wonderful Sky Objects to See and Explore*. New York: Springer, 2010. Print.

"Karl Jansky and the Discovery of Cosmic Radio Waves." *National Radio Astronomy Observatory*. Assoc. U, 16 May 2008. Web. 27 Apr. 2015.

Kaufman, Marc. "Mars Rover Finds Ancient Streambed—Proof of Flowing Water." *National Geographic*. Natl. Geographic Soc., 28 Sept. 2012. Web. 5 June 2015.

Kaufmann, William J. *Universe*. 4th ed. New York: W.H. Freeman, 1994. Print.

Keeton, Charles. *Principles of Astrophysics: Using Gravity and Stellar Physics to Explore the Cosmos*. New York: Springer, 2014. Print.

Kemsley, Jyllian. "Space Dust Science." *Chemical and Engineering News* 19 Apr. 2010: 36–38. Print.

"Keto." *Theoi Project: Greek Mythology*. Aaron J. Atsma, n.d. Web. 20 Mar. 2015.

Khan, Amina. "Trojan Asteroid Tags along on Earth's Orbit." *Los Angeles Times*. Los Angeles Times, 28 July 2011. Web. 11 Mar. 2015.

King, Andrew. *Stars: A Very Short Introduction*. Oxford University Press, 2012.

King, Bob. "New Binocular Nova Discovered in Sagittarius." *Universe Today*. Universe Today, 16 Mar. 2015. Web. 23 Apr. 2015.

Kirkman, Thomas W. "Constellations: Frequently Asked Questions." *Tools for Science*. Coll. of St. Benedict & St. John's U, 1 May 2006. Web. 24 Mar. 2015.

"Kleopatra Gave Birth to Twins . . . Moons." *Cosmic Log*. NBC News, 23 Feb. 2011. Web. 16 Feb. 2015.

Klioner, Sergei A. "Lecture Notes on Basic Celestial Mechanics." *Department of Astronomy / Lohrmann Observatory*. Technical U Dresden, 2011. Web. 14 May 2015.

Kluger, Jeffrey. "Gagarin's Golden Anniversary: The High Price Paid by the First Man in Space." *Time*. Time, 12 Apr. 2011. Web. 4 June 2015.

Knowles, J. W. "Crystal Diffraction Spectroscopy of Nuclear $\gamma$-Rays." *Alpha-, Beta- and Gamma-Ray Spectroscopy*. Vol. 1. Ed. Kai Siegbahn. Rpt. Burlington: Elsevier, 2012. Digital file.

Koberlein, Brian. "What Are Comet Tails?" *Universe Today*. Universe Today, 21 July 2014. Web. 9 Mar. 2015.

Kornreich, Dave. "What Is a Nova?" *Ask an Astronomer*. Cornell U Astronomy, 24 Jan. 1999. Web. 31 Mar. 2015.

Koupelis, Theo. *In Quest of the Stars and Galaxies*. Sudbury: Jones, 2011. Print.

Krco, Marko. "What Are Constellations Used For?" *Ask an Astronomer*. Curious Team, Oct. 2002. Web. 24 Mar. 2015.

Krebs, Robert. "Fowler's Theory of Stellar Nucleosynthesis." *Encyclopedia of Scientific Principles, Laws, and Theories*. Vol. 1. Westport: Greenwood, 2008. 195–96. Print.

"Kuiper Belt and Oort Cloud." *Solar System Exploration*. National Aeronautics and Space Administration, 19 Nov. 2014. Web. 10 May 2015.

"Kuiper Belt." *Cosmos: The SAO Encyclopedia of Astronomy*. Swinburne Centre for Astrophysics and Supercomputing, Swinburne University of Technology, n.d. Web. 10 May 2015.

"Kuiper Belt Objects: Dwarf Planets." *Laboratory for Atmospheric and Space Physics*. U of Colorado at Boulder, Aug. 2007. Web. 24 Mar. 2015.

Kuzmin, Andrey. "Meteorite Explodes over Russia, More Than 1,000 Injured." *Reuters*. Thomson, 15 Feb. 2013. Web. 9 Mar. 2015.

Lacey, Sheridan, and Sarah Maddison. "Binary Asteroids." *SAO Astro News*. Swinburne Astronomy Online, 13 Apr. 2014. Web. 20 Mar. 2015.

Lamb, Robert. "Why Explore Mars?" *Discovery News*. Discovery Communications, 4 May 2010. Web. 5 June 2015.

Landau, Elizabeth. "'Bright Spot' on Ceres Has Dimmer Companion." *Jet Propulsion Laboratory*. California Inst. of Technology, 25 Feb. 2015. Web. 22 Apr. 2015.

Lang, Kenneth R. *A Companion to Astronomy and Astrophysics: Chronology and Glossary with Data Tables*. New York: Springer, 2006. Print.

———. *The Cambridge Guide to the Solar System*. 2nd ed. Cambridge: Cambridge University Press, 2011. Print.

———. *The Life and Death of Stars*. Cambridge University Press, 2013.

"Largest Asteroid May be 'Mini Planet' with Water Ice." *HubbleSite*. NASA, 7 Sept. 2005. Web. 22 Apr. 2015.

"Laser Ablation Experiments." *University of Strathclyde Engineering*. U of Strathclyde, 2015. Web. 4 June 2015.

"Laser Bees Concept." *University of Strathclyde Engineering*. U of Strathclyde, 2015. Web. 4 June 2015.

"Launching Galileo." *European Space Agency*. ESA, 9 Mar. 2015. Web. 26 June 2015.

Launius, Roger D. "It All Started with Sputnik." *Air & Space*. Smithsonian Inst., July 2007. Web. 4 June 2015.

Lawrence, Andy. "Spectroscopy." *Astronomical Measurement: A Concise Guide*. Heidelberg: Springer, 2014. 121–44. Print.

Lemaître, Georges. "The Beginning of the World from the Point of View of Quantum Theory." *Nature* 127 (1931): 706. Rpt. in *The Book of the Cosmos: Imagining the Universe from Heraclitus to Hawking*. Ed. Dennis Richard Danielson. Cambridge: Perseus, 2000. 407–8. Print.

Lemonick, Michael D. "Kuiper Belt Missions Could Reveal the Solar System's Origins." *Scientific American*. Nature America, 14 Oct. 2014. Web. 9 Mar. 2015.

———. "Winds Blasting from Black Holes Shut Down Star Growth." *National Geographic*. Natl. Geographic Soc., 19 Feb. 2015. Web. 31 Mar. 2015.

Leverington, David. *Babylon to Voyager and Beyond: A History of Planetary Astronomy*. Cambridge: Cambridge University Press, 2003. Print.

Levy, David H., and Wendee Wallach-Levy. *Cosmic Discoveries: The Wonders of Astronomy*. New York: Prometheus, 2001. Print.

"Life Cycle of a Star." *National Schools' Observatory*. Natl. Schools' Observatory, 2004–15. Web. 21 May 2015.

"LightSail Launch Success!" *EarthSky*. Earthsky Communications, 20 May 2015. Web. 5 June 2015.

Lissauer, Jack J., and Imke de Pater. *Fundamental Planetary Science: Physics, Chemistry and Habitability*. New York: Cambridge University Press, 2013. Print.

Longair, Malcolm S. *High Energy Astrophysics*. 3rd ed. New York: Cambridge UP, 2011. Print.

MacRobert, Alan. "The Spectral Types of Stars." *Sky & Telescope*. F+W Media, 1 Aug. 2006. Web. 10 June 2015.

Magrass, Yale R. "Stephen Hawking." *Salem Press Biographical Encyclopedia* (2014): n. pag. *Research Starters*. Web. 3 June 2015.

Mahoney, T.J. *Mercury*. New York, NY: Springer Science+Business Media, 2014. Print.

Maisel, Merry, and Laura Smart. "Annie Jump Cannon." *Science Women*. San Diego Supercomputer Center, 1997. Web. 10 June 2015.

Majewski, Steven R. "Stellar Populations and the Formation of the Milky Way." *Globular Clusters*. Ed. C. Martinez Roger, I. Pérez-Fournon, and F. Sanchez. New York: Cambridge University Press, 1999. Print.

Major, Jason. "Recommissioned NEOWISE Discovers Near-Earth Asteroid." *Discovery News*. Discovery Communications, 7 Jan. 2014. Web. 8 May 2015.

Malik, Tariq. "NASA Maps Dangerous Asteroids that May Threaten Earth." *Space.com*. Purch, 14 Aug. 2013. Web. 4 June 2015.

Malin, David. "The Tail of the Corona Australis Reflection Nebula." *Australian Astronomical Observatory*. Commonwealth of Australia, 25 July 2010. Web. 10 Apr. 2015.

Mangum, Jeff. "The Earth-Moon Spin-Orbit Resonance." *Ask an Astronomer*. Natl. Radio Astronomy Observatory, 24 Oct. 2013. Web. 24 Apr. 2015.

Mann, Adam. "Voyager 1 Becomes First Man-Made Object to Reach Interstellar Space." *Wired*. Condé Nast, 12 Sept. 2013. Web. 26 Mar. 2015.

Marchione, Demian. *The Heriot-Watt Astrochemistry Research Group*. Heriot-Watt U, 26 Mar. 2014. Web. 14 Apr. 2015.

"Mars Curiosity Navigation." *Jet Propulsion Laboratory at the California Institute of Technology*. NASA, n.d. Web. 7 May 2015.

Marsden, Brian. "Eugene Shoemaker (1928–1997)." *Jet Propulsion Laboratory*. NASA, 18 July 1997. Web. 5 May 2015.

Marshall, John. "Introduction: Black Holes." *New Scientist*. Reed Business Information, 6 Jan. 2010. Web. 30 Apr. 2015.

Martin, Ronald. *Earth's Evolving Systems:. The History of Planet Earth*. Burlington, MA: Jones & Bartlett, 2013. Print.

Matzner, Richard A., ed. *Dictionary of Geophysics, Astrophysics, and Astronomy*. Boca Raton: CRC, 2001. Print.

McFadden, Lucy-Ann, Paul R. Weissman, and Torrence V.Johnson, eds. *Encyclopedia of the Solar System*. 2nd ed. San Diego, CA: Academic Press, 2007. Print.

McKee, Maggie. "Cosmic Rays Originate from Supernova Shockwaves." *Nature*. Nature, 15 Feb. 2013. Web. 20 Mar. 2015.

"Messier 74." *SEDS Messier Database*. SEDS-USA, 14 Aug. 2013. Web. 8 May 2015.

Miglio, Andrea, Josephina Montelban, and Arlette Noels, eds. *Red Giants and Probes of the Structure and Evolution of the Milky Way*. Springer, 2012.

Miller, Ron. *Mars*. Minneapolis, MN: Twenty-First Century Books, 2006. Print.

_____. *Uranus and Neptune* Minneapolis, MN: Twenty-First Century Books, 2003. Print.

Miralda-Escudé, Jordi. "How do Quasars Really Work?" *Department of Astronomy and Meteorology*. University of Barcelona, n.d. Web. 11 May, 2015.

"Molecules in Space." *University of Cologne Institute of Physics*. U of Cologne, 2015. Web. 26 May 2015.

"Molecules in the Interstellar Medium." *Department of Astronomy and Astrophysics*. Pennsylvania State U, n.d. PDF file.

Morgan, W. W., Philip C. Keenan, and Edith Kellman. *An Atlas of Stellar Spectra*. Chicago: University of Chicago Press, 1943. Digital file.

Mosher, Dave. "Space Shuttle in Extreme Detail: Exclusive New Pictures." *National Geographic*. Natl. Geographic Soc., 16 Apr. 2012. Web. 1 July 2015.

Moskowitz, Clara. "Hottest Particle Soup May Reveal Secrets of Primordial Universe.'" *LiveScience*. Purch, 13 Aug. 2012. Web. 2 June 2015.

_____. "Right Again, Einstein! New Study Supports Cosmological Constant." *Scientific American*. Scientific Amer., 17 Jan. 2013. Web. 5 June 2015.

_____. "Searching for Alien Life? Try Failed Stars." *Space.com*. Purch, 31 Mar. 2011. Web. 27 Mar. 2015.

Mottola, Stefano, et al. "Rotational Properties of Jupiter Trojans." *Astronomical Journal* 141.5 (2011): 170–201. Print.

Mullen, Leslie. "Star Bright . . . and the Sub-Brown Dwarfs." *Astrobiology Magazine*. Astrobio.net, 4–6 Aug. 2003. Web. 27 Mar. 2015.

Murayama, Hitoshi. "The Origin of Neutrino Mass." *Physics World* May 2002: 35–39. Print.

National Optical Astronomy Observatory. "Supergiant Census Supports Stellar Evolution Theory."*Astronomy Magazine*. Kalmbach, 23 Apr. 2012. Web. 30 Mar. 2015.

"Navigating in Space." *National Air and Space Museum*. Smithsonian Institution, n.d. Web. 14 May 2015.

Neal-Jones, Nancy. "NASA's LRO Spacecraft Captures Images of LADEE's Impact Crater." *Lunar Reconnaissance Orbiter*. NASA, 28 Oct. 2014. Web. 9 Mar. 2015.

"Near Earth Asteroid Rendezvous." *Discovery is NEAR*. JHUAPL, 2001. Web. 10 Mar. 2015.

*Near-Earth Object Program.* Caltech, 6 Feb. 2015. Web. 27 Mar. 2015.

Nelson, Marcia L., Daniel T. Britt, and Larry A. Lebofsky. "Review of Asteroid Compositions." *Resources of Near-Earth Space.* Ed. John S. Lewis, Mildred Shapley Matthews, and Mary L. Guerrieri. Tucson: U of Arizona P, 1993. 493–522. Print.

"NEOWISE: The Wide-Field Infrared Survey Explorer (WISE)." *Near Earth Object Program.* NASA, 7 May 2015. Web. 6 May 2015.

Neuman, Scott. "New Research Affirms That Milky Way Has Four Spiral Arms." *The Two-Way.* NPR, 17 Dec. 2013. Web. 16 Mar. 2015.

"New Asteroid Families Discovered." *Science News.* NASA, 29 May 2013. Web. 6 May 2015.

Newcomb, Simon. "Apparent Motion of the Sun." *Elements of Astronomy.* 1900. *Tools for Science.* Coll. of St. Benedict & St. John's U, n.d. Web. 27 Mar. 2015.

"New Insights on How Spiral Galaxies Get Their Arms." *Harvard-Smithsonian Center for Astrophysics.* Harvard U, 2 Apr. 2013. Web. 15 May 2015.

"New Rocky Planet Found in Constellation Leo." *ScienceDaily.* ScienceDaily, 10 Apr. 2008. Web. 9 Apr. 2015.

"Newton's Laws of Motion." *NASA.* Natl. Aeronautics and Space Administration, 5 May 2015. Web. 28 May 2015.

"Nobel Prize in Physics 1927: Arthur H. Compton—Biographical." *Nobel Prize.* Nobel Media, 2015. Web. 15 Apr. 2015.

Noll, Keith S., et al. "Binaries in the Kuiper Belt." *The Solar System beyond Neptune.* Ed. M. A. Barucci et al. Tucson: U of Arizona P, 2008. 345–63. Print.

Nordberg, John T. *Grand Unification.* John T. Nordberg, n.d. Web. 6 May 2015.

"Northern Circumpolar Constellations." *Windows to the Universe.* Natl. Earth Science Teachers Assn., 2012. Web. 1 May 2015.

North, Gerald. *Astronomy in Depth.* Rev. and expanded ed. London: Springer, 2003. Print.

Noyes, Dan. "Cosmic Rays Discovered 100 Years Ago." *CERN.* CERN, 7 Aug. 2012. Web. 12 June 2015.

Oakes, Kelly. "On the Origin of Chemical Elements." *Basic Space.* Scientific Amer., 2 Aug. 2011. Web. 16 July 2015.

Odenwald, Sten. "What Is Grand Unification Theory?" *Astronomy Cafe.* Sten Odenwald, 1997. Web. 6 May 2015.

Olcott, William Tyler. *Star Lore of All Ages.* New York: Knickerbocker, 1911. *AAVSO.* Web. 1 Apr. 2015.

O'Meara, Stephen James. *Deep-Sky Companions: The Messier Objects.* 2nd ed. Cambridge: Cambridge University Press, 2014. Print.

O'Meara, Stephen James. *Southern Gems.* New York: Cambridge UP, 2013. Print.

O'Neill, Ian. "Clouds of Water Possibly Found in Brown Dwarf Atmosphere." *Discovery News.* Discovery Communications, 26 Aug. 2014. Web. 30 Mar. 2015.

O'Neill, Ian. "Why Are Quark Stars So Strange?" *Discovery News.* Discovery Communications, 19 Jan. 2010. Web. 6 Apr. 2015.

Ouellette, Jennifer. "The Brilliance behind the Plan to Land Curiosity on Mars." *Smithsonian.com.* Smithsonian Inst., Dec. 2013. Web. 5 June 2015.

Overduin, James. "Einstein's Spacetime." *Gravity Probe B: Testing Einstein's Universe.* Stanford U, Nov. 2007. Web. 5 May 2015.

Paat-Dahlstrom, Emeline. "A Giant Leap for Curiosity, a Small Step for Robot-Kind." *Forbes.* Forbes.com, 25 July 2012. Web. 22 May 2015.

Pais, Abraham. *Inward Bound: Of Matter and Forces in the Physical World.* New York: Oxford UP, 1986. Print.

Pál, A., et al. "TNOs Are Cool: A Survey of the Trans-Neptunian Region—VII. Size and Surface Characteristics of (90377) Sedna and 2010EK139." *Astronomy & Astrophysics* 541 (2012): n. pag. Web. 7 Apr. 2015.

Palmer, Brian. "Mars Rover Gets Instructions Daily from NASA via a Network of Antennae." *Washington Post.* Washington Post, 29 Oct. 2012. Web. 5 June 2015.

Parry-Hill, Matthew J., et al. "Spatial Resolution in Digital Images." *Molecular Expressions Optical Microscopy Primer.* Florida State U, 26 Mar. 2014. Web. 24 Feb. 2015.

"Part I. Historical Context." *Space Transportation System.* Denver: Historic American Engineering Record, Natl. Park Service, US Dept. of the Interior, Nov. 2012. PDF file.

Peeters, Z., et al. "Astrochemistry of Dimethyl Ether." *Astronomy & Astrophysics* 445.1 (2006): 197–204. Print.

Perryman, Michael. *The Exoplanet Handbook.* New York: Cambridge University Press, 2011. Print.

Peterson, Carolyn Collins. *Astronomy 101: From the Sun and Moon to Wormholes and Warp Drive, Key Theories, Discoveries, and Facts about the Universe.* Avon: Adams, 2013. Print.

Peterson, Kit A. "Introduction to Basic Measures of a Digital Image for Pictorial Collections." *Library of Congress Prints & Photographs Division.* LOC, June 2005. Web. 7 Apr. 2015.

Pevtsov, Alex. "Explain the Coronal Heating Problem, Please." *Stanford Solar Center.* Stanford U, n.d. Web. 29 May 2015.

"Phorkys." *Theoi Project: Greek Mythology.* Aaron J. Atsma, n.d. Web. 20 Mar. 2015.

"Photometry." *Encyclopaedia Britannica.* Encyclopaedia Britannica, 2015. Web. 22 May 2015.

"PHYS 3.2: Gravitation and Orbits." *PPLATO: Promoting Physics Learning and Teaching Opportunities.* University of Reading, 1996. Web. 4 June 2015.

"Pierre Simon de Laplace (1749–1827)." *High Altitude Observatory.* UCAR, n.d. Web. 24 Apr. 2015.

Plait, Phil. "The Upper Limit to a Planet." *Discover.* Kalmbach, 7 Sept. 2006. Web. 27 Mar. 2015.

Plataforma SINC. "Trans-Neptunian Objects Suggest That There Are More Dwarf Planets in Our Solar System." *ScienceDaily.* ScienceDaily, 15 Jan. 2015. Web. 10 May 2015.

"Pluto and the Developing Landscape of Our Solar System." *International Astronomical Union.* IAU, n.d. Web. 22 Apr. 2015.

Popova, Olga, and Ivan Nemchinov. "Bolides in the Earth Atmosphere." *Catastrophic Events Caused by Cosmic Objects.* Ed. Vitaly Adushkin and Nemchinov. Dordrecht: Springer, 2008. 131–62. Print.

Pössel, Markus. "Elementary Einstein." *Einstein Online.* Max Planck Inst. for Gravitational Physics, n.d. Web. 5 May 2015.

"Post–Main Sequence Stars." *Australia Telescope National Facility.* Commonwealth Scientific and Industrial Research Org., n.d. Web. 21 May 2015.

Prantzos, Nikos, and Sylvia Ekström. "Stellar Nucleosynthesis." *Encyclopedia of Astrobiology.* Ed. Muriel Gargaud et al. Vol. 3. Heidelberg: Springer, 2011. 1584–92. Print.

Privett, Grant, and Kevin Jones. *The Constellation Observing Atlas.* New York: Springer, 2013. Print

"Protostar." *Space Book.* Las Cumbres Observatory Global Telescope Network, 2012. Web. 1 Apr. 2015.

"Radiometric Resolution." *Natural Resources Canada.* Govt. of Canada, 29 May 2013. Web. 24 Feb. 2015.

Randall, David A. *Atmosphere, Clouds, and Climate.* Princeton: Princeton University Press, 2012. Print.

Rapp, Donald. *Human Missions to Mars: Enabling Technologies for Exploring the Red Planet.* 2nd ed. Chichester, UK: Praxis/Springer, 2016. Print.

Ratcliffe, Martin, and Alister Ling. "The Deep Sky: The Region Surrounding Orion's Belt Is Studded with a Broad Array of Celestial Wonders." *Astronomy* Mar. 2004: 63. Print.

Rayman, Marc D., et al. "Dawn: A Mission in Development for Exploration of Main Belt Asteroids Vesta and Ceres." *Acta Astronautica* 58 (2006): 605–16. Print.

Razumovskaya, Olga. "GPS-Only Devices to Be Hit with Tax." *Moscow Times.* Moscow Times, 28 Oct. 2010. Web. 6 May 2015.

Reddy, Francis. "NASA Satellites Catch 'Growth Spurt' from Newborn Protostar." *WISE: Wide-Field Infrared Survey Explorer.* NASA, 23 Mar. 2015. Web. 1 Apr. 2015.

"Report of the Space Task Group, 1969." *NASA Headquarters.* NASA, n.d. Web. 26 June 2015.

"Responding to Potential Asteroid Redirect Mission Targets." *Jet Propulsion Laboratory.* California Institute of Technology, 14 Feb. 2014. Web. 9 Mar. 2015.

Reynolds, Mike D. *Falling Stars: A Guide to Meteors & Meteorites.* Mechanicsburg: Stackpole, 2010. Print.

Richardson, James, and James Bedient, comps. "Fireball FAQs." *American Meteor Society.* Amer. Meteor Soc., n.d. Web. 11 Mar. 2015.

Ridpath, Ian, and Wil Tirion. *Stars and Planets: The Most Complete Guide to the Stars, Planets, Galaxies, and the Solar System.* 4th ed. Princeton: Princeton University Press, 2007. Print.

"Robot Systems." *Rover Ranch.* NASA, 3 Apr. 2003. Web. 22 May 2015.

"Rocket Propulsion." *Glenn Research Center.* NASA, 12 June 2014. Web. 21 Apr. 2015.

Romanishin, W. "An Introduction to Astronomical Photometry Using CCDs." *Astrophysics Research at OU.* University of Oklahoma, 22 Oct. 2006. Web. 28 May 2015.

Rothery, David A. *Planet Mercury: From Pale Pink Dot to Dynamic World.* New York, NY: Springer, 2015. Print

Rothstein, Dave. "What Can We Learn from the Color of a Star?" *Ask an Astronomer.* Astronomy Dept., Cornell U, 17 Mar. 2002. Web. 28 May 2015.

Rousseaux, Charles. "Dark Energy Cam: Fermilab Expands Understanding of Expanding Universe." *Energy.gov.* US Dept. of Energy, 12 Mar. 2012. Web. 5 June 2015.

Rozitis, B., et al. "A Thermophysical Analysis of the (1862) Apollo Yarkovsky and YORP Effects." *Astronomy & Astrophysics* 555 (2013): N. pag. Web. 7 Apr. 2015.

Russell, C. T., et al. "Dawn: A Journey to the Beginning of the Solar System." *Proceedings of the Conference on Asteroids, Comets and Meteors, July 29–August 2, 2002.* Berlin: Technical University of Berlin, 2002. Web. 16 Mar. 2015.

Russian Institute of Space Device Engineering. *Global Navigation Satellite System Interface Control Document.* Moscow: Russian Institute of Space Device Engineering, 2008. *Spacecorp.ru.* Web. 6 May 2015.

Sample, Ian. "Life aboard the International Space Station." *Guardian.* Guardian News and Media, 24 Oct. 2010. Web. 13 May 2015.

Sandage, Allan. "Classification and Stellar Content of Galaxies Obtained from Direct Photography." *Galaxies and the Universe.* Ed.

Sanders, Ray. "When Stellar Metallicity Sparks Planet Formation." *Astrobiology Magazine.* Astrobio.net, 9 Apr. 2012. Web. 29 Mar. 2015.

Sanders, Robert, and Tim Stephens. "Astronomers Announce the Most Earth-Like Planet Yet Found outside the Solar System." *Currents Online.* UC Santa Cruz, 13 June 2005. Web. 13 Apr. 2015.

———. "How Cleopatra Got Its Moons." *UC Berkeley News Center.* UC Regents, 22 Feb. 2011. Web. 16 Feb. 2015.

"Satellite Navigation—Global Positioning System (GPS)." *Federal Aviation Administration.* FAA, 13 Nov. 2014. Web. 14 May 2015.

Schmadel, Lutz D. *Dictionary of Minor Planet Names.* 6th ed. 2 vols. Heidelberg: Springer, 2012. Print.

Schmude Jr., Richard *Uranus, Neptune, Pluto and How to Observe Them* New York, NY: Springer Science+Business Media, 2008. Print

———. *Uranus, Neptune, Pluto and How to Observe Them.* New York, NY: Springer Science+Business Media, 2008. Print

Schneider, Howard. *National Geographic Backyard Guide to the Night Sky.* Washington, DC: Natl. Geographic Soc., 2009. Print.

Schultheis, M., et al. "Mapping the Milky Way Bulge at High Resolution: The 3D Dust Extinction, CO, and X Factor Maps." *Astronomy & Astrophysics* 566 (2014): n. pag. Web. 10 Apr. 2015.

Schultz, Colin. "Gliese 581g, the First Exoplanet Found That May Have Been Able to Host Life, Doesn't Actually Exist." *Smithsonian Magazine.* Smithsonian.com., 7 July 2014. Web. 24 Mar. 2015.

Seeds, Michael A., and Dana E. Backman. *Foundations of Astronomy.* 12th ed. Boston, MA: Brooks/Cole, Cengage Learning, 2013. Print.

Seitz, David E. "NASA Announces Next Steps on Journey to Mars: Progress on Asteroid Initiative." *Asteroid Redirect Mission.* NASA, 25 Mar. 2015. Web. 4 June 2015.

Setton, Dolly. "Neutrinos: Ghosts of the Universe." *Discover.* Kalmbach, 31 July 2014. Web. 15 Apr. 2015.

Shimada, Toru. "Hybrid Rocket Will Go for the Next Generation of Space Transportation." *Institute of Space and Astronautical Science.* Japanese Aerospace Exploration Agency, 2012. Web. 21 Apr. 2015.

Shimizu, Mikio. "The Hydrogen Clouds of Comets." *Comets in the Post-Halley Era.* Ed. Ray L. Newburn Jr., Marcia Neugebauer, and Jurgen Rahe. Vol. 2. Norwell: Kluwer, 1991. 897–905. Print.

Simpson, Phil. *Guidebook to the Constellations: Telescopic Sights, Tales and Myths.* New York: Springer, 2012. Print.

Smith, Nick. "Geospatial Technology: Mapping the Future." *E & T.* Inst. of Engineering and Technology, 20 Aug. 2012. Web. 24 Feb. 2015.

"Smithsonian Selects NEAR Mission for 2001 Aerospace Trophy." *Applied Physics Laboratory.* Johns Hopkins U, 14 Nov. 2001. Web. 5 May 2015.

Somov, Boris V. *Plasma Astrophysics, Part II: Reconnection and Flares.* 2nd ed. New York: Springer, 2013. Print.

Soper, Davison E. "Cepheid Variable Stars as Distance Indicators." *Institute of Theoretical Science.* U of Oregon, n.d. Web. 30 March 2015.

———. "Homestake Gold Mine Experiment." *The Electronic Universe.* University of Oregon, 22 Oct. 2007. Web. 6 April 2015.

"Spacecraft Design, Structure, and Operations." *AU Space Primer.* Air U, 2003. *Air University.* Web. 9 June 2015.

"Space Race." *Royal Air Force Museum*. Trustees of the Royal Air Force Museum, n.d. Web. 13 May 2015.

Sparrow, Giles. *Constellations: A Field Guide to the Night Sky*. London: Quercus, 2013. Print.

"Spectral Classes." *Australia Telescope National Facility*. CSIRO, n.d. Web. 10 June 2015.

"Spectral Classification of Stars." *University of Nebraska-Lincoln*. Astronomy Education, University of Nebraska-Lincoln, n.d. Web. 28 May 2015.

"Spectroscopy." *Cosmos: The SAO Encyclopedia of Astronomy*. Centre for Astrophysics and Supercomputing, Swinburne University of Technology, n.d. Web. 29 Mar. 2015.

Sprawls, Perry. "Radiation Penetration." *Sprawls Educational Foundation*. Sprawls Educ. Foundation, n.d. Web. 6 Apr. 2015.

"Sputnik 1." *National Space Science Data Coordinated Archive*. NASA, 26 Aug. 2014. Web. 4 June 2015.

"Sputnik and the Dawn of the Space Age." *NASA History Program Office*. NASA, 10 Oct. 2007. Web. 4 June 2015.

"Star Classification." *Astrophysical.org*. Instituto Scientia, 2004–14. Web. 10 June 2015.

"Star Clusters." *Australia Telescope National Facility*. CSIRO, n.d. Web. 30 Mar. 2015.

"Star Formation & Molecular Astrophysics." *Blake Research Group*. California Inst. of Technology, n.d. Web. 18 May 2015.

"Stellar Populations." *Astrophysics*. Amer. Museum of Natural History, n.d. Web. 29 Mar. 2015.

Stenmark, John. "Precise to a Fault: How GPS Revolutionized Seismic Research." *Earth: The Science behind the Headlines*. Amer. Geosciences Inst., 30 Apr. 2014. Web. 10 July 2015.

Stern, David P. "How Orbital Motion Is Calculated." *From Stargazers to Starships*. Author, 6 Apr. 2014. Web. 14 May 2015.

Stiennon, Patrick J. G., and David M. Hoerr. *The Rocket Company*. Reston: AIAA. 2005. Print.

Stolte, Daniel, and Christian Veillet. "Supermassive Black Hole Discovered with Mass of 12 Billion Suns." *Scientific Computing*. Advantage Business Media. 27 Feb. 2015. Web. 29 May 2015.

Stott, Carole, Robert Dinwiddie, David Hughes, and Giles Sparrow. *Space from Earth to the Edge of the Universe*. New York, NY: DK Publishing, 2010. Print.

"Strömgren, Bengt George Daniel." *A Dictionary of Astronomy*. Ed. Ian Ridpath. 2nd ed. rev. New York: Oxford University Press, 2012. 453. Print.

Strom, Robert G. and Ann L Sprague. *Exploring Mercury: The Iron Planet*. Chichester, UK: Praxis/Springer, 2003. Print.

"Structure of the Moon." *Natural History Museum*. Trustees of the Natural Hist. Museum, London, n.d. Web. 9 Mar. 2015.

Sturrock, Peter Andrew. *Plasma Physics: An Introduction to the Theory of Astrophysical, Geophysical and Laboratory Plasmas*. New York: Cambridge University Press, 1994. Print.

"Summary." *XMM-Newton*. European Space Agency, 19 July 2011. Web. 18 May 2015.

"Super-Cool Brown Dwarf Helps Distinguish Dwarfs and Giant Planets." *Marie Curie Actions*. European Commission, 28 Apr. 2014. Web. 30 Mar. 2015.

Sutton, George P., and Oscar Biblarz. *Rocket Propulsion Elements*. Hoboken: Wiley, 2010. Print.

Tate, Jean. "Sirius B." *Universe Today*. Universe Today, 15 Nov. 2009. Web. 17 Apr. 2015.

Tate, Karl. "Brown Dwarfs: Strange Failed Stars of the Universe Explained (Infographic)." *Space.com*. Purch, 29 Jan. 2014. Web. 10 June 2015.

Taylor, Frederic W. *The Scientific Exploration of Venus*. New York, NY: Cambridge University Press, 2014. Print.

"Temporal Resolution." *Natural Resources Canada*. Govt. of Canada, 22 Apr. 2014. Web. 24 Feb. 2015.

Than, Ker. "Densest Matter Created in Big-Bang Machine." *National Geographic*. Natl. Geographic Soc., 26 May 2011. Web. 16 July 2015.

———. "Trojan Asteroid Found Sharing Earth's Orbit—A First." *National Geographic News*. Natl. Geographic Soc., 28 July 2011. Web. 6 Mar. 2015.

"The Benefits of Particle Accelerators for Society." *Accelerators for America's Future*. US Dept. of Energy, 30 June 2010. Web. 12 June 2015.

"The Case Files: Dr. Edwin Hubble." *The Franklin Institute*. Franklin Inst., 2015. Web. 26 May 2015.

"The Death of Stars II: High Mass Stars." *Australia Telescope National Facility*. CSIRO, n.d. Web. 18 Mar. 2015.

"The Electromagnetic Spectrum." *Imagine the Universe!* NASA, Mar. 2013. Web. 14 Apr. 2015.

"The EPIC Instrument." *XMM-Newton*. Astrophysics Science Division, NASA Goddard Space Flight Center, 18 June 2012. Web. 18 May 2015.

"The European Photon Imaging Camera (EPIC) Onboard XMM-Newton." *XMM-Newton.* European Space Agency, 29 Sept. 2009. Web. 18 May 2015.

"The Hertzsprung-Russell Diagram." *Australia Telescope National Facility.* CSIRO, n.d. Web. 30 Mar. 2015.

"The Hubble Tuning Fork: Classification of Galaxies." *European Space Agency.* ESA, 15 Aug. 2013. Web. 26 May 2015.

"The Interstellar Medium." *University of New Hampshire Experimental Space Plasma Group.* U of New Hampshire, n.d. Web. 26 Mar. 2015.

"The Kuiper Belt." *New Horizons.* Johns Hopkins U Applied Physics Laboratory, n.d. Web. 5 Mar. 2015.

"The Large Hadron Collider." *CERN.* CERN, n.d. Web. 12 June 2015.

"The Myths of Ursa Major, the Great Bear." *AAVSO: American Association of Variable Star Observers.* AAVSO, 16 Apr. 2010. Web. 30 Apr. 2015.

Theodossiou, Efstratios, et al. "Sirius in Ancient Greek and Roman Literature: From the Orphic *Argonautics* to the *Astronomical Tables* of Georgios Chrysococca." *Journal of Astronomical History and Heritage* 14.3 (2011): 180–89. *NARIT.* Web. 17 Apr. 2015.

"The Origins of the Universe: The Big Bang." *The Stephen Hawking Centre for Theoretical Cosmology.* U of Cambridge, n.d. Web. 22 Apr. 2015.

"Theorists Propose a New Way to Shine—and a New Kind of Star." *Astronomy Magazine.* Kalmbach, 15 Dec. 2009. Web. 29 May 2015.

"The Space Race." *History.com.* A&E Television Networks, n.d. Web. 4 June 2015.

"The Sunyaev-Zeldovich Effect." *NASA Science.* NASA, 6 Apr. 2011. Web. 6 Apr. 2015.

"The Surveyor Program." *Lunar and Planetary Institute.* Lunar and Planetary Inst., n.d. Web. 22 May 2015.

"The Third Way of Galaxies." *Hubble Space Telescope.* European Space Agency, 12 Jan. 2015. Web. 11 May 2015.

Thompson, Andrea. "Major Space Telescopes." *Space.com.* Purch. 18 May 2009. Web. 28 May 2015.

Thorne, Kip. "The Man Who Imagined Wormholes and Schooled Hawking." Interview by Susan Kruglinski. *Discover.* Kalmbach, 9 Nov. 2007. Web. 10 July 2015.

"Tides and Tidal Forces." *Royal Museums Greenwich.* Natl. Maritime Museum, n.d. Web. 24 Apr. 2015.

Tielens, A. G. G. M. "The Galactic Ecosystem." *The Physics and Chemistry of the Interstellar Medium.* Cambridge: Cambridge UP, 2005. 1–25. Print.

Timerson, Bradley, et al. "Occultation Evidence for a Satellite of the Trojan Asteroid (911) Agamemnon." *Planetary and Space Science* 87 (2013): 78–84. Print.

Tirion, Wil. *The Cambridge Star Atlas.* Cambridge University Press, 2011.

"Top Ten WISE Factoids." *Jet Propulsion Laboratory.* California Inst. of Technology, n.d. Web. 6 May 2015.

Trachet, Tim. "Delporte, Eugène-Joseph." *Biographical Encyclopedia of Astronomers.* Ed. Thomas Hockey et al. New York: Springer, 2007. 288–89. Print.

"Transit Photometry: A Method for Finding Earths." *Planetary Society.* Planetary Society, n.d. Web. 29 May 2015.

"Transmitted Intensity and Linear Attenuation Coefficient." *NDT Education Resource Center.* Iowa State U, n.d. Web. 6 Apr. 2015.

"Two Companion Moons Found Near Dog-Bone Asteroid." *SETI Institute.* SETI Inst., n.d. Web. 3 Mar. 2015.

Tyson, Neil deGrasse. *Death by Black Hole and Other Cosmic Quandaries.* W. W. Norton & Company, 2014.

United Nations. *Education Curriculum: Global Navigation Satellite Systems.* Vienna: Author, 2012. PDF file.

United Nations. Office for Outer Space Affairs. *Global Navigation Satellite Systems.* Vienna: United Nations, 2012. *UNOOSA.* Web. 6 May 2015.

United States. Natl. Aeronautics and Space Administration. *Near-Earth Object Survey and Deflection Analysis of Alternatives: Report to Congress.* N.p.: Author, 2007. *NASA.* Web. 7 Apr. 2015.

Valtonen, M. J., and Hannu Karttunen. *The Three-Body Problem.* Cambridge: Cambridge University Press, 2005. *eBook Collection (EBSCOhost).* Web. 9 June 2015.

Van der Hucht, Karel. "Near Earth Asteroids: A Chronology of Milestones." *IAU.* International Astronomical Union, 7 Oct. 2013. Web. 9 Mar. 2015.

Vergano, Dan. "Dwarf Planet Discovery Hints at Hidden World Orbiting Solar System." *National Geographic.* Natl. Geographic Soc., 26 Mar. 2014. Web. 7 Apr. 2015.

Villaneuva, John Carl. "VY Canis Majoris." *Universe Today*. Universe Today, 8 Sept. 2009. Web. 17 Apr. 2015.

"Violent Origins of Disc Galaxies: Why Milky Way–Like Galaxies Are So Common in the Universe." *ScienceDaily*. ScienceDaily, 17 Sept. 2014. Web. 11 May 2015

Viti, Serena, et al. "The Making of *Stars 'R' Us!*" *Astronomy & Geophysics* 45.6 (2004): 6.22–24. Print.

Wall, Mike. "Deflecting Killer Asteroids away from Earth: How We Could Do It." *Space. com*. Purch, 7 Nov. 2011. Web. 4 June 2015.

Ward, Peter D., and Donald Brownlee. *Rare Earth: Why Complex Life is Uncommon in the Universe*. New York, NY: Springer-Verlag, 2000. Print.

Ward-Thompson, Derek, Helen Fraser, and Jonathan Rawlings. "The Chemistry of Star Formation." *News and Reviews in Astronomy and Geophysics* 43.4 (2002): 26–27. Print.

Wayman, Erin. "Cryovolcano." *Science News*. Soc. for Science and the Public, 2 Dec. 2013. Web. 22 Apr. 2015.

Weintraub, David. *Is Pluto a Planet? A Historical Journey through the Solar System*. Princeton: Princeton UP, 2007. Print.

Weiss, Achim. "Big Bang Nucleosynthesis: Cooking Up the First Light Elements." *Einstein Online*. Max Planck Inst. for Gravitational Physics, 2006. Web. 2 June 2015.

"What Are the Types of Comet Tails?." *HubbleSite*. Space Telescope Science Inst., n.d. Web. 12 Mar. 2015.

"What Is a Comet?" *Qualitative Reasoning Group*. Northwestern U, n.d. Web. 5 Mar. 2015.

"White Dwarfs: Aging Stars." *Science and Space*. Natl. Geographic Soc., n.d. Web. 31 Mar. 2015.

"White Dwarf Supernovae Are Discovered in Virgo Cluster Galaxy and in Sky Area 'Anonymous.'" *SMU Research*. Southern Methodist U, 26 Feb. 2013. Web. 6 May 2015.

WhiteHorne, Mary Lou. "How Far Are the Constellations?" *Space.com*. Purch, 24 Jan. 2007. Web. 20 Mar. 2015.

White, Martin. "Big Bang Nucleosynthesis." *Martin White, Professor of Physics, Professor of Astronomy, UC Berkeley*. Regents of the University of California, 27 Mar. 2006. Web. 6 May 2015.

Wilcox, Roger M. "Stellar Evolution." *The Internet Stellar Database*. Author, n.d. Web. 6 May 2015.

Williams, David A., and Thomas W. Hartquist. "The Chemistry of Star-Forming Regions." *Accounts of Chemical Research* 32.4 (1999): 334–41. Print.

Williams, Forman A. "Flame Extinguishment Experiment (FLEX)." *International Space Station*. NASA, 21 Oct. 2014. Web. 10 Apr. 2015.

Woo, Marcus. "Will We Ever...Travel in Wormholes?" *BBC Future*. BBC, 26 Mar. 2015. Web. 10 July 2015.

"XMM-Newton Reveals Origin of Elements in Galaxy Clusters." *XMM-Newton*. European Space Agency, 11 May 2006. Web. 18 May 2015.

"X-Ray Mass Attenuation Coefficients." *Physical Measurement Laboratory*. Natl. Inst. of Standards and Technology, n.d. Web. 6 April 2015.

"X-Ray Test Facility Calculations." *Extreme Universe Laboratory*. Moscow State U, n.d. Web. 6 Apr. 2015.

Yeomans, Donald K. *Near-Earth Objects: Finding Them before They Find Us*. Princeton: Princeton University Press, 2013. Print.

Young, Monica. "Cooking Up High-Mass Stars." *Sky & Telescope*. F+W Media, 4 June 2014. Web. 18 Mar. 2015.

Zak, Anatoly. "Uragan Satellites." *RussianSpaceWeb. com*. RussianSpaceWeb.com, 30 Nov. 2014. Web. 6 May 2015.

Zeeberg, Amos. "How to Avoid Repeating the Debacle That Was the Space Shuttle." *Discover*. Kalmbach, 22 July 2011. Web. 26 June 2015.

Ziurys, L. M., and A. J. Apponi. "Interstellar Molecules as Probes of Prebiotic Chemistry: A Laboratory in Radio Astronomy." *Arizona Radio Observatory*. U of Arizona, 2006. PDF file.

Zubrin, Robert. and Richard Wagner,. *The Case for Mars: The Plan to Settle the Red Planet and Why We Must*. New York, NY: The Free Press, 2011. Print.

——. *Mars Direct:. Space Exploration, the Red Planet, and the Human Future*. New York, NY: Penguin, 2013. Print.

Zuckerman, Catherine. "Hubble Pictures: Top Five Hidden Treasures." *National Geographic*. Natl. Geographic Soc., 4 Sept. 2012. Web. 7 Apr. 2015.

# WEB RESOURCES

**Amazing Space**
Home of the Hubble and James Webb Space Science Missions
http://amazingspace.org/

**Astrogeology Science Center**
U.S. Department of Interior U.S. Geological Survey
http://astrogeology.usgs.gov/

**Australian Astronomical Observatory**
Commonwealth of Australia
https://www.aao.gov.au/

**Blake Research Group**
California Inst. of Technology
http://web.gps.caltech.edu

**CERN**
CERN, the European Organization for Nuclear Research
http://home.cern/

**Constellation Guide.**
http://www.constellation-guide.com/

**Cosmic Evolution**
Harvard-Smithsonian Center for Astrophysics
https://www.cfa.harvard.edu

**Cosmos: The SAO Encyclopedia of Astronomy**
Swinburne University of Technology,
http://astronomy.swin.edu.au/cosmos/

**Division of Geological and Planetary Sciences**
California Institute of Technology
https://www.gps.caltech.edu/

**EarthSky**
Earthsky Communications
http://earthsky.org/

**Essential Radio Astronomy.**
National Radio Astronomy Observatory
https://science.nrao.edu/

**Everyday Cosmology**
Observatories of the Carnegie Institution for Science
http://cosmology.carnegiescience.edu/

**Galileo's Pendulum.**
Rice University
http://galileo.rice.edu/index.html

**HyperPhysic**
Department of Physics & Astronomy, Georgia State University
http://phy-astr.gsu.edu/hyperphysics/

**International Astronomical Union. IAU**
http://www.iau.org/

**Johnston's Archive**
Asteroids with Satellites Database: Johnstons's Archive
http://www.johnstonsarchive.net/index.html

**Kavli Institute for Particle Astrophysics and Cosmology (KIPAC)**
Stanford University
https://kipac-web.stanford.edu/

**LiveScience**
http://purch.com/tag/live-science/

**Math Open Reference**
http://www.mathopenref.com/

**NASA**
*National Aeronautics and Space Administration*
http://www.nasa.gov/
*Jet Propulsion Lab*
http://www.jpl.nasa.gov/
*SpaceFlight*
https://www.nasaspaceflight.com/
*PDS: The Planetary Data System*
https://pds.jpl.nasa.gov/
*PDS Small Bodies Node*
http://pds-smallbodies.astro.umd.edu/

**Palomar Observatory**
California Institute of Technology
http://www.astro.caltech.edu/palomar/homepage.html

**Particle Adventure**
Particle Data Group
http://www.particleadventure.org/

**Physicsworld.com**
Institute of Physics (IOP)
http://physicsworld.com/

**RussianSpaceWeb.com.**
http://russianspaceweb.com/

**Science and Space**
National Geographic
http://science.nationalgeographic.com/science/

**SETI Institute.**
http://www.seti.org/

**Space Book**
Las Cumbres Observatory Global Telescope Network
https://lcogt.net/

**Space Telescope Science Institute.**
http://www.stsci.edu/portal/

**Space.com**
Science & Astronomy
http://www.space.com/science-astronomy

**StarDate**
The University of Texas McDonald Observatory
https://stardate.org/

**Students for the Exploration and Development of Space.**
SEDS USA
http://seds.org/

Sydney Observatory
Museum of Applied Arts & Sciences Sydney Observatory
https://maas.museum/sydney-observatory/

**The Constellations and Their Stars**
University of Wisconsin-Madison Department of Astronomy
http://www.astro.wisc.edu/~dolan/constellations/

**The Munich Astro Archive**
http://www.maa.mhn.de/

**The Stephen Hawking Centre for Theoretical Cosmology**
University of Cambridge
http://www.ctc.cam.ac.uk/

**Time and Navigation**
Smithsonian
https://timeandnavigation.si.edu/

**Universe Today**
http://www.universetoday.com/

# TIMELINE OF SPACE EXPLORATION

**1957**

October 4 - The Soviet Union launched the first satellite, *Sputnik*, into space.

November 3 - The Soviet spacecraft *Sputnik 2* was launched with a dog named Laika on board. Laika did not survive the voyage.

**1958**

January 31 - *Explorer 1* was the first satellite launched by the United States when it was sent into orbit on January 31, 1958. It was designed and built by the Jet Propulsion Laboratory (JPL) of the California Institute of Technology. The satellite was sent aloft from Cape Canaveral in Florida by the Jupiter C rocket that was designed, built, and launched by the Army Ballistic Missile Agency (ABMA) under the direction of Dr. Wernher Von Braun.

**1960**

August 19 - The Soviet craft *Sputnik 5* was launched, carrying the dogs Strelka and Belka. They became the first living beings to survive a trip into space.

**1961**

April 12 - Russian cosmonaut Yuri Gagarin became the first human in space.

May 5 - Astronaut Alan Shepard became the first American in space.

May 25 - President Kennedy challenged the country to put a man on the moon by the end of the decade.

**1962**

February 20 - Astronaut John Glenn became the first American in orbit.

June 16 - Valentina Nikolayeva Tereshkova became the first woman in space.

**1965**

March 18 - While tethered to his spacecraft, cosmonaut Alexi Leonov became the first man to walk in space.

June 3 - Astronaut Ed White became the first American to walk in space.

July 14 - The spacecraft *Mariner 4* transmitted the first pictures of Mars.

**1966**

February 3 - The Russian spacecraft Luna 9 became the first spacecraft to land on the moon.

June 2 - Surveyor 1 became the first American spacecraft to land on the moon.

**1967**

January 27 - Astronauts Gus Grissom, Ed White, and Roger Chaffee were killed in an accidental fire in a command module on the launch pad.

April 24 - Cosmonaut Vladimir M. Komarov was killed in a crash when the parachute on his Soyuz 1 spacecraft failed to deploy.

October 18 - A descent capsule from the Soviet probe *Venera 4* collected data about the atmosphere of Venus.

**1968**

September 15 - The Soviet spacecraft *Zond 5* was launched and later became the first spacecraft to orbit the moon and return to Earth.

December 21 - *Apollo 8* was launched, and later her crewmembers became the first men to orbit the moon.

**1969**

July 20 - Neil Armstrong and "Buzz" Aldrin became the first men on the moon.

**1970**

April 11 - *Apollo 13* was launched.

September 12 - The Soviet craft Luna 16 was launched and became the first automatic spacecraft to return soil samples of the moon.

November 17 - The Soviet automatic robot Lunokhod 1 landed on the moon with *Luna 17*.

December 15 - The Soviet *Venera 7* became the first probe to land on Venus.

### 1971
April 19 - The Soviet space station Salyut 1 was launched.

July 30 - The moon rover was driven on the moon for the first time.

November 13 - The *Mariner 9* probe became the first craft to orbit another world - Mars.

### 1972
December 11 - Eugene Cernan and Harrison "Jack" Schmitt became the last men to walk on the moon.

### 1973
May 14 - The U.S. launched its first space station, Skylab.

### 1975
July 17 - The American Apollo 18 and Soviet *Soyuz 19* dock in the Apollo-Soyuz Test Project.

### 1976
September - The American probe *Viking 2* discovered water frost on the Martian surface.

### 1977
August and September - *Voyagers 1* and *2* were launched. (*Voyager 2* was launched before Voyager 1, but Voyager 1 was on a faster trajectory.)

### 1979
March and August - *Voyagers 1* and *2* began transmitting images of Jupiter and her moons.

September - The U.S. probe Pioneer 11 reached Saturn and began transmitting images.

### 1980
November 13 - *Voyager 1* reached Saturn and began transmitting images.

### 1981
April 12 - *Columbia* became the first Space Shuttle to be launched.

August 26 - *Voyager 2* reached Saturn and began transmitting images.

### 1983
April 4 - The second Space Shuttle, *Challenger*, was launched.

June 19 - Sally Ride became the first American woman in space on *Challenger's* second mission.

August 30 - Guion Bluford became the first African American in space.

### 1984
February 3 - Astronaut Bruce McCandless became the first man to take an untethered space walk.

August 30 - The third Space Shuttle, *Discovery*, was launched.

October - Kathryn Sullivan became the first American woman to walk in space.

### 1985
October 3 - *Atlantis*, the fourth Space Shuttle, was launched.

### 1986
January 24 - *Voyager 2* began transmitting images from Uranus.

January 28 - The Space Shuttle *Challenger* exploded seconds after liftoff.

February 20 - The core section of the Space Station Mir was launched.

### 1989
August - *Voyager 2* began transmitting images from Neptune.

### 1990
August 10 - The *Magellan* spacecraft began mapping the surface of Venus using radar equipment.

August 24 - The Space Shuttle *Discovery* deployed the Hubble Space Telescope.

### 1992
May 7 - The Space Shuttle *Endeavor* was launched on her maiden voyage.

September 12 - Mae Jemison became the first African-American woman in space.

### 1993
December - The Space Shuttle *Endeavor* made the first servicing mission of the Hubble Space Telescope.

### 1994
February 3 - Sergei Krikalev became the first Russian cosmonaut to fly on a Space Shuttle.

### 1995
February 2 - Eileen Collins became the first female Shuttle pilot.

December - The Galileo probe began transmitting data on Jupiter.

### 1997
July 4 - The *Mars Pathfinder* arrived on Mars and later began transmitting images.

### 1998
October 29 - John Glenn became the oldest man in space.

### 1999
July 23 - Eileen Collins became the first female Shuttle Commander.

### 2000
February 14 - The U.S. *Near Earth Asteroid Rendezvous* (*NEAR*) spacecraft began transmitting images of the asteroid Eros.

### 2001
February 12 - *NEAR* landed on the surface of Eros.

April 28 - American Dennis Tito became the first tourist in space after paying the Russian space program $20,000,000.

### 2003
February 1 - The Space Shuttle *Columbia* broke up on re-entry into the Earth's atmosphere.

February 13 - An investigative panel found that superheated air almost certainly seeped through a breach in space shuttle *Columbia's* left wing and possibly its wheel compartment during the craft's fiery descent, resulting in the deaths of all seven astronauts.

August 25 - NASA launched the largest-diameter infrared telescope ever in space, the Spitzer Space Telescope.

September 21 - NASA's Galileo mission ended a 14-year exploration of the solar system's largest planet and its moons with the spacecraft crashing by design into Jupiter at 108,000 mph.

### 2004
January 14 - President Bush proposed a new space program that would send humans back to the moon by 2015 and establish a base to Mars and beyond.

July 1 - The *Cassini* spacecraft sent back photographs of Saturn's shimmering rings.

### 2005
July 3 - A NASA spacecraft collided with a comet half the size of Manhattan, creating a brilliant cosmic smashup designed to help scientists study the building blocks of life on earth.

July 26 - Space Shuttle *Discovery* was launched with seven astronauts aboard; this was America's first manned space shot since the 2003 *Columbia* disaster.

### 2006
January 15 - NASA spacecraft *Stardust* returned safely to Earth in a desert near Salt Lake City with the first dust ever collected from a comet.

## 2007

August 4 - NASA launched its Phoenix Mars Lander.

August 8 - Space Shuttle *Endeavour* and a crew of seven blasted off with teacher-astronaut Barbara Morgan aboard as a crewmember. Morgan was the first teacher in space since the *Challenger* disaster in 1986.

## 2008

January 14 - The NASA space probe Messenger skimmed 124 miles above Mercury.

May 25 - NASA's *Phoenix Mars Lander* landed safely and began sending images home after a 10-month, 422 million-mile journey. Scientists later reported that *Phoenix* discovered chunks of ice.

## 2009

March 6 - The NASA spacecraft *Kepler* was launched. Its mission is to search for planets outside our solar system, in a distant area of the Milky Way.

June 18 - NASA launched the Lunar Crater Observation and Sensing Satellite, also known as LCROSS. The mission is to confirm the presence or absence of ice on the moon. On November 13, 2009, NASA scientists announced the discovery of a "significant amount" of ice in a crater near the moon's South Pole.

## 2010

October 10 - Virgin Galactic, a private company, announced the successful first manned glide flight of the *VSS Enterprise*. This vehicle is a suborbital plane designed to take private citizens on suborbital space flights.

October 11 - President Barack Obama signed legislation focusing NASA's efforts on exploring Mars and the asteroids.

December 8 - A private company named SpaceX launched a spacecraft into orbit and returned it to earth safely. It was the first non-government organization to accomplish this.

## 2011

July 8 - The space shuttle *Atlantis* became the last American space shuttle to be launched into space. Mission STS-135 and its 4-member crew brought much-needed supplies and equipment to the International Space Station (ISS).

July 16 - NASA's *Dawn* spacecraft became the first man made craft to orbit an asteroid.

November 26 - NASA launched *Curiosity*, the biggest, best equipped robot ever sent to explore another planet. It will reach Mars in 2012.

## 2012

May 22 - SpaceX, a commercial space company, launched its Dragon C2+ mission to resupply the International Space Station (ISS).

August - NASA's *Voyager 1* probe, launched in 1977, entered interstellar space.

August 6 - NASA's *Curiosity* rover successfully landed on Mars. As large as a car, it carried an array of advanced new instruments and experiments.

## 2013

September 7 - NASA launched the unmanned LADEE spacecraft from NASA's Wallops Flight Facility in Virginia. It was the U.S. space agency's third lunar probe in five years.

December 24 - NASA astronauts wrapped up successful repairs at the International Space Station after a rare Christmas Eve spacewalk to fix an equipment cooling system.

## 2014

March 11 - Two Russian cosmonauts and an American astronaut landed back on Earth in Kazakhstan after a stay aboard the International Space Station (ISS) of over half a year.

May 12 - A NASA study said the West Antarctic ice sheet is starting a slow collapse in an unstopable way.

May 29 - Elon Musk, CEO of SpaceX, unveiled *Dragon V2*, a new spacecraft designed to carry up to seven astronauts to the Int'l. Space Station (ISS).

Jul 2, NASA's Orbiting Carbon Observatory-2 reached orbit after launch from Vandenberg Air Force Base.

Jul 10 - Arianespace launched a rocket from French Guiana carrying four satellites that will help provide Internet and mobile connectivity to people in nearly 180 countries.

Sep 21 - NASA's *Maven* spacecraft entered orbit around Mars to study the planet's atmosphere.

Oct 31 - A Virgin Galactic *SpaceShipTwo* blew apart over southern California after being released from a carrier aircraft killing one pilot, Michael Alsbury (39) and seriously injuring the other, Peter Siebold (43). The suborbital vehicle was undergoing its first powered test flight since January over the Mojave Desert.

### 2015

Mar 6 - NASA confirmed that its *Dawn* spacecraft has arrived to orbit the dwarf planet Ceres for a 16-month exploration.

Mar 28 - NASA said a Russian Soyuz spacecraft with three crew on board successfully docked with the International Space Station after blasting off from Kazakhstan. An American and two Russians floated into the International Space Station, beginning what is to be a year away from Earth for two of them.

Apr 17 - The SpaceX supply ship arrived at the International Space Station, delivering the world's first espresso machine designed exclusively for astronauts.

Apr 30 - Blue Origin, a private space company founded by Amazon CEO Jeff Bezos, launched an unmanned spaceship. Its New Shepard capsule reached 58 miles and landed in the west Texas desert.

Jun 11 - In Kazakhstan NASA astronaut Terry Virts, Samantha Cristoforetti of the European Space Agency and Russia's Anton Shkaplerov returned to Earth after 199 days on the Int'l. Space station, nearly a month longer than planned.

Sep 2 - In Kazakhstan the first Dane in space accompanied by 26 custom-made figurines from Danish toymaker Lego blasted off in a *Soyuz* spacecraft as part of a three-man team to the International Space Station.

Sep 12 - A Russian space capsule landed safely in Kazakhstan, bringing home a three-person crew from the International Space Station, including a record-breaking Russian cosmonaut. Gennady Padalka completed his fifth mission for a world record of 879 total days in space.

# FAMOUS FIGURES AND EVENTS IN ASTRONOMY

*Thales* (624-547 B.C., Ionian). A philosopher who traveled widely in Mesopotamia and Egypt, and brought astronomical records from these cultures back to Greece. He believed that the Earth is a disk floating on an endless ocean.

*Anaximander* (611-547 B.C., Ionian). A philosopher who made the first detailed maps of the Earth and the sky. He knew that the Earth was round, and believed that it was free-floating and unsupported. He measured its circumference, and was the first to put forward the idea that celestial bodies make full circles in their orbits. He was the first to conceptualize space as having depth.

*Pythagoras* (569-475 B.C., Ionian). A mathematician who stated that the universe is made of crystal spheres that encircle the Earth. According to him, the Sun, the Moon, the planets, and the stars travel in separate spheres. When the spheres touch each other, they create the "music of the spheres.".

*Aristotle* (384-322 B.C., Greek). A philosopher who proved that the Earth is spherical, and believed that it was at the center of the universe. His reason for believing this was that if the Earth revolved around the Sun, then we should see the stars shift position throughout the year, but unable to detect the shift since he did not have the technology of today, he concluded that Earth rests at the center of the universe with the Sun, planets, and stars located in spheres that revolved around the Earth.

*Aristarchus* (310-230 B.C., Greek). Put forward the idea that the Sun was at the center of the universe, about 1,750 years before Copernicus. He also attempted to measure the relative distances between the Earth and the Sun and the Earth and the Moon. He used a reasonable method, but his results were not very accurate, lacking the technological equipment necessary for a precise measurement.

*Hipparchus* (190-120 B.C., Greek). He compiled the first known star catalog to organize astronomical objects, and also came up with a scale to define the brightnesses of stars. A version of this magnitude system is still used today. He measured the distance from the Earth to the Moon to be 29.5 Earth diameters. Another of his important discoveries was the precession, or wobble, of the Earth's axis, caused by the gravitational pull of the Sun and Moon.

*Claudius Ptolemy* (85-165 A.D., Greek). Ptolemy used Hipparchus' extensive observations to develop a model that predicted the movements of the Sun, Moon, planets, and stars. The Ptolemaic system put forth an Earth-centered universe and assumed that all astronomical objects move at constant speeds in circular orbits. The Ptolemaic model was the cornerstone of astronomy for 1,500 years.

*al-Khwarizmi* (780-850, Islamic). He developed algebra, in words, not mathematical expressions, based on what we today call Arabic numerals. His work was translated into Latin hundreds of years later, and served as the European introduction to the Indian number system, complete with its concept of zero. Al-Khwarizmi performed detailed calculations of the positions of the Sun, Moon, and planets, and did a number of eclipse calculations. He constructed a table of the latitudes and longitudes of 2,402 cities and landmarks, forming the basis of an early world map.

*Omar Khayyam* (1048-1131, Persian). was a great scientist, philosopher, and poet. He compiled many astronomical tables and performed a reformation of the calendar which was more accurate than the Julian and came close to the Gregorian. An amazing feat was his calculation of the year to be 365.24219858156 days long, which is accurate to the sixth decimal place!

*Nicolaus Copernicus* (1473-1543, Polish). He concluded that the Sun was the center of the universe instead of the Earth. Copernicus kept the orbits circular, even though he though Ptolemy's model was contrived. Other astronomers such as Brahe and Galileo helped to eventually prove that this model of the universe more accurately portrayed reality.

*Tycho Brahe* (1546-1601, Danish). He made the most accurate astronomical observations up to that time in his observatory, which contained sophisticated equipment for mapping star positions. He believed that the universe was a blend of the Ptolemaic and Copernican models, and created his own model in which the planets orbit the Sun and the Sun orbits the Earth

*Galileo Galilei* (1564-1642, Italian). The father of observational astronomy. In 1609, he built a telescope and was able to make a number of discoveries that changed astronomy: craters, mountains, and valleys of the Moon, the number of stars making up the Milky Way. Galileo kept precise records of sunspot activity and the phases of Venus, and discovered four moons orbiting Jupiter (the Galilean Moons). He publicized the fact that other astronomical bodies were clearly revolving around something other than the Earth. Galileo was put on trial in 1633 and forced to renounce his views by the Catholic Church. He died under house arrest.

*Johannes Kepler* (1571-1630, German). Kepler inherited his Brahe's astronomical records and used them to develop three laws of planetary motion. He used the idea of elliptical orbits to describe the motions of the planets, which became known as Kepler's first law. His second law states that a line from the Sun to a planet sweeps out equal areas in equal amounts of time. The third law states that the square of the number of years of a planet's orbital period is equal to the cube of that planet's average distance from the Sun.

*Giovanni Cassini* (1625-1712, Italian). First discovered the division in the rings of Saturn, today known as the Cassini division. He also found four moons orbiting Saturn, and measured the periods of rotation of Mars and Jupiter.

*Isaac Newton* (1643-1727, British). was a mathematician who developed extensive mathematics to describe the astronomical models of Copernicus and Kepler. His Theory of Universal Gravitation was the foundation of Kepler's laws of planetary motion, but it also went further: Newton showed that the laws governing astronomical bodies were the same laws governing motion on the surface of the Earth. Newton's scientific ideas are so complete that they still offer an accurate description of physics today, except for certain cases in which 20th century physics must be used.

*Edmond Halley* (1656-1742, British). Famous for predicting the 1682 appearance of a comet called Halley's Comet, he proved that the orbit of comets is periodic.

*Charles Messier* (1730-1817, French). Published a list of 110 astronomical objects that should not be mistaken for comets. This list includes objects visible through small telescopes, including galaxies, nebulae, and star clusters. The M objects, as they are now called, are used today to identify the most brilliant objects in the sky.

*William Herschel* (1738-1822, British). Discovered Uranus and two of its moons as well as two more moons of Saturn and several asteroids. He made a catalog of 2,500 astronomical objects and found the polar ice caps on Mars.

*Johann Bode* (1747-1826, German). Published a law now known as Bode's Law, that predicts mathematically the distances of the planets from the Sun. Using his law, he was able to determine that there should be another planet between Mars and Jupiter.

*Joseph von Fraunhofer* (1787-1826, German). Discovered dark lines in the spectrum coming from the Sun. He carefully measured the positions of over 300 of these lines, creating a wavelength standard that is still in use today.

*Joseph Lockyer* (1836-1920, British). Discovered the element helium while studying the Sun's atmosphere. He made detailed records of sunspot activity and studied solar flares and prominences. He conducted several tours to places where solar eclipses would be visible. He was one of the first archaeoastronomers: his book, *The Dawn of Astronomy*, investigates the astronomy of ancient cultures, in particular Egypt.

*Annie Jump Cannon* (1863-1941, American). A member of the a group of Harvard astronomers called "Pickering's Women." She single-handedly classified 400,000 stars into the scheme we use today (O B A F G K M), and discovered 300 variable stars.

*George Hale* (1868-1938, American). Discovered that sunspots have localized magnetic fields, He founded three important observatories: Yerkes, Mt. Wilson, and Palomar.

*Henrietta Swan Levitt* (1868-1921, American). Another member of "Pickering's Women" who discovered that a particular type of variable star known as a Cepheid could be used as a distance marker. This discovery made it possible to determine astronomical distances to objects.

*Ejnar Hertzsprung* (1873-1967, Danish). One of the inventors of the Hertzsprung-Russell diagram, showing the relationship between the absolute magnitude and the spectral type of stars. He determined the distance to the Small Magellanic Cloud, a galaxy visible from Earth's southern hemisphere.

*Karl Schwarzschild* (1873-1916, German). The first to study the theory of black holes. The Schwarzschild radius is the distance from a black hole at which bodies would have an escape velocity exceeding the speed of light and therefore would be invisible. He wrote on the curvature of space, based on Einstein's theory of relativity.

*Henry Russell* (1877-1957, American). One of the inventors of the Hertzsprung-Russell diagram describing the spectral types of stars. He measured the parallax of the stars photographically, allowing them to be properly placed on the H-R diagram.

*Albert Einstein* (1879-1955, German). His special theory of relativity, proposed in 1905, extends Newtonian mechanics to very large speeds close to the speed of light and describes the changes in measurements of physical phenomena when viewed by observers who are in motion relative to the phenomena. In 1915, Einstein extended this further in the general theory of relativity to include the effects of gravitation. According to this theory, mass and energy determine the geometry of spacetime, and curvatures of spacetime manifest themselves in gravitational forces.

*Arthur Eddington* (1882-1944, British). Proved observationally that Einstein's prediction of light bending near the extreme mass of a star is scientifically accurate. He also explained the behavior of Cepheid variables, and discovered the relationship between the mass of a star and its luminosity.

*Edwin Hubble* (1889-1953, American). Discovered that faraway galaxies are moving away from us. Known as Hubble's Law, the theory states that galaxies recede from each other at a rate proportional to their distance from each other. This concept is a cornerstone of the big bang model of the universe.

*Jan Oort* (1900-1992, Dutch). The first to measure the distance between our solar system and the center of the Milky Way Galaxy and to calculate the mass of the Milky Way. He proposed a large number of icy comets left over from the formation of the solar system, now known as the Oort Cloud

*George Gamow* (1904-1968, Russian-born American). First to put forward the idea that solar energy comes from the process of nuclear fusion.

*Karl Jansky* (1905-1950, American). Discovered that radio waves are emanating from space, leading to the science of radio astronomy.

*Gerard Kuiper* (1905-1973, Dutch-born American). Discovered a large number of comets at the edge of the solar system beyond Pluto's orbit, known as the Kuiper belt as well as several moons in the outer solar system and the atmosphere of Saturn's moon, Titan.

*Clyde Tombaugh* (1906-1997, American). Discovered Pluto, which he found photographically in 1930, using the telescope at the Lowell Observatory in Arizona.

*Subramanyan Chandrasekhar* (1910-1995, Indian-born American). Made important contributions to the theory of stellar evolution. He found that the limit, now called the Chandrasekhar limit, to the stability of white dwarf stars is 1.4 solar masses: star larger that cannot be stable as white dwarves.

*James Van Allen* (1914-, American). Discovered the magnetosphere of the Earth and gave his name to the belts of radiation surrounding the planet, the Van Allen belts, which moderate the amount of solar radiation hitting Earth.

*Fred Hoyle* (1915-2001, British). Although a proponent of steady-state model of the universe, and he coined the term "big bang." He believed that early life forms were transported to Earth by comets.

*Robert Dicke* (1916-1997, American). Believed it was possible to detect radiation left over from the big bang and invented the microwave radiometer to detect this radiation, which has a wavelength of one centimeter.

*Alan Sandage* (1926-, American). Calculated the ages of many globular clusters, and discovered the first quasar.

*Roger Penrose* (1931-, British). Expanded the physics of black holes by showing that singularities in space were responsible for their existence.

*Arno Penzias* (1933-, German-born American). Co-discoverer (with Wilson) of the cosmic microwave background radiation left over from the big bang.

*Carl Sagan* (1934-1996, American). Popularized astronomy with the general public and championed the search for extraterrestrial intelligence, which continues today with a number of missions to Mars to search for signs of life on that planet.

*Robert Wilson* (1936-, American). Co-discoverer (with Penzias) of cosmic microwave background radiation left over from the big bang.

*Kip Thorne* (1940-, American). Contributed to the understanding of black holes.

*Stephen Hawking* (1942-, British). Combined the theory of general relativity and quantum theory to prove that black holes emit radiation and eventually evaporate.

*Alan Guth* (1940-, American). Developed the theory called the inflationary universe as an addition to the big bang model, which predicts that the universe is flat and infinite.

# SUBJECT INDEX

**Symbols**

1 Ceres v, 29, 73, 74, 114
2 Pallas 29, 326
3C 273 254, 328
3 Juno 29
4 Vesta vi, 29, 113, 114, 325, 326
5 Astraea 29
51 Pegasi 207
55 Cancri 58
55 Cancri A 58
70 Virginis 328
80 Ursae Majoris 318
81P/Wild 2 17
(216) Kleopatra v, 169, 170, 171
243 Ida v, 29, 106, 107, 172, 173, 174
(243) Ida I Dactyl v, 29, 106
253 Mathilde 210
433 Eros v, 16, 29, 127, 128, 209, 210, 211
588 Achilles 315
911 Agamemnon 11, 12
951 Gaspra 29
1036 Ganymed 13
1221 Amor v, 12, 13, 14
1862 Apollo v, 18, 19
1866 Sisyphus 19, 21
1932 EA1 13, 14
1950 DA 170
1991 BA 28
1992QB1 175
2003 UB313 124, 125
2003 VB12 271
2010 TK7 316
2062 Aten v, 38, 40, 41
4450 Pan vi, 233, 234
5535 Annefrank v, 16, 17
10199 Chariklo 30, 72
25143 Itokawa 29
42355 76
65489 Ceto v, 75, 76, 237, 238
(65489) Ceto I Phorcys vi, 237
90377 Sedna vi, 270, 271
136199 Eris v, 29, 124, 125
(136199) Eris I Dysnomia v, 117
(NEO) Observations Program 212
α (alpha) Tauri 88

**A**

Abell 2065 95
ablation 1, 2, 178, 179
ablation shield 1
absolute magnitude 13, 152, 153, 227, 240
absolute zero 139, 142, 206
absorption 100, 101, 165, 192, 206, 255, 287, 288, 294
accelerometers 274
accretion disk 3, 36, 49, 167, 168, 255, 256, 290, 291
Acrab 268
Acrux 99, 100
Acta Mathematica 310
Actas 318
active galactic nuclei (AGNs) 51, 52, 254, 333
active planet 119, 250
Acubens 56, 58
Adams 216
adaptive optics x, 117, 118, 169
Adhafera 183
Adhara 60
Adrasteia 321
Advanced Research Projects Agency (ARPA) 116
AE Aquarii 235
Aeetes 26
Aeolis Mons 103
Aeolis Palus 103, 105
aerodynamic 1, 181, 182
aerospace engineering 2, 4, 7, 8, 80, 122, 131, 161, 177, 180, 200, 259, 272, 275, 278, 284, 333
Age of Aquarius 243
Age of Pisces 243
albedo 72, 188, 194, 215, 249, 324
Albrecht, Andreas 96
Alcor 318
Alcyone 305
Aldebaran 259, 305, 306, 307
Alexandria 122, 238, 260
Alexhelios v, 169, 170, 171, 172
al Farkadain, Akhfa 320, 321
al Farkdain, Anwar 320, 321
Alfecca Meridiana 91, 92
Alfvén, Hannes 22, 247
Algedi 64, 65
Algieba 183, 185
Alhena 144
Alioth 318
al-Jauza 232

Alkaid 318
Allen, James Sircom 217
Allen, James Van 248
ALMA array 130
Almagest ix, 59, 61, 238
Alnasl 264
Al Nasl 264
Alnilam 231
Alnitak 231
Al Niyat 269
alpha 1 64
alpha 2 64, 65
alpha 2, beta, and omega Capricorni 65
alpha Aquarii 23
alpha Arietis 26, 243
alpha Boötis 328
alpha Cancri 56, 58
alpha Canis Majoris 62
alpha Canis Minoris 62
Alpha Capricorni 64
alpha Cassiopeiae 67
alpha Centauri 206, 207
alpha Centauri B 206
alpha Coronae Borealis 94
alpha Crucis 99
Alpha Geminorum 144
alpha Leonis 183
alpha Librae 186
alpha Orionis 61, 231
alpha Piscium 242
alpha Sagittarii 264
alpha Scorpii 268
alpha Scorpiids 269
alpha stars 64
alpha Tauri 305
alpha Ursae Minoris 320
alpha Ursa-Majorids 319
alpha Virginis 328
Alphecca 93, 94, 95
Alpherg 242
Alpher, Ralph 37, 222
Alrescha 242
Alrisha 242
Alshat 64, 65
Al Tarf 56, 58
Alula Australis 318
Amalthea 65
Ames Research Center 276
ammonia crystals 82, 140, 303

Amor v, 12, 13, 14, 15, 16, 28, 41, 127, 128, 188
Amor asteroid group v, 12, 188
Amor asteroids v, 12, 13, 14, 15, 127
Andromeda 68, 133, 196, 199, 242, 256
Andromeda constellation 196
Andromeda galaxy 133, 196, 256
Andromeda Nebula 133, 199
Angara launch vehicle 181
angular momentum 2, 3, 219, 244, 245
anomalies 229
Antares 206, 259, 268
Antecanis 62
antiquark 252
aphelion 12, 18, 41, 172, 193, 266, 267, 270, 271
Aphrodite 243, 324
Apis 306
*Apollo* v, 14, 15, 18, 19, 20, 21, 22, 28, 41, 68, 127, 200, 232, 233, 274, 276, 278, 286
*Apollo 11* 286
*Apollo 12* 276
Apollo asteroid v, 18, 21, 233
*Apollo* lunar landings 278
Apollo mission 200, 286
Apollo Navigation and Guidance System 274
Aquarius v, 22, 23, 24, 25, 64, 65, 242, 243
Aquila 64, 92
Archytas 260
Arcturus 259, 293, 328
Arecibo Observatory 19, 40
Ares 190, 268
argon atoms xii
Argonauts 26
Ariadne 95
Arich 328
Aries v, 25, 26, 27, 28, 242, 243
Arkab Posterior 264
Arkab Prior 264
Artemis 232
Ascella 264
asterism 59, 61, 264, 305, 317, 320
asteroid impact avoidance 12, 18, 28, 40, 169, 177, 211
Asteroid Initiative 30, 180
Asteroid Redirect Mission (ARM) 14, 16, 19, 20, 22, 30, 31, 40, 41, 180, 213
Asteroth 94
Astraia 329
astrobiology 177, 223, 296, 300

381

astrochemistry  31, 32, 33, 77, 82, 139, 141, 164, 176, 205, 246
astrogeology  234
astronautics  1, 4, 7, 88, 161, 180, 182, 200
astronomical unit (AU)  40, 115, 119, 127, 188, 209, 213, 272
astrophotography  293
astrophysics  vi, 35, 36, 38, 46, 47, 55, 96, 143, 167, 168, 191, 214, 215, 216, 217, 218, 247, 248, 293, 307, 308, 309
Atargartis  243
Aten  v, 28, 38, 39, 40, 41, 127
Aten asteroid  40, 41
Aten objects  40
Atens  40
Athamas  26
Atira  41
Atlantis  280, 281
Atlas family  180
*Atlas of Stellar Spectra, An*  205, 207
Atlas rocket  104
Atlas V  11
atmospheric carbon  324
attenuation  vi, 191, 192
attenuation coefficient  191, 192
attitude control  4, 6
Aulus Gellius  260
Auriga  68, 319, 320
auroras  248, 332
Australis  v, 90, 91, 92, 93, 94, 183, 264, 318
AutoNav  116
autumnal equinox  186
Avdeyev, Sergei  201
average distance of Neptune from the Sun  214
axion  110
axis of rotation  119, 244, 291
Azure Dragon  269

# B

Background Imaging of Cosmic Extragalactic Polarization (BICEP2)  97
Bahcall, John N.  218
Baikonur Cosmodrome  284
ball of light particle model  46, 47
barred  134, 135, 156, 157, 158, 197
barred spiral  134, 135, 156, 157, 158, 197
barred spiral galaxy  135, 156, 197
barycenter  119, 310
baryon  252

BASS-II  81
Baten Algiedi  65
Becrux  99
Beehive Cluster  57
BeiDou Satellite Navigation System (BDS)  147
Belka  285
Bellatrix  231
Bell, Jocelyn  219
Benetnash  318
Berenices  196, 328
beta 1 and beta 2 Sagittarii  264
beta Aquarii  23
beta Arietis  26
beta Cancri  56, 58
beta Canis Minoris  62
beta Capricorni  64
beta capture  217
beta Cassiopeiae  67
Beta Crucis  99
beta decay  217
Beta Geminorum  144
beta Leonis  183
beta Librae  186
beta Orionis  231
beta Piscium  242
beta Scorpii  268
beta Tauri  305
Beta Taurids  306
beta Ursae Minoris  320
Betelgeuse  59, 61, 206, 231, 232, 240, 259, 293
Bethe, Hans Albrecht  78
big bang nucleosynthesis  37, 220, 222
big bang theory  35, 36, 37, 46, 47, 96, 97, 216, 217, 220, 221, 223, 307, 308, 309
Big Dipper  99, 100, 317, 318, 319, 320, 328
binary centaur  76, 238
binary pair  335
binary star  2, 3, 26, 58, 60, 64, 65, 91, 94, 185, 268, 305, 318, 328
binary system  3, 50, 60, 62, 75, 76, 237, 245, 298
binary trans-Neptunian object  75, 76, 237, 249, 250
blackbody  96
blackbody spectrum  96
Black Eye  196
blazar  51, 52, 255
Blaze Star  94
blink comparator  251
BL Lacertæ objects (BL Lacs)  52
blue giant  335

Bode, Johann Elert 29
Bode's galaxy 318
Bode's law 29
Bok, Bart 291
Bok globule 290, 291
bolide 52, 53, 129, 130
Boötes 62, 94, 293, 328
Boötes and Coma 328
Boötes constellation 94
bottom quarks 252
Bouvard 216
Brachium 186
Bragg, Sir W. H. 102
Bragg, Sir W. L. 102
Bragg's law 102
Bragg spectrometer 84
Bragg, William Henry 84
Brahe, Tycho x, 69, 159, 228
braid structure in a planetary ring 215
braking radiation 54
bremsstrahlung 54, 55, 192
bremsstrahlung process 54
bremsstrahlung radiation v, 54
Bright Target Explorer (BRITE) 240
Broucke-Hénon family 311
Broucke, Roger 310, 311
Brouwer, Dirk 71
brown dwarf 37, 110, 299, 300
Brown, Michael E. 117
Brown, Mike 124
Bruns, Heinrich 311
bulge 49, 133, 134, 135, 156, 157, 158, 197, 198, 245, 295, 296
Bull of Heaven 231
Burning and Suppression of Solids II (BASS-II) 81
Bush, George H. W. 280
Bush, George W. 280

# C
Caer Arianrhod 95
California Institute of Technology 11, 22, 111, 114, 117, 124, 267, 331
Callisto 193, 318
Caloris Basin 194
CalTech 14, 33, 76, 113, 121, 124, 126, 213, 271, 300, 304, 326, 327
Camelopardalis 68, 320
Canada-France-Hawaii Telescope 169
Canadian Space Agency 161, 162

Canadian Space Agency (CSA) 162
canali 190
cancer 56, 120, 236
Canes Venatici 196
Canis Major v, 58, 59, 60, 61, 63, 206, 231, 321
Canis Minor v, 56, 58, 59, 61, 62, 63, 231
Canis Minorids meteor shower 63
Cannon, Annie Jump 205, 207, 293, 294
Cape Canaveral 74, 104, 114, 326
Caph 67
Capricorn 65, 66
Capricorni 64, 65
Capricornid meteor shower 66
Capricornus v, 63, 64, 65, 66, 242
carbonaceous chondrites 129
carbon dioxide atmosphere 188, 323
carbon-nitrogen-oxygen (CNO) cycle 154
Carina 99
Cartesian coordinate graph 229
Cartesian coordinate system 229
Cassiopeia v, 66, 67, 68, 69, 320
Cassiopeia's Chair 68
Castor 56, 144, 145
Castor and Pollux 56, 144, 145
cathode rays 235
CCD 34, 271, 301, 333
celestial mechanics v, 69, 70
celestial pole 90, 93, 320
centaur 72, 75, 76, 92, 237, 265
Centaurus 92, 240, 265
Cepheid 34, 99, 133, 134, 144, 145, 146, 240
Cepheid variable 34, 99, 133, 134, 144, 145
Cepheus 68, 320
Ceto v, vi, 75, 76, 237, 238
Cetus 26, 68, 242
Challenger 280
Chamaeleon 320
Chandra X-Ray Observatory 25, 66, 69, 95, 198, 239, 241, 265, 270, 307, 319
Chaos 121
charge-couple devices (CCDs) 34, 241, 333
Chariklo 30, 72
Charlois, Auguste 127
charm quarks 252
Charon 249, 250
Chelae Scorpionis 186
chemical propulsion 8, 10
Chenciner, Alain 311
Chernushka 285

Chiron 72
chi Tauri 305
chondritic meteoroids 129
Christopher C. Kraft, Jr. Mission Control Center 273
chromosphere 287, 288
Chrysomallus 26
Cigar galaxy 318
Circlet of Pisces 242
circular accelerator 235
circumpolar constellation 66, 90, 93, 98, 99, 319, 320
circumpolar stars 66, 90
circumstellar disks 299
Classical KBOs 174, 266
classical mechanics 69, 70
Cleopatra Selene II 169, 171
Cleopatra VII 169, 171
Cleoselene v, 169, 170, 171, 172
Cluster II probes 331
CMB 37, 38, 96, 97, 308
CNO cycle 77
Coalsack 99
Colchis 26
cold classical KBOs 174
Columbia 7, 280, 281
coma v, 79, 83, 111, 135, 196, 328
Coma Berenices constellation 196
Coma Cluster 135
combustion 8, 10, 80, 81, 82, 131, 262
Combustion Integrated Rack (CIR) 80, 81, 82
comet v, vi, 73, 79, 82, 83, 84, 160, 224, 225, 267, 303, 304
cometary meteoroids 129
command and control interface 275, 276
common center of mass 2, 26, 29, 49, 60, 62, 75, 76, 91, 94, 99, 119, 121, 237, 249, 264, 310, 311, 318, 328
companion star 60, 298
Compass system 89, 147
composite remnants 302
Compton, Arthur H. 84, 85, 86, 167, 168
Compton effect 167
Compton scattering v, 84, 85, 167, 168, 191
Compton scattering equation 167
Computational Sciences Division 276
Concerning an Heuristic Point of View toward the Emission and Transformation of Light 85
conservation of energy 84, 85
conservation of momentum 84

contact iv, 91, 103, 140, 146, 176, 191, 210, 215, 291, 313
continuous spectrum 54, 167, 206, 294
convection 214, 215, 258, 287, 288, 323, 324
convective zone 287, 288
Copernican revolution 69
Copernicus ix, xii, 69
Copernicus, Nicolaus 69
Core navigation system v, 88
Corinna 92
corona v, 90, 91, 92, 93, 94, 95
Corona Australis v, 90, 91, 92, 93, 94
Corona Borealis v, 91, 93, 94, 95
Corona Borealis galaxy cluster 95
Coronae Australis 91, 92
coronal mass ejections 245
Corvus 328
cosmic dust 31, 92
cosmic inflation v, 96, 97
cosmic microwave background radiation 37, 38, 96, 308
cosmic rays 34, 54, 55, 123, 141, 164, 165, 234, 235, 301, 302
cosmological constant 107, 108, 221
cosmological principle 221
cosmology xi, 35, 36, 96, 191, 216, 221, 223, 308
*Cosmos: A Personal Voyage* 308
coulomb force 247
crab-like supernova remnant 301
Crab Nebula 196, 301, 304, 305
crab-type remnant 301
Cretaceous-Tertiary extinction 41
Cronus 121, 250
Crotus 265
crust of Mercury 194
Crux v, 87, 98, 99, 100
cryovolcanism 250
cryovolcanoes 74
crystal spectrometer 101
crystal spectroscopy 101, 102
C-type asteroid 209, 210
Cupid 243
*Curiosity* v, 85, 103, 104, 105, 116, 117, 277
Cygnus X-1 49
Cynosura 321

## D

D/1993 F2  233
Dabih  64, 65
Dactyl  v, 29, 106, 107, 172, 173
dark energy  v, 107, 329
Dark Energy Camera  109
Dark Energy Survey  109, 110
dark halo  197, 199
dark nebula  99, 290
da Vinci, Leonardo  275
Davis detector  78
Davis, Raymond Jr.  78, 218
Davis telescope  78
Dawn mission  v, 22, 40, 73, 74, 112, 325, 326
Dawn Science Center  113
*Dawn* spacecraft  113, 114
de Bort, Léon Teisserenc  42, 43
DECam (Dark Energy Camera)  109
deep-sky object  99, 195
deep-space exploration  166
deep-space navigation  115
Deep Space Network (DSN)  116, 117
deep-space objects  230, 305, 318
Defense Advanced Research Projects Agency (DARPA)  116, 151
Deimos  13, 29, 190
Delisle, Joseph-Nicolas  195, 196
Delphinus  24
Delporte, Eugène  13, 14, 23, 25, 183, 185, 268
delta Capricorni  65
delta Cassiopeiae  67
Delta family of launch vehicles  180
Delta II  209
delta Orionis  231
delta Scorpii  268
delta Ursae Minoris  320
Demeter  328
Deneb Algedi  65
Denebola  183
DES  109
detached objects (resonant)  312
deuterium  37, 96, 222, 299
deuteron  222
differential rotation  215
diffraction gratings  294
Dike  329
Dionysus  27, 62, 92, 95
dioxide accumulation  324
Discovery Program  17, 31, 112, 113, 114, 128

*Discovery Stardust* spacecraft  16, 17
Dmitrašinović, Veljko  310, 311
DNA  236
Dog Star  59, 60, 61, 293
Doppler radar  115, 117
double star  65, 67, 91, 196, 318
Draco  87, 320
drag coefficient  181
Dragon's Wing  320
Draper, Henry  205, 293, 294
Drizzle algorithm  34
Drizzling  34
Dschubba  268
Dubhe  317, 318
duct propulsion  8, 10
dumbbell-shape  169
dust coma  79, 82, 83, 224
dust-grain model  176
dust storms  188, 189
dust tail  83, 224, 303
dwarf elliptical galaxies  135
Dwarf galaxy  321
dwarf spherical galaxy  196
dwarf star  60, 205, 207, 235, 257, 297, 328, 329
dynamic planet  119
dynamo effect  120
Dysnomia  v, 117, 118, 125, 126

## E

E0  133, 134, 157, 158
E3  133
E5  158
E7  133, 134, 157
Earth-imaging satellite  122
earthquakes  120, 151, 203
Earth's moon  14, 19, 22, 28, 40, 102, 125, 126
Eastern Fish  242
eccentric orbit  174, 193, 249, 266, 267, 271
Echidna  76
ecliptic  25, 56, 64, 86, 144, 174, 183, 242, 249, 268, 304, 305, 320, 328
Eddington, Arthur  222
Edgeworth, Kenneth Essex  71
Edgeworth-Kuiper Belt  71, 174, 214, 249, 266, 312
Edwards Air Force Base  262, 280
Eiffel Tower  234
Einstein, Albert  xi, xiii, 35, 36, 48, 70, 77, 85, 107, 110, 111, 221, 281, 307, 308, 330
Einstein-Rosen bridge  331

Einstein-Rosen model 331
Eisenhower, Dwight D. 113, 286
ejecta 179, 203
electric field 46, 47, 54, 247
electromagnetic (EM) energy 167
electromagnetic (EM) radiation 168, 297, 298
electromagnetic (EM) spectrum 165
electromagnetic spectrum x, xi, 32, 51, 101, 140, 164, 165, 238, 239, 290, 295, 301, 302
electromagnetic wave 100, 191
electromagnetic wavelengths 44, 165, 212
electron xii, xiii, 32, 54, 55, 78, 85, 167, 168, 217, 218, 246, 247
electron density 246
electron neutrino 78
electrostatic force 247
electrostatic ion propulsion (EIP) 10
electroweak star 289
elementary particles 46, 54, 222, 252, 289
Eley-Rideal mechanism 176
Eley-Rideal model 177
elliptical galaxy 133, 135, 156, 157, 158, 321
elliptical orbit xi, 14, 82, 119, 233, 245, 325
elliptical path 12, 14, 29, 228, 310
Elnath 305
Emden, Robert 307, 309
emission spectra 52, 226, 257
Enceladus 102
Endeavor 280
Endurance 103
Enki 65
Enterprise 7
EPIC Radiation Monitor System (ERMS) 334
epsilon Arietis 26
epsilon Canis Majoris 60
epsilon Orionis 231
epsilon Sagittarii 264
epsilon Tauri 305
epsilon Ursae Minoris 320
epsilon Virginis 328
Eratosthenes 122, 185, 270
Eridanus 24, 65
Erigone 62
escape velocity 181, 189, 194, 259, 260
escape velocity of Mars 189
Eskimo Nebula 145
ET 7, 280
Eta Cassiopeiae 67
eta Piscium 242, 243

eta Tauri 305
eta Ursae Minoris 320
Euler, Leonhard 310, 311
Euphrates River 243
Europa 59, 305
European Geostationary Navigation Overlay Service (EGNOS) 138
European Organization for Nuclear Research (CERN) 235, 236, 237
European Photon Imaging Camera (EPIC) 333, 334, 335
European Space Operations Centre (ESOC) 334
Euryale 231
event horizon 48, 49, 308, 330, 335
exoplanet 183, 184, 185, 241
exotic matter 330, 331
exotic star 289
exploration systems technologies 276
*Explorer 1* 8, 116, 123, 248
*Explorer 6* 123, 124
*Explorer I* 261
external tank 7, 278, 279
extinction 14, 16, 19, 21, 39, 41, 178, 213, 295
Extragalactic astronomy 50, 115, 133, 136, 156, 195, 254
extravehicular activity (EVA) 200, 201, 202
Eyes galaxies 328

## F

Fadeout Star 94
failed star 287
False Cross 99
Farquhar, Robert 209
Federal Aviation Administration (FAA) 89, 90, 274
Fermi, Enrico xii, 217
Fermi Gamma-Ray Space Telescope 52, 112, 283
figure-eight family 311
figure-eight solutions 311
fireball 52, 53, 96, 129, 130
First Big One, The 269
first law of motion 228
First Point of Aries 27
Flame Extinguishment Experiment (FLEX) 81, 82
Flamm, Ludwig 330
flat-spectrum radio quasars (FSRQs) 52
Fleming, Williamina 205, 293
flow information 182
flux xiii, 238, 239, 332
flux transfer events 332

focus xi, 8, 36, 116, 177, 178, 202, 212, 229, 230, 239, 241, 285, 325
force balance 182
Frank, Anne 16, 17
freedom of space 286
free-free radiation 54
frequency division multiple access (FDMA) 147
Friends of Amateur Rocketry 262
fuel injection 131
Full Operational Capability (FOC) 137, 138
Fum al Samakah 242
fusion reaction 247, 257

# G
Gacrux 99
Gaea 121
Gagarin, Yuri A. 285, 286
Gaia 231, 269
galactic nucleus 2, 3, 50
galaxy classification v, 133, 156, 157, 158
galaxy-classification scheme 133
galaxy cluster 95, 111, 328
Gale Crater 103, 105
Galilei, Galileo 137
Galileo v, ix, x, 29, 89, 106, 107, 136, 137, 138, 139, 147, 172, 173, 190, 194, 274, 324
Galileo global satellite navigation 89
Galileo Mission 106
Galle 216
gamma and rho Cassiopeiae 67
gamma Aquarii 23
gamma Arietis 26
gamma Capricorni 65
gamma Cassiopeiae 62, 68
gamma Coronae Australis 91
gamma Crucis 99
gamma Geminorum 144
gamma Librae 186
gamma Orionis 231
gamma Piscium 242
gamma ray 51, 78
gamma-ray spectrometer 114, 211
gamma star 67
gamma Tauri 305
gamma Ursae Minoris 320
gamma Virginis 328
Gamow, George 35, 36, 37, 222, 308
Ganymede 24, 193
Garvey Spacecraft Corporation 262

gas coma 79, 82, 83, 224
gaseous atmosphere 120
gas giant 17, 75, 127, 214, 215, 249, 313, 326
gas-grain model 139
gas-phase astrochemistry model 142
gas-phase model v, 139, 140, 141, 142
gas-phase reaction 176
Gauss 190
Gemini v, 56, 63, 144, 145, 146, 232, 242, 319
Gemini South Observatory 232
Gemma 94
general theory of relativity xi, 48, 107, 221
geocentrism 69
Geodetics 146
geographical information systems (GIS) 122
geomagnetic storms 332
geometry xi, 121, 122, 325
geospatial technology 122, 123
giant blue stars 328
giant red star 65
Giffard, Jules Henri 5
Gilgamesh 231
Glashow, Sheldon 47
Gliese 581d 187
Gliese 581g 187
Gliese 876 23, 24, 241
global navigation satellite systems (GNSS) 90, 147, 272, 274
globular cluster 65, 133, 196
Globular cluster Messier 2 (M2) 23
GLONASS v, 89, 136, 137, 146, 147, 148, 272, 274
gluons 222
Gnosia 94
Goat Horn Constellation 64
Goddard, Robert H. 5, 8, 261
Gomeisa 62, 63
Gorbachev, Mikhail 280
Gorgons 76
GP-B spacecraft 283
Graffias 268
Grandblaise, Françoise 195
grand-design spiral galaxy 243
grand unification equation 46
grand unification theory 46, 47
gravitational attraction 37, 193
gravitational collapse 297
gravitational field xiii, 46, 47, 48, 106, 193, 215
gravitational lens 3, 110, 111
gravitational lensing 110

Gravity Probe B (GP-B) 283
grayscale 33, 34
Great Bear 317
Great Orion Nebula 196
Great Red Spot 215
Great Square of Pegasus 242
greenhouse effect 44, 323, 324
greenhouse gases 120
Grissom, Gus 68
Groombridge 1830 319
ground-based navigation 115, 116
Guardians of the Pole 320
Guth, Alan 96
GX 339-4 binary system 3
gyroscope 283

# H
H2 165, 176, 177, 227
HA 1932 19
Hades 121, 250, 328
hadron 223, 235, 236, 252, 289
Hale-Bopp 82, 83, 159, 160, 224, 303
Halley, Edmond 159, 196, 197
Halley's Comet 82, 224, 225
Hamal 26
Harch, Ann 173
harmonic standing spherical wave 46
Harold II 304
Harvard College Observatory 133, 205
Harvard Computers 293
Harvard Observatory 18, 226
Harvard scheme 292, 293
Harvard-Smithsonian Center for Astrophysics 25, 136, 141, 223, 241
Hawking, Stephen 38, 50, 98, 308, 309, 332
Hayabusa 29
HD 19445 27
Heidelberg Observatory 11
Hektor 315
Helin, Eleanor Francis 16, 22, 38, 41, 42
heliocentrism 69
Helios, Alexander 169, 171
helium fusion 257, 289
Helix Nebula 23
Helle 26
Hellespont 26
Hénon, Michel 310, 311
Henry Draper Catalogue 205, 293
Hera 26, 57, 318

Heracles 56, 305
Herbig-Haro (HH) object 290, 291, 292
Hercules 56, 57, 92, 94, 184, 305
Hercules constellation 92, 94
Herman, Robert 37
Hermes 194
Heron of Alexandria 260
Herschel x, 27, 29, 74, 140, 141, 143, 239, 271, 272, 320
Herschel, Sir William 27, 29, 74, 320
Herschel Space Observatory 140, 141, 143, 239, 271, 272
Hertz, Heinrich 85
Hertzsprung, Ejnar 152
Hertzsprung gap v, 152, 153
Hertzsprung-Russell diagram 77, 152, 153, 227, 257, 287, 294
Hesperides 321
Hess, Victor 234, 235
Heze 328
Higgs boson 235, 236
high-mass star 154, 155, 219, 252
high-resolution spectroscopy 164
HII region 226, 227
Hinshelwood, Sir Cyril Norman 176
Hipparchus 205, 238
Homestake experiment 218
Homestake Mine 218
Hooker Telescope 134
HOPS 383 291, 292
horizon 48, 49, 96, 98, 197, 231, 268, 308, 330, 335
Horizon 116
hot classical KBOs 174
hot core region 31
hot-wire velocity probes 182
Hoyle, Fred 222
H-R diagram 152, 153
Hubble classification 158, 197
Hubble, Edwin Powell 108, 110, 133, 134, 136
Hubble sequence 133, 197
Hubble tuning-fork diagram 133
Hyades 153, 304, 305, 306
Hyadum I 305
Hydra 56, 57, 63, 92, 186
hydrogen cloud 82, 83, 159, 160
hypergiant 60, 207

## I

I Alexhelios v, 171
Icarius 62
Ichthyes 243
Ida v, 29, 106, 107, 172, 173, 174, 321
II Cleoselene v, 171
Ikhthyes 243
impact basin 203, 326
impact craters 120, 193, 203, 204, 210
Indian Regional Navigation Satellite System (IRNSS) 147, 274
inertial navigation systems (INS) 274
infinite density 308
infinite gravity 38
Infrared Astronomical Satellite (IRAS) 212, 213
infrared energy 120, 140
infrared mapping spectrometer 114
infrared wavelengths 239
inner planets 13, 15, 18, 20, 39, 188
In-Orbit Validation (IOV) 138
Institute of Physics Belgrade 311
International Council of Scientific Unions 284
International Geophysical Year (IGY) 284, 285
International Space Station Intergovernmental Agreement (IGA) 162
International Space Station (ISS) v, 43, 80, 82, 161, 162, 163, 164, 202, 273, 275, 277, 279, 280, 281, 286
International Union for Cooperation in Solar Research 205
Interplanetary Kite-craft Accelerated by Radiation of the Sun (IKAROS) 10, 11
interstellar chemistry v, 164, 165, 166
interstellar dust 31, 32, 139, 142, 176, 232, 239
interstellar medium (ISM) 31, 32, 133, 139, 141, 142, 164, 165, 176, 177, 198, 199, 223, 301, 302
interstellar space 116, 164, 165, 166, 248, 270
inverse Compton scattering 167, 168
ionization 140, 165, 246, 247, 248
ionosphere 43, 44, 246, 248, 284
ion propulsion 8, 10, 112, 113, 114
ion tail 83, 224, 303
iota Arietis 26
Iota Cancri 57
iota Cassiopeiae 67
iota Orionis 231
irregular galaxy 26, 133, 156, 157, 158
Ishtar 329
Italian Combustion Experiment for Green Air (ICE-GA) 81
Ivan 53, 68

## J

Jabbah 269
Jansky, Karl Guthe 43
Japanese Aerospace Exploration Agency (JAXA) 132, 162, 235
Jason 26
jet propulsion 8, 10
Jewel Box 99
Jewitt, David 175, 312
Johns Hopkins University Applied Physics Laboratory 210, 211
Johnson Space Center 181

## K

K9 rover 276
Kajita-Macdonald theory 78
kappa Coronae Australis 91
kappa Orionis 231
kappa Tauri 305
Kaus Australis 264
Kaus Borealis 264
Kaus Media 264
Kaus Meridionalis 264
Keck II 169, 171
Keck Observatory 118, 126
Keenan, Philip 205, 226
Keenan, Philip Childs 206
Kellman, Edith 205, 206, 207, 226
Kelvin, Baron 206
Kennedy, John F. 200, 286
Kennedy Space Center (KSC) 7, 276, 280, 281
Kepler, Johannes x, 70, 190, 228
Kepler's first law 228, 229
Kerberos 250
Kerss, Tom 320
kinematic properties 296
kinematics 295
kinetic energy 54, 55, 78, 84, 246
King Arthur 319
King Arthur's Chariot 319
King Cepheus 68
King David 319
King Minos 305
King Oscar II 310
Kleopatra v, 169, 170, 171, 172

Kochab 320
Korabl-Sputnik 285
Korabl-Sputnik 2 285
Korabl-Sputnik 4 285
Koronis v, 172
Koronis asteroid 172
Koronis family 172
Kristall 201
K-shell emission 54
Kuffner, Moriz von 173
Kuiper Belt object (KBO) 124, 126, 174, 175, 266, 312
Kuiper, Gerard 71, 174, 266
Kullat Nunu 242
Kvant II 201

## L

Laboratory for Planetary Studies at Cornell University 308
Lacerta 68
Lagoon Nebula 264
Lagrange-Euler family 311
Lagrange, Joseph-Louis 310, 311, 315
Lagrangian points 314, 315
Laika 285
lambda Arietis 26
lambda Orionis 231
lambda Sagittarii 264
lambda Scorpii 269
lambda Tauri 305
land masses 120
Lane-Emden equation 307
Lane, Jonathan Homer 307
Langley 182
Langley Research Center 182
Langmuir-Hinshelwood mechanism 176
Langmuir-Hinshelwood model vi, 176, 177
Langmuir, Irving 176, 177, 247
Langmuir waves 247
Laplace, Pierre Simon de 244, 246
large dark blue spot 215
Large Hadron Collider (LHC) 223, 235, 236
Large Magellanic Cloud 33, 35, 78, 158
laser 1, 2, 104, 118, 177, 178, 179, 180, 182, 211, 236, 247, 248
Launch vehicle aerodynamics vi, 180
lava 74, 120, 194, 326
law of cosines 167
law of universal gravitation x, 70, 221, 228

laws of motion x, 8, 9, 36, 69, 70, 228, 259, 261
Leda 145
Lemaître, Georges 36
Lenard, Philipp 85
lenticular galaxy 133, 156
Leo vi, 56, 183, 184, 185, 242, 328
Leonid meteor shower 129, 184
Leonid shower phase 129
Leonids-Ursids 319
Leo Ring 184
Leo Triplet 184
lepton 290
Lepus 59, 62, 232
LeVerrier 216
Libra vi, 185, 186, 187, 242, 328
libration points 314, 315
light element 96
light quanta 85
LightSail 11
Linde, Andrei 96
linear accelerator 235
linear and circular accelerators 235
linear attenuation coefficient 191
liquid hydrogen 131, 278, 280
Little Dipper 319, 320, 321, 322
low-density comets 129
low-Earth orbit 181, 202
Lowell Observatory 250, 321
Lowell, Percival 190, 250, 251, 321
low-mass star 290
L-points 314, 315
luminescence 100, 101
luminosity class 205, 206
luminosity classification 257, 258, 292, 293
Lunar Atmosphere and Dust Environment Explorer (LADEE) 204
lunar missions 105, 200
Lunar Reconnaissance Orbiter (LRO) 204
Lupus 92, 186
Luu, Jane X. 175, 312
Lynx 56, 320

## M

machine vision and compensation 275
Maera 62
magma 120
magnetic field reversal cycle 324
magnetic monopole 96
magnetic poles 193

magnetic portals 332
magnetic reconnection 331
magnetohydrodynamics 247, 248
magnetohydrodynamic waves 247
magnetosphere 248, 331
Magnetospheric Multiscale Mission (MMS) 332
main-belt asteroid 172
main sequence 37, 77, 152, 153, 223, 227, 257, 287, 288, 293
main sequence star 77, 227, 293
main-sequence star 62, 287, 291
major axis 174, 229, 325
mantle 114, 120, 139, 140, 142, 324, 326
Maraldi, Jean-Dominique II 196
March equinox 242
Marchis 169, 170, 171
Marchis, Franck 169
Marconi, Guglielmo 42, 43
Mariner 3 103
Mariner 4 103
Mariner 9 29
Mars 188
Mars *Curiosity* Rover 116
Mars Exploration Program 117
Mars Global Surveyor 103
Mars mission 213, 277, 278
Mars *Observer* 103
Mars *Pathfinder* mission 103
Mars rover 104, 277
Mars Science Laboratory mission 104, 105
Martian day 189
mass attenuation coefficient 191, 192
mass extinction 14, 16, 19, 21, 39, 178
mass fraction 259, 261
mass fraction (MF) 259, 261
mass imbalance 193, 194
massive compact halo objects (MACHOs) 110, 111
mass of Mercury 194
*Mathematical Principles of Natural Philosophy, The* 70
Mathilde 210
matrix 33, 34
Maui 269
Maury, Antonia 205
MDCA/Cool Flames experiment 81
M dwarf 23
Méchain, Pierre 243
medium-Earth orbit (MEO) 138, 146, 149, 274
Medusa Nebula 145
Megrez 318

Meissa 231
Mekbuda 145
Merak 318
Mercury 193
Mercury transit 196
Mesarthim 26
meson 252
Messier 1 (M1) 196, 305
Messier 2 (M2) 23, 196
Messier 4 (M4) 269
Messier 6 (M6) 269
Messier 7 (M7) 269
Messier 30 (M30) 65
Messier 31 (M31) 133
Messier 41 (M41) 60
Messier 42 (M42) 232
Messier 44 (M44) 57
Messier 49 (M49) 328
Messier 58 (M58) 328
Messier 60 (M60) 328
Messier 65 (M65) 184
Messier 66 (M66) 184
Messier 67 (M67) 57, 58
Messier 74 (M74) 242, 243
Messier 77 (M77) galaxy 35
Messier 80 (M80) 269
Messier 81 (M81) 318
Messier 87 (M87) 135, 328
Messier 104 (M104) 158, 328
Messier Album 196
Messier Catalog 195, 196, 197
Messier, Charles xi, 57, 195, 197, 243, 328, 329
Messier Marathon 197
Messier, Nicolas 195
Messier object 195, 196
metallicity 295, 296
metal oxide semiconductors 333
meteoroid 2, 52, 53
meteor shower 63, 66, 92, 129, 184, 321
methane-containing atmosphere 250
methane crystals 215
Michell, John 48
microgravity 80, 81, 163, 200, 202, 277
Microscopium 64
microwaves 32, 42, 43, 116, 167, 168, 212, 297, 301
microwave signals 165
microwave wavelengths 239
Mikheyev-Smirnov-Wolfenstein (MSW) effect 218
Mimosa 99, 100

Minkowski, Hermann 283
minor planet 52, 233, 271, 313
Minotaur 95, 305
Mintaka 231
*Mir* vi, 200, 201, 202, 280, 286
mirror bee 177, 178
*Mir* space station 200, 280, 286
Mission Control 273
mission payload 4, 6
MIT (Massachusetts Institute of Technology) 293
Mizar 318
MK vi, 205, 226
MKK vi, 205, 206, 207
Moeckel, Richard 311
molecular hydrogen 139, 141, 142, 143, 176, 227
molecular radioastronomy 164, 165
Monoceros 62, 63
Montgomery, Richard 311
Moon impacts vi, 203
moon of Saturn 193
moons of Jupiter 29, 193
moons of Mars 190
Moore, Cristopher 311
Morgan-Keenan Classification System (MK or MKK) vi, 205, 226
Morgan, William 205, 206, 226
Morgan, William W. 205
morning star 324
MOS CCDs 333
Mount Ida 106, 173, 321
Mount Stromlo Observatory 108
Mount Wilson Observatory 133, 134
moving wave 46
MSW 218
Mt. Palomar Observatory 254
M-type (metal rich) asteroids 28
multi-ring basins 203
Multi-User Droplet Combustion Apparatus (MDCA) 81
muon xiii, 78, 217, 218
Musca Borealis 26
Muscida 318

## N
Na'ir al Saif 231
NanoSail-D 11
nanosatellite 240
NASA Dawn mission 74, 326, 327
Nash 264
Nashira 65
National Aeronautics and Space Agency (NASA) 34
National Aeronautics and Space Association (NASA) 5
natural satellite 106, 119, 121, 172
Navigation Signal Timing and Ranging 89, 137, 150
Navigation Signal Timing and Ranging Global Positioning System (NAVSTAR GPS) 137, 150
Navstar Global Positioning System 274
n bodies 70
n-body problem 69, 70, 71
NEA (near-Earth asteroid) 12, 13, 14, 15, 16, 18, 19, 20, 21, 22, 38, 39, 40, 41, 127, 209, 211, 233
NEAR 16, 22, 29, 40, 127, 128, 209, 210, 211
near-Earth asteroid (NEA) 1, 12, 14, 18, 20, 29, 38, 127, 209, 211, 233
Near-Earth Asteroid Rendezvous (NEAR) mission 128
Near-Earth Asteroid Rendezvous Shoemaker (NEAR Shoemaker) mission 209
Near Earth Network 116
near-Earth object (NEO) 14, 16, 22, 30, 42, 130, 212, 233
Near Earth Object Search Program (NEOP) 16
NEAR mission 209
NEAR Project 128
NEAR Shoemaker mission 16, 40, 127, 128, 210
*NEAR Shoemaker* spacecraft 128, 209, 210
nebula 57, 92, 99, 196, 232, 290, 305, 306, 317, 318, 319
Nemean lion 184
NEOWISE Missions 212
Nephele 26
Neptune 214
Neptune's mass 214
Neptune's rotational axis 215
neutralino 110, 111
neutrino xii, xiii, 78, 110, 216, 217, 218
neutrino oscillation 78
neutrino telescope xii, 78
neutron capture 221
neutron degeneracy limit 289
neutron degeneracy pressure 289
neutron star 35, 154, 155, 218, 219, 252, 257, 259, 289, 297, 298, 301, 335
New General Catalog 697 (NGC 697) 26
New General Catalog 772 (NCG 772) 26
New General Catalog 877 (NGC 877) 26
New General Catalog 1156 (NCG 1156) 26

New General Catalog 1952 (NGC 1952)  305
New General Catalog 3628 (NGC 3628)  184
New General Catalog 4435 (NGC 4435)  328
New General Catalog 4438 (NGC 4438)  328
New General Catalog 4567 (NGC 4567)  328
New General Catalog 4755 (NGC 4755)  99
New General Catalog 4874 (NGC 4874)  135
New General Catalog 4889 (NGC 4489)  135
New General Catalog 5101 (NGC 5101)  158
New General Catalog 6541 (NGC 6541)  92
New General Catalog 6872 (NGC 6872)  135
New Horizons mission  249, 314
Newtonian mechanics  69
Newton, Isaac  ix, 8, 9, 35, 36, 48, 69, 70, 107, 131, 159, 165, 221, 228, 244, 259, 261, 310, 333
Newton's law of universal gravitation  70
Newton's laws of motion  69, 70, 259, 261
Newton's second law  9, 228, 261
Nixon, Richard  7, 278
Nobel Peace Prize  167
non-dwarf planet  117
nonexotic matter  331
Nordberg Fusion Technologies  46
Nordberg, John T.  46, 48
North Pole  63, 87, 242, 305, 320, 321
North Star  87, 99, 320, 322
nozzle  131, 260
nu Capricorni  64
nuclear fusion  77, 154, 155, 217, 223, 246, 247, 255, 290, 297, 301
nuclear propulsion  8
nucleosynthesis  35, 37, 220, 221, 222, 223
nucleus  xii, 2, 3, 37, 50, 54, 55, 77, 78, 79, 82, 83, 140, 157, 159, 220, 221, 222, 224, 225, 257
Nusakan  93, 95
nu Scorpii  269
Nushaba  264

# O

OAO-2 (Orbiting Astronomical Observatory 2)  160
Oberth, Hermann  5
Object D  284
Object M44  57
OB star  226
OB star association  226
Odyssey  105
Olbers, Heinrich Wilhelm  114
Olympus Mons  189
OM  334

omega Capricorni  65
Omega Nebula  264
Omicron Capricorni  65
ONS  115
Oort Cloud  82, 175, 224, 225, 270, 272, 314
Oort, Jan Hendrik  270
opacity  v, 42, 43, 44, 323, 324
open atmosphere  120
Ophiuchus  186, 242, 269, 289
opportunity  105, 179, 200
optical double star  65, 91
optical monitor  333
optical navigation system  115
optics  x, 81, 117, 118, 165, 169, 238, 239
orange giant  26, 144, 259, 320
Orbital Maneuvering System (OMS)  278, 280
orbital mechanics  69, 119, 122, 131, 188, 193, 214, 228, 230, 249, 275, 284, 310, 323
orbital radius  13, 244
orbital resonance  119, 244, 312, 313
orbital robotics  119, 121, 275, 277
Orbital Technologies Corporation (ORBITEC)  262
orbit plotting  vi, 228
ordinary chondrites  129
O ring  280
Orion B1  226
Orion Bullets  232
Orion Nebula  165, 196, 226, 231, 232, 293
Orion Nebula Star Cluster  232
Orion OB1  226
Orion's Belt  60, 61, 63, 227, 231, 232
Orion Spur Arm  198
Osiris  306
outer planets  72, 75, 76, 188, 237
Owl Nebula  318
ozone (O3)  120

# P

P/2013 P5  30
Pabilsaĝ  265
pair production  191
Palisa, Johann  169, 173
Palitzsch, Johann Georg  196
Pallas  29, 74, 326
Palomar  x, 16, 22, 38, 41, 42, 124, 233, 254, 270, 271
Palomar Planet-Crossing Asteroid Survey (PCAS)  41
Palomar QUEST large-area CCD camera  271
Pan  vi, 65, 233, 234
Parsons, William (third earl of Rosse)  318

particle acceleration vi, 234, 236
particle accelerators 223, 234, 235, 236
Pasiphaë 305
Patroclus 315
Pauli, Wolfgang xii, 217
payload 4, 5, 6, 7, 104, 105, 180, 182, 260, 261, 279, 280
Payload Operations Center 80
Pegasus 68, 180, 242
Penzias, Arno Allan xi, 37, 308
periodic comet 266
period of rotation 193, 215, 323
periphelions 173
Perseid meteor shower 129
Persephone 328
Perseus 26, 38, 68, 320
Personal Satellite Assistants 276
perturbation 69, 70, 215
PHA 41
Phad 318
Phecda 318
Pherkad 320
*Philosophiæ Naturalis Principia Mathematica* 70
phi Tauri 305
PHO 18
Phobos 13, 29, 190
Phorcys vi, 75, 76, 237, 238
photoelectric effect 84, 85, 86, 167, 191
photoionization 246
photometers 239
photometry 240, 241
photon 54, 55, 78, 85, 100, 167, 333
photosphere 287, 288
Phrixus 26
Piazzi, Giuseppe 29, 74, 114
pi Capricorni 65
Pickering, Edward Charles 205, 293, 294
Pindar 92
Pinwheel galaxy 318
Pioneer Venus spacecraft 160
Pisces 242
Piscis Austrinus 64
pi Scorpii 268
pixel 33, 34, 212
Planck constant 168
Planck-Einstein relation 85
Planck, Max 84, 85, 223, 284, 332
Planck mission 309
Planck Observatory 239

Planck relationship 84, 85
Planck satellite 97, 140
Planck satellite telescopes 140
Planck scale 332
planetary astronomy 16, 112, 172, 270, 325
Planetary Data System 212, 238
planetary nebula 318
planetary nebulae 258
planetary rings 215
planetary rover 188, 190, 275
planetary science 117, 308
Planetary Society 11, 179, 180, 241
planetoid 28, 31, 271, 272
plasma 44, 47, 168, 222, 223, 245, 246, 247, 248, 287
plasma density 246, 247
plasma oscillations 247
plasma physics 247, 248
Pleiades 153, 196, 232, 304, 305, 306
plerion 301, 302
plerionic composite 301
plutino 312, 313
Pluto 249
Pluto-Charon pair 250
Pluto's moon 249
pn-CCD 333
PNT 150, 151
Pogson, Norman Robert 239
Poincaré, Henri 282, 310, 312
Poincaré, Jules Henri 310, 312
Polaris 87, 99, 100, 319, 320, 321, 322
polarization 46
pole star ix, 66, 320
Pollux 56, 144, 145
Polyakov, Valeri 201
polytrope 307
Pontus 121
population III stars 207
Population II stars 207
population I star 207
Porrima 328
Poseidon 68, 121, 216, 231, 305
positrons 51, 78, 234
potentially hazardous asteroids (PHAs) 41
potentially hazardous object (PHO) 18
power sources and recharging 275
Praesepe 57
Precise Positioning Service (PPS) 150
pressure rakes 182
Prima Giedi 64, 65

Princeton University 40, 45, 244, 259, 314, 331
Priroda 201
Procyon 59, 61, 62, 63, 293
Procyon A 62
Procyon B 62, 63
Project Apollo 278
propellant 7, 8, 9, 10, 260, 261, 278, 279, 280
propulsion v, 8, 9, 10, 80, 112, 113, 114, 177, 213, 334
protoplanet 36
protostar 36, 37, 142, 287, 290, 291, 292
Ptolemy ix, xii, 24, 57, 59, 61, 65, 69, 92, 93, 95, 145, 238, 243, 269, 318, 321, 328
pulsar 218, 219, 235, 236, 289, 301
pulsar 3C 58 289
Putin, Vladimir 147
Pyramus 184

## Q

Q2343-BX442 135
Q-type asteroid 18, 19
Q-types 19
quantum foam 332
quantum hypothesis 85
quantum mechanics 50, 85, 298
quantum physics 107
quantum theory 84, 308
quark 222, 223, 252, 253, 289, 298
quark-gluon plasma 222, 223
quark-nova 252, 253
quark star 252
quasar vi, 2, 3, 254, 255, 256, 328
quasars 3, 38, 52, 254, 255, 256, 290
quasi-constellation 99
quasi-stellar vi, 2, 3, 254
quasi-stellar radio source vi, 3, 254
Quasi-Zenith Satellite System 274
Queen Cassiopeia 68

## R

R2 277
R-7 rocket 284
radiation hardening 188, 190, 275
radiative zone 287, 288
radioastronomy 164, 165
radio emission 254
radioisotopes 236
radioisotope thermoelectric generator (RTG) 104
radiometric resolution 122, 123
radio telescope x, 219, 254

Ram Constellation 25
Ras Elased Australis 183
Ras Elased Borealis 183
Rayleigh scattering 191, 250
R Coronae Australis 91
R Coronae Borealis 94, 95
R CrB 94
R Crucis 99
Reagan, Ronald 280
Reconnaissance 105, 204
red dwarf 241
red giant 257
red giant star 257, 305
red hypergiant star 60
red planet 103, 105, 188, 277
redshift 35, 36, 38, 52, 254
red supergiant 61, 259, 268
Reflection Grating Spectrometer (RGS) 334
Regulus 56, 183
Reinmuth, Karl 17, 18
Reinmuth, Karl Wilhelm 11
relativistic energy expression 84, 85
Relativistic Heavy Ion Collider (RHIC) 223
remnant 112, 193, 196, 205, 252, 297, 301, 302, 305, 330
remote sensing 103, 122, 123
remote sensing (RS) 122, 123, 164
Resolution 5A 74
Resolution B5 125, 271, 327
Resonance KBOs 174
retrograde rotation 323
Reverse Nova 94
Rheasilvia 326
rho Capricorni 65
rho star 67
Rigel 206, 231, 240, 292
Rigil Kentaurus 240
ring system 215
Robonaut 2 277
rocket propulsion 8
Röntgen, Wilhelm C. 54
Roscosmos 162
Rosen, Nathan 330, 331
Rosetta 79, 83, 84, 160, 175, 225, 304
Royal Observatory Greenwich 320
Royal Observatory of Belgium 13
Royal Society of London 196
RR Lyrae variables 153
RROxiTT 276

rubble-pile asteroid  169, 170
Rubin, Vera  111
Ruchbah  67
Rukbat  264
Russell, Henry Norris  152
Russian Federal Space Agency  137, 161, 162
Rutherford, Ernest  77

## S
Sa  134, 157, 158
Sadachbia  23
Sadalmelik  23
Sadalsuud  23
Sagan, Carl  179, 308, 331
Sagittarii  264
Sagittarius  vi, 64, 65, 92, 198, 242, 263, 264, 265, 269, 328
Sagittarius A*  198, 264, 265
Saiph  231
Salam, Abdus  47
Salyut  200
Salyut 1  200
Samuel Oschin Schmidt Telescope  271
Sargas  269
satellite navigation  88, 89, 136, 137, 138, 146, 147, 148, 149, 150, 273, 274
Saturn  ix, 23, 29, 72, 102, 116, 121, 132, 180, 182, 193, 216, 244, 245, 250, 315
Saturn Nebula  23
Saturn V  180, 182
Sb  134, 157, 158
SB  157
SBa  134, 157, 158
SBb  134, 157, 158
SBc  134, 157, 158
Sc  134, 157, 158, 190, 195, 216, 325
scattered disk objects  71, 72, 174, 266, 312, 313
scattered KBOs  72, 174, 266
Schilt, Jan  239
Schmidlap, Johann  260
Schmidt, Maarten  254, 256
Schmidt photographic telescope  41
Schwarzschild, Karl  330
Scorpiids  269
Scorpio  206, 265
Scorpius  vi, 92, 186, 231, 242, 267, 268, 269, 270
S Crucis  99
Scutum  92
Scylla  76

SDO  313
Secchi  190
Secunda Giedi  64
Sedna  vi, 267, 270, 271, 272, 313
seismic activity  249
semiconductors  236, 333
semimajor axis  13, 40, 228, 229, 325
semiminor axis  229
SEP  213
Serpens Caput  94
Seven Sisters  305
Seyfert galaxies  51, 254, 255
Shapley, Harlow  xi, 133, 199
Shaula  269
Shedir  67
shell-type remnant  301
Shen  232
shepherd moons  214, 215
Sheratan  26
Sherpa  329
Shoemaker, Carolyn  233
Shoemaker, Eugene  210, 211, 233
Shoemaker-Levy 9  233
short-period comet  266
Siamese Twins  328
sigma Librae  186
sigma Scorpii  268
sigma Tauri  305
silicate materials  120
silicate (S-type) asteroid  106
Simó, Carles  311
simple crater  203
single quantum  37
singularity  2, 38, 48, 49, 50, 235, 255, 298, 308, 330
Sirius  58, 59, 60, 61, 62, 206, 293
Sirius A  60
Sirius B  60
sister planet  323
Sleeping Beauty galaxy  196
small angle formula  216
Small Dog Constellation  61
small solar system body (SSSB)  28, 52, 76, 159, 237
Smithsonian National Air and Space Museum  160, 209, 281
SN 2002ap  243
SN 2003gd  243
Sojourner  103, 277
solar cycle  ix, 245
solar eclipse  xi, 195

solar electric propulsion 213
solar flares 334
solar masses 49, 78, 223, 288, 289, 335
solar neutrino 216, 218
solar neutrino problem 218
solar propulsion 8, 10
solar radiation pressure 115, 273
solar sail 6, 8, 10, 11
solar wind 10, 82, 83, 224, 245, 277
solid-propellant rocket motors 7
solid rocket boosters (SRBs) 7, 278, 279, 280
Sombrero galaxy 158, 328
Sommerfeld, Arnold 54
sonic boom 129, 130
Southern Cross 87, 98, 100
southern crown 91
South Pole 60, 87, 99, 100
*Soyuz* 148, 161, 162, 163, 181, 200, 286
*Soyuz-Apollo* mission 200, 286
Space Communications and Navigation Program (SCaN) 116
spacecraft propulsion 8, 80, 131, 259, 278
Space Launch System (SLS) 182, 262
Space Network 40, 103, 105, 116
space race 5, 8, 116, 162, 286
space shuttle program 116, 278, 280
Space Task Group 7, 180, 278, 280, 281
space-time continuum 282
space-time theory 283
space-time vortex 283
space transportation systems 278
spatial resolution 122, 123
special theory of relativity 167
spectral class 153, 205, 206
spectral classification 152, 226, 257, 258, 292
spectral emissions 254, 257, 292
spectral pattern 165, 205, 206
spectral resolution 123
spectral type xii, 77, 152, 205, 226
spectrometer 84, 100, 101, 104, 114, 211
spectroscopic analysis 250
spectroscopy v, 100, 101, 102, 164, 165, 166, 293, 294, 295, 333
Spektr 201
spheroid galaxies 134
Spica 328, 329
Spica A 328
Spica B 328
spin axis 193

spinning galaxies 112
spin-orbit coupling vi, 119, 121, 244, 245
spiral galaxy 133, 135, 156, 157, 184, 196, 197, 199, 242, 243
*Spirit* 85, 103, 277
Spitzer Space Telescope 141, 232, 239, 291
Spitzer Space Telescope mission 141
spring equinox 242, 305
*Sputnik* vi, xi, 5, 8, 116, 123, 200, 261, 284, 285, 286
*Sputnik 1* 116, 123, 200, 284, 285, 286
*Sputnik 2* 285
*Sputnik 3* 284, 285
*Sputnik I* xi, 5, 261, 284
Sputnik missions 285
Sputnik space program 284
Square of Pegasus 242
staged combustion 80, 81
Standard Positioning Service (SPS) 150
standing wave 46
Stardust 16, 17, 179
Stardust mission 17
Star in the Jaws of the Dog 59
star nurseries 290
steady state theory 223
Steinhardt, Paul 96
stellar classification 292
stellar evolution 35, 37, 290
stellar halo 197, 199
stellar mapping 295
stellar neutrino 216
stellar nucleosynthesis 37, 220, 223
stellar nursery 291
stellar population vi, 295, 296
stellar remnants 297
step rocket 261
Stetson, Harlan 239
strange star 253
Strelka 285
Strömgren, Bengt Georg Daniel 227
Strömgren sphere 226
STS 7, 180, 278, 280
S-type asteroid 127, 209, 211
Sub-brown dwarfs (SBDs) vi, 299, 300
subdwarf star 293
sublimation 1, 2, 79, 83, 159, 178
sub-planetary astronomy 16, 112, 172, 270, 325
sulfuric acid 323, 324
sulfuric acid droplets 324
sum x, 85, 229, 281

sunspots 244, 245
Sunyaev, Rashid Alievich 168
Sunyaev-Zel'dovich effect 167, 168
supergiant 61, 145, 205, 206, 259, 268, 320
Supergiant Fast X-Ray Transient IGR J18410-0535 335
supermassive black hole 3, 48, 49, 51, 135, 198, 254, 255
supernova 2012ha 329
supernovae 301, 302, 329
supernova remnant 196, 301, 302, 333
supersymmetry theory 111
*Surveyor* 103, 276, 278
*Surveyor 1* 276
*Surveyor 3* 276
Surveyor missions 276
Šuvakov, Milovan 310, 311
Suzaku 235
Swan, Horseshoe, or Lobster Nebula 264
Syene 122
Sylvia 169, 171, 223
synchrotron radiation 301, 302
Synopsis of the Astronomy of Comets, A 196
SZE 168

# T
Talitha 318
Tammuz 329
Tania Australis 318
Tania Borealis 318
tau1 Arietis 26
tau Ceti 207
tau Librae 186
Taurid meteor 306
Taurus vi, 26, 59, 62, 87, 88, 144, 180, 196, 231, 232, 242, 304, 305, 306, 307
Taurus A 305
Taurus constellation 88, 196, 305, 307
tau Sagittarii 264
tau Scorpii 269
T Coronae Borealis 94, 95
T Crucis 99
Teapot asterism 264
Teapot of Sagittarius 265
tectonic plates 120, 151, 189, 194, 324
telemetry 4, 6
teleoperation 275, 276, 277
Telescopium 92
Tellus 121

Tempel-Tuttle 184
temporal resolution 123
terrestrial planet 23, 114, 119, 188, 325
terrestrial planets 73, 74, 119, 188, 193, 249, 323, 325, 326
terrestrial plasma 248
Thales of Miletus 321
THEMIS 331
theoretical astronomy 31, 96, 110, 220, 244, 254, 281, 330
theoretical astrophysics 42, 84, 191, 307, 310
theoretical physics 308

theory of general relativity 36, 48, 70, 108, 221, 282, 308, 309, 330
theory of gravity 109
theory of special relativity 36, 281, 282
thermal composite 301
thermal considerations 188, 190, 275
thermal control 4, 6
thermal energy 287, 288, 289, 299
thermal equilibrium 97, 309
thermal radiation 37, 96
thermal spectrum 167, 168
Theseus 95, 305
theta Scorpii 269
theta Tauri 305
thick disk 197, 199
thin disk 197, 199
third law of motion 131
Thisbe 184
Thomson, William 206
Thorne, Kip 331, 332
three-body problem 70, 310, 311, 315
three-body system 310
thrust 1, 8, 9, 10, 80, 113, 131, 179, 180, 181, 228, 260, 262, 278, 279, 280
tidal evolution 69, 71
tidal force 237, 244
Time History of Events and Macroscale Interactions during Substorms (THEMIS) 331
time travel 309, 330, 332
Tisserand, Félix 70
Titan 121, 180, 193, 232, 250, 306
Titan Cronus 250
Titan family of launch vehicles 180
Titans 121
Titius-Bode law 29
Titius, Johann 29

Tombaugh, Clyde 250, 312
top quarks 252
torque 6, 193
tracking and commanding 4, 7
Traité de mécanique céleste 70
trajectory 1, 115, 116, 203, 273, 274, 292
trans-Neptunian Asteroid Belt 313
trans-Neptunian objects (TNOs) 28, 214, 249, 266, 271, 312, 314
trans-Neptunian region 312
Transonic Dynamics Tunnel (TDT) 182
Trapezium 232
Treatise on celestial mechanics 70
Triangulum 26, 242
Triassic period 329
Trifid Nebula 264
triple star system 320
triplet production 191
tritium 222
Trojan vi, 11, 12, 28, 314, 315, 316
Trojan asteroid 11, 314, 315, 316
tropical year 242
Tropic of Capricorn 65
Tsiolkovsky, Konstantin 5, 261
T Tauri 290, 291
T Tauri star 290
TT&C 6
turbomachinery 8, 10
twin paradox 282
Tyndarus 145
type 1 fireballs 129
type 2 fireballs 129
type 3a fireballs 129
type 3b fireballs 129
type II galaxies 136
type I irregular galaxies 136
Typhon 76, 243
Typhon (Typhoeus) 243

## U

ultraviolet radiation 44, 60, 140, 159, 206, 226, 271, 302
Ulysses 277, 278
Ulysses mission 277
unique atom 37
United States Geological Survey 210
University of California, Berkeley 169
University of Vienna 240
up quark 252

upsilon Librae 186
Urodelus 320
Ursae Majoris 318
Ursa Major 317
Ursa Minor 319
Ursa Minor Dwarf 321
Ursa Minor Dwarf galaxy 321
Ursid meteor shower 321
Ursids 319, 321
US Air Force Advanced Upper Stage Engine Program 262
U Scorpii 269
US Department of Energy 235
US Explorer 6 123
US space program 278
US space shuttle 181
UV radiation 120, 165, 226, 227

## V

vacuum 43, 80, 141, 179, 235, 282
Valles Marineris 189
Van Allen belts 248
Van Allen radiation belt 248
Vanguard satellite program 285
van Maanen, Adrian 243
Van Maanen's Star 242, 243
variable specific impulse magnetoplasma rocket (VASIMR) 10
Vasile, Massimiliano 178
vector x, 47, 244
Vega 240
Vela 99
Venera 277, 324
vernal equinox 25, 27, 91, 98, 230, 242, 243, 263, 268, 304, 317, 320, 327
Vespucci, Amerigo 100
Vesta vi, 22, 29, 40, 73, 74, 75, 112, 113, 114, 115, 325, 326, 327
Victoria 104
Vienna Observatory 173
*Viking* 44, 103
*Viking 1* 103
*Viking 2* 103
Villard, Paul 235
Vindemiatrix 328
Virgin 327, 328
Virgo vi, 62, 183, 186, 242, 327, 328, 329
Virgo Cluster 328, 329
visible light spectrum 34, 226, 238, 239, 240, 297

visible mapping spectrometer 114
volcanoes 74, 102, 120, 203, 324
von Zeppelin, Ferdinand 5
vortex engine 262
vortex liquid-fuel rocket 262
Vostok 285, 286
Vostok 1 285, 286
Voyager 1 116, 177
Voyager 1 and 2 116
VY Canis Majoris 60

## W

Wang Ganchang 217
weakly interacting massive particle (WIMP) 110
Weinberg, Steven 47
Western Fish 242
Wheeler, John Archibald 331
white dwarf star 60, 235, 297, 329
white hole 331
white main-sequence dwarf 321
Wide-Field Infrared Survey Explorer 34, 52, 76, 214, 292, 316
Wide-field Infrared Survey Explorer telescope 212
Wide-field Infrared Survey Explorer (WISE) mission 72, 211, 212
William the Conqueror 304
Wilson, Albert George 321
Wilson, Robert xi, 37
Wilson, Woodrow 308
WIMP 110
wind tunnel test 182
Winter Triangle asterism 59, 61
WISE mission 73, 212
Witt, Gustav 127
Wolf, Max 315
work iv
wormhole 307, 330, 331, 332
Wright brothers 131
Wright, Orville 5
Wright, Wilbur 5

## X

xi Scorpii 268
Xi Ursae Majoris 318
XMM vi, 333, 334, 335
*XMM-Newton* 333
XMM-Newton mission 334
XMM observatory system 334
X-points 332
x-ray astronomy 167
X-Ray Multi-Mirror Mission 333
XTE J1739-285 289

## Y

Yamaha rotorcraft RMAX 276
Yarkovsky-O'Keefe-Radzievskii-Paddack (YORP) effect 19
yellow giant star 328
yellow star 166
yellow supergiant star 320
yellow-white main-sequence dwarf 321
Yerkes Observatory 226
Yerkes spectral classification 226
Yildun 320
YORP 18, 19, 20
YORP effect 20

## Z

Zel'dovich, Yakov Borisovich 168
Zenelgenubi 186
zenith 122
zeta Geminorum 145
zeta Orionis 231
zeta Sagittarii 264
zeta Ursae Minoris 320
zeta Virginis 328
Zib-ba An-na 186
Zubenelakrab 186
Zubenelgenubi 186
Zubeneschamali 186
Zwicky, Fritz 111